Textbook of ANIMAL BEHAVIOUR

Fourth Edition

FATIK BARAN MANDAL
Principal, Department of Zoology
Bankura Christian College
Bankura, West Bengal

PHI Learning Private Limited
Delhi-110092
2025

In fond memory of ***Shri Asoke K. Ghosh*** *(October 1942 – February 2024), Founder Chairman and Managing Director of PHI Learning, whose vision endlessly inspires.*

The Legacy Continues....

Published by Pushpita Ghosh, PHI Learning Private Limited, Rimjhim House, 111, Patparganj Industrial Estate, Delhi-110092 and Printed by Multi Colour Services, I-45, DLF Industrial Area, Sector-32, Faridabad, Haryana-121003.

₹1295.00

TEXTBOOK OF ANIMAL BEHAVIOUR, Fourth Edition
Fatik Baran Mandal

ISBN-978-93-5443-949-0 (Print-book)
ISBN-978-93-5443-888-2 (e-book)

To

My Mother

Contents

Preface

This is indeed a pleasure for me to finalize the fourth edition of the book "Textbook of Animal Behaviour." In fact, I have a passion for knowing and sharing information. Nowadays, information is being created at a significantly high rate. It is very difficult to acquire all this information and present it to the readers. We must think about their needs, and capabilities and not overburden them with information. Keeping all such information in mind, I have tried my level best to present the information to the readers in a very simple way. In the present edition, I have incorporated some recent understandings of animal behaviour that were previously not included. For example, animal culture, the learning of animal species through socially transmitted behaviour is now widely accepted. The Convention on the Conservation of Migratory Species of Wild Animals has been trying to use the knowledge of animal culture to protect endangered wildlife. Whales, dolphins, elephants, and primates acquire their knowledge and skills through social learning. Some animals learn from adults or peers about various behaviours. The nut-cracking chimpanzees highlight the species' unique culture. The species can crack open different types of nuts by using stones and pieces of wood. Cultural capacity enables animals to survive in unfavourable environments. Recent understandings in behavioural biology have revealed how various animal groups use environmental cues through complex processes.

I do believe that with the expertise of PHI Learning Pvt. Ltd., this edition must attract more readers.

FATIK BARAN MANDAL

Preface to the First Edition

Behavioural biology is a fascinating subject. Significant advances have been made in this field with the knowledge of molecular biology, endocrinology, and neurobiology. It has gained significant importance along with increasing realization of human responsibility towards animal welfare. Animals are dependent on their families for care, food, and protection, for varying lengths of time, before they survive on their own. Each species has its own instincts, family structure, and social rules. A wealth of knowledge about behaviour accumulated during past few decades is really astounding. Behaviour evolves by natural selection. Behaviours simply are the survival strategies which keep the organisms alive, and have been a practical matter for early trappers, hunters, traditional shepherds, and herdsmen.

Behavioural studies help in enhancing quality of our societal life, in understanding links between the genes and the brain, explaining susceptibility to diseases and stress, and designing strategies for biodiversity conservation, animal welfare, and pest control. Reproduction, the sole method for a species' continuation, manifests itself in numerous deviations throughout the nature. The ultimate goal of each species is to produce maximum offsprings with exerting least energy and increasing the fitness. The genes are parts of DNA that carries codes for feature(s), such as behaviour. Natural selection acts on these features, more accurately on genes. Parents' genes are passed to their children through sexual reproduction. A child gets on average half the genes of the mother and half of the father. Caring for one child is equivalent to caring for 2 grandchildren or 2 nephews, or 4 cousins, and so on. Genetically, we can express any class of relatives as offspring equivalent, and then compare the fitness of individuals with various propensities for rearing offspring or helping relatives. A gene that causes an organism (bearer of the gene) to help its relatives do well in future, if cost to the organism is smaller than benefit to its relative, divided by probability that the relative shares the same gene. The genes are understandably crucial to physiology, morphology, and biochemistry. However, the contribution of genes in behaviour seemed outrageous. Presently, our understanding about relationship between the genes and the behaviour is becoming informative and productive over the past several decades. Predation pressure, resource accessibility, and competition for attracting mates greatly influence the likelihood of an individual within a species in producing viable offspring, and thereby donating genes to the subsequent generations. Different fish species learn unique reproductive strategies employed from fertilization tactics to parental care methods. Females typically exert much energy and time into reproduction than males, and females are usually more selective in choosing their mates.

Ecology, natural selection, microevolution, macroevolution, and gene constitute the foundation of animal behaviour. The natural selection is a competitive process, which helps those individuals that help themselves. The various animal groups exhibit various strategies for their survival and reproduction. These interesting strategies are discussed in behavioural sciences. Biomimicry is a relatively new branch of science that deals with the nature, its models, systems, processes and elements, and initiates to solve the problem sustainably in nature's way. This is a new way of viewing and valuing nature with immune potential for developing novel products and technologies, as ideas come directly from the nature.

A number of books have been published to cater to the needs of the students, teachers, as well as researchers. However, there is a gap in the information containing in these books and student's needs, particularly, in the Indian context. Present endeavour has been undertaken to fill up the gap, with up-to-date information in a student friendly way, which would have particular value to the students. The book is about the concepts needed to understand animal behaviour coherently, and scientifically. It grows from a concern that we have become so fascinated with our individual lives that we do not pay enough attention to the lives of other species. Hope, this book would have significance in creating interest among students and other interested readers in the above-mentioned field.

FATIK BARAN MANDAL

Acknowledgements

While teaching animal behaviour to B.Sc. (Hons.) students and hence going through a number of books for fulfilling the student's interest, I became gradually attracted to the field of animal behaviour. Perhaps, this book took its origin from my discussions with students. Therefore, I am thankful to my students. I am grateful to the Department of Psychology, University of Plymouth for allowing me to use some figures from their resource materials. Thanks are due to my wife, Mrs. Amita Mandal, for her constructive criticisms. I am thankful to my elder brother, Dr. Asit B. Mandal, Project Director, Directorate of Seed Research, Kushmaur, Mau (U.P.) for his support. I am also grateful to Professor S.K. Maitra, Department of Zoology, Visva-Bharati, Shantiniketan, West Bengal, and to Dr. Richard R.N. Bajpai, Principal, Bankura Christian College, Bankura, West Bengal, for their encouragement.

I also wish to express my sincere thanks to Editor, PHI Learning, Delhi, their constructive suggestions for improving the draft manuscript. The professionalism of the editorial and production teams at PHI Learning Private Limited is also thankfully acknowledged. I wish to express my sincere thanks to Mr. Tarak Nath Ghosh for his help in improving some figures, Mr. Himanish Prasad Jana, an M.Sc. student in Conservation Biology, Durgapur Government College, Durgapur and Mr. Madhusudan P. Srinivasan, a Ph.D. candidate of Kentucky University, USA for providing me with a number of photographs. A few photographs collected from *www.fotosearch.com* and *www.yourwallpapers.com* are thankfully acknowledged.

I wish to express thanks to the Editor, Apis-UK Newsletters, for allowing me to use three figures in the second edition of the book.

FATIK BARAN MANDAL

Chapter 1

Introduction

Behaviour is an observable response of an organism to external or internal stimulus. Any evolved adaptive action or activity of an organism that interacts with its environment, not a simple by product or side effect of activity, is a behaviour. This is a fundamental property of living organisms and ranges from simple response in bacteria to intricate social interactions in humans. Behaviour may be conscious or unconscious, overt or covert, and voluntary or involuntary. Blinking, eating, walking, running, flying, etc. are examples of behaviour. Animals behave for four basic reasons: to get food, to interact in social groups, to avoid predator, and to reproduce. Behaviour is the outcome of multiple interacting forces like genes, physiology, development and environment. Genetic variation is related to survival and reproduction of individual. As no 2 individuals face and feel the same internal and external environments, the outcome varies in the behaviour of the individual. Successful behaviours pass on to the next generation and evolve. Unsuccessful behaviours are generally not transmitted to the next generation. In the presence of sufficient variation and heritability (Hager 2012) in populations, behaviour evolves. Threats, sounds, and smells from other animals as well as hunger, and fear prompt behaviour.

Environmental flexibility helps to adjust to change during the lifetime of organisms. Behaviour ranges through habitat selection to social organization. Some animals sacrifice their own survival and reproduction to help others.Bacteria move in response to chemical gradients. Amoebas alter body shape to wrap around food (Etnier and Vogel 2000). Pathway leading from genes to behaviour with many intermediate steps is long. Dynamic genetic and physiological processes influence each other, specifically genetic switches in regulatory region. Causes of behaviour are internal and external environments influencing the expression of genes, as well as physiological and developmental pathways. Specific and standardized methods exist to study behaviour (Martin and Bateson 2007). Behavioural research ranges from species to genes and can be undertaken in field, laboratory, or, both, assessed in the context of ecology and evolution to gain insight into animal's interact action with environment and their change over time.

Behavioural strategies confer an adaptive advantage to an organism with success that is measured as fitness. Natural selection favours individuals, who utilize habitats in which the greatest number of successful offsprings can be raised. Species with highly fixed genetically programmed habitat selection require considerable time to evolve mechanisms for selection of a new habitat. A more comprehensive prediction of an animal's most profitable behaviour requires consideration of other's acts. Animal behaviour is intrinsically fascinating, and applies equally to domestic pets, and exotic species. Understanding animal behaviour is essential for development of animal welfare standards in farms, zoos, and in conservation areas. Relative number of foragers in each patch should match the relative available amount of resources obtained there. Individuals distribute themselves among habitats in a way so that every individual maximizes its return. Costs and benefits may depend on number of other users using the resource. A variety of factors determine the presence or absence of a species at a particular site. A number of factors play role in decision making among animals.

Evolutionary changes over long time result in speciation. Behaviour may be partly responsible for the speciation. *Sylvia atricapilla* commonly called *blackcap* breed in central Europe and winters in southern Europe and northern Africa. Variation in migratory behaviour results in a small percentage of the bird that consistently over winters in Britain (Rolshausen et al., 2009). Non-British and British blackcaps (Hager 2012) migrate back to same breeding forests in central Europe, except that individual that winter in Britain arrive and begin nesting earlier and produce large clutches of eggs. British-wintering blackcaps predominantly breed with one another and not with members of other population. Blackcaps are reproductively isolated from one another despite breeding proximity. Natural selection and reproductive isolation cause speciation through accumulation of genetic and behavioural differences.

A member of a male-female pair kills and consumes other during copulation in sexual cannibalism as seen in Praying Mantis. In Lycosa tarantula, a wolf spider sexual cannibalism is adaptive for female (Rabaneda-Bueno et al., 2008). Cannibalistic female mated previously with male achieved higher reproductive rate and produced high quality offspring relative to non cannibalistic female. Female Lycosa face difficulty in finding and consuming quality prey. Successful females behave for increasing the chance to cannibalize the males. Such behaviours are inherited to the offspring.

Appetites fall into category called *behavioural homeostasis*, meaning the tendency of organism to maintain internal equilibrium. Thirst, sleep, hunger, body temperature drives significant behaviours. Animal forages for food to obtain their energy and to obtain macro-, and micro-nutrients for sustaining life. Foraging involves inherent risks, as animal often moves from the protected shelter in search of food. Vertebrates need sleep. Two hypothesized adaptive functions for sleep exist. First, sleep allows a period for neural repair and consolidation of memory. Second, sleep in a protected location keep animal from the risk of predation. Some birds and mammals sleep with one eye open during which one hemisphere of brain remains in sleep, and the other remains alert. Sleep is related to circadian clock of about 24 hour which is regulated by melatonin. The clock is being set by exposure to natural (sun, moon) light cycles. Behavioural thermoregulation is important for ectotherms and endotherms. To warm the body and, for the free movements of muscles and faster metabolic processes, the ectotherms bask in the sun. When overheated they search for shade. Endotherms produce heat internally, and maintains body temperature physiologically. As heat production requires vast energy, birds and mammals bask to reduce calory expenditure. Behaviour provides the flexibility to respond to change to environmental conditions and to move about in their habitat to find resources essential for survival.

Pain is a sensation associated with physical damage of body providing feedback that allows avoidance of further injury. Responses of mammals and birds to physical injury suggest similarity of pain perception in such organisms. All vertebrates share physiological pathways to prevent bodily damage. In response to experimental stimuli the groups like Crustacea, insects, and molluscs suffer from physical discomfort like mild electrical shocks.

Agitated animals exhibit displacement behaviour. Grooming, typical displacement behaviour can be seen in humans. In socially uncertain condition we dress our hair or show other grooming movements. Lack of manifestation of natural behaviour are behavioural pathologies like repetitive pacing or grooming causing physical damage in captivity. Birds and mammals exhibit symptoms similar to obsessive-compulsive disorder of humans under behavioural stress through pull out their feathers or fur or groom constantly.

A behaviour may be rewarding when it is rare in a population, but may not be nearly as advantageous to its actor when it becomes common. Game theory models consider value of frequency-dependent fitness benefits. Originally developed as a tool to predict rational human economic behaviour, its application in many evolutionary problems has improved understanding of situations where fitness consequences of a behaviour depend on types and frequencies of behaviours exhibited by other animals. Predictable fluctuations in environmental conditions impact behaviour and movement patterns of many animals. When northern temperate ponds freeze over in winter, bald eagles cutoff from their prey, and forced to travel in search of open water. Animals respond by travelling to seasonal changes and increase carrying capacity by seeking out abundant resources in summer, and by emigrating from limited and harsh resource conditions. Migration refers to these regular seasonal journeys which alternate between breeding and non-breeding locations. Many species move larger distances, and excel at an ability to find their way. A combination of cues and mechanisms is employed to complete some of amazing achievements. A series of questions can be addressed to explore the causation of migration in animals.

Some animals are 'nice' to each other. Vampire bats share meals. If a bat fails to find a meal, a well-fed bat spares some of its meal through regurgitation to the starved one. Bats within a colony are not necessarily relatives. They show a behaviour which simply means that they lend each other favours, in expectation that favours will be repaid in future. By being generous one day at little cost to self might save the starved one. There is a game called 'Prisoner's Dilemma' that can be used to understand conditions under which this process works.

Animals communicate with each other through signals which serve different purposes. Some are big announcements and others are subtle whispers to a listener. Animals filter out which signal is relevant to them. Some animals can hear a wide range of sounds. Others are attuned to a very specific frequency, such as call of their mate or echo from their own echolocation. Many insects have bold red and black markings which birds avoid quickly. Moths have night-time predators, like bats. Bats cannot see these signs and use sound for a similar warning effect. Tiger moths make a loud noise perfectly within bats' range of hearing, and timed to be in between bats echolocation calls, so that they are as unmistakable as possible. Some predators learn to exploit prey. One bat species, from South America, has 'cracked the frog code'. Their main prey, frogs call each other to attract mates. Bat has tuned hearing to hear these calls, and can distinguish different species of frog in order to capture edible one and avoid poisonous one. The same is true for signals, between males and females. Male displays are designed to attract attention of choosy females. A female spends some time deliberating over who to mate with. Each species occupies its own habitat (Figures 1.1 and 1.2) or biological niche based on interaction with other species and with natural processes in the environment. If one's 'most desired' niche is taken by someone else, competition arises. Individual that is more determined, and has most resources to hold that niche wins. This competition can take both visible and invisible forms. The 'loser' then either finds a different behavioural strategy in group or distances himself from the group in hope of finding a different society or niche.

FIGURE 1.1 Atlantic walruses in marine habitat.

FIGURE 1.2 A *Squilla* inhabits marine habitat.

Each species has a set of 'survival strategies'. Specialization also occurs within a species. Sex is the most basic form of intra-species specialization, but other kinds are possible. Ants are a perfect example of extreme specialization. Ant's colony typically contain different 'castes'—a queen or queens, drones whose only function is mating, nurses, workers, and soldiers. Some species have complex group interaction and rely on teamwork to survive. The more a species relies on teamwork to survive, the greater the degree of specialization among its individuals. Members of such species often divide up into guards, explorers, hunters, warriors, nurses, helpers, workers, etc. sometimes temporarily, and sometimes permanently. Species that rely on cooperation for survival have internal mechanism(s) that motivates them to seek out other members of their species for cooperation and interaction. If one baboon in a pack is highly aggressive, this affects behaviour of other baboons.

Social animals are constantly adapting to changes in group structure and changing roles around them. Cooperation without competition flourishes where individuals pursue different niches and gain benefits from each other's niches. If one individual is very aggressive, he/she might form a partnership with another individual that is not at all aggressive, but who likes to reap the benefits of being around aggressive individuals—much like hyenas benefit from lions' hunting skills. This 'helper' would also need to bring benefit to aggressive partner, perhaps by offering useful 'advice' by copying other's actions in critical situations or by befriending those offended by his partner, so that they do not rebel against him. Thus, symbiotic relationships form, based on nonconflicting, mutually beneficial niches. If an individual is forced by circumstances to occupy a less natural niche, he/she switches to a more desirable niche as soon as circumstances permit. If two or more individuals 'found themselves' in their respective mutually beneficial niches, their cooperation and friendship might continue indefinitely increasing each other's chances for success within community and for survival in environment.

Within complex animal societies, group and individual specialization is common. A society might have groups of hunters, explorers, caretakers, and guards. There would be intense competition in such groups for the same niches. When specialized groups are formed, not every-one in a group must necessarily be a hunter, guard, caretaker, explorer, etc. 'at heart'. For a group to cooperate effectively, some individuals occupy roles that are auxiliary to main task that group performs. Nature tends to maintain a 'dynamic equilibrium' of opposing behaviour strategies, within communities of social animals. High concentration, of one behavioural trait, naturally creates a high demand, for opposing or complementary traits. If too many members of a group are aggressive and dominant, that group falls apart, and individuals have less chance of survival, thus having fewer chances to pass on their aggressive genes.

Societies of social animals remain in a constant state of flux, with elements of both competition, and cooperation, with a multitude of both 'harmonious' and 'forced' partnerships, and with individuals occupying different niches in different groups—some more natural, others less so. Within such imperfect societies, separate individuals may be able to 'get away with' feeling satisfied with their overall situation, experiencing fulfilling relationships and realizing their desires to a great extent. To experience this, one must among other things manage to consistently find satisfying behavioural niches, and co-operative partnerships, where one's natural traits are in demand.

Behaviours are controlled by the endocrine system, nervous system, environment, and genetic make up of the animals. Behavioural complexity is related to complexity of nervous system. Generally, organisms with complex nervous system show a greater capacity to learn. Behavioural studies involve not only what an animal does, but also when, how, why, and where the behaviour occurred (Lehner 1979). Narrow range environmental conditions are generally favourable to organisms. Particular characteristic of particular animal helps to survive in a particular environment. These characteristics are called *adaptations*. Adaptations are structural and functional or behavioural. Internal stimulus triggers a number of functional behavioural responses. For example, flight behaviour is associated with the structure like wing (Figures 1.3 and 1.4). Most organisms place themselves in a favourable environment, through orientation that occurs through taxis, and kinesis. Taxis is deliberate, more or less automatic movement, towards or away from stimulus. Kinesis is a random movement, not orientated towards or away from a stimulus, and are seen when animals are exposed to light, sound, touch, heat, or chemicals. Broadly, two forms of behaviour are innate or inborn, and learned behaviour. Behaviours that change in response to input, form an animal's social and physical environment, are learned. Innate behaviours are determined by nervous pathways and are inherited involving reflex actions. Same stimulus produces

FIGURE 1.3 Flight in a bird.

FIGURE 1.4 Flight in a butterfly.

same response, due to 'built-in' nervous pathways. Learned behaviour depends on both short- and long-term memory, requires habituation, and includes imprinting, classical and operant conditioning, and insight learning. In complex animals, specific behaviour involves integration of innate and learned behavioural patterns, although distinctions between innate and learned behaviours are not always clear cut.

European school was founded in 1930's by Lorenz (Figure 1.5). He collaborated with Tinbergen, to establish 'ethology', as 'biological study of behaviour'. Tinbergen's (Figure 1.6) book, '*The Study of Instinct*', perhaps provides the best introduction, to ethological approach. American approach to animal behaviour roots in Watson's work. Watson (1925) introduced experimental behaviour approach in the book, '***Behaviourism***'. He was influenced by

FIGURE 1.5 Konrad Lorenz.

FIGURE 1.6 Niko Tinbergen.

Pavlov's work on classical conditioning. Watson's ideas were adopted by experimental psychologists interested in learned behaviour. Skinner believed that behaviour is shaped by reward. Ethologists were concerned, with identifying, and describing species-specific behaviours, with little variability between members of same species, understanding genetic basis for behaviour, and preparing a documentation of behavioural patterns. Lorenz believed that species-specific behaviour develops without experience of stimuli to which it responds or without practice of motor patterns that it performs. Watson tried to develop a psychology, for utilization by 'educator, physician, jurist, and business man'. Acquired and learned behaviour, associated with the view, is that behaviour is the result of 'nature'. Ethological approach typified by Lorenz, Tinbergen, and von Frisch was concerned with behaviour of organisms as it expresses in their environment. Some scientists emphasized on mechanistic underpinnings of behaviour on Norway rat in a controlled laboratory setting. B.F. Skinner led to development of use of learning paradigms. Skinner Box remains an important tool in animal psychology. Learning theorists sought similar mechanisms in animals that allow animals to respond to their environment. Comparative psychology includes many developments in psychological sciences and spanned through behaviours of infra-human animals:

Many organisms live in groups, like herd, shoal (Figure 1.7), school, and flock. Some groups are temporary groups. Frogs gather together at a pool during mating season. Birds, migrate in large numbers, are temporary groups. The largest group, a swarm of locusts is also a temporary group. Some groups are highly organized. Herds of gnu and zebra stay in large groups, for their own protection. One gnu eating on its own, has a 100% chance of being caught by a predator. In case of two gnus, each has a 50% chance, and in case of 100 gnus, each has 1% chance. Pigeons flock together, which helps them to find a mate. Reasons for being in a group are completely selfish. Social animal groups show cooperation between its members. Some animals perceive geophysical stimuli, seismic, or acoustic waves at low frequency (below 50 Hz), electric field changes and olfactory stimuli, which may precede earthquakes. Some birds and fishes are sensitive to sound with frequencies below 40 Hz. Many animals perceive low frequency vibrations through skin. Certain fishes are sensitive to electric field changes as small as 10–5 V/m. Some mammals respond significantly to weaker fields. Release of gases from small cracks is perceived by some animals, before earthquakes. Animal behaviour reported in post earthquake period resembles fear or escape reactions, and ranges from mild response to bizarre behaviour. Such abnormal behaviour is species-typical behaviour, which may be triggered by a variety of stimuli not necessarily related to earthquakes. Many behaviours reported before geophysical events, other than earthquakes, such as thunderstorms, or volcanic eruptions are somehow similar. In 1950s, Chinese found that domestic animals sense earthquakes in advance. Rats coming out of holes, snakes rushing of their pits, goats refusing to eat or go into pens, chicken dashing out of coops, pigs squealing strangely, birds flying out at night, insects congregating in large numbers near sea shores, and cattle rushing to higher ground much ahead of earth tremors. Animals in Yala National Park in south Sri Lanka, might felt signs of tsunami 3 or 4 days ahead of strike on December 26, and there is no report of animals' death from there.

FIGURE 1.7 Shoal of fish.

Studies to relate matter to mind, via inner experiences of feelings towards defined physical stimuli were conducted. Main focus laid on way the subjects perceive experimental stimuli that were objectively measured, such as pure tones varying in intensity, or light varying in luminance. Senses including vision, hearing touch, taste, smell, or sense of time were studied. Freud believed that psychopathologies stemmed from culturally unacceptable, repressed and unconscious desires and fantasies of sexual nature. He stressed presumed value of dreams as sources of insight into unconscious desires. Influential approach to psychology understood psyche through exploring worlds of dreams, art, mythology, world religion, and philosophy. Behaviourism, a movement in psychology advocates use of strict experimental procedures to study behaviours in relation to environment. Behaviourists regard learning and environmental conditions as dominant effects on heredity. Learning was explored using artificial tasks in strictly controlled environments. Comparative psychology viewed science for explaining events, predicting, and controlling them. External environment was thought as main determining influences on behaviour. Pavlov realized that behaviour changes. Thorndike, using a puzzle box, demonstrated that behaviours followed by a positive outcome are often repeated, while those followed by a negative outcome are not at all. He postulated *law of effect*, where behaviour changes result from an action proved of consequences. Stimulus-response theory holds that emotional reactions are learned in the same way as other skills. Skinner contends that nearly all behaviours are shaped by complex patterns of reinforcement in a person's environment.

Late 19th century witnessed rise of modern neuroscience. Hypotheses that distinct brain functions were contained within specific regions of brain, emerged with evidence from patients with epileptic seizures, head trauma, and stroke. Extensive characterizations of neurons throughout the brain led to propose neuron doctrine, i.e. function of brain as a product of electrical and chemical activity of individual neurons. Further support came from discovery that muscles and neurons show electrical excitability which could be transferred to neighbouring cells. Synapses, as sites of communication between neurons, were identified. Ethology emerged in diverse roots and aimed to provide a more comprehensive study of animal behaviour than previously available. Heinroth focussed on adaptive value of behaviour. Holst showed that behaviours could be elicited experimentally even in absence of any sensory input. Interest in genetically programmed elements was contrasted to those subject to learning. Causation of behaviour is viewed in context of a metaphor, psychohydraulic model. An animal that has had a good meal will have a low drive to eat more. As it goes for a while without eating, the drive builds up. If food is then encountered, animal will try to consume it. If no good food is encountered, less preferred items will be accepted. This model faced much criticism, as performance of many behaviours appears to increase tendency to perform it again, rather than bring about a decrease. Frisch used classical conditioning to show that fish could hear and bees are able to perceive colours. Three ethologists shared the 1973 Nobel Prize for Physiology or Medicine in recognition of their work over 30 years as founders of ethology.

Behavioural genetics explores heritability of behavioural traits. Studies on *Drosophila* in particular have revealed many aspects of behavioural genetics. Individuals live and die, but genes survive. The ones that survive the best are those that programme their survival mechanics. Genes are not visible to selection, they achieve their effect by influencing phenotypes. Genomes for any two individuals, except for those of identical twins or cloned animals are unique. Such differences, i.e. polymorphism produce different versions of same genes. Despite these differences in genetic makeup, major commonalities across much of code, are shared by all forms of life. Individual's phenotype emerge as interaction of genetic information with environmental conditions. "A man's natural abilities are derived by inheritance, under exactly same limitations, as are form, and physical features of whole organic world". Behavioural ecology is the study of ecological and evolutionary basis, for animal behaviour. Sociobiology aims to explain behaviour, in all species, by considering evolutionary advantages of social behaviours, based on idea that humans will tend to act in ways to improve their own reproductive success and this will result in social processes conducive to genetic fitness. Evolutionary psychology attempts to explain 'useful' mental and psychological traits, like memory, perception or language as adaptations. Neuroethological approach stems from idea that nervous systems have evolved to address problems of sensing and acting in certain environmental conditions. Mechanisms of sensory processing have utilized a variety of model systems including moth's acoustic startle response to bat sounds, jamming avoidance in electric fish (Watnabe and Takeda 1963) or sound localization in owls (Knudsen and Konishi 1978). Advances in our knowledge of hormonal control of behaviour have profited from work on division of labour in bee colonies, changes in sexual behaviour due to altered level of sex hormones (Carter 1992) and aggression in lizards (Greenberg 1977).

Mimicry is a kind of adaptive resemblance in which an animal animate other animal or objects in form, colour, and attitude, to escape from eyes of a predator, or exhibit an apparent aggressiveness which is not at all real. Biomimicry is a relatively new branch of science that studies nature, its models, systems, processes and elements to solve human problem sustainably. Biomimicry came as an ethical stance in late 20th century. One major application of biomimetics is the field of biomaterials involving synthesizing natural materials and applying this to practical design. Animal models are being used as inspiration for different type of robots. Biomimicry would be a part in ensuring long-term sustainability of human development. Currently, behavioural studies collect information on species-by-species, to solve conservation problems.

In biology, two types of casual explanations are generally recognized—**proximate and ultimate** explanation. Former explanation appeals to motivational variables, experience, and genotype as the case of behaviour. Latter explanation refers to selection pressures, and other factors that cause evolution of behaviour. These two explanations are independent, complementary, serve different purposes, and one can not replace other. Rather than distinguishing just between proximate and ultimate causes, Tinbergen (1963) argued that four questions—causation (mechanisms), function, development (ontogeny), and evolution must be answered to understand behaviour. Dichotomy between Ethology and Comparative Psychology with their concerns for adaptation and mechanism can be succinctly described as a concern for ultimate versus proximate causes. The pursuit of ultimate causes for "Why Questions" is described (Ernst Mayr 1961). The proximate causes are concerned with the way the world works on "How Question". Tinbergen's 4 study areas block out into ultimate versus proximate causation. Tinbergen's causation is concerned with Proximate Causation. Development is the proximate cause. The function of adaptive value or fitness is directly related to evolution (Curio 1994).

Behaviour has no clear beginning or end. Analysis of behaviour starts innocently to describe interactions of organisms with the environment. However, trends in animal behaviour research, during 1968–2002, shows the rise of sexual selection and mating behaviour, as prominent areas of research. Researchers had been studying magnetism in smaller animals, and looking for a way to extend their work to larger species. Researchers have long known that certain bacteria, birds, fishes, whales, and even rodents have minute organs in their brains containing particles of magnetite, that act like a compass. New results are the first hint that larger land-based mammals may also have such organs. Findings are 'very interesting, and not at all implausible'. Cows are known to align their bodies facing uphill, facing into a strong wind, to minimize heat loss, or broadside to sun on cold mornings, to absorb heat, but pictures taken at many locations at different times of the day, and in generally calm weather minimized the effect of environmental factors. 'We have to remember that whales are descended from a common ancestor of cows, so this is not a surprise given what we know about whales'. Bats, birds, bees, and whales, all use their magnetic sense to help navigate. Kirschvink recently reported that if a pulsing magnetic field is applied to bats perpendicular to Earth's field, the animals will change the direction of their flight by 90 degrees. What the benefit could be for cows, however, remains a mystery. It might help them find their way to home, experts said, or perhaps, it is simply a vestigial sense that is no longer used for any purpose.

Cause, Development, Evolution, and Function

Tinbergen's viewed genes, ecology, physiology, development and learning, evolution,(http://bio.research.ucsc.edu/~barrylab/classes/CHAPTER_PDFS/Chap_1_History.pdf) and sociality which provide a schema to know the adaptation in behaviours at various temporal scales. Paul Sherman (1988) would add Cognition to the list. Cognitive theory is an outgrowth of development and learning, and should be included in those categories. Behavioural Ecology is changing to generalize the classically-developed ideas of Psychology and Cognitive Processes into wild populations (Real 1994). Genes and Ecology cover basics of Darwinian Natural and Sexual selection as they apply to animal behaviour. To copeup with variation in environment, organism acquires adaptation in physiology to promote successful survival or reproduction. Such changes act at level of metabolism, endocrinology, neurophysiology, or any proximate mechanisms which help organism to cope with both abiotic and biotic environmental factors. Developmental changes and learning occur from ovum or sperm to maturity that is adaptations to a particular way of life unfold during lifespan of an organism. By understanding genetic, developmental, ecological, physiological and cognitive processes, the concepts of behavioural evolution could be realized.

Phylogeny and Constraints in Behavioural Evolution

Organisms are constrained by their own phylogenetic effects. In evolution of a lineage, adaptations pile on top of each other as result of which closely related organisms share similar features and further constraining the acquisition of new adaptation. Functional and structural constraints arise from material properties of organism. More development constraints originates during embryogenesis. Constraints are found at the level of proximate causation. Consider a phylogenetic example taken from 2 lineages—birds and mammals. All birds lay eggs, as the common ancestor of birds, reptile-like dinosaur laid eggs. Most mammals bear live young because in remote past a new kind of mammal-like reptile evolved different mode of life and passed this novel trait onto all subsequent species in lineage or phylogeny. Exception to mammalian generalization includes egg laying mammals like *Echidna* and *Platypus*. The monotremes branched off from main stock of mammals so early in past that they retain the more ancestral mode of egg-laying. Such differences in reproductive mode (egg-laying versus live-bearing) constrain both birds and mammals as in terms of parental care that evolve in each group. Extra adaptations in mammals may constrain evolution of parental care. Evolution of mammary gland as primary source of nutrition tends to lead to species of mammals displaying a preponderance of maternal care. There are few examples of male care in mammals as compared to birds. Many bird species have evolved male and female parental care behaviours so that rearing the young can be accomplished by both parents. Some bird species secretes a milky substance in part of their digestive system called crop. Because male and female birds have crop, in theory, both parents evolve to produce a milky substance as a form of parental investment. Phylogeny difference for male versus female care between mammals and birds leads to additional differences in evolution of mating systems in these 2 groups. To understand phylogeny constraints that operate on other traits, knowledge of the proximate mechanisms, and natural selection is required.

Societal and Cultural Evolution

Social evolution and advent of culture introduce another mode of long-standing transmission of behavioural traits between generations. One need to walk into library to realize impact of mass storage of human culture. Library is a storehouse from where information is passed onto subsequent human generation, but there is no genetic basis to information in libraries. Theory of cultural evolution holds that many behavioural changes in humans might have a largely non-genetic component arising from cultural transmission of information. Behavioural genetics studies heritability of behavioural traits in animal behaviour. The field has emerged due to overlap of ethology, psychology and genetics.

Nature versus Nurture

Darwin shifted the way we view animal behaviour. However, bevavioural biology has a tradition that were present before Darwin (Drickamer and Vessey 1986). Ethology deals with the evolution and functional significance of behaviour. C.O. Whitman in 1800's coined the term *instinct* to denote display patterns in pigeons. The ethogram, a graph of time course or switch points in of the behavioural sequence, became way of categorizing species-typical behaviours. Many instincts are triggered by environmental stimuli. von Uexkull termed such triggers of instinctive stereotyped behaviours sign stimuli. Courtship display of male three-spined sticklebacks fish is triggered by a classic stimulus. The zig-zag dance in male stickleback is triggered by enlarged belly of the female. Males dance to entice the female to enter nest built by that male. Early ethological work was mostly due to Konrad Lorenz and Niko Tinbergen. Lorenz is famous for work on genetically programmed behaviours in young for imprinting. The young geese forming an image of parent after hatching is a case of classic imprinting. If hatchlings first encounter a human like Lorenz, they imprint and follow him around as if he was their mother. Karl von Frisch studied communication in bee. Tinbergen formulates a method for studying animal behaviour (Tinbergen 1963). Ethological approach had a strong Darwinian tradition underlying its development. Most study in ethology aimed at understanding the ultimate evolutionary reasons for behaviour. Tinbergen listed 4 areas of inquiry that understands issues of behaviour. Ethology and comparative psychology formed 2 different schools of thought on the causes of behaviour. Originating in Europe, Ethology looked to genetic underpinnings of behaviour. The field of comparative psychology originated in America, has seen behaviours as product of environment. Differences between ethology and animal psychology causes debate in behaviour that captured in "nature versus nurture" which must be looked in terms of interaction nurture between gene and environment.

Behavioural Ecology and Sociobiology

Behavioural Ecology attempts to synthesize both evolutionary traditions of Ethology, and mechanistic studies of Comparative Psychology. This is a new movement compared to traditions of ethology and psychology and has grown over the last 3 decades. Behavioural ecology looks at interaction between organisms in natural environments (Krebs and Davies 1987). Both mechanistic underpinnings of behaviour and fitness consequences of behavioural traits are important which can be traced back to Tinbergen's 4 areas of study namely causation, development, evolution and function. Behavioural ecology draws issues of energetics and physiology (Calow 1987), measures differences in survival and reproduction of behavioural traits and is broad than study of behaviour. Behavioural ecologists use behavioural traits to maximize energy acquisition as proxy for fitness traits. Development of optimal foraging during the 70's and 80's added a distinct theoretical perspective to the Behavioural Ecology. New approach to study behaviour involves a study of social systems in diverse group of organisms. This field has taken off since publication of Sociobiology by E.O. Wilson (1980). As some ideas were applied to humans, theory is controversial. Sociobiology with strong Darwinian tradition attempts to explain evolution of social systems. The field of Evolutionary Psychology has co-opted the approaches of behavioural ecology and sociobiology to explain diverse human behaviours like foraging and siblicide. The same "organic rules" that shape other organisms are also applicable to human. This area is ripe for debate as researchers attempt to derive explanations for behaviours displayed by humans in modern society.

1.1 HISTORY OF ANIMAL BEHAVIOUR AND BIOMIMICRY

Study of animal behaviour has burgeoned over past 4 decades. Natural selection theory of Darwin (Figure 1.8) set the stage for considering animal behaviour in evolutionary terms. In his book, *The expression of the emotions in man and animals* (1872), Darwin elaborates some facts of animal behaviour. Major theoretical developments in Britain in second half of 19th century had led at first, to empirical studies of animal behaviour that were original and significant. Acceptance of experimental behaviour theory, to serve societal objectives was soon demonstrated to Pavlov's work, after Russian revolution (Wilson, D.S. 2002). Spalding examined relationship between instinct and the environment. In early 1870s, he established existence of inborn behaviour (Spalding 1873). Lubbock's analysis and explanation of insect societies, as in ants, bees and wasps was carefully set before as wide public as possible, so that lessons might be learned from insects about social organization so that achievement of science in gleaning this information could be properly acknowledged. Romanes (1884), published anecdotal accounts of animal behaviour. Hackle described Romane as the first to recognize psychology as ultimate expression of Darwinism.

FIGURE 1.8 Darwin.

Lloyd Morgan was a strong advocate of an evolutionary approach in comparative psychology, and later set out a doctrine of emergent evolution of consciousness. Experimental work described in Morgan's '*Habit and Instinct*' (1896), illustrated his theory of imitation and also approached the problem of habit formation, and learning in birds by 'trial and error'. His studies bear important contribution in application of laboratory methods for behaviour of higher vertebrates. Thorndike initiated animal experimentations that led him to form his 'connectionist' theory, which he later retained in face of behaviourism. Between 1900 and 1940, animal psychology represented an opportunity for British women academics, to make further inroads into academic science and a high proportion of them engaged in animal psychology (Mayers 1961). In early 20th century, naturalists and zoologists increasingly using the term 'ethology', who stepped in to fill the vacuum created by others, who became enmeshed in environmental conditioning, and learning theories. According to Durant "it may be said that ethnologists shared a distinctive view of animal's behaviour and of the way in which it should be studied". This view held that animal's possess specific innate characters, which can be understood often by direct analysis with human character on the basis of prolonged and sympathetic observation (Durant 1986). Design of Thorndike's best known experiment, that involving cats in puzzle boxes (Thorndike 1911) was at fault, because it did not allow animals to apply their full problem solving potential, but his procedure was excellent.

Early history of international study reveals a series of oscillations in manner in which investigations were conceived, and carried out. In relatively quiescent period of 1900–1940, laboratories for animal behaviour were formed in Britain. Opportunities for physiological psychology as a biological science, through collaboration with neurophysiologists, such as Sherrington and then Adrian were ignored partly because of conservation or philosophical orientation of psychologists themselves. Behaviourism was methodologically strong although theoretically weak, and filled to elucidate complex behaviour. Until 1940s, experimental studies were limited and received little official, institutional, or public recognition. Lorenz and others, in late 1970s, stressed that behaviour is something that an animal "has", as well as, what he or she "does", and is a phenotype on which natural selection can act. Recent advances in behavioural biology has been made with contribution from workers, like McFarland (1993), Pearce (1997), Dewsbury (1992), Martins (1996), Huntingford (2003) and Stamps (2003).

Homo sapiens is increasingly being studied within evolutionary adaptationist, selectionist framework, with various labels including evolutionary psychology, human behavioural ecology, and human sociobiology, collectively called Human Evolutionary Psychology (HEP). Sexual selection and sex differences are especially prominent in recent

HEP research. Many other topics are parent–offspring relations, reciprocity and exploitation, foraging strategies and spatial cognition. Inherent adaptationism of psychological science is obvious in study of sensation and perception, where proposed mechanisms and processes are labelled in terms of information processing problems that they solve, like edge detection, sound localization, olfactory discrimination, image stabilization, light-level compensation, and face recognition. Behavioural ecologists, and sociobiologists have been making real progress, in understanding non-human social psychology and behaviour, apparently because they have partitioned the subject along lines of discrete and real-world problem domains, and this partitioning carves the psyche more nearly at its joints.

Many prominent contributors to development of HEP came to the subject from back-grounds in non-human animal behaviour, ecology, and evolutionary biology. Such founders of human ethology as Jones, Eibl-Eibesfeldt, Hinde, and Tinbergen published major works on behaviour of other vertebrates before turning to human animal. In 1970s, Alexander was an important investigator of anthropological field-work within an adaptationist, selectionist framework, while DeVore, Trivers and Wilson inspired a number of students to research on both human beings and other animals. Symons, the author of the book, *The Evolution of Human Sexuality*, wrote an earlier monograph on rhesus monkey play. In recent work on sexual selection and sex differences, researchers have taken current concepts from theoretical biology and from studies of other animals, and have applied them without essential modification to study of *H. sapiens*. Workers have investigated sex differences in polygamous inclinations, sexual jealousy, intrasexual competition, and mate choice criteria from this perspective. Issue that has been developed in non-human sexual selection research and applying it to human case include explorations of sex differences in spatial information processing in relation to sexually selected ranging behaviour (Gaulin and Hoffman 1988); application of 'polygyny threshold' model to human marriage transactions, investigation of whether shared MHC alleles have same effects on sexual attraction in humans that they have in mice (Wedekind et al., 1996); efforts to elucidate prevalence and consequences of facultative polyandry, extra-pair paternity and sperm competition (Scheib 1991); and a spate of recent work on fluctuating asymmetry as an indicator of 'developmental stability' and phenotypic quality, especially, but not solely with reference to the problem of assessing whether women's mate choice psychology exhibits adaptation for gaining 'good genes' benefits independent of material benefits.

Sexual selection has not been the sole preoccupation of HEP, and works on other topics, like violence against children (Daly and Wilson 1996) in which risk factors have been predicted from a general theory of evolved facultative variations in parental solicitude. Orians and Heerwagen's approach to environmental aesthetics was viewed as a reflection of psychological adaptations for habitat selection, efforts to understand peculiar human phenomena as language, artistic production and appreciation, humour and governance routinely invoke the concepts of sexual selection, evolutionary game theory, kin selection, Zahavian handicaps, and other theoretical staples of contemporary animal behaviour research (Miller 1998). Human linguistic ability greatly facilitates the collection of information on everything from matrilineal kinship links, marital and sexual histories, and bride prices paid for women of different reproductive values, desires, preferences, attention priorities, beliefs, and grievances. Unfortunately, articulateness of human animal is a double-edged sword. It has made possible the investigation of questions that one can scarcely imagine how to address in other animals, but it has also tempted researchers down a variety of garden paths (Daly and Wilson 1999). In simple terms, social Darwinism involves the idea that it is possible to apply some of Darwin's original observations about the nature of animal behaviour/evolution in relation to human social groupings.

Biomimicry (from bios meaning life, and mimesis meaning to imitate) refers to studying nature's most successful developments, and then imitating these designs, and processes to solve human problems. It is referred as "innovation inspired by nature." For most of our history, humans have treated nature as a gigantic warehouse and commissary. Natural world has been a source of raw materials, food, and ideas for architects, artists, and engineers. Rapid advances in biological science and growth of nano-technology have resulted in a new field of intentional biology, along with its two main subfields, biomimicry, and synthetic biology. Biomimicry views nature as a source of new materials, structures, and processes, and it aims to create new lightweight technologies and organizational forms, based upon natural analogs. Synthetic biology views nature as code, as sets of instructions, guiding the lifestyle of living organisms. Many aboriginal Alaskans stalk seals in a similar fashion to the way polar bears hunt smaller marine mammals. The "sciences of the nervous system" did not see the light of day until the late 19th century, with the combined development of physiology, histo-anatomy, and anatomical clinical science. The reflex theory which associates every action with a stimulus, the neuronal theory which takes the view that the nervous system consists of an articulation of discrete elements or neurons, and the theory of locations which links the mental functions and faculties with specific areas of the brain, all influenced the materialist philosophies. Later, the growth of ethology based on rigorous observation of animal behaviour caused attention to turn once again to the nerve structures underlying stereo-typed and innate forms of behaviour. A first synthesis was made in the late 1950s between experimental psychology, ethology and physiology, largely as an outcome of the development of techniques to study the brain directly by fitting intracerebral electrodes to record the electrical activity of the structures or to stimulate them.

The 1960s brought a veritable revolution with the discovery of the chemical messengers used by the nerve cells to communicate between themselves and the identification of their receptors. Computer science provides a theoretical model for understanding the brain; the analogy with the computer serves as a frame of reference for the cognitive sciences. The techniques of cerebral imaging, made possible by computer science, permit an increasingly precise visual display of the activities of the different regions of the brain and provide an anatomical foundation for the study of complex functions.

Genetics and molecular biology have been the fundamental advancements of the last ten years. It is generally accepted that the genetic patrimony of a human being comprises some 100,000 genes; the brain structure alone uses more than 50,000 of them. Although their functions are not always known, new genes expressed in the nervous system are constantly being discovered. A final phase involves the study of the development of the brain during the evolutionary processes (phylogenesis) and in the individual (ontogenesis). The discovery of regulating genes, guiding and adhesion molecules, cell growth and death factors, enables us by explaining the structure of the brain, to understand how it works, and opens up demiurgic prospects for its subsequent transformation.

At the molecular level, in addition to identifying the structure of the channels, neuro-transmitters and their receivers, the identification of abundance of molecules which interact in cascades enable us to understand the functional and molecular substrates of fundamental phenomena such as pleasure, suffering, dependence, and memory. The discovery of regulating genes, which are responsible for the development of the brain according to the plans of the species exposed to the influences of its environment, holds out considerable prospects not only for the understanding of the phylogenesis of the brain, but also of the way in which it works.

1.2. HUMAN BEHAVIOUR

Human behaviour is broadly determined by three factors—genetic endowment, environmental impact, and the way the individual collates every-thing to determine his long-term personality. Genes, as hardware shape entire physiology. Software is not genetically written in form of gene sequences. Software is knowledge and perception that one gathers from environment. Six unique human traits found in animals are:

Culture

Culture is simply the sum of characteristics, ways of living of a particular group learned from one another and passed down the generations and others. Primate species have unique practices, such as greeting each other. Killer whales fall into two distinct groups, resident and transients. Both live in same water with different social structures, lifestyles, communication, tastes in food, characteristic hunting technique, and interbreed. Non-vocal forms of sperm whale are different cultural clans of sperm whales, affected very differently by EI Nino events with different reproductive rates. In sperm whales and likely in other whales and dolphins, culture affects population biology.

Mind reading

Great apes, and some monkeys (Figure 1.9) understand deception and their understanding of others' mind is probably implicit, rather explicit, as in adult humans. Studies point seeing as a form of knowing, reading intentions, and goals.

FIGURE 1.9 A monkey can read the mind.

Tool use

Some chimps use rocks to crack nuts. A gorilla gauges the depth of water with equivalent of a dipstick. No animals perhaps wield tools with quite alacrity of New Caledonian crow. To extract tasty insects from crevices, they craft a selection of hooks and long barbed tapers called **stepped-cut tools**, made by cutting a pandanus leaf with their beaks. Animals understand function of tools, deploy creativity and planning to construct them.

Morality

Hungry rhesus monkeys would not take food offered by another monkey which received an electric shock in collecting food. The same is true for rats. Social mammals learn right and wrong of social interaction, moral norms that are extended to other situations, such as sharing food, defending resources, grooming, and giving care. Animals' ability to make social evaluation is foundation for animals' moral behaviour.

Emotions

Emotions allow bonding with others by regulating social interactions, behaving flexibly in different situations, and include elephants caring for a crippled herd member, a funeral ritual performed by magpies, a male baboon called nick to take revenge on a rival by urinating on her. Divers who freed a humpback whale caught in a crab line described its reaction as one of gratitude. Chimps perform excited dance when faced with a waterfall.

Personality

Individuals of same species living under same conditions vary in their degree of boldness. Personality ranges from coward spiders and reckless salamanders, to aggressive songbirds and fearless fishes. Personality traits evolve to help individuals survive in a wider variety of ecological niches. A review on social animals, like birds and bees, suggests that social interactions can alter gene expression in brain, to influence behaviour. Insight from a study of songbirds led by Clayton, showed that expression of a specific gene increases in forebrain of a zebra finch just after it hears a new song from a male of same species. The gene egr1, codes for a protein which regulates expression of other genes. Clayton's team emphasizes potential of social interactions, which alter gene expression in brain. Foragers produce a pheromone that signals other bees about food sources. If foragers are removed from hive, young bees develop into foragers much earlier in life than usual. Social information alters expression of same gene over a much shorter timescale within lifespan of a honey bee. Differences in gene expression occur over vastly different time scales, helps understand some of complex relationships between genes, brain, and behaviour.

Our early ancestors had understanding of habitat preference, movement patterns and sensory biology of prey and predators. Egyptians had a keen interest in temperaments and sensory abilities of animals. Plato believed human mind as an entity pre-existed somewhere in heavens before being sent down to a body on earth. Hippocrates believed that brain was the seat of sensation and intelligence. Aristotle viewed human nature to be constituted by formed habits, and individuals are born with no innate or built-in mental content rather their entire knowledge is built up gradually from experiences and sensory perceptions of outside world. Herophilus studied nervous systems and distinguished between sensory and motor nerves. Erasistratus distinguished between cerebrum and cerebellum. Galen told about behavioural deficits which resulted from acute head trauma, paralysis, and problem in spinal cord. Arabian scholars described cranial and spinal nerves, along with surgical procedures for neurological disorders. Descartes defines 'thought' as every activity of which a person is immediately conscious. He viewed body as a machine, and mind as a non-material entity. Leonardo da Vinci conducted studies of human anatomy. Empedocles seems to be first in recording universal and fundamental nature of cooperation, which underlies group action and struggle for existence. Shaftesbury, who lived before and after 1700, seems to be the first in modern period to recognize racial drives that go beyond personal advantage and can be explained by their advantage to group. Patten made general statement of fundamental nature of cooperation, when in 1920; he gave it a central place in his analysis of grand strategy of evolution. Deegener's distinctive contribution was a classification of different social levels, from the simplest sorts of artificial collections of animals to parasitism, and truly social life. His rating of these different aspects of sub-social and social life in one long outline has the great merit of showing no hard and fast lines which can be drawn between social and sub-social organisms, and social communities are natural outgrowth of sub-social groupings. Morton Wheeler's book on social life among insects appeared in 1923, is a noteworthy general summary of insects including termites, bees, wasps, ants (Figure 1.10), and beetles (Figure 1.11). Social habit has arisen some 24 distinct times, in about one-fifth of known major divisions of insects. Animals do not always act for their own best interests; still they do so to a certain degree, or are exterminated in long-run. Breeding aggregations of worms, crustaceans, fishes, frogs, snakes, birds, and mammals have long attracted attention. Locusts in migration swarm out of sky in the Sahara borderlands in southern Russia, in South Africa, and on Malay Peninsula in terrorizing numbers. Migrating army worms and chinch bugs present impressive aggregations pressure to cause evolutionary changes.

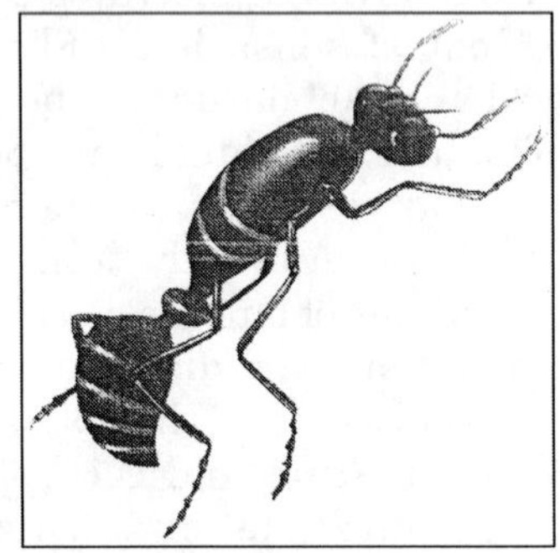

FIGURE 1.10 An ant.

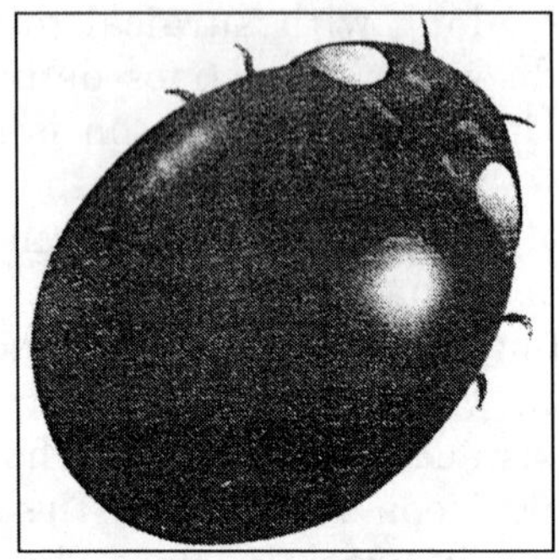

FIGURE 1.11 A lady bird beetle.

Many problems in human society are related to interaction between environment and behaviour, or genetics and behaviour. Socio-ecology, behavioural genetics, and behavioural ecology deal with such problems. Studies on chimpanzees and monkeys have illustrated the importance of cooperation and reconciliation in social groups, and provide new perspectives to view, and ameliorate aggressive behaviour among human beings.

Methodology of studying animal behaviour has tremendous impact in psychology, and social sciences. Behavioural study of humans would be much diminished today without the influence of animal research. Charles Darwin's work on emotional expression in animals has important influence on study of human emotional behaviour. Comparative study of behaviour, over a wide range of species, can provide insights into influences affecting human behaviour. We might learn how to minimize aggression, through understanding woolly spider monkey's way of avoiding aggression. Understanding the differences in adaptability between species that live in a variety of habitats, versus those are restricted to limited habitats, we can lead to improving our adaptability with our changing environment. Basic research, on circadian and other endogenous rhythms in animals, has led to research relevant to human factors, and productivity in areas, such as coping with jet-lag, or changing from one shift to another. Research on animals has developed many important concepts relating to coping with stress.

1.3 SIGNIFICANCE OF ANIMAL BEHAVIOUR RESEARCH

Behaviour helps in understanding vexing conservation problems, like saving endangered species, assessing environmental quality, and designing nature preserves and evaluating the importance of human-related threats to survival in other animals. Birds collide with windows for various reasons. Visual cues, migration, territoriality, and habitat preference explains birds suffering and high mortality from the urban threat (Hager et al., 2013). Animal behaviour aids in assessing the fate of climate change (Parry et al., 2007). Climate changes have already altered the timing of animal migration, pollination, and breeding (Hager 2012). Studies with the non-human primates (Hager 2012) offers valuable perspectives into causes and evolution of social, and reproductive human acts (Breed and Sanchez 2010). Research on the behaviour of non-human animals (Hager, 2012) has improved the human condition including technological invention (Hager 2012) (flight via aircraft) and minimizing health risks. Earthquakes could be successfully predicted through toads (Bufo bufo), which detect and respond to a seismic event before it happened (Grant and Halliday 2010). Perceptions of external world through hearing, vision, and smell give information that lead to change in behaviour. Perceptions of internal world including (Breed and Sanchez 2010) hunger, sex, pain, and provide motivations for behaviour. Animals integrate external and internal inputs to fix behavioural priorities.

Animal behaviour, the bridge between molecular and physiological aspects and ecology is the most important property of animal life that link between organisms, environment and nervous system. It plays critical role in the biological adaptations. Beauty of an animal includes its behavioural attributes. Studies by de Waal on chimpanzees and monkeys showed the importance of cooperation and reconciliation in social groups and provided new perspectives to ameliorate aggressive behaviour in human. Methods used in animal behaviour has tremendous impact in psychology and social sciences. Jean Piaget began his career with study of snails and observed carefully about behavioural observations on human cognitive development. J.B. Watson began his study of behaviour biology with gulls. Experimental design, observation methods, and attention to nonverbal communication signals were developed in animal behaviour before application to human behaviour. Charles Darwin's work on emotional expression in animals has important influence on many psychologists, like Paul Ekman's study on human emotional behaviour. Harry Harlow's work on social development in rhesus monkeys and work of Overmier, Maier and Seligman have importance to theories of child development and psychiatry. Comparative study of species gives insights into influences affecting human behaviour. The woolly spider monkey found in Brazil exhibits non aggressive behaviour among group members. Avoidance of aggression in these monkeys could help minimize human aggression. Bat uses sonar to locate objects that has led to the use of sonar techniques in applications from military to diagnostics of foetus. Knowledge of chimpanzees using language analogues have developed new technology that has applied to teaching language to disadvantaged human populations. Research on circadian and endo-genous rhythms is relevant to human and productivity in areas like coping with jet-lag.

1.3.1 Animal Behaviour and Neurobiology

Sir Charles Sherrington, an early Nobel Prize winner, developed a model for structure and function of nervous system, based only on a close behavioural observation, and deduction. Seventy years of subsequent neurobiological research has supported the inferences, Sherrington made from behavioural observation. Neuroethology, the integration of animal behaviour and neurosciences, provides important frameworks, for hypothesizing neural mechanisms. Careful behavioural data allow neurobiologists to narrow the scope of their studies, and to focus on relevant input stimuli, and attend to relevant responses. Animal studies show that quality of social and behavioural environment have a direct effect on immune system functioning.

1.3.2 Animal Behaviour and Conservation

The behaviour of animals often provide the first clue for environmental degradation. Changes in sexual and other behaviour, occur much sooner, and at lower levels of environmental disruption. Field studies of natural behaviour are vital to provide baseline data for future environmental monitoring. Basic research on how salmon migrate back to their home streams, started more than 40 years ago, by Arthur Hasler. This has taught us much about the mechanisms of migration. Understanding predator-prey relationships can lead to introduction of natural predators on prey species. Knowledge of honeybee foraging behaviour can be applied to mechanisms of pollination, which in turn is important for plant breeding, and propagation. An understanding of foraging behaviour in animals can lead to an understanding of forest regeneration. Many animals serve as seed dispersers and thus are essential for the propagation of tree species, and habitat preservation. Conservation of endangered species requires knowledge about their behaviour (migratory patterns, home range size, interactions with other groups, foraging demands, reproductive behaviour, communication) in order to develop effective reserves, and effective protection measures. Reallocation or re-introduction of animals is not possible without detailed knowledge of a species' natural history. Basic behavioural studies on reproductive behaviour, have led to improved captive breeding methods for whooping cranes, golden lion tamarins, cotton-top tamarins, and many other endangered species.

1.3.3 Animal Behaviour and Animal Welfare

Animal welfare without knowledge of behaviour is difficult. Animal behaviour biologists look at the behaviour and well-being of animals in laboratory, and field.

1.4 APPLIED ANIMAL BEHAVIOUR

Traditionally, humans dominate domesticated animals. Our role in their social environment varies with the species in addition to frequency and duration of the relationship. This influence has reached another phase when animals are kept in husbandry systems. Hediger (1964) proposed a model of human–animal relationships with a behavioural element involving knowledge of animal behaviour which was later modified by McBride (1978). The interface (McBride 1980) is defined as all the contacts, interactions, transactions, and relationships between domestic animals and environment with the man. The active, direct contact occurs when humans and animals come into physical contact and interact with each other. The passive, direct contact occurs where humans are not in physical contact with them. Automatic services, usually feeding and the provision of water, are active, indirect interfaces. Fences, shelters and their design features are the passive, indirect interfaces.

Dominance Concept

The first study of dominance relationships in groups of animals was due to Schjelderup–Ebbe who studied the organization of flocks of chickens (1935) and developed a concept of simple dominance where one individual had preferential rights over another. This is called *peck-right* of one hen over another. Allee (1949) noted that irregularities such as triangular relationships often occurred in a simple straight line hierarchy. A hen with medium or even low status might have the peck-right over some individual that out-ranked her in the linear hierarchy. Horses use coalitions so that affiliated pairs in a herd have an accumulative power to displace a third horse that normally out-ranks both of them on an individual basis. However, linear hierarchies remain a useful construct. A dominance hierarchy is the system of space sharing in a group arranged on a priority basis that keeps friction to a minimum. Once the division of space is complete, there is no further challenge to the order unless a young member of the group matures significantly or an aged member becomes senescent (McBride 1971). To maintain a hierarchy, each animal must recognize all the members of the group and follow the dominance–subordinate relationship. In flocks of poultry (n = 80 birds), individuals did not move freely but remained attached to a site, so they learned to recognize a small subgroup (McBride and Foenander 1962). Such territorial behaviour should be encouraged in large flocks, such as are commonly found on modern broiler units (e.g. n = 10,000), since it reduces the number of conflicts when strangers meet.

Flight Distance

A tame animal has a flight distance of zero (Hediger 1964). The concept is of considerable importance in round-pen gentling of horses. Animals have a 'zone of safety' around them and when this zone is penetrated, the animals will move away. Retreating from the flight zone causes the animal to stop moving. If an animal starts to turn back, the handler should retreat. In round entraining, the flight zone is repeatedly penetrated.

Crowding and Over-crowding

Crowding begins, not when animals jostle each other, but when they are forced into the personal spaces of their neighbours (McBride 1971). Animals need space to walk on, space to lie on and also a personal space around them. Spacing systems operate in natural populations and crowding does not occur. In domestic animals kept close together, crowding may be an important problem, often affecting fitness and productivity. At this point, it is called over-crowding. As animals are forced into each other's personal space, interactions of a violent nature may occur. At high densities, it is difficult for the animals to avoid such intrusions as they cannot move away. Some adapt to the intrusion while others become so stressed that productivity is affected.

Social Facilitation

This behavioural phenomenon can cause problems in intensively housed animals. It is the tendency of animals to join in an activity like feeding. This activity requires a large space, which causes competition for the key resources leading to crowding stress. It may increase the amount of food being eaten. In nursing sows housed together, social facilitation increases the number of suckling bouts that has to be undertaken by all sows. Grunting by a nursing sow in one pen causes piglets in a neighbouring pen to approach their own dam and initiates a suckling bout.

Stress and Its Measurement

Selye (1976) commented that things such as cold, heat, drugs, sorrow and joy would provoke an identical biological reaction. These agents are known as *stressors*. Stress is often associated with effects due to "weaning stresses", and "transport stress" or it can refer to 'behavioural stress', which are problems associated with intensification. McBride (1971) used the definition of Lee (1966), who described stress as the pressures acting on individuals to cause strain. Fraser et al., (1975) suggested that in a veterinary context 'stress' is a profound physiological change in the condition of an animal. If an animal can avoid an unpleasant stimulus, it can remove itself from the stress. Unsuccessful animals try to adapt by the process of habituation. If they do not habituate themselves to the aversive conditions, they enter a state of learned helplessness or apathy in which they remain distressed but no longer attempt to make appropriate responses to improve their plight.

In the animals, continued arousal generates the General Adaptation Syndrome (GAS) of Selye (1976). There is a general pattern of alarm leading to homeostatic resistance to change and the stage of exhaustion follows if the stressor is severe enough and is applied for a sufficient length of time. Strain may develop which is characterized by a susceptibility to a variety of infectious or other environmental stressors. In other words, the animal has not been able to adapt to the environment. At this level, problems arise affecting the welfare of the animals. Often, there are comparisons of the stresses to which free-living and domestic animals are exposed. The sources of these stresses are different, but in both conditions there will be environmental challenges that may contribute to the well-being of an animal. It can either adapt or be successful or fail. The physiological reactions of animals to stressors are difficult to measure. Some measurements may require the slaughter of animals (e.g. weighing of adrenal glands), withdrawal of blood, etc. The use of telemetric transmitters taped to an animal, e.g., pig's back (Mayes 1982), can send signals of the animal's heartbeat. The refinement of microchips is allowing scientists to record large amounts of physiological data from animals wearing implanted and telemetric devices. Meanwhile, the use of ACTH stimulation tests and cortisol assays in samples ranging from blood, saliva, faeces and even eggs is becoming more commonplace as a means of plotting trends in physiological stress responses. However, the importance of diurnal rhythms and the possibility of physiological fatigue as a result of chronic stress mean that such measurements should not be interpreted in isolation from behavioural observations. Changes in behaviour, most notably the appearance of placement or stereotypic behaviours, may be an early indication of a stressful situation.

Review Questions

Short Answer Questions

1. What do you understand by the term behaviour?
2. Why do we study animal behaviour?
3. What are adaptations?
4. What are physiological adaptations?
5. Define taxis.
6. Define kinesis.
7. Define learned behaviour.
8. Define innate behaviour.
9. What is ethology?
10. Mention the reason for being in a group in case of animals.
11. What is meant by the term culture?
12. What constitutes the foundation of moral behaviour of animal?
13. What is meant by the term "personality"?
14. What is behaviourism?
15. Name the ethologists who shared the 1973 Nobel Prize.
16. Define behavioural genetics.
17. What is behavioural ecology?
18. What are the aims of sociobiology?
19. Define mimicry.
20. What is biomimicry?
21. Comment on proximate explanation.
22. Comment on ultimate explanation.
23. Name the four questions raised by Tinbergen to understand behaviour.
24. Who examined the relationship between the instinct and environment?
25. What is social Darwinism?
26. Define stressors.
27. What is the full form of GAS?
28. What do you understand by social facilitation?
29. What do you understand by "zone of safety"?
30. Define peck-right.

Long Answer Questions

1. 'Ants show a perfect example of extreme specialization of survival strategies'. Discuss this statement.
2. Write a note on significance of animal behaviour studies.
3. Describe the history of animal behaviour in brief.
4. Give examples of 5 survival strategies.
5. Write an essay on history of biomimicry.
6. Write an essay on applied animal behaviour in light of modern study.

Socioeconomic Improvement and Absolute Fitness

Fatik B. Mandal

Reproduction occurs in a behavioural ecological context to maximize the absolute fitness of many animals and human. Life history evolution attempts to maximize the survival of genes of a species in the next generation.

We care for our sons or daughters as they possess half of our own genetic materials, which have been explained in the "selfish gene" concept. The concept states that the gene increases its copy in the next generation by increasing reproductive success of its bearers and bearer's relatives. In the process of caring our children, we care for our genes. Caring for one child is equal to care for 2 grandchildren, or 2 nephews, or 4 cousins.

The ultimate goal of each species is to produce maximum number of offspring with exerting least energy and increasing the fitness. In fact, each organism attempts to produce maximum number of offspring, to pass on their genes to the next generation. This is genetically hard wired. Thus, biological meaning of life is the production of offspring through reproduction.

Reproductive effort requires that cost as time, energy, and risks to life, social reputation, or material resources in human. Reproductive effort includes mating effort, and *parenting* effort, which together contribute directly to fitness, besides nepotistic effort which concerns the investment made in the reproduction of genealogical sidelines. Marriage as an institution legalizes heterosexual *relationships* and reproduction in human, in the interest of a rapid harvest of one's own personal fitness.

Socioeconomic success is believed to secure reproductive success. It has been reported that socioeconomic status plays an important role in mate choice besides other factors like vigour, beauty, facial signals, intelligence, attractiveness, etc. so the fitness increases, when humans attempt to improve the socioeconomic basis of their existence and reproduction as this in turn results in finding a suitable mate. Thus, continual struggle for socioeconomic improvement may be explained from the evolutionary point of view.

Practically, in modern society, a woman prefers to marry an intelligent and wealthy man who would provide best care for their future offspring.

Source: http://www.associatedcontent.com/pop-print.shtml/content-type=article

Chapter 2

Trends in Animal Behaviour Studies

Aristotle, a great philosopher, offered the earliest recorded account of life about 2400 years ago. We have gathered vast information in the last decade about living organisms than over the past several centuries. The modern definition of life was recorded in a book, *What is Life* published in 1943. The author, a physicist and a Nobel Laureate Erwin Schrödinger, observed that life defies the laws of physics. Schrödinger's observation stimulated interests of James Watson, Francis Crick, and many other biologists. Life was defined by its ability to convert molecules into complex compounds and to store genetic information to pass down to next generation. As a self-sustaining chemical system, life followed evolution. Chemicals are organized into the basic unit of life, the cells, and the cells are organized into individuals, individuals into species that evolved over many million years. DNA depends on metabolism for its replication. Membranes depend on DNA for proteins that build them. The cell stores energy for metabolism and the genes encode the necessary enzymes.

2.1 SURVIVAL OF THE FITTEST OR NATURAL SELECTION

2.1.1 The Central Idea of Animal Behaviour

Darwin thought that in struggle for existence, individuals with favourable variations meet the conditions of life more successfully, survive and propagate their kind. Herbest Spencer termed the process "the survival of the fittest". The organisms lacking such variations perish, and their characters are eliminated from the population. In consequent generations, the process continues and results in gradually adapting animals suitable to a particular environment. Natural selection is based on inherent variation that Darwin called characteristics of species. In a natural population, selective forces, like environmental factors or biotic factors, favour fit individuals than less fit individuals. The variation conferring a selective advantage would be diminished in future generations and ultimately eliminated.

Central to Darwin's theory is the idea that evolution precedes by accumulation of small, heritable changes; not large and sudden changes, and selective forces act on individuals. Evolution acts without design-heritable randomly accumulated traits and natural selection depends on the prevailing conditions. Darwin's theory of evolution forms the foundation of evolutionary theory, and his book *The Descent of Man and Selection in Relation to Sex* provides the core concept of animal behaviour. Darwin attempted to explain animal behaviour including emotions and thought in humans. He noticed that traits related to mate acquisition and mate choice were different from traits under natural selection. De Vries theory along with the re-discovery of Mendel's laws began confronting with the theory of natural selection. Mutations theory states that evolution occurs by production of new variants, or species suddenly.

Natural selection was not accepted widely in the-then scientific community. The Neo-Darwinian Synthesis considers the opposing views together. New heritable variation in which natural selection acts arises by mutation. The weakest point of natural selection is the lack of understanding about mechanism of heredity. Enrichment of Darwin's theory by perspectives of population genetic as well as the findings of Paleontology and Biogeography is called the modern synthesis (Neo-Darwinism).

2.2 MODERN SYNTHETIC THEORY

Modern synthesis is a merger of Darwinian selection and genetic theory. Point mutations and genetic recombinations are the sources of variation. Evolution, in terms of genetics, means changes in gene frequency; it proceeds generally in small steps and is the result of operation of natural selection on genetic variation. Evolution explains the origin of higher taxa from the simple one as a result of its act over prolonged duration. This evolution is known as *microevolution* or *phylogenetic gradualism*.

Modern understanding in cytology, genetics, population genetics and evolution formulated a coherent theory called the modern synthesis around the 1930s by scientists like Wright, Muller, Dobzhansky, Goldschimdt, Huxley, Fisher, Haldane, Mayr, and Stebbins to fulfil the gap of Darwinian theory of natural selection. Synthesis theory of evolution explains transformation of a species by natural selection and the splitting of a species into reproductively isolated subgroups called the *speciation*.

This theory recognizes five basic types of processes, namely, gene mutation, changes in chromosome structure and number, genetic recombination, natural selection, and reproductive isolation. The first three processes provide genetic variability without which changes can not occur. Natural selection and reproductive isolation guide population into adaptive channels. Mutation, genetic recombination, structural changes in chromosomes (which affect linkage and epistatic interaction of genes), and natural selection interact together to produce a pro-gressive change in the population which keeps it adapted in the changing environment. Three accessory processes affect the operation of these five basic processes. Migration of individuals from one population to another, as well as the hybridization between races or closely related species, increases the amount of genetic variability available to a population. Effects of a chance, acting on small populations, may alter the way in which the natural selection guides the course of evolution.

Mutation, genetic recombination, and natural selection are equally indispensable for the evolutionary change to occur. Evolution can be compared to an automobile being driven along the highway. Mutation corresponds to gasoline in tank. Since it is the only possible source of new genetic variation, it is essential for continued progress, but it is not the source of motive power. This source is genetic recombination, shuffling of genes and chromosomes, which operate during sexual cycle. Since this process provides an immediate source of variability on which the selection exerts its primary action, it is comparable to the engine of an automobile. Natural selection, which directs genetic variability towards adaptation to environment, is comparable to the driver of the vehicle. Structural changes in chromosome can have profound effects on the interrelationships between genetic recombination and natural selection and so can be compared to transmission and accelerator of the automobile. Finally, reproductive isolation, which includes all the barriers of exchanging gene between populations, has a canalizing effect similar to the highway with its limits and directive signs which are exerted on the driver of the automobile, thus permitting several vehicles to drive in the same direction at the same time.

Decision-making rules or 'Darwinian algorithms' are an aspect of behaviour. To process information gathered from their environment, organisms depend on such rules with the resultant behavioural output that guides key behavioural- and life-history decisions. Darwinian algorithms involve sensory and cognitive processes that perceive and prioritize cues, which are then transformed into motor outputs along with a stimulus threshold or dependence on the cue related with a fitness-enhancing outcome. Darwinian algorithms are shaped over evolutionary time by the selective regime of each population. Darwinian algorithms are necessarily complex to produce adaptive outcomes under a species' environment, but not so complex to cover all the induced contingencies.

2.3 DARWINIAN SENSE AND PERSPECTIVE OF ANIMAL BEHAVIOUR

Since Hamilton's formulation of kin selection, many studies of animal behaviour yielded results that are consistent with the Darwin's theory of sexual selection. Parental care for offspring, as observed in birds and mammals, is one obvious form. It appears to be difficult for parents to recognize their offspring, but communal cave nurseries of Mexican free-tailed bat may contain thousands of young pups, where mothers are reported to find and feed their own offspring greater than 80% of the time. In certain birds, young receive assistance from non-parents. Hymenopterans live in societies with a strict division of labour. Workers diligently care for queen's offspring and yet are sterile and unable to have offspring of their own—certainly an extreme form of altruism. These species are haplodiploid, meaning that each female receives normal half of its mother's genes, but all of its father's genes. Because of this genetic quirk, sterile female workers are actually more closely related to their siblings than they would be to their own offspring.

Kin selection theory predicts that such altruistic individuals show a very high degree of genetic relatedness to each other, so that altruistic genes they carry have a high probability of also being present in individuals they assist, even though they have no descendants of their own. This fact was found for naked mole-rat (Reeve et al., 1990). Reciprocal altruism in which one individual assists another that is not closely related in order to receive some benefit in return is widely reported and exemplified by olive baboons (*Papio anubis*). Sexually receptive females of this species are usually closely attended by a single male consort on lookout for opportunities to mate. A rival male may solicit aid of an accomplice male who engages consort in a fight. While distracted, rival has uncontested access to a female. What is in this for accomplice who fights but does not mate? He will likely get his chance at mating next time, when his buddy will take his turn in distracting another consort (Packer 1977).

Huge quantity of sperm cells produced by a male means gaining access to as many mates as possible, which increases his reproductive success. But this is usually not the case for a female whose reproductive success depends more on fate of her fertilized eggs. Male should be more eager to mate and less discriminating in his choice of mates than females are expected. Females should be more restrained and choosier in their selection of mates. Discriminative mate choice is exemplified by female insects that demand a "nuptial gift" from male before allowing copulation. Female black-tipped hanging fly (*Bittacus apicalis*) rejects advances of any male that does not first offer a morsel of food. And the larger the male's gift, the better the male's chances of inseminating female, since a quickly consumed tidbit may lead the female to cut short mating process and seek another gift-bearing male (Thornhill 1976). Such behaviour puts selection pressure on males, to provide larger bits of food, since males with little or no gifts are not likely to have their "stingy" genes represented in next generation, whereas those with larger gifts are more likely to reproduce. Many female birds mate only with males that control a food-producing territory (Figure 2.1). Female bullfrogs prefer mating with largest males, and it is not likely coincidental that the largest male usually controls breeding locations that are best suited to development of fertilized eggs. Female birds often select males based on their song repertoire, plumage, size, or courtship ritual which

FIGURE 2.1 A bird in the act of defending territory.

are indicators of health, strength, and parental ability as well as the likely mating success of male offspring fathered by male (Alcock 1993).

But there are some fascinating exceptions. In some species, a complete reversal of typical sex roles is seen. *Syngnathus typhle* male receives eggs from female; he has fertilized and keeps them in his brood pouch, until they hatch. Since females can produce eggs more quickly than males can rear them, brooding pouches are in great demand among females. Thus, male pipefish is picky about his mates, preferring large well-decorated females who appear to be able to provide many high-quality eggs for him to carry. In Mormon cricket, male produces for his mate a large nutritious meal in form of spermatophore. Since spermatophore may weigh as much as 25% of his body weight, he can usually produce only one in his short lifetime, thereby limiting his mating opportunity to just one female. Since he invests so much in his single mating he is choosy, preferring to mate with large females who carry a greater number of eggs, and females compete for access to him. Thus, it is not gender itself, or any intrinsic property of egg or sperm cells that normally makes males competitors for, and female selectors of mates. Rather, it is gender with higher reproductive costs that is choosy in selecting a mate, whereas gender with lower costs is less discriminative and more competitive.

Hanuman langurs, found in India, live in bands and consist of one sexually active male and a harem of females with their young. Occasionally, resident male is expelled from the group by another male after a series of violent confrontations. When this occurs, incoming male attacks and kills infants that were fathered by previous resident male. Infanticide by male also occurs in lion (Figure 2.2). Perhaps, high testosterone levels left over from fighting result in heightened aggression, and attacks on easy victims or may be the new male use infants as a source of high-protein food after great physical exortion or infanticide may be a pathological reaction to high stress accompanying artificially high population densities of langurs in many locations where they are fed by humans. Nursing females provide resources to offspring of previous male. Lactating females do not ovulate and cannot be impregnated by new male. So, by killing infants, incoming male both eliminates reproduced genes of his male rival and makes females sexually receptive once again. Those male langurs have never been observed to eat infants they kill and that infanticide occurs also in areas of low population density, support hypothesis that infanticide is a means of achieving reproductive advantage (Hrdy 1977). Also consistent with this interpretation is observation of infanticide in similar conditions by other animals including lion (Pusey and Packer 1987) and jacana (Emlen, Demong, and Emlen 1989), a water bird. Of course, male langurs need not be conscious of reasons for their killing ways, any more than they are conscious of why they have a tail or fingers. It is extremely unlikely that they have figured out that lactating females do not ovulate and that killing infants will make their mothers fertile and sexually receptive. It may be that incoming males simply have an instinctive desire to eliminate from their band all infants, a goal that was repeatedly selected in past generations because of reproductive advantages it conveyed.

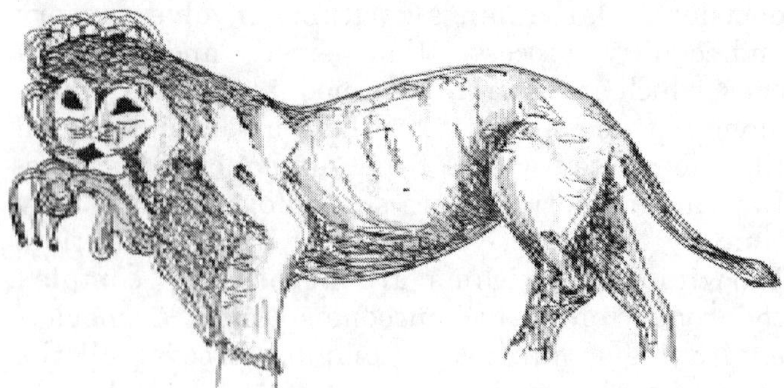

FIGURE 2.2 Infanticide in lion.

These are a few examples of how an evolutionary perspective focusing on reproductive success provides answers to the ultimate why questions concerning a wide range of animal behaviours. Many examples could be given showing survival and reproductive function of behaviours those animal's use to find and make places to live, obtain food, defend themselves from predators, cooperate with other animals, mate, and care for offsprings (McFarland 1993). Thus, evolutionary theory now provides core explanatory framework for studies of animal behaviour in natural settings. It is strongly supported by countless experiments in both field and laboratory settings (Alcock 1993). But when invoking evolutionary answers to these ultimate why questions, we must be on guard against tendency to see specific behaviours as being selected for their survival and reproductive benefits. Instead, we know that what are selected and inherited are not fixed patterns of action, but rather goals in form of reference levels and physical means to achieve them, despite continual and unpredictable disturbances provided by an uncaring mother nature.

Let us consider a spider spinning its web. The webs of any given species of orb-weaving spider are all of same basic design, but actual dimensions must vary because of variations in locations where they are installed, such as branches of a tree or bush. So, it is obvious that no invariant sequence of actions will be successful in installing a web in all locations. Instead, each web must be custom designed for site it is to occupy. A spider (Figure 2.3) is able to fit the web to site not by engaging in a fixed pattern of actions, but by varying its behaviour for each stage of web building, until certain goals are met, before it proceeds to next stage. First, spider perched on a branch, releases a strand of silk into wind until it catches on another branch. Since distance to other branch is different for each site, spider cannot release a fixed length of silk each time, and therefore, it has no fixed sequence of behaviour. It must continually let out silk, until it feels that its sticky end is attached to another branch. Amount of silk it will then pull back in for proper tension which also vary from one web to another depending on distance between branches and stiffness of branches themselves. After tying near end to its branch spider uses this first strand to drop a looser second strand and then a third to form a Y configuration with three stands meeting at what is to become centre of the web. Spider then begins to construct additional radials, like spokes of a bicycle wheel, checking angles between radials with its outstretched legs and continuing to add radials until angle between each spoke and its neighbour falls below a certain value. Each radial is also carefully cinched in so that it has proper tension.

FIGURE 2.3 A garden spider.

Next, spiral portion of web is constructed. Using a temporary nonsticky strand as a scaffold, spider works first from centre outward and then from periphery back towards the centre, laying down permanent sticky silk that will trap its future meals. Spider again carefully controls spacing, between spirals, since too much space would allow insects to pass through the web, and too little would be wasteful of precious silk. Finally, spider determines how much web sways in breeze. If sway is excessive, it may attach weights in the form of small pebbles or twigs to one of web's lower corners. If after all this work, spider judges web to be unsatisfactory, it will abandon site and construct another web else-where. Due to nature of web building and varied conditions in which it occurs, sensory feedback is essential to all stages of construction. It is only by varying its behaviour as required to achieve each sub goal that spider is successful in recreating same basic design that evolved over millions of years for its prey-catching ability. In the case of the spider's web, the fixed end can be brought about only by achieving a number of sub goals in a particular order. It is these sub goals and means for achieving them not spider's actions themselves that evolved because of their value in providing the spider with a means for its livelihood.

2.4 STUDYING BEHAVIOUR

The function of behaviour can be studied through observing the operation of natural selection on organisms under natural conditions by observing differential reproduction. This can be done by considering the naturally occurring variations among individuals. Sometimes, the researcher experimentally increases the behavioural variations where little variation exists in nature with disadvantage of involving unnatural variants, but with advantage of revealing how differences among individuals cause variation in reproductive fitness. Each great tit male has a repertoire of up to eight songs for advertising his presence. Krebs and his colleagues recorded the singing behaviour of breeding male to determine repertoire size. They recorded the egg-laying date, and the size of clutch and brood. The survival of each male's young to the time of breeding was noted. Individual tits had different repertoire sizes. Males with large repertoires had heavy chicks at fledging; most of such chicks survive to breed than offspring of males with small repertoires. Males with large song repertoires acquired superior territories. So it is clear that the function of a great tit male's singing of multiple songs helps him to secure a top-quality breeding territory and mate. Perhaps songs are learned over time and only the old males possess a large repertoire. Alternatively, there are costs to singing multiple songs, and only the strong males can sing many songs.

2.4.1 Adaptive Design

Echolocation in bats, nest-building of weaver birds and alarm calls of ground squirrels serve definite purposes. Mechanisms to perform such behaviours are similar. Such adaptive behaviours have arisen through natural selection which favours some traits, including behavioural ones, which enable the organism to propagate its genes into future generations. Creating a formal optimality model is one way to infer the adaptive design or function of behaviour. If an optimality model includes an exact understanding of the function of behaviour, it may predict the behavioural type in nature. Foraging honeybees return to the hive with less than a full load of nectar. The fuller bee shows slower fly. Carrying a full load may decrease the rate of nectar collection. Trained bees forage from an array of artificial flowers with each flower containing a fixed amount of nectar and the time spent in flying between flowers was varied to alter the duration and cost of foraging, but the size of the bee's load did not maximize the net rate of energy delivery to the hive. Clearly, bees behave for the highest foraging efficiency than the highest food-delivery rate.

A second method is the comparative method of studying the adaptive design. Peter Jarman first used this method to study the mating systems among the African antelopes. In species like dik-dik, individuals are solitary and cryptic. During mating season, they form monogamous pairs. In some species, during the breeding season, only a few males control sexual access to a group of females in a polygynous system. Jarman found that body size, group size, and mating system were interrelated in the African antelopes. Smaller species with relatively high metabolic rates consume high-quality food while concealing themselves from predators. Because of the sparse distribution of food and the need to remain solitary and cryptic to avoid capture by predators, the smaller species are widely dispersed, leaving no opportunity for a male to monopolize access to many females.

Small-bodied species are monogamous. The large size species graze where food is generally abundant and they are clearly visible to the predators. Wildebeest live in large herds that migrate seasonally. Group living creates the chance for the male to monopolize several females, and polygamy in the large-bodied species. Thus, the mating system is derived from the selection pressures related with food and predation. The selection pressures influence the spatial distribution of females and their defensibility by individual males.

2.4.2 Character Mapping

Character mapping constructs a phylogenetic tree which means the evolutionary history of one or a group of interrelated species. The tree can be drawn by reducing the number of characters or changes. The shortest evolutionary path along with considering any character from its origin to the present is known to be the 'most parsimonious' and the most probable. The behaviours of extant species have remained the same since the last speciation event in their lineage and that the shortest evolutionary path is most likely can be formulated about the relative timing of the origin of various behaviours and their subsequent loss.

2.4.3 Phylogenetic Grading

Phylogenetic grading deals with comparisons among extant species, with respect to a particular behaviour, and then arraying the various behaviours. Complexity increases over evolutionary time. Simple behaviour is therefore ancestral. Existing species with a simple behaviour probably have not experienced the selection pressures. Karl von Frisch, who decoded the 'dance language' of honeybees reportedly said: We cannot believe that the bee dance of the European bees has come from heaven as it is and, since the Indian honeybees and the stingless bees there live in a more primitive social organization, we should expect some phylogenetically primitive stages of the bee dance. According to this view, bees (*Melipona*) might not even possess a dance language, as they live in small, less-organized colonies. Studies of stingless bees provide evidence that successful foragers communicate distance, direction, height, and smell of food to their mates. Stingless bees do every work like the 'advanced' honeybees. Stingless bees have a communication system that is different but not primitive than the communication system of honeybees.

2.4.4 The Comparative Approach

This approach reconstructs the behavioural history and studies its fitness consequences. If the existing behaviour offers higher fitness than its alternatives, it is obvious that the natural selection acting in similar antecedent environments caused its initial spread. This approach assumes that the present selective pressures are similar to those that operated in the past. The European starling and the English sparrow were imported to the United States during the second half of the 19th century. Some aspects of their new environment were different, whereas the other environmental aspects like nesting sites and social environment did not change. As a result, the birds' reproductive and communicative behaviours closely resemble those of starlings and sparrows living in Europe today.

2.5 PSYCHOLOGY AND BEHAVIOURISM

The ethological approach contributed by Lorenz, Tinbergen, and von Frisch was concerned with the behaviour of organisms. Some scientists focused on the mechanistic underpinning of behaviour in controlled conditions. Skinner leads to the development of the use of learning paradigms. Scientists sought the similar mechanisms in all animals that allow them to respond to their environment. The defined field of comparative psychology spanned areas as follows:

- *Perceptual psychology:* This branch deals with the reception of environmental stimuli through the senses, and the subjective perceptual interpretation of these stimuli.
- *Physiological psychology:* This branch attempts to relate physiological properties of an organism to external behaviours.
- *Functionalism:* This branch deals with the study of the mind.
- *Behaviourism:* This branch deals with the study of accumulated experiences in shaping the organism's behaviour. Central to behaviourism is the idea that an organism is born in a blank slate in which experience accumulates to shape the behaviour.
- *Animal psychology:* This branch was initially related to the study of learning in model systems, although now the field includes a large body of work related to cognition in diverse animals.

2.6 CLASSICAL ETHOLOGY

Through much of the 20th century, European and American scientists were sharply divided over the animal behaviour methodology. European scientists stressed on imprinting, innate-releasing mechanisms, communication, development of behaviour, and value of comparative studies of specific behavioural patterns, such as mating across species, to gain knowledge about the evolution of these behaviours. American scientists focused on learning and conditioned responses. Two divergent approaches, nature versus nurture, stimulate the rethinking about the roles of nature and nurture in behaviour. Well-characterized four key subjects were chosen by Tinbergen: causation, survival value, ontogeny, and evolution.

Contemporary animal behaviourists combine the biological context and apply techniques from the fields of genetics, statistics, and mathematical modelling to solve problems in behaviour. Drive stands out as the interesting concept in behaviour. Early ethologists thought about energy supply in animal. This energy accumulates, and is released through the behavioural act. Energy models of the drive explain prioritization of behaviour. Excess energy, particularly if an intended behaviour is blocked, might be dissipated in redirected behaviour. If a conflict occurs between allocating energy to two or more behaviours, energy might be dissipated. This is called *displacement behaviour*.

2.7 MODERN CONCEPTS OF HOMEOSTATIC REGULATION

The hormonal and neurobiological regulation of behaviour has tremendously increased. Biologists have shifted away their thinking from motivation and drive. Behaviour, the result of physiological needs, is regulated by neurochemical and hormonal mechanisms. Many behaviours improve an animal's physiological condition and internal homeostasis. Thermoregulation is an example of interaction between behavioural and physiological homeostases. Strategies used in prioritizing behaviour remain an important area of inquiry. An animal's ultimate goal is to successfully reproduce. Animals maintain themselves by feeding, protecting and finding appropriate environmental conditions for survival. They also need to store food reserves.

2.8 BEHAVIOURAL ECOLOGY AND SOCIOBIOLOGY

Behavioural ecology studies the interaction of organisms in their natural environments. Both the mechanistic underpinnings of behaviour and the fitness consequences of behavioural traits need understanding. This can be traced back to causation, development, evolution, and function. Ecologists use traits that maximize energy acquisition as proxies for fitness traits. The development of optimal foraging during the 1970s and 1980s has added a different theoretical perspective in the field of behavioural ecology. The recent approach involves social systems in various organisms.

Sociobiology has a strong Darwinian tradition as it attempts to develop rules for explaining the evolution of social systems. Evolutionary psychology has co-opted the methods of behavioural ecology and sociobiology to explain various human behaviours. Animal behaviour results from the interaction of several forces. Individual's gene, physiology, development along with the internal and external environments interact to influence the behaviour. Genetic variations among individuals are related to survival. The interaction between the nervous system and the environment is manifested through the behaviour. Behaviour includes learning, communication, navigation, foraging, mating, parenting, and social cooperation as well as social conflict. Motivation, cognition and emotion shape an animal's decisions. Such phenomena have neurobiological basis. Evolution over generations plays an important role in shaping the behaviour. An animal's choice of strategy is influenced by the genetic background, experience, and the evaluations of existing conditions. A strategy that cannot be dislodged by other strategies is termed 'evolutionarily stable strategy' (ESS).

Behaviour is fundamental to living organisms. Bacteria move in response to chemical gradients. Animals respond to both internal and external stimuli. Dynamic genetic and physiological processes influence each other. The nervous system is the interface between molecular and cellular functions. Variation among individuals of the same species arises from mutations and crossing-over during meiosis. Behaviour enhances survival. If variation and heritability are high in populations, behaviour evolves over many generations. Animal behaviour is shaped by natural selection. Our own behaviours are shaped by survival need. Some behaviours with survival value became instinctive. Behaviour has adaptive value.

2.9 THOUGHTS OF DIFFERENT SCHOOLS AND PRESENT STATE OF BEHAVIOURAL STUDIES

Behavioural science, comparative psychology, and neurobiology developed independently. In the early stages, behavioural science formed and developed in Europe. Studies of comparative psychology proceeded in the America. Before the year 1950, studies of behavioural science focused on the nature and animal behaviour of specific species, seldom involved in learning, human, and mammal behaviour. Comparative psychologists concentrated on learning. Comparative psychologists wondered whether the simple reflex is different from conditioned reflex, or both are the same one. Besides reflex, they also showed much interest in imprinting learning. Different domains caused the confliction of opinions between comparative psychology and behaviour. During 1930s, a big divergence between comparative psychology and behaviour occurred. Theodore C. Scheirla introduced a modern thought of learning behaviour, which padded the gulf between comparative psychology and behaviour.

Neurobiology which evolved from classical medicine, anatomy, and physiology developed independently. Neurobiologists were absorbed in the studies of instantaneous and direct response related to neural morphology. On the other hand, since it was difficult to study the nerve alive, for a long time, the development of neurobiology relied on scientific and technological progress as well as observation of cases. Since the 1960s, behavioural studies began to develop through ecological causes related to evolution and neuro-physiological causes. Four main features in modern behavioural studies are designing experiment, and studying the mechanism of ecology in animal behaviour, seeking theoretical model and developing comprehensive theory, and application of high technology.

In 1977, Emlen and Oring introduced concept of mating system which suggests that space, distribution of food, and safe breeding ground affect animals. In 1996, Reynolds bought about the concept of breeding system that contains mating system and parental care system. This concept insists that breeding system lies on the mate choice and the expense as well as benefit of competition and courtship display. Presently, behavioural science use application of high technology, such as detecting short tandem repeat DNA by DNA fingerprinting technique, application of dynamic analysis by computer, detecting mitochondrial DNA sequence which is relative conservation to identify the relationship between species and subspecies, and detecting D-loop region to check the genetic variation in populations. Behaviours that are helpful to others, but costly to originator should simply not evolve as instincts. Altruistic acts, like sharing food, reduce reproductive success of altruistic donor while increase that of its recipients and genetic competitors.

A ground squirrel warns other squirrels by emitting alarm call upon noticing a predator and put itself at greater risk of predation. A vampire bat regurgitates blood for a starved neighbour.

Such altruistic acts attracted attention of biologists in 1950s through 1970s. Aldine noted that a gene predispose an animal to save another animal from some danger with potential "hero" running a 10% risk of being killed in attempt. Such gene could spread in population through natural selection if animal thus saved was a close relative of hero. A closely related individual would have a good chance of sharing same altruistic gene as hero, so that a copy of same gene would likely be saved, even if the hero perish by his actions.

Aldine noted that such a gene could even spread, although not as quickly if saved individual was more distantly related to hero, such as a cousin, niece, or nephew. "I am prepared to lay down my life, for more than two brothers, or more than eight first cousins" was his way of summarizing this phenomenon. This was the beginning of formulation of kin selection. The idea states that a gene is not "judged" by natural selection solely on its effects on individual who carries it, but also on its effects on genetically related individuals, who are also likely to carry a copy of gene. The altruistic behaviour towards kin can be understood as a form of selfishness on part of a gene. Since related individuals who receive assistance are likely to carry a copy of same gene and pass it to their offspring. There are different degrees of relatedness. Thus, they make sense for altruistic behaviour, to be scaled according to degree of relatedness so that it may be directed toward closest relatives.

Hamilton noted that evolution should be expected to bias the altruistic behaviour towards close relatives, and therefore, also select for ability of altruistic animals to discriminate close relatives from more distantly related individuals and their acts could be preferentially directed toward the former, and not latter. Kin selection is an important factor in evolution of behaviour, and altruistic acts are directed toward unrelated individuals. The modern answer was first attempted by Williams (1966) in *Adaptation and Natural Selection*, a book that became a classic in evolutionary biology. He suggested that beneficent behaviour towards another unrelated individual that was initially costly for donor could in the long-run be advantageous if favour was later returned. This idea was further developed by Trivers (1971), with theory of reciprocal altruism. Here, cooperative and seemingly altruistic behaviour was shown to evolve among individuals who are not closely related. It also account for mutually advantageous relationships between different species, such as between cleaner-fish and larger fish that they clean. During cleaning, cleaner-fish obtains a meal, and cleaned fish gets rid of troublesome parasites, but only as long as it refrains from gobbling down much smaller cleaner fish. Through such symbiotic behaviour both cleaner and cleaned profit, in ways that would not be possible without mutual cooperation. Sexual reproduction involves a mate. Offspring of many species require parental care. Finding a mate and factors determining mate selection were first pointed out by Darwin (1871) in "*The Descent of Man and Selection in Relation to Sex*". He observed that males often compete with each other for females, and females tend to be choosy in selection of partners often with alluring courtship displays or some physical characteristics. He also suggested that such selection pressure was responsible for elaborate "ornaments" by males, such as bright and striking plumage of paradise bird and peacock. Sexual selection and its consequences for animal behaviour were largely ignored for the next century.

Trivers (1972) drew attention to the fact that sex cells produced by males are much smaller and more numerous than those produced by females. An individual male may provide enough sperm cells during a single mating, theoretically, to impregnate every female of the species. Females of most species produce a much smaller number of much larger eggs. This discrepancy in potential reproductive potential should have consequences for differences in sexual behaviour.

A glimpse of history in behavioural science is given in Table 2.1.

2.10 HISTORY OF THE STUDY OF ANIMAL BEHAVIOUR

In the history of animal behaviour study, the focus has remained on the development of key concepts in the field. It is difficult to pinpoint the precise beginnings of animal behaviour studies. Let us consider some highlights in the development of the discipline. The idea of intellectual continuity among animals shaped some earliest views of animal behaviour, which were summarized in Principles of Psychology (Spencer, 1855). However, its roots can be traced back to the ideas of the ancient Greek philosophers, who focused on continuity in mental states between lower and higher animals. It was based on a picture of evolution like Aristotle's *Scala naturae*, the great chain of being, in which the evolution of species was linear and continuous. This classification system was hierarchical, with animals ranked according to their degree of relationship. The highest point of

TABLE 2.1 A Glimpse of History in Behavioural Science

1859	Charles Darwin published the "*Origin of Species by Means of Natural Selection*" in 1859. Darwin's theory of natural selection prepared the stage to consider animal behaviour in an evolutionary light. With this foundation, Konrad Lorenz, Niko Tinbergen, and Karl von Frisch began to practice ethology.
1872	Darwin published *"The Expression of the Emotions in Man and Animals*".
1873	Spalding conducted research on imprinting.
1898	Whiteman studied the behaviour of pigeons and doves.
1898	Thorndike reported trial and error learning.
1904	Pavlov demonstrated the conditional reflex.
1910	Heinroth studied the behaviours of ducks and geese.
1913–1937	Kohler studied insight learning.
1934	Von Vexkull studied the perceptual world of animals as Umwelt.
1937	Konrad Lorenz had done research on the instinct.
1950	Frisch, Kramer, and Pittendrigh independently discovered compelling evidence for internal clocks in animals.
1952	Tinbergen and his co-workers identified the social releaser.
1952	Kramer worked on the orientation of birds.
1960	Goodall became the first person to study chimpanzee behaviour by living with the chimps in Chimpanzee Reserve in Tanzania, Africa. Goodall's difficult and time-consuming observations broke apart old ideas in animal behaviour.
1963	A landmark paper by Niko Tinbergen laid out the aims and methods of ethology.
1966	Reynolds brought about the concept of breeding system.
1970	A book "*The Social Contract: A Personal Inquiry into the Evolutionary Sources of Order and Disorder*" written by anthropologist, Robert Ardrey was published.
1972	Trivers drew attention to fact that sex cells produced by males are much smaller than those produced by females.
1973	Niko Tinbergen shared the Nobel Prize in Medicine with Karl von Frisch and Konrad Lorenz for their contributions to animal behaviour.
1977	Emlen and Orien introduced the concept of mating system.
1989	Clogan reviewed the process of animal motivation.
1994	The process of sexual selection was reviewed by Andersson.
1997	Benys gave a detailed account of biomimicry.
2001	Cloned genes and their influences on complex behaviour was reviewed by Sokolowski.
2003	Emmons and Lipton summarized various gene products and their role in behaviour.

evolution is humans (Hodos and Campbell 1969). At the bottom of the scale were sponges, insects, fish, amphibians, reptiles, birds, nonhuman mammals, and humans. In evolution each higher species evolving from a lower one until the emergence of humans. It was thought that the animal mind and the human mind were points on a continuum. Darwin published his thoughts on evolution by natural selection in On the Origin of Species (1859). He provided a conceptual framework for development of animal behaviour which can be summarized as follows: 1. Individuals in a population vary. They differ in physiology appearance, behaviour, and other part of their phenotype. 2. Some variation is inherited and passed on from mother to offspring. Most of the offspring produced do not survive to reproduce. Some individuals survive longer and produce more offspring. Natural selection is the differential survival and reproduction of individuals. This results from genetically based variation in their morphology, behaviour, physiology, and so on. Evolutionary change occurs as the heritable traits of successful individuals which are spread. In these volumes, he recorded his observations on animal behaviour. He believed that careful observations were useless unless they were associated to a general theory. Darwin's general theory was evolution by natural selection. Because humans evolved from other animals, he considered the minds of humans and animals to be similar in kind and differ only in complexity. Darwin described the animal behaviour by using terms that denote human emotions and feelings.

Darwin's opinion was influential, and others who became interested in animal behaviour followed his lead. Ethologists and comparative psychologists trace the beginnings of their fields to the ideas of Darwin. After crystallizing casual observations of animal behaviour, Tinbergen (1963) identified four questions: The mechanisms that cause behaviour. How does it develop? What about its survival value? How does it evolve? Tinbergen believed that ethology should "give equal attention to each of them and to their integration." Tinbergen's four questions were later condensed into two categories. "How" questions which deal with causation and the development of behaviour. "Why" questions which deal with function and evolution of behaviour. Mechanisms of behaviours, in some cases, include cognitive and emotional processes (Emery and Clayton 2005). Let's consider the seasonal migration of songbirds between northern and southern latitudes. As new birds appear daily in the spring, one may be curious about migration. Each of us may ask different questions. How do the birds "know" it is time for migration? How do they find their way during migration? Such questions point to the mechanisms of the behaviour. The first-time learner birds learn the route from experienced travellers. When we ask why an animal behaves in a certain way, some of us ask about immediate causes, while others ask about evolutionary causes. These are not competing avenues of investigation, rather, they are complementary. Each may provide feedback on the others, deepening our understanding and broadening our avenues of investigation (Armstrong 1991; Halpin 1991; Stamps 1991). Animal requires answers to four types of questions, about (1) its immediate mechanisms, (2) its development, (3) its survival value, and (4) its evolution. The study of animal behaviour begins with observation. The next step is the tentative explanation, called hypotheses. Each hypothesis should produce testable predictions. Such predictions support or refute the hypothesis. Romanes charted the evolution of the mind. He listed the emotions that follow their evolutionary appearance. In Animal Intelligence, Mental Evolution in Animals, and Mental Evolution in Man, he considered the Darwinian view on the continuity of species. Loeb (1918) believed that all behaviours are tropisms, physiochemical reactions with stimuli. Jennings emphasized the variability and modifiability of behaviour, but he disagreed with Loeb's ideas. Differences in opinion developed with the expansion of the subject with two major approaches. These are ethology in Europe, and comparative psychology in the United States. Ethologists often observe animals in nature and study innate behaviour in birds, fish, and insects. They focus on the function and evolution of behaviour. Comparative psychologists focus on the mechanisms and development of behaviour. They focus on learned behaviour in species such as the Norway rat, with the belief that learning could be best studied in the laboratory. They searched for general laws of behaviour. Some on to study the effects of hormones on behaviour.

2.11 BEHAVIOURAL BIOLOGY

Behavioural biology describes research that includes more than one of Tinbergen's four questions (Taborsky 2006). Studying behaviour at the level of the whole organism and integrated studies of the four questions are described in the papers celebrating the 40th anniversary of Tinbergen's classic paper on the four questions. Ryan (2005) argues that animal behaviour that integrates Tinbergen's four questions—cause, development, survival value, and evolution—is needed to provide a complete understanding of behaviour. Ryan illustrates how knowledge of the evolutionary history of the calls of túngara frogs helped understand the mechanisms of male calling and female response, and how they develop and increase fitness. Sherry (2005) suggests that knowledge of the survival value of the behaviour assists research on the causes of the behaviour. An animal's mental capabilities are viewed as a product of natural selection. The field of study began with Griffin's controversial 1976 book, The Question of Animal Awareness. Gryphon later (2001) named the field cognitive ethology. It is an interdisciplinary area that brings Tinbergen's four questions to bear on the study of animals' mental experiences. Three areas of research with rapid progress are animal communication, seed caching and recovery, and navigation and orientation (Balda et al., 1998). Applied animal behaviourists focus on captive animals. Some work with companion animals is now continuing.

2.12 ANIMAL BEHAVIOUR IN THE PRESENT WORLD: CHALLENGES AND PROSPECTS

We are now living in the era of the Anthropocene. The activity of our species is the major influencing factor on the earth. We are changing the planet in a variety of ways. Anthropogenic activities that have changed the earth include climate change, land-use change, pollutants, the emergence of disease, groundwater depletion, and the loss of biodiversity at an unprecedented rate. We have also introduced invasive species and converted natural areas into urban and agricultural landscapes. These changes cause animals to experience novel environments with new threats. To survive, animals must adjust to such changes. Animals often respond by changing their behaviour. We should explore the behavioural responses of animals in changing scenarios to determine the fate of animal populations and ecosystems and ensure that current research contributes to mitigating large-scale negative changes. A suitable way of addressing the present global change can only ensure the future fate of our species.

2.13 IMPORTANCE OF CHANGING ANIMAL BEHAVIOUR

Over the past three decades, the increasing need for understanding animal behaviour and its relevance for conservation of natural resources has been the major focus of animal behaviour studies. Although it is late, we have now realized the importance of our natural resources, such as air, land, water, flora, and fauna. It has been realized that animals can adjust their behaviour to modulate their responses to environmental change (Gunn et al., 2022), which can change the persistence and evolution of species in a changing world (Tuomainen and Candolin, 2011). Animals can change their behaviour to times or locations that are habitable to them (Gilbert, 2023; Brown et al., 2023) (Templeton et al., 2023). Light pollution negatively affects nocturnal animals and interferes with their seasonal movements, such as migration (Burt et al., 2023) and regular movements (Brown et al., 2023).

Changes in animal behaviour affect the role species play in ecosystem functioning and thereby affect the provision of key ecosystem services (Wilson et al., 2020), like seed dispersal (Mortelliti, 2023). For example, in iScience, (Bose et al., (2022) demonstrate that pharmaceuticals can alter the predation efficiency of odonate larvae, a top predator found in many stream systems. Understanding animal behaviour is crucial for effective conservation actions (Caro, 1999). Greggor and Goldenberg (2023) have discussed the need for integrating information from social networks into conservation translocations. This being a key intervention, it can benefit from better integration with animal behaviour.

2.14 EMERGING ISSUES

Animals are often neophobic, frightened, or uncomfortable in new situations. Sih et al., (2023) have correlated a fear response to one factor with a response to a second or third factor. In a changing world, animals face multiple new threats. It is important to understand whether animals show consistent differences in novel and dangerous situations. This has implications for conservation initiatives such as translocating individuals, as discussed by Greggor and Goldenberg (2023). Underlying mechanisms behind behavioural responses related to the evolutionary history of the species (Gilbert et al., 2023), we are now equipped to develop efficient management strategies. Such strategies could reduce the negative effects of human activities on populations and ecosystems (Mortelliti, 2023). The disruption of behaviours due to drastic environmental change requires the rewiring of animal spatial, social, and community networks. Animal social networks are believed to be the drivers of population dynamics. Thus, these are relevant to conservation (Snijders, et al., 2017). Blumstein et al., (2023) provide a framework to understand the environmental change-induced destabilization of these networks. He assessed both indirect and direct effects on social structure and interactions, with ramifications for population persistence. Gilbert et al., (2023) emphasize that changes in the daily activity patterns of animals can rewire potential interactions across species. Similar kinds of rewiring can occur across space as land-use change continues (Jacoby and Freeman, 2016). Mortelliti (2023) emphasizes how interaction rewiring can have downstream effects on many ecosystem services.

As animal behaviour changes, we need to develop systems and techniques to track animals and monitor their behaviour. Tracking individuals has historically been highly labour-intensive. However, the combined efforts of advances in sensing technology with machine learning and artificial intelligence (Tuia et al., 2022) would enable us to detect and analyze the natural complexity of animal behaviour. This in turn moves beyond pattern description to a better understanding of causal processes (Couzin and Heins, 2023). Burt et al., (2023) and Gilbert et al., (2023) have shown the application of emerging technologies to real-world systems. Many of the outstanding questions for conserving animal behaviour in a changing world are interdisciplinary. Interdisciplinary approaches can only translate our understanding of the effects of light pollution on migration to policy changes, like the adoption of 'lights-out' programs (Burt et al., 2023). Building sustained interdisciplinary efforts to develop predictive capacities in behavioural science for conservation biologists has also been challenged (Couzin and Heins, 2023). Sensory ecology shows how behaviour can impact human well-being (Mortelliti 2023). Thus, the future of animal behaviour in a changing world is an interdisciplinary one.

2.15 INSTINCTIVE LEARNING

Animal adjusts its behaviour based on experience, as experience provides useful information at a later time. Learning is a tool for survival and reproduction. An animal needs to know about good food, when and where the good food will be available, whom to avoid and approach, potential mate. When such things are not genetically preprogrammed, the animal learns them. Most animals are genetically predisposed to acquire information for developing behaviour. Most learning is instinctive and is conspicuous in the flower-learning behaviour of honeybees. Since the time of Aristotle (384–322 BC), worker bees show 'flower constancy', although honeybees are generalist foragers capable of exploiting at least 700 others. They distinguish colour from yellow into the ultraviolet. A bee learns the flower's colour during the final few seconds before beginning to feed, and odour learning during feeding. The time course of learning is highly adaptive. 'General process learning theory' is accounts for learning with a single set of principles, but the unconstrained 'associative learning' is studied in instrumental (operant) conditioning and classical (Pavlovian) conditioning. Associative learning occurs when an animal changes its behaviour to form association between an environmental event and its own response to the event. In operant conditioning, it learns to associate a voluntary activity with specific consequences. In classical conditioning, it learns to associate a novel stimulus with a familiar one. Pavlov demonstrated that by exposing a dog to a particular sound and simultaneously placing meat powder in its mouth the dog could salivate in hearing the sound even without the meat stimulus.

John Garcia discovered several puzzling phenomena that indicated adaptive limits of learning and contradicted the general principles of conditioning. These anomalies include flavour aversion learning. When rats and humans become ill due to a flavour, they learn to avoid that flavour. Westermarck noted that arranged marriages between children growing together are far more likely to fail than the arranged marriages between individuals not raised together. The failures are due to sexual incompatibilities. Children are genetically guided to learn to treat as siblings with whom they are raised. As siblings tend to avoid sexual contact due to a detrimental consequence associated with inbreeding, marriages between these individuals tend to fail. Instinctive learning has ended the centuries-old debate about whether 'nature' or 'nurture' is the source of adaptive behaviour. Animals are shaped by their experiences. The interpretation of each experience is governed by the genes in each species. Learning that prevailed during most part of the 20th century was based on the assumption that learning is always beneficial, and animal learning abilities are like human learning abilities.

Lorenz referred to 'the innate schoolmarm' for expressing the reality that animals possess adaptive predispositions in their learning. Functional interpretations of behaviour were usually made in terms of goodness of the behaviour for the species. Social behaviours that excluded some individuals from reproducing were viewed as adaptations. George C. Williams and David Lack revealed the underlying problem with the view that animals behave in ways that limit their reproduction for the favour of their species. Individuals who maximize their own reproduction will have greater genetic success than those who behave in ways that limit their reproduction. Over time reproduction-reducing behaviours will be replaced by reproduction-enhancing ones.

Lack's study of the reproductive behaviour of the European swift is mentionable. At first glance, swifts appear to voluntarily restrict their own reproduction. When Lack removed the eggs laid each day from a pair's nest, the female could lay up to 72 or more eggs in a season, although she usually lays just two or three eggs. Individuals are 'selfish', who behave in ways for their own reproduction regardless of its long-term effect on the survival of their species. Sometimes animals exhibit altruism. Female Belding's ground squirrels give staccato whistles that warn nearby squirrels but also attract the predator's attention to the caller. Worker honeybees perform suicidal attacks on intruders to defend their colony. Female lions nurse cubs that are not their own. The key insight to the evolution of such self-sacrificial behaviour was provided by William D. Hamilton in the mid-1960s.

2.16 LEARNED BEHAVIOUR

A spider's web catches prey, and spider must custom-build each web to fit its site. But the design it uses is the same basic one that has been successful over many thousands of years. If this design now turns out to be unsuccessful, for a particular spider in securing food, spider cannot make another kind of web, like more productive one being used nearby by another species. It is stuck with design of its species, in much the same way that it is stuck with having eight legs, a hairy body, and an appetite for juicy insides of insects.

Other animals show more flexibility in being capable of learning. Whereas an insect-eating spider eats only insects, rats nibble on just about anything that might be edible and learn to distinguish what is nutritious from what is not. Thus, individual rats of the same species may have very different diets and food preferences according to their dining experiences.

One way of looking at such learning is to see it as a behavioural adaptation to environmental changes that happen too quickly to be tracked by natural selection. Gradual changes in climate or the gradual appearance and extinction of pathogens, prey, and predators can affect instinctive behaviour, through differential survival and reproduction of organisms with adaptive behaviours. But more rapid environmental changes taking place from one generation to next or even within a generation cannot be tracked by evolution. As Skinner (1981) observed, "contingencies of survival cannot produce useful behaviour if environment changes substantially from generation to generation, but certain mechanisms have evolved by virtue of which the individual acquires behaviour appropriate to a novel environment during its lifetime." These "mechanisms" refer to ways of learning that allow animals to adapt their behaviour to unpredictably changing environments.

Skinner included both one-way cause-effect and selectionist components in his theory of how animals acquire new behaviours in much the same way that Lorenz included both of these in his account of instinctive behaviour. Selectionist component for Skinner had to do with the learning process itself; that is, how new behaviours are first emitted (random variation) with certain ones selected by environment according to their consequences. It is for this reason that Skinner emphatically rejected frequently applied characterization that was a stimulus-response theory, because of "unstimulated" nature of originally emitted novel behaviours.

But despite his protests, an important one-way cause-effect component of Skinnerian theory comes into play after a new behaviour has been learned. This is because the new behaviour is then elicited or caused by sensory stimuli that are same as or similar to environmental stimuli experienced when behaviour was originally selected. Skinner designed a box for his experiment (Figure 2.4). Rat may stumble upon pushing lever to obtain food in a haphazard random way, but after it learns this new way of feeding itself it will immediately approach and push the bar (if hungry), when placed into same or similar box in which behaviour was learned. It is for this Newtonian inspired one-way cause-effect conception of performance of already learned behaviours, Skinner's theory was and still is characterized by many behavioural scientists as a stimulus-response theory. The notion is understandable if not completely justified, when it is realized that Skinner repeatedly referred to "stimulus control" of behaviour. Although he understood stimulus broadly as the cumulative effect of all previous sensory stimuli experienced by an organism, he emphasized that "the environmental history is still in control" (Skinner 1971). By control, he actually meant cause. This view of behaviour is in striking contrast with circular causality of perceptual control

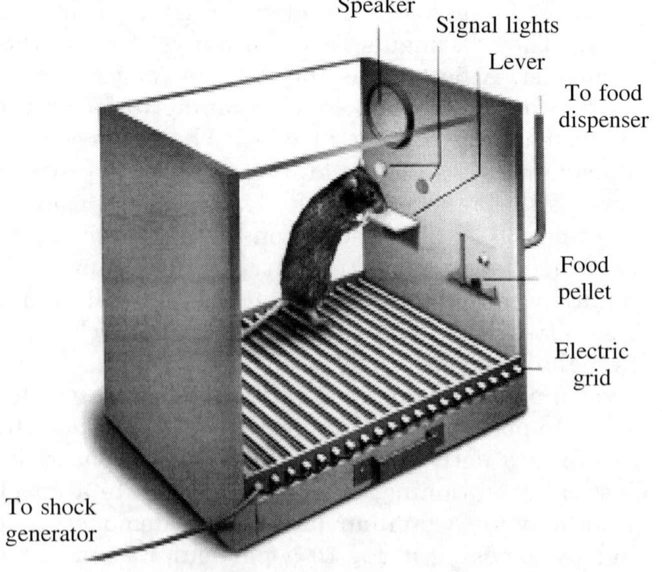

FIGURE 2.4 The skinner box.

theory, which sees organisms purposefully varying their behaviour to control perceived environmental consequences of those behaviours. In other words, instead of Skinner's selection by consequences, we have Power's selection of consequences.

A good way to contrast difference between these theories of how organisms modify their behaviour, is to consider an intriguing pattern of behaviour, Skinner observed. He found that he could obtain very high rates of behaviour by gradually decreasing the rate of reinforcement. These high rates could be obtained by starting out with a relatively generous reinforcement schedule that provided a grain of food for each key peck and then using progressively more stingy schedules, requiring more and more pecks for each reward. Skinner was thereby "able to get the animals to peck thousands of times, for each food pellet over long enough periods to wear their beaks down to stubs. They would do this, even though; they were getting only a small fraction of the reinforcements initially obtained".

But if according to this theory of operant conditioning, reinforcement increases the probability of the behaviour that resulted in the reinforcement, how it could be that reducing the reinforcement leads to an increase in rate of behaviour? This puzzle is solved when we see reinforcement not as an environmental event, but rather as a goal that organism achieves by varying its behaviour as required. If circumstances are arranged, so that the hungry Skinner-box-trained rat must perform more bar presses to be fed, and as it has no other way to obtain food, it will adapt its behaviour by increasing rate of bar pressing. If rate of reinforcement is increased to the point at which rat can maintain its normal body weight, a control-system model of behaviour based on circular causality would predict that further increases in reinforcement should lead to decrease in rate of behaviour. Skinner also believed that in any behaviour, an animal was physically capable of could be brought about through contingencies of reinforcement. He took particular delight in demonstrating games that he taught pigeons to play, such as the one in which bird used its beak to roll a midget bowling ball down a miniature alley to a set of tiny pins (Skinner 1938).

But other research on animal learning has discovered clear constraints on types of behaviours that animals can learn, and that instinctive behaviours can often interfere with learning new ones. Keller and Marian Breland, who worked for many years in training animals for commercial purposes, reported several such examples, in their informative and entertaining 1961 paper "The Misbehaviour of Organisms." Included in their report are accounts of chickens that could not learn to stand on a platform for twelve to fifteen seconds without vigorously scratching it; raccoons that could learn to put one coin in a container, but when given two coins would spend minutes rubbing them together and refuse to deposit them; and pigs that after having learned to pick up and place large wooden coins in a piggy bank would after several weeks or months begin repeatedly to drop coin push it with their snout and pick it up again, taking up to ten minutes to transport four coins over a distance of about six feet. Other researchers reported that male three spined sticklebacks were successfully trained to swim through a ring to gain access to a female, but they could not learn to bite a glass rod for same reward, since they attempted instead to mate with the rod (Sevenster 1973).

In all these cases, we see animal's normal instinctive behaviours related to eating and reproduction interfering with new behaviour the researcher wanted to learn, a phenomenon referred to by Brelands as "instinctive drift." Other interesting evolutionary constraints on learning were investigated in the laboratory rat. For example, rats are quite handy with their front paws, and so a hungry rat normally learns quite quickly to press a bar to obtain food. But it is very difficult to get a rat to press a bar to avoid a shock (Slater 1981). This seems due to rat's freezing in response to fear, an instinctive behaviour incompatible with bar pressing. Rats also can make certain associations between stimuli and their effects, but not others. If a rat is made sick after consuming a food with a certain taste, it will consequently avoid all foods having the same taste. And if a sound or visual stimulus regularly precedes an electric shock, a rat will associate this as a signal of the impending shock and will learn to make an appropriate avoidance response. But rats cannot learn to associate taste with electric shock, or use auditory or visual cues to learn that a food is noxious (Garcia et al., 1968). These findings may be puzzling for the psychologist who has no appreciation of the evolutionary past of the rat, but they make quite good sense from an evolutionary perspective. For rats, which often scurry about in dark places and eat an amazing variety of foods, taste is a better indicator of the quality of food than its visual appearance or the sounds they make while eating. In contrast, physical dangers are usually accompanied by visual and auditory signals as produced by birds (Figure 2.5), not gustatory ones. So, it makes sense that evolution would have selected rats that learn what is bad to eat by taste and what

FIGURE 2.5 The birds can warn danger.

is physically dangerous by sight and sound. That rat can learn food aversion based on taste, is itself a quite remarkable adaptation that led psychologists to seriously revise their theories about learning. It was once widely believed that two stimuli had to be presented several times, and within a very short time, if one was to become associated with the other. But John Garcia and his associates fed rats a harmless substance with a characteristic taste and later made the animal sick using radiation. Contrary to expectations, rats would learn to avoid the new food even if they were made sick several hours after ingesting it. And this food-avoidance learning appeared permanent.

Findings of this and several similar studies were quite surprising to psychologists at the time, although this type of learning ability, again, makes good evolutionary sense. Rats live in a wide variety of rapidly changing environments and consume a wide range of foods; often those intended for humans or discarded by them. Since they cannot know beforehand whether a new food is toxic or nutritious, they are very cautious, and at first take only a small quantity of it. And since, it may take a few hours for food poisoning to take effect; they have evolved a learning mechanism that can operate over an interval of hours, so that they forever avoid the taste of a food that has made them ill just once. This well-adapted learning is why rat poisons have limited success. On the other hand, a rat whose normal diet is deficient in an essential nutrient has a stronger inclination to try a new food. If new addition happens to be followed by recovery from dietary deficiency, rat will develop a marked preference for it as recorded elsewhere.

But what exactly is learned when an animal escapes from a type of puzzle box that Thorndike used? Animal presses a bar to obtain food in a Skinner box, or develops a preference for a food that contains some essential nutrient. Learning can be adaptive only, if learned behaviours remain flexible and permit the organism to obtain its goals in the face of these disturbances. Hierarchy of controlled perceptions provides a quite different perspective on learning. It will be recalled that it shows how higher-level goals are achieved through manipulation of combinations of lower-level goals (sub goals). A spider is able to catch prey only by achieving a rather large number of sub goals that involve spinning a web, catching prey, and injecting its venom to kill or paralyze is prey. Fortunately for spider, it inherits a control system hierarchy, in which these goals and sub goals are specified, and so, it requires no learning to be able to feed itself. This is what is referred to as instinctive behaviour. But other animals are more adaptable. A rat inherits certain taste preferences and as long as it can find sufficient quantities of these foods, it may live its entire life without having to try new ones. But a starving rat must try new foods, if it is to survive. It will then come to prefer tastes associated with feelings of wellness and avoid those associated with sickness. Rat is not learning specific new eating behaviours, but rather to reset reference levels for lower-level perceptions based on consequences for higher-level goals.

Rat placed in Skinner box also demonstrates learning, but this involves learning which patterns proprioceptive, auditory, and visual perceptions lead to the delivery of a food pellet. A rat's behaviour is more flexible than that of a spider in that rat is able to reset reference levels based on experience, whereas the spider's reference levels are less modifiable. That evolution allows certain types of flexibility, but not others; recall that a rat quickly learns in a single trial which taste leads to nausea and which sounds are followed by skin pain. In perceptual-control-theory terms, the rat learns to set a very low or zero reference level for these tastes and sounds to avoid the nausea and pain that follow them. But its behavioural flexibility is limited, in that it cannot change its reference level for taste based on sound or for a certain sound based on nausea.

Learning from a hierarchical-perceptual-control-theory perspective is actually finding out by a form of trial and error, which combinations of lower-level perceptions are successful in bringing about a higher-level goal. This process is referred as reorganization. Reorganization is a process akin to rewiring or microprogramming a computer, so that those operations it can perform are changed. Reorganization alters behaviour, but does not produce specific behaviours. It changes the parameters of behaviour, not the content. Reorganization of a perceptual function results in a perceptual signal altering its meaning owing to a change in the way it is derived from lower-order signals. Reorganization of an output function results in a different choice of means, a new distribution of lower-order reference signals as a result of a given error signal.

Animals act on their world based on what they perceive, and thereby change their environment and what they consequently perceive of it. Animals also change how they act on the world, when old ways are no longer effective in getting what they want. But this change in behaviour is based on a Darwinian process involving spontaneous variation and selection; not variation and selection of specific behaviours, as conceived by Skinner and his behaviourist followers, but rather variation and selection of goals as the organism discovers, which new combinations of controlled lower-order perceptions lead to the attainment of higher-level goals.

2.17 CONCLUSION

During the 15th to the 19th centuries, with the development of science, a better comprehension of the nature was achieved. Meanwhile, the study of animal behaviour has also advanced a lot. Three sources of current animal behaviour science are: medical anatomy and physiology, biological evolution, and psychology. These have constituted three disciplines, which are behavioural science, comparative psychology, and neurobiology (Figure 2.6).

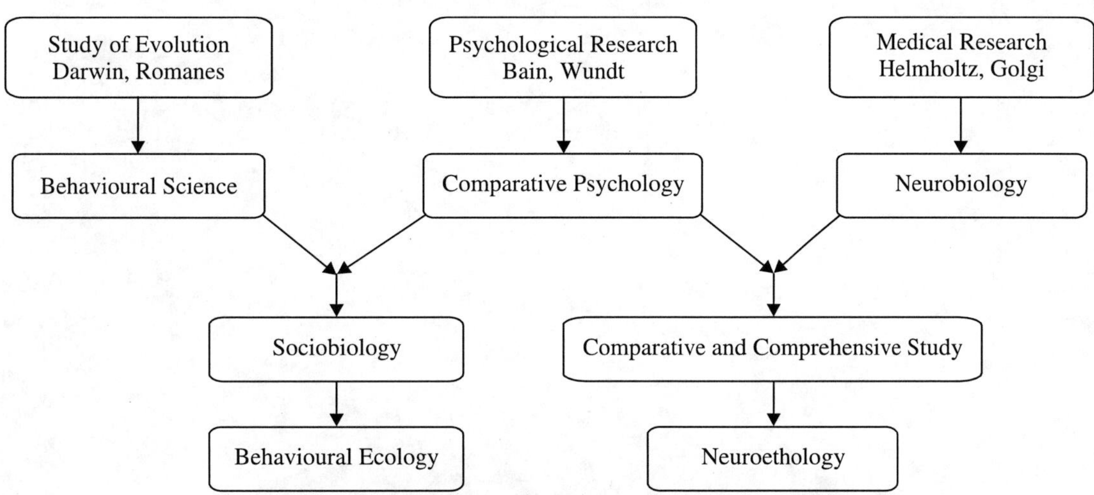

FIGURE 2.6 Diagramatic representation of evolution of behavioural science.

Review Questions

Short Answer Questions

1. State four main features of modern behavioural studies.
2. Who introduced the concept of mating system?
3. Who introduced the concept of breeding system?
4. State the idea of kin selection.
5. Who developed the theory of reciprocal altruism?
6. Give an example of symbiotic behaviour.
7. Define sexual selection.
8. Comment on the reversal of typical sex role.
9. Write a note on the contribution of Skinner.
10. What is reinforcement?
11. Give two examples of instinctive behaviour.
12. What is reorganization?
13. Name the author of the book, *What is Life*.

14. What do you understand by natural selection?
15. Define mutation.
16. What is microevolution?
17. Define speciation.
18. Define reciprocal altruism.
19. What is kin selection?
20. What are adaptive behaviours?
21. What do you understand by character mapping?
22. When does associative learning occur?
23. Define behavioural ecology.
24. Define displacement behaviour.

Long Answer Questions

1. Describe the contribution of Darwin in ethology.
2. "Infanticide is a means of achieving reproductive advantage". Discuss.
3. Discuss the web construction in spider in the light of modern science.

Chapter 3

Interaction between Gene, Neuron, Hormone and Environment

Behavioural observations on Mediterranean monk seals date back to Aristotle, who in his ground breaking work; "*Natural History*" described herding and social behaviour of seals on coasts of Lesbos in fourth century B.C. Behavioural studies originate in very early days, but advances in our understanding in this field are very recent. Recent understanding in behavioural biology suggests that behaviour is the result of interaction between gene, neuron, hormone, environment, and subfields like, behavioural genetics, neuroethology, and behavioural ecology have emerged very recently. With the development of molecular biology and cellular biology, much have revealed about animal behaviour. Behaviour is controlled by the genetic regulation, neuroregulation, hormonal regulation, and environmental regulation. Nervous system controls behaviour in a direct way, but hormones show a slower process with longer duration. Stimulus from the environment influence endocrine physiology, and neuroendocrine physiology may interfere with the functioning of gene and ultimate manifestation of behaviour. Gender difference of animal behaviour probably accounts for the relationship among the behaviour, hormone, and gene.

3.1 BEHAVIOURAL GENETICS

Genes bear the evolutionary responses of prior populations to selection on behaviour. Genetic characteristics are largely determined by genes. Genes may play role in many behaviours, but they never determine and directly code for a behaviour-genes code only for proteins. A change in single protein cause many effects and even brings about distinct phenotype. External environment strongly influence the expression of genes in behaviour via nervous and hormonal mechanisms. Phenotype emerges from an interaction of its genotype with environmental factors. Gene composed of DNA is the smallest functional unit of heredity which specifies codes for amino acid chains that make up proteins. Locus is a section on chromosome which is related in a significant way to a function. Monomorphic loci contains single common segment of code and are found in 95–99% individuals. Polymorphic loci is found in form of 2 or more alleles with a combined frequency > 0.05. Diploid organisms carry 2 copies of each gene, 1 derived from each parent. In Homozygous condition both copies for gene are identical. In heterozygous condition chromosomes contain 2 different alleles of gene. Sometimes phenotype of heterozygous individuals follows one allele and not other. When one gene pattern is well established in particular organism, it is called *wild type allele*. A mutant allele is a new and less common modification. Pleiotropy is the situation where there is no clear 1:1 correspondence between one particular gene and particular behavioural trait. Fruitflies with mutant period gene display many behavioural effects, from differences in daily activity rhythms, time of day at which adults hatch, and male courtship song pattern. Heritability is an estimate of variation in trait that is attributable to genetic differences in individuals. Number of human arms is heritable and can be expressed as a proportion. An individual's fingerprint and height are strongly heritable, while variables like musical abilities are much less so.

Classical Genetics

Classical (Mendelian) genetics examines the distribution of hereditary characteristics of behaviours from one generation to next. With selective breeding, presence/absence of specific behaviours can be tracked through outcomes of sexual reproduction. Individual is diploid with one half of the genes contributed by each parents. Some genes dominate phenotype over others. Genetics of many diseases follows simple Mendelian rules. Most morphological/behavioural traits (body size, intelligence, boldness) occur across a continuous phenotypic range. Seymour Benzer discovered behavioural mutants including single gene disruptions of circadian rhythms in fruit flies, and genetic causes in neuro-degeneration and aging.

Hygienic behaviour in honeybees involves the ability to detect and remove diseased, larval and pupal brood from nest before pathogen becomes infectious. Mechanism of resistance to bee diseases and parasitic mites consists of 2 components: uncapping a cells and removing its content. These 2 traits were controlled in simple Mendelian manner by 2 recessive loci (Rothenbuhler 1964). Molecular evidence from Quantitative Trait Loci (QTL) linkage mapping has identified multiple specific stretches of DNA with genes that underlie variation in this trait suggesting that genetic basis of hygienic behaviour is complex. Seven QTLs are associated with hygienic behaviour, each controlling only 9–15% of observed phenotypic variance. Quantitative genetics explores the inheritance of traits which are subject to significant environmental influences and involve several underlying genes. Although, each gene may follow a simple Mendelian pattern, overall results are not characterized in such terms. Quantitative approaches study the extent of relatedness among individuals which is matched by resemblances in behavioural characteristics. Aside from cases where multiple genes affect single traits, changes in one individual gene affect several traits. Such instances can be characterized through measures of phenotypic covariance across multiple traits.

Environment, Genetics and Cognitive Development

Cognition separate animals from immediacy of environment and to solve future problems reflected on past. Cognition make novel associations once thought to define humanity from animals. Cognitive ability is not confined solely to humans. Learning through cognition may be removed from genetic constraints than other types of learning. Cognitive problem solving ability varies among different individuals within species. Variation in cognitive ability is inherited and there is a genetic element underlying such abilities. Cognition gives high level of flexibility in social and physical environments, but cognition is constrained by genetic limits. It allows an animal to distinguish itself as a distinct identity. Evidence of cognition is found if an animal look at its own image in mirror and recognizes "self". Dying a patch of hair of an animal and then observe the reaction of animal to its mirror image is a test of cognition. If it touches the dyed patch this is taken that animal has a concept of "self." Apes, some monkeys, elephants and dolphins respond positively in mirror tests, emphasizing that cognition is important in behavioural development (Plotnik et al., 2006). Social cognition, the ability of animal to predict how its own actions affect its future relationships within social group is found in chimpanzees. Social cognition makes animals to be more calculating and manipulative in social relationships. Chimpanzees perform exact revenge against group members that exhibit selfish behaviour (Jensen et al., 2006).

Living organisms perhaps are not consciously aware of the purpose of their survival and they attempt to survive with comfort and produce offsprings. Survival is regarded as instinct. We are genetically hardwired to survive. Organisms suffer a lot to survive. For example, they fight for the limited food, shelter, and potential mates. Plants, animals and microbes have evolved various mechanisms over millions of years to survive and to propagate depending on changing environmental conditions. Plants produce beautiful flowers, sometimes with attractive smell, for attracting pollinators with the ultimate aim of their survival through offspring. Diverse anatomical, physiological and morphological features help organisms to live in diverse environment. Darwin, the noted naturalist and evolutionary biologist, first developed a coherent account of the purpose of survival. His theory of evolution is now fully accepted, although not fully realized in most cases. Evolution may be biological or cultural. Biological evolution is the key in understanding life, life forms and their attainment of complexity from the simplicity in structure and functions. Cultural evolution exerts significant influence to shape the life.

Sociobiology including sociogenomics has recently emerged as a potential field in explaining many behaviours. Contrasting behavioural phenomena are often seen. For example, both cooperation and conflict are common. Emigration and immigration, hibernation and aestivation, infanticide and parental care, monogamous relationship and polygamous relationship run side by side. Thus, the nature of all behaviours with a common formula is difficult to predict.

Darwinian concept of evolution is important in explaining and understanding nature. Man is dominated by culture, language and intelligence. Some put emphasis on culture than genes in explaining human nature creating the debate—"nature versus nurture". Cultural transmission is analogous to biological transmission. Human language is a tool of cultural evolution. Fashions, ceremonies and customs, art and architecture, as well as various rituals in various human societies evolve through the course of cultural evolution, and exhibit both the similarities and dissimilarities. In fact, cultural homogenization is a matter of concern for human civilization. Life evolves by the differential survival of the gene. Replication or copying of gene and passing the copies to the next generation through natural selection constitute the foundation of biological evolution.

The replicator of human culture is meme which means imitation. Examples of memes are tunes, ideas, and fashions. New and novel idea spreads from one human to another and from one human society to another. All genes do not replicate successfully. All memes do not spread equally. Some memes are more successful than others in propagation. Religious meme complex is the faith. Memes and genes may reinforce each other, although sometimes oppose each other.

Natural selection favours memes that exploit their cultural environment. The meme pool has a role to play in evolution. The gene, a length of chromosome, serves as a unit of natural selection. Each gene competes with its own alleles. Memes are similar to the early replicating molecules than genes. Co-adapted gene complexes sometimes arise in the gene pool. A set of genes causing mimicry in butterflies remain closely linked together on the same chromosome. Suitable teeth, claws and sense organs in carnivore gene pools have evolved from herbivore gene pools. Genes are passed onto the next generation to preserve the species. Perhaps, this is hardwired. As generation ends, the contribution of genes becomes halved and through successive generations it ultimately becomes insignificant. Idea, tune and invention may remain intact after the mixing of genes in the common pool. Einstein may or may not have a single gene or two today but his ideas are strong.

We have the conscious foresight capacity and true altruism. We produce the gene to pass on to the next generation. DNA, a self-replicating structure, is different from rival pieces of DNA. DNA is not the lone factor for evolution. Two distinct units of natural selection are the gene and the organism. Genes, the functional part of DNA, remain inside the cells, store and carry biological information from parent to offspring. Each living organism possesses inherited genetic differences. Living organism is the repository of a history of a specific gene. Genetic diversity confers species the ability to adapt to the changing environments, including new pests and diseases and new climatic conditions. Intra- specific genetic diversity represents the range of heritable differences of a trait or a set of traits among individuals within a species as well as variation among different population. The term is synonymous with "genetic variation" or "genetic variance". Previously existed population shapes the genetic diversity. Physical, chemical, and biological environments are constantly changing over short- and long-term time scales. To survive in the changing environments, species must adapt or face extinction. Species' ability to adapt to alter environment is directly related to quantity of genetic diversity available to natural selection. The populations with much genetic diversity for fitness traits adapt more quickly in the environment. Maintenance of genetic diversity is a key for the long-term survival of most species. The main forces that determine genetic diversity within species are the mutation, migration, selection, and genetic drift.

3.1.1 Gene Concept

In the pre-1940 beads on a string concept, the gene as the basic unit of inheritance was defined by three criteria, namely, function, recombination and mutation. As a unit of function, the gene was thought to control the inheritance of one character or one attribute of phenotype. The gene as a unit of structure could be operationally defined in two ways: by recombination and by mutation. In 1941, G.W. Beadle and E.L. Tatum, based on their discovery on biochemical mutations in *Neurospora*, proposed "one gene–one enzyme" hypothesis. This hypothesis states that a gene controls a biochemical reaction by directing the production of a single enzyme.

Evolution of gene concept

As per the classical concept, a gene is not subdivisible and cross over does not occur within a gene but occurs between two separate genes. But Oliver's studies show that a gene is more complex than a bead on a string. Operator and promoter genes do not code for a polypeptide or an enzyme. The gene as the unit of function controls the synthesis of one polypeptide. Extensive study on the fine structure of gene was undertaken by Seymour Benzer who defined the functional unit as cistron and confirmed operationally closely to what we commonly think as gene. This cistron may be thought of as the gene at the functional level. There can be over a hundred points within a functional unit wherein a mutation can take place and cause a detectable phenotypic effect. This means that a cistron is over hundred nucleotide pairs in length. Some cistrons may be as long as 30,000 nucleotide pairs in length. Each cistron is part of a gene which is responsible for coding of only one polypeptide chain of an enzyme that has two or more different polypeptide chains in its complete enzymatic unit.

The muton is the smallest unit of DNA which, when altered, causes mutation. The study of genetic code makes it clear that an alteration of a single nucleotide pair in DNA may result in a missence codon in transcribed mRNA (e.g. AGC-AGA) or nonsense (e.g. UGC-UGA). A cistron may consist of many mutable units or mutons. The term *muton* was given by Benzer. The recon is the smallest part of DNA which is inter-changeable through crossover and recombination. Recombination studies in *E. coli* indicate that a recon consists of not more than two pairs of nucleotides. A recon may occur within a cistron. A gene classically is made up of several functional units, the cistrons, which consist of several recons and mutons. Separable mutable sites can exist within a single gene. The unit structure not divisible by recombination was the single nucleotide pair as first demonstrated experimentally by Yanofsky and others. Convincing evidence for colinearity between sequence of nucleotide pairs in gene and sequence of amino acids in polypeptide gene product is obtained in other studies, in both prokaryotes and eukaryotes. Demonstrations of non-coding segments, introns within certain eukaryotic genes, show that the concept of an uninterrupted sequence of triplet base pairs in gene specifying a collinear sequence of amino acids in polypeptide gene product often needs modifications. A polypeptide which showed that independent mutations that alter the same amino acid may either affect the same mutable site and not recombine or occur at different mutable sites. Moreover, data show that independent mutation at the same mutable site is not always identical.

Fine structure of genes and complex loci in eukaryotes

Intragenic recombination was first detected in *Drosophila*. Genetic fine structure maps have been constructed for many genes of *Drosophila*. Fine structure maps have been worked out for several other higher animals and plants. In almost every case where an extensive analysis of a gene has been carried out, mutable sites separable by recombination have been detected. Clearly then, the genes of eukaryotes are sequences of nucleotide pairs which specify the amino acid sequences of polypeptides. In eukaryotes, some genes have interesting structural features which are not observed in any prokaryotic genes. The most complete picture for a locus in eukaryotes is rosy locus in *Drosophila*, which codes for enzyme xanthine dehydrogenase. Besides rosy locus, the fine structure maps of gene of many other eukaryotes have been prepared such as white eye, notch wing, lozenge eye, zest loci of *Drosophila*, waxy and other loci of maize.

3.1.2 Concept of Genetic Diversity

Variations in the living world are due to the variations in the sequences of the four base pairs of DNA and their interactions with the environment. The pool of genetic diversity of a species exists at three levels, viz., genetic diversity within individuals often called *heterozygosity*, genetic diversity between individuals within a population, and genetic variation between populations. Genetic diversity is the heritable variation within a population and between populations. New genetic variations are produced through gene or chromosomal mutations. In sexually reproducing organisms, recombination also contributes in genetic diversity. Genetic diversity in a population enables natural evolutionary changes to occur. The rate of such changes is proportional to the amount of available genetic diversity. Genetic diversity at the individual and population levels determines the ability of the species to survive and adapt to the environmental conditions. This measure of fitness is fundamental to the process of natural selection and evolution.

Higher genetic diversity offers more opportunities for survival. Under natural conditions, a degree of genetic diversity is retained as populations of a species interact over time and space. The environment exerts its effect on an organism at genetic level through regulating the transcription of a specific gene. Out of 10^9 genes in the world biota (10^5 are found in higher organisms), not a single one duplicates the contribution of others. Genes control fundamental processes like photosynthesis and respiration, and are highly conserved across different taxa and generally exhibit little variations (Groombridge 1992). Genetic diversity of micro-organisms is higher than micro-organisms because the former has originated first and existed on the earth for a longer time.

Of the 95 phyla of living organisms (Margulis and Schwartz 1988), 52 belong to microorganisms excluding viruses. A total of 159 mutant lines in *Chlamydomonas reinhardii* and more than 3000 mutants in *Neurospora crassa* (Woese et al., 1990) exemplify genetic diversity of microbes. Most genetic variability within species over a single generation occurs due to shuffling of homologous chromosomes, crossover, and fusion of gametes during sexual reproduction. Mutations are rare, occurring once in 10^5 to 10^6 genes per individual in a generation. Evolution would stop without mutation although mutation does not take place in anticipation of environmental demand.

It simply happens bringing structural and functional changes of organisms. Environmental conditions determine whether such changes are beneficial, harmful or neutral for the present as well as for the future. When a single species population invades several new habitats and evolves due to differing environmental pressures, then many new species would evolve in a relatively short period through adaptive radiation. This happens due to superior adaptation in a species that displaces a less-adapted species from a variety of habitats. Natural selection may favour the spread of a mutation or not, but it cannot directly detect an organism's genotype which acts on phenotypes. Mutation, migration, selection and drift, individually and collectively, alter allele frequencies, bring about evolutionary divergence and cause the formation of species.

Wild relatives often provide genes that are unavailable in domesticated plants. Such genes afford resistance to diseases/pests and other environmental stresses and have been acquired by the wild relatives through their long period of co-evolution with microbes/pests as well as survival for a long time under stressed environments of various sorts. Many resistant genes control specific resistance. The primary gene pools (GP1) represent the true biological species including all its cultivated (cultigens), wild and weedy forms; hybrids among these forms are fertile; and gene transfer to the crop is simple, direct and poses no problem. Most primary gene pools show at least 80% genetic closeness to the crop species. The secondary gene pools (GP2) represent the species group that can be artificially hybridized with the crop but gene transfer may not be easy. The hybrids, if produced, are usually weak or partially sterile. The secondary gene pools show around 60% closeness to the crop species. The tertiary gene pools (GP3) include all species that can be crossed to the crop species but with some difficulty. These gene pools show around 40% closeness to the cultivated species. There is also a quaternary gene pool (GP4); its constituents are incompatible with related species (Krishnamurthy 2004).

3.1.3 Mutations

Mutations are heritable changes in genetic materials producing alternate forms of any gene (alleles). Of the two broad types of mutation, one affects gene, and is called *gene mutation*; the other affects the whole chromosome or chromosomal segment called *chromosomal aberration*. Gene mutation at nucleotide level is generally known as *point mutation*. Chromosomal mutation may be due to changes in chromosome structure (deletion, duplication, inversion, and translocation) or number.

Chromosomal mutation

In eukaryotic organisms, the DNA molecules in the nucleus are combined with proteins, mainly histones, to make chromosomes. Each chromosome represents a single DNA molecule. Chromosomes are usually clearly visible and identifiable only when cells are dividing, when the chromosomes have already been divided into daughter chromatids. However, the daughter chromatids retain connection to one another at a position called the *centromere*. The number of chromosomes, their lengths and the positions of their centromeres are unique to each species and these characteristics are used as descriptors for the species karyotype. Chromosomes themselves mutate and evolve (Beaumont and Hoare 2003).

Inversion

When a region of a chromosome breaks off and rotates 180° before rejoining the chromosome, then the phenomenon is called *inversion*. When the centromere is not included in inversion, it is called *paracentric inversion*. When the inversion includes centromere, it is called *pericentric inversion*. Crossover in a heterozygous pericentric inversion results in deletion and duplication and produces rod-shaped acrocentric chromosomes. A crossover in the inverted region of a heterozygous paracentric inversion produces a dicentric chromosome.

Translocation

The breakage of a chromosomal region and rejoining either at the other end of the same chromosome or another non-homologous chromosome is termed *translocation*. This may be simple-, shift- and reciprocal translocation. In the simple translocation, a single break in a chromosome occurs and the broken piece gets attached to one end of a non-homologous chromosome. In the shift translocation, the broken segment of one chromosome gets inserted interstitially in a non-homologous chromosome. In the reciprocal translocation, a segment from one chromosome is exchanged with a segment from another non-homologous one. In *Oenothera*, a rare series of reciprocal translocation is reported involving all its seven chromosome pairs. If each chromosome end is labelled with a different number, the normal set of seven chromosomes would be represented as 1-2, 3-4, 5-6, 7-8, 9-10, 11-12 and 13-14.

Likewise a translocation set would be represented as 2-3, 4-5, 6-7, 8-9, 10-11, 12-13 and 14-1. Such a multiple translocation heterozygote would form a ring of 14 chromosomes during meiosis. Various lethals in each of two haploid sets of seven chromosomes administer the structural heterozygosity. Since only alternate disjunction from the ring can form viable gametes, each group of seven chromosomes behaves as a single large linkage group with recombination confined to the pairing ends of each chromosome. Each set of chromosomes inherited as a unit is called a *Renner complex*.

Deletion

Deletion involves loss of a chromosomal region. A terminal portion of a chromosome is lost due to only one break in the terminal deletion. An intermediate portion of a chromosome is lost due to two breaks in the intercalary deletion. Here, the chromosome is broken into three pieces, the middle one of which is lost and the remaining two pieces get joined again. Notch-wing mutation is an example of deletion. Notch-wing is a recessive allele and when this allele behaves like a dominant allele and is phenotypically expressed, then this phenomenon is called *pseudo dominance*. Pseudo dominance (deletion) is found in Cri-du- chat syndrome where human individuals lose a portion of the short arm of chromosome 5.

Duplication

The presence of a chromosomal part more than the normal chromosomal complement is called *duplication*. This may be tandem-, reverse tandem-, displaced- and transposed duplication. In tandem duplication, the duplicated region is situated by the side of the normal corresponding chromosome section. Here, the gene sequences are similar in both normal and duplicated regions of the chromosome. In the reverse tandem duplication, the sequence of genes in the duplicated chromosomal region is reverse to that of the normal chromosome. In the displaced duplication, the duplicated region is not located adjacent to the normal chromosome. Such duplication may be homobranchial or heterobranchial. In the transposed duplication, the duplicated chromosomal portion becomes attached to a non-homologous chromosome. In extra-chromosomal duplication, the duplicated chromosomal part acts as an independent chromosome in the presence of centromere.

Changes in chromosome morphology

Morphological changes in chromosome are observed in isochromosome, ring chromosome, and Robertsonian translocation. In isochromosome, both the chromosome arms are identical due to the division of a centromere in the wrong plane, producing two daughter chromosomes, each of which carries the information of one arm only but is present twice. In the case of ring chromosome, breakage at each end of the chromosome occurs and the broken ends are joined to form ring chromosome. Robertsonian translocation is believed to play a role in the evolution of human beings. Humans and great apes have 46 and 48 chromosomes, respectively. Humans would have evolved from the ape ancestor due to centric fusion of two acrocentric chromosomes to produce a single large one. Actually such translocation results in the reduction of chromosome number due to the whole arm fusion in the non-homologous chromosome.

Changes in chromosome number

The gametic chromosome number constitutes a basic set of chromosomes called genome. Chromosomal aberrations may include whole genomes and the entire single chromosome. Changes in chromosome number are called heteroploidy. Heteroploidy may be either aneuploidy or euploidy. The chromosome numbers of both aneuploidy and euploidy are given in Table 3.1.

TABLE 3.1 Chromosome Numbers of Aneuploidy and Euploidy

Aneuploidy	*Chromosome number*
Monosomy	$2n-1$
Trisomy	$2n+1$
Tetrasomy	$2n+2$
Pentasomy	$2n+3$
Euploidy	
Diploidy	$2n$
Triploidy	$3n$
Tetraploidy	$4n$
Pentaploidy	$5n$
Polyploidy	$3n$, $4n$, $5n$
Autopolyploidy	Multiplication of the same genome
Allopolyploidy	Multiplication of different genomes

Aneuploidy

Monosomy: A 2n – 1 complement due to the loss of one chromosome is called *monosome.* Monosomy for the X chromosome is found in humans 45, X Turner syndrome.

Trisomy: A $2n + 1$ complement due to the addition of an extra chromosome is called *trisomy.* In the primary trisomic, the extra chromosome is identical to its homology. In the secondary trisomic, the extra chromosome is an isochromosome. In tertiary trisomic, the extra chromosome is the product of translocation. In human beings, trisomy is found in Down's syndrome, Edward's syndrome, and Patau syndrome.

Tetrasomy: Organisms with two extra chromosomes are known as tetrasomics ($2n + 2$). Tetrasomics are found in wheat.

Euploidy: When the genome contain chromosome in multiples of some basic number (x). Euploids are organisms with balanced set or sets of chromosomes in any number. A basic set of chromosome numbers is called *monoploid.* When more than two sets of chromosomes are present, the term polyploid applies. Only even-number polyploids are fertile because the odd number of chromosomes cannot be divided into two at reductional meiotic division. The addition of one or more extra sets of chromosomes, identical to normal haploid complement of the same species, results in autopolyploidy. The combination of chromosome sets forms different species because interspecific mating results in allopolyploidy. When a diploid gamete is fertilized by a haploid gamete or when two sperms fertilize an ovum, then triploids are produced. Polyploids with the equivalent of four haploid genomes derived from two separate species are called allotetraploid. Endopolyploidy is the condition in which only certain cells of diploid organisms are polyploids. In *Gerris,* 1024 to 2048 copies of each chromosome are found in salivary gland cells.

Although chromosome variations are no longer used as markers in population genetic studies, they play an important role in evolution. Chromosome duplications provide an extra copy of a block of DNA that may contain complete gene sequences. As might be expected, duplications are less harmful than deletions and when duplications contain complete gene sequences, natural selection can operate independently on both the new and the old sequences to produce divergent roles for the genes. This is the principal process for the evolution of new genes (Beaumont and Hoare 2003).

Gene mutation

A gene is a unit of information which is held as a code in a discrete sequence of DNA. DNA is a polymeric molecule consisting of chains of nucleotide monomers. The complete DNA molecule actually consists of two plain-clothes chains wrapped around one another as a double helix. The sugar phosphate backbones are outside the molecule while the bases point towards the middle of the structure; the two strands of the molecule run in opposite direction. The fundamental beauty of the DNA molecule is a result of complementary base pairing where G can only bond with C, and A can only bond with T, at the middle of the molecule.

The replication produces daughter DNA molecules, each of which has one parental strand and one copied strand. This is called semi-conservative replication. During replication, various proofreading activities take place and almost all errors are corrected by removing the incorrect base and inserting the correct one. Despite proofreading, a few errors are inevitable and it is estimated that about one in every 3 billion bases is incorrectly inserted. Such errors are called *point mutations*. Without such errors, no genetic change at the DNA level would take place, but with too many errors daughter cells would too often be non-viable and the organisms carrying that DNA would soon become extinct. Gene mutation may be point mutation or gross mutation.

Point mutation: Point mutation results from base substitution in a gene that codes for a polypeptide. This may be missence, nonsense or same sense mutation. Missence mutation results in the replacement of one sense codon for another. Sickle-cell anemia is a good example of such a mutation. A nonsense mutation creates one of the three terminating codons (UGA, UAA, UAG). Such a mutation produces polypeptides that are shorter than the normal ones and are unable to function properly and therefore represent a loss of biological characteristic.

Frame shift mutations result due to insertion or deletion of one or more bases causing the reading frame of the gene to be altered and a different set of codons to be read downstream (3′) of the mutation. Several examples have been observed in individuals with haemophilia A. These include a deletion of four boxes that results in a change of reading frame after codon 50 and an insertion of 10 bases which alters the reading frame beyond codon 38. Splice site mutations alter the signal sequences for splicing that occur at the 5′ and 3′ ends of introns leading to a failure of RNA transcribed from the mutated gene to be spliced properly. Mutations in introns also result in the creation of splice sites, again causing splicing to occur abnormally. Promoter mutations affect the way through which gene transcription is regulated. Such a mutation is associated with haemophilia.

Gross mutation: Gross mutation causes substantial alterations to the DNA often involving long stretches of sequence. Such mutation may be categorized into deletion, insertion, re-arrangement, and trinucleotide repeat mutation. Deletion mutations involves the loss of a portion of the DNA sequence. Insertion mutation occurs as a result of insertion of extra bases and usually forms another part of a chromosome. Rearrangement mutation involves segments of DNA sequences within or outside a gene exchanging portion with each other. A usual form of gene mutation, trinucleotide repeat mutation, has been described that involves unstable trinucleotide repeat sequences. During meiosis, a dramatic increase in the number of copies of the trinucleotide repeat occurs, a dramatic increase of which leads to the development of disease in subsequent offsprings. Gross mutation involves major alterations to gene sequences and invariably has a serious effect on the encoded protein and is frequently associated with a mutant phenotype (Figures 3.1 to 3.3).

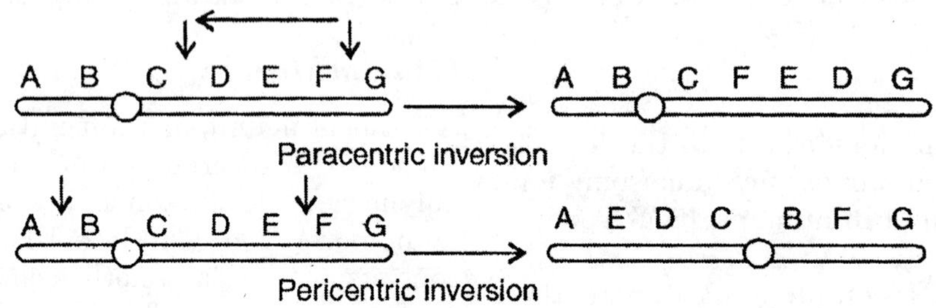

FIGURE 3.1 Paracentric and pericentric inversion.

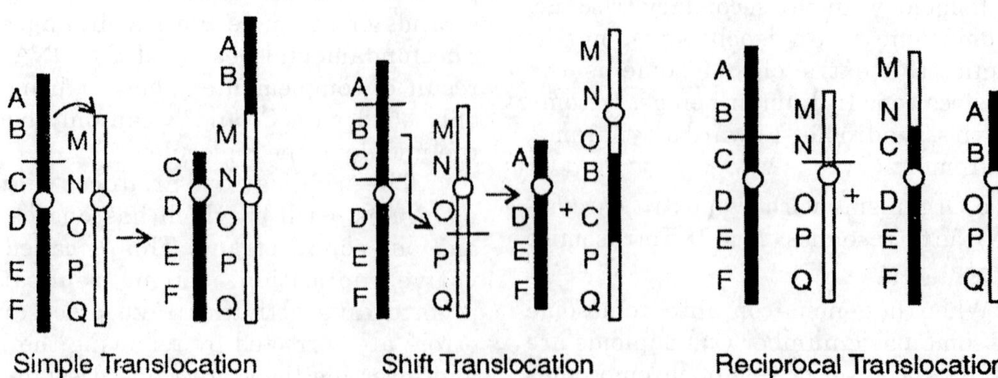

FIGURE 3.2 Various types of translocation.

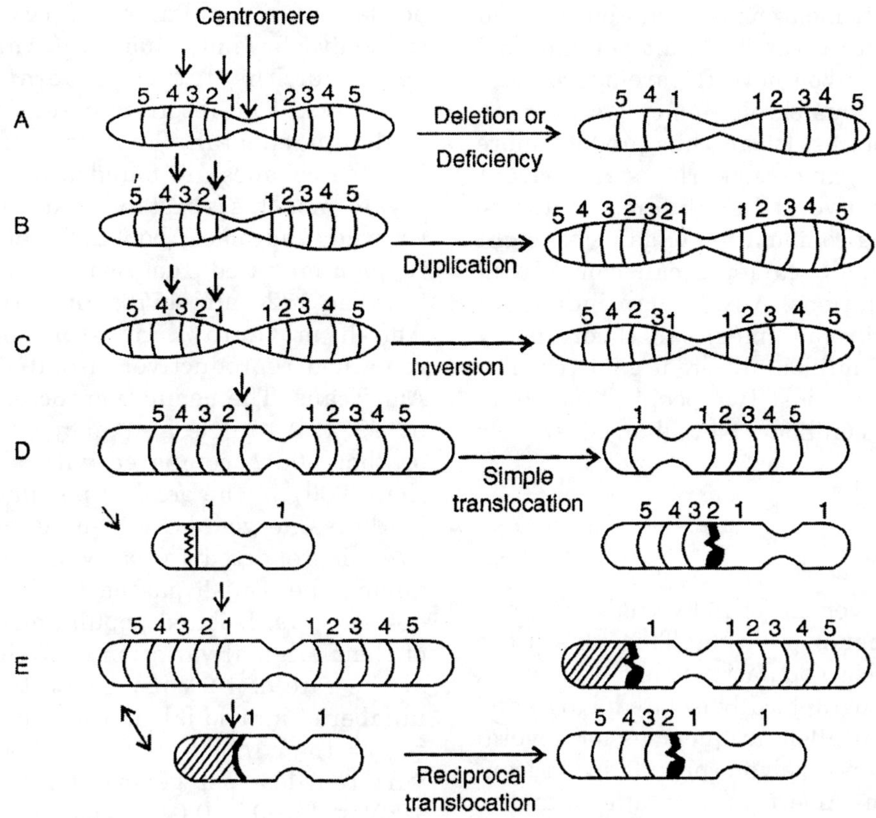

FIGURE 3.3 Different types of chromosomal aberrations.
Source: Chatterjee and Chakroborti, 2005.

Molecular basis of gene mutation

The sequence of nucleotides determines the structures of rRNA, mRNA, and tRNA as well as proteins that are required by living cells. Genetic information should be accurately replicated and passed on to the future generations for the survival of species. Insertion, deletion or substitution of one or more bases in mutation results from the error(s) in replication of a gene within the DNA molecule. One spontaneous mistake generally occurs in 10^4 to 10^7 replications. Such a mistake leads to a heritable change in the sequence of a DNA molecule. Many mutations are due to the instability of the nucleotide bases in the DNA. Nucleotide bases on absorbing sufficient energy from bombarding water molecules undergo structural changes called *tautomeric shifts*. A tautomeric shift causes the redistribution of electrons and protons in the bases so that they no longer pair normally.

Actually, the spontaneous movement of a hydrogen atom from one position to another within a base converts it to an alternative isomeric form called a *tautomer*. The process itself involves a tautomeric shift. All purines and pyrimidines can exist as tautomers. When a tautomeric shift occurs, the altered base tends to pair with the wrong complementary base, thus creating a mutation that would cause a different phenotype. It is important to mention in this connection that some genes contain regions that are more likely to mutate than others. These sites are called *hot spots*. Benzer first demonstrated hot spots during his landmark study of the topology of the rII locus of bacteriophage T4. Benzer mapped a total of 1612 spontaneous mutations in the rII locus and found 60 sites within the locus exhibited a larger than expected number of mutations.

Shuffling of homologous chromosome, crossover and fusion of gametes

At meiosis I, maternal and paternal homologues are randomly shuffled to the daughter cells. An organism with three pairs of homologous chromosomes can align in form configurations at metaphase I. Anaphase I can therefore produce eight possible sets of chromosomes. Each of these chromosome clusters will enter meiosis II to produce two gametes and the hypothetical parent organism with three pairs of homologous chromosomes could produce gametes with eight different chromosome sets. Each time the hypothetical parent produces gametes, its homologous chromosomes crossover in new and different places. It can thus produce more than eight types of gametes. Perhaps the parent never produces any two identical gametes due to crossing over. During combination of gametes, eight male and eight female gametes can combine in 64 different combinations. Any human individual (with 23 pairs of chromosomes) can theoretically produce 2^{23} (about 8 million) different gametes. The fusion of gametes from just two people thus could produce 8 million × 8 million = 64 trillion genetically different children.

3.1.4 Migration

Migration means movement of individuals from one population to another population. This causes the flow of genes between populations that were once, but are no longer, geographically isolated. Parra and co-workers measured allele frequencies for several different DNA sequence polymorphisms in African-American, European- American populations, and in African and European populations representative of the ancestral human populations from which the two American populations are descended. One locus they studied, a restriction site polymorphism called FY-NULL has two alleles, FY-NULL* 1 and FY-NULL* 2. The frequency of this allele is 0 in the three African populations and 1.0 in the three European populations, but lies between these extremes in all African-American and European-American populations. Such a result can be attributed to the mixing of genes between American populations with predominantly African ancestry and American population with predominantly European ancestry (Klug and Cummings 2004).

3.1.5 Genetic Drift

Chance events mostly alter the allele frequencies in a small population than in a large population by genetic drift. Audesirk and Audesirk (1999) noted two important points about genetic drift:

1. Genetic drift tends to reduce genetic variability within a small population.
2. Genetic drift tends to increase genetic variability between populations.

Two special cases of genetic drift are population bottleneck and founder effect. In population bottleneck, a population reduces to such an extent that only a few individuals are available to contribute genes for the future generations. Population bottleneck is on record in case of northern elephant seal and cheetah. Founder effect occurs when isolated colonies are founded by a small number of organisms. This is best illustrated in human Ellis-van Creveld syndrome. About 200 members of the Amish religion migrated from Switzerland to Pennsylvania between 1720 and 1770. Since then, Pennsylvania Amish move to Lancaster country and have remained reproductively isolated from non-Amish Americans. The population increased to about 8000 by 1964. An allele frequency for Ellis-van creveld of about 0.07, compared with a frequency of less than 0.001 in the general population, was recorded in the same year. One couple who immigrated in 1744 is reported to carry the allele. Inbreeding among the Amish passed the allele along to their descendents. Isolated populations lose a percentage of their original variation over time, approximately at the rate of $1/2N$ per generation, where N = the number of individuals in the populations. Thus, if N = 500, then $1/2N$ = 1/1000 = 0.001 or 0.1% genetic variation lost per generation and if N = 50, then $1/2N$ = 1/100 = 0.01 or 1% genetic variation lost per generation (Stiling 2004). Thus, a population of 500 will retain 99.9% of its genetic variation in a generation, whereas a population of 50 will retain 99.0%. After 20 generations, a population of 500 will thus still retain over 98% of it original variation, but a population of 50 will retain only 81.79%. Such effect becomes more severe as the population size decreases. A population of 500 is necessary to reduce the effects of genetic drift. Thus, the rule has entered the conservation theory literature as a magic number (vide Simberloff 1988).

3.1.6 The Double-helix and Behavioural Genetics

All animals, plants, bacteria, and viruses produce copies of their DNA. Different organisms exist, although in fundamental chemistry they are the same. They bear the same gene, and DNA works in them in complex ways. A DNA with known shape is made up of nucleotides. A pair of nucleotide chains are twisted together in a spiral. The nucleotides of four different kinds (Adenine, Thymine, Guanine, and Cystosine) are shortened to A, T, C and G which are the same in all eukaryotes. In the nucleus of every human cell, there are 46 chromosomes. Meiosis produces the sperms or eggs and they contain only 23. A sperm is made in the testicle. Single gene codes for single polypeptide chain. The gene, just a length of chromosome, may overlap with other genetic units and may include smaller units, and form part of larger units. A large cluster of separate genes comes together in a tight linkage group on a chromosome by the inversions and other accidental rearrangements of genetic material.

The butterfly mimicry cluster is a good example. As the cistrons leave one body and enter the next, they find their close neighbours from the bodies of distant ancestors. Neighbouring cistrons on the same chromosome form a tightly-knit troupe of travelling companions who seldom separate during meiosis. A gene travels intact from grandparent to grandchild, passing straight through the intermediate generation without being merged with other genes. The genes are immortal. A population is not a discrete entity to be a unit of natural selection. Each individual is unique. The genes are not destroyed by crossover; they merely change partners. The lifespan of a DNA molecule is short—a month, definitely not more than one lifetime. But a DNA molecule could live on as copies of itself for a hundred million years. Copies of a particular gene may be distributed all over the world.

The gene's potential immortality makes it a good candidate for natural selection. Genes have an effect on the embryonic development. The effect of a gene depends on other genes. Sometimes a gene has one effect in the presence of another gene, and a completely different effect in the presence of another set of companion genes. The whole set of genes in a body constitutes a genetic climate or background, modifying and influencing the effects of any particular gene. A gene for sexuality manipulates all the other genes for its own selfish interests. Mutator genes manipulate the rates of copying-errors in other genes.

A copying-error is the disadvantage of the gene, if it is the advantage of the selfish mutator that induces it. The mutator can spread through the gene pool. Sex and crossover plays a fundamental role in evolution. The true "purpose" of DNA is to survive. The immediate manifestation of natural selection is always at the individual level. Non-random individual death and reproductive success change gene frequencies in the gene pool. Because of sex and crossover, the gene pool is kept well stirred, and the genes partially shuffled.

The vast majority of interactions between genes take place within cells, notably the cells of developing embryos. Well-integrated bodies exist as they are the products of stable set of genes. Animals exploit the plants, either by eating them, or by other animals. Evolution has caused the immense diversity of animals and plants. Both animals and plants evolved into many celled complex organisms. Natural selection favours genes that cooperate with others. The currency used in the evolution is the survival, strictly gene survival, but for many purposes individual survival is a reasonable approximation. The genes predict the sweet taste in the mouth as eating sugar is beneficial to gene survival.

Compelling evidence suggests that genes influence behaviour in all animals, including humans. Now biomedical researches are devoted to human behavioural problems such as alcoholism, obesity, schizophrenia, and Alzheimer disease. These studies are pursued using animal models. There are genes for coat colour in guinea pigs or horses which means that genetic variation in the guineapig or horse causes some variation in coat colour. Identifying a gene that influences a behaviour does not imply that the behaviour is inevitable. Considerable variations in behaviours of the individual's genetic make-up and its environment exist for the expression of the behaviour. Occasionally, the possession of a gene does consistent result in the individual having behaviour. The influence of genes on the behaviour is quantified by a genetic measure called *heritability,* which is defined as the fraction of the total variation in a trait of individuals in a population. The remaining source of the variation is the environment. Values of heritability range between zero and one. The smaller the environmental variation in the individuals, the greater will be the fraction of the total variation in the behaviour. The degree of genetic influence on a particular behaviour is not a fixed characteristic. Heritability varies greatly depending on how much the individuals experienced environmental variation in the specific population.

Behavioural genetics attempts to explain various aspects of behaviour from genetical view point. Sir Francis Galton was the first to study heredity and human behaviour systematically. The term "genetics" did not appear until 1909, only 2 years before Galton's death. With or without a formal name, study of heredity has always been dealt with study of biological variation. Human behavioural genetics, a relatively new field, seeks to understand both genetics and environmental contributions to individual variations in human behaviour. Behaviours, like all complex traits involve multiple genes, which complicates search for genetic contributions. Studies on genes and behaviour require analysis of families and populations for comparison of those who have trait in question, with those who do not. Result often is a statement of "heritability," that estimates amount of variation in a population that is attributable to genetic factors. If population or environment changes, heritability is likely to change as well. Heritability statements provide no basis for predictions about expression of trait in question in any given individual. In humans, some behaviours run in families. Behaviour has an evolutionary history that persists across related species. Chimpanzees are our closest relatives, separated from us by a mere 2% difference in DNA sequence. Genes are evolutionary glue, binding all life in a single history that dates back about 3.5 billion years. Conserved behaviours are part of that history which is written in language of nature's universal information molecule, DNA. Some important works (Hamilton 1964, Neiderhiser 2001, Moehring and Mackay 2004) in behavioural genetics are mentionable.

Genes, brain and behaviours are essential to study. Nowadays, the important development in the brain and cognitive science is coming from molecular genetics and cell physiology. *Drosophila*, as an important biological model, has a relatively simple brain. Genes and function of genes as well can easily be studied in *Drosophila*. *Drosophila* has many different kinds of behaviours. They are able to learn and suffer from disease symptoms, like senile dementia. *Drosophila* needs sleep. All the behaviours are accomplished through the complex control mechanism.

Traditional research in behavioural genetics includes studies of twins and adoptees. Recently, pieces of DNA associated with particular behaviours are also investigated, which has been most productive in identifying potential locations for genes associated with mental illnesses, like schizophrenia and bipolar disorder. Search for genes for characteristics, like sexual preference and basic personality traits are frustrating. Genetics and molecular biology have provided insights into behaviours associated with inherited disorders. For example, extra-chromosome 21 is known for mental retardation that accompanies Down's syndrome, although processes for disrupting brain function are not yet clear. Steps from gene to effect for a number of single-gene disorders that result in mental retardation including phenylketonuria is known.

Behavioural genetics have asserted claims for a genetic basis of numerous physical behaviours including homosexuality, aggression, impulsivity, and nurturing. Focus on genes and behaviour has contributed to a resurgence of behavioural genetic determinism. Defining a specific endpoint that characterizes a condition, identifying, and excluding other possible causes of condition, are obstacles to correlate genotype, and behaviour. No single gene determines a particular behaviour. Behaviours are complex traits, involving multiple genes that are affected by a variety of factors. A variety of genetic and environmental factors are involved in development of trait. Presence of certain genetic factors can express or repress other genetic factors. Genes are turned on and off. Protein encoded by a gene can be modified in ways that affect its ability to carry out its normal function. Genetic factors also influence role of certain environmental factors in development of a particular trait. A person may have a genetic variant to increase risk for emphysema from smoking, an environmental factor. If that person never smokes, then emphysema will not develop.

Frequent news reports claim discovery of gene for aggression, intelligence, criminality, homosexuality, feminine intuition and even bad luck. Such reports tend to suggest usually incorrectly, a direct correspondence between a mutatant gene and a disorder. Single genes do not determine most human behaviours. Rare disorders, like Huntington's disease, have a simple mode of trans-mission in which a specific mutation results in developing disorder. Most behaviours have no such clear-cut pattern and depend on interplay between environmental factors and multiple genes. Genes in such multiple-gene systems are called Quantitative Trait Loci (QTLs), as they are likely to result in continuous (quantitative) distributions of phenotypes that underlie susceptibility to common disorders. Behaviour is unique as a product of our brain. Sequencing of multiple human genomes and identification of several million DNA base pairs that differ among us would have significance. These DNA variations are responsible for ubiquitous genetic influences on individual differences in behavioural dimensions and disorders. With human genome sequence available, it is possible to locate QTLs with aid of a greatly improved map based on hundreds of thousands of Single-Nucleotide Polymorphisms (SNPs), and very high-throughput genotyping as discussed by Peltonen and McKusick. Understanding QTLs effects will take much more time.

Using fruit fly as a model system, Benzer identified "behavioural mutants". Animal with a mutation in a gene was incapable of performing specific behavioural tasks (Benzer 1973). Brenner later introduced *C. elegans* as a model system, for behavioural genetics. Benzerian-type mutant analysis and simple rationalization that genes are required to build neurons and that it is activity of neurons (Figure 3.4) that produce behaviour, suggest that individual genes are required for correct expression of behaviour. Traditional two conceptually different perspectives on how to tackle behavioural genetics are "Hirschian" and "Benzerian" perspectives (Tully 1996). Former approach is based on study of naturally occurring polygenic variations (Hirsch 1967). Benzerian approach is based on induction of mutations and identification of their molecular nature.

Behavioural genetics deals with heritability of behavioural traits. The field comprises genetics, ethology and psychology. Genes do not directly code for a behaviour. Genes only code for proteins. A change in a single protein causes a host of downstream effects and even brings about a distinct phenotype. External environment exerts strong influences on gene expression behaviour via development of neuronal and hormonal mechanisms. Phenotype emerges from interaction of genotype with environmental factors. Classical genetics deals with distribution of hereditary characteristics of behaviours from one generation to next. With selective breeding, presence or absence of specific behaviours can be traced through outcomes of sexual reproduction. Genetics of many diseases follow simple Mendelian rules. Benzer discovered a series of behavioural mutants in fruit flies, including single gene disruptions of circadian rhythm, and various genetic causes in neurodegeneration and aging. Hygienic behaviour in honeybees involves ability to detect and remove diseased larval and pupal brood from nest before pathogen becomes infectious. Genetic basis of hygienic behaviour is considerably more complex.

Quantitative genetics deals with traits which are often subject to significant environmental influences and usually involve a great number of underlying genes. Quantitative approaches aims to estimate proportion of total phenotypic variance that is explained by relatedness. Aside from cases where multiple genes affect single traits, changes in one individual gene may impact a number of different traits. Such instances can be identified and characterized through measures of phenotypic covariance across multiple traits. Population genetics is concerned with mechanism that changes relative occurrence of genes within population. It examines gene frequencies changes within population through sexual selection, mating systems, dominance, or territoriality. Developmental genetics explores interference of genetics on ontogenetic processes in behaviour.

Genome mapping and genetic analyses identifies genes for behaviour. Genes are cloned and swapped between species to access their role in control of behaviours. Knowledge of genomic structure combined with environmental factors, like intrauterine effects, social environment, and feeding and rearing habitat help in understanding the origin of individual variation in behaviour. Mechanisms underlying the control of sensitivity help in understanding origin and treatment of behavioural disorders. Behaviour is obligatory gateway to examine brain functioning. Mechanism of brain functioning helps in understanding disease, like Alzheimers. Learning and imprinting reveal new insight into neural plasticity and developmental influences on brain. MRI scanning provides potential for linking dynamic changes in brain with specific behavioural events. Exploration of cognitive capacities reveals ways in which different species navigate in space and perceive time extent to which they predict future events, recognize other individuals, and solve abstract reasoning problems.

Many gene regulatory systems include environmental contributions at molecular levels. When *E. coli* grow in presence of glucose and lactose, it uses glucose first and then switches to lactose. In doing so, it activates genes for lactose metabolism only, when lactose is present and glucose is absent via interactions among several types of sugar and protein molecules. Such complex interaction with food molecules in *E. coli*, suggests extensive interactions with environment in human beings leading to diverse changes in physiological functions. Most genetic and environmental effects on behaviour are mediated by the nervous system. Environmental conditions are known to affect growth of nerve cells, number and strength of connections amongst neurons, both during development and in adults. Adult rats develop more structural elaboration in neurons in a complex environment full of toys, than in a blank cage. If a cat is reared with one eye closed, during first several weeks of life, organization of inputs to a visual portion of cerebral cortex is permanently altered and neurons that would normally respond to an object seen by either eye now respond only via an eye that has remained open throughout (Purves and Lichtman 1985).

In 1991, Sinsheimer affirmed "in deepest sense we are who we are because of our genes". Genes for schizophrenia, manic depression, alcoholism, and a gene site for bed wetting are reported. Interactions of genes, proteins, hormones, food,

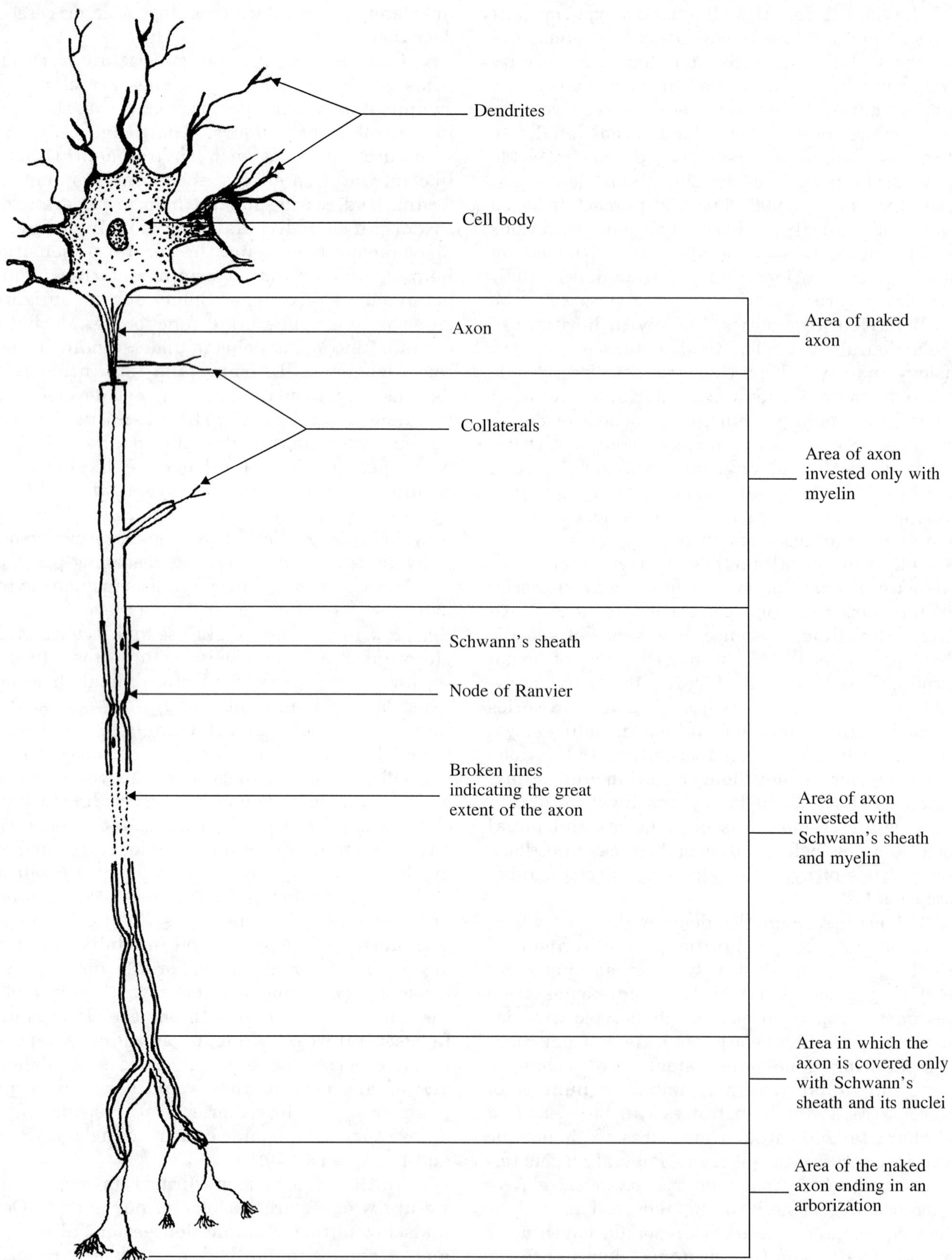

FIGURE 3.4 A motor neuron.

life experiences, lead to cognitive and behaviour functions. Behavioural genetics is traditionally studying differences between people, differences in intelligence, personality, and tries to figure out the differences between people due to genetic or environmental differences. Genetic differences are very powerful. Serious unresolved question is the nature of human intelligence. General intelligence (G-factor) exists. This G-factor is a statistical abstraction, which captures the ways that people who are good at one thing are also good at another thing. People who tend to be good at certain kinds of mental things, like having a large vocabulary also tend to be good at other mental tasks, such as mathematics or spatial navigation.

The honeybee (Figure 3.5) is a living engine of selfless domesticity, which form caretaking colonies of kin. The honeybee workers go through behavioural changes with each new assignment in the hive, transforming from housebound nest nurses into field explorers that require travel of 550 miles in a lifetime for pollen and nectar. Much of this social life is coordinated by activity of a single gene. One protein made by a gene controls several important traits for worker behaviour, including division of labour. Gene makes a protein called **vitellogenin**, a reproductive biochemical which is crucial for formation of yolks in many egg-laying insects. Researchers created bees that expressed higher or lower levels of protein. Level of vitellogenin controls worker bees foraging, affecting preference for pollen or nectar for hive. Higher levels of protein early in bee's life favoured pollen, and lower levels favoured nectar. Protein levels also affect life span. Higher level makes the life span of worker bees longer. This gives the idea that evolution of community is founded on basic molecular biology of reproduction, with those genes and proteins being used as building blocks to assemble ever more complex behaviour. They affect sensory system of animal in a new way from which emerges social behaviour that took its origin 65 million years ago.

FIGURE 3.5 A honeybee.

A "phenotype" termed "phenocopies" can be caused entirely by non-genetic factors. Fruit flies develop, with four wings instead of two in a mutation in bithorax gene, or after exposure to heat or chemical stress at a critical phase of embryonic development (Capdevila and Garcia–Bellido 1978). Environmentally induced effects are phenocopies of bithorax. Role of gene in a behavioural outcome includes protein it codes for, function(s) of this protein and mechanism through which protein influences nervous system. A human's foetus exposure to sex hormones effects gene regulation, brain, and external sex differences. When foetus or mother produces too little or much of testosterone, or other androgen, or foetus lacks appropriate receptor molecules for hormone, problem arises in the newborn. In case of androgen-insensitivity syndrome, a genetically male human becomes a complete female, except her internal reproductive organs are inadequately formed. In such cases, foetus possesses a defective gene for androgen receptor protein. Higher amount of circulating androgen causes prenatal masculinization of genetic females, transforms clitoris partly or completely into a penis by a genetic defect, or treatment of pregnant women with progestin. Deletion or alteration of a gene interferes with amount of protein coded by that gene, effects several other proteins as it may interact with other DNA, RNA, proteins, hormones and dietary substances to mediate gene regulation, which mediates behaviour. Each neuron affects many neurons. Some conditions caused by gene can be prevented or reversed by non-genetic means, such as providing a phenylalanine-free diet to children with phenylketonuria. Alcohol abuse by pregnant women has permanent effects. Obesity is a psychological problem with physiological regulation. That genes, proteins, hormones, food, and life experiences shape most behaviours in a human are recorded.

In 1869, Galton introduced "nature versus nurture" concept following publication of his book, "*Hereditary Genius*" in 1869. Watson and his followers believed that we have been shaped from birth by external influences and learning, and not by inherited innate differences. A debate is ongoing over last several decades on importance of nature (genes) as opposed to nurture (environment), to explain human development, diseases and behaviour. Ridley (1996) concluded that genes influence response to external factors, while environmental factors determine genes functioning. He writes "The discovery of how genes actually influence human behaviour and how human behaviour influences genes, is about to recast the debate entirely. No longer is it nature versus nurture, but nature via nurture." Genes take their cues form nurture. Ridley emphasized importance of genes, and that 30,000 genes of man can account of billions of permutations underlying every aspect of our physiology, diseases, behaviour and personality.

Genetic similarities are shared between man and apes also with mice, frogs, flies, worms, and even plants. Differences generate in differential gene expression. During evolution "the same gene can be reused in different places and at different times". Differential gene expression resolve dilemma of 'similar, but different', and allows extrapolation that "nurture can start to express itself, through nature". Galton in his book published in 1874, *English Men of Science: Their Nature and Nurture*, emphasized that scientific geniuses are born, not made. Galton studied identical and non-identical twins. Former remained similar throughout the life in their appearance, health, and personality, while latter became increasingly dissimilar, even though twins were exposed to same external influences, he concluded that "nature prevails enormously over nurture". Studies on twins reinforced emergence of behaviour genetics.

A mutation in a gene is necessary for development of normal grammar and speech in humans, and fine motor control of larynx. Monkeys and mice also have this gene, but have not evolved a vocal language, which Ridley explains by speculating that at some time after 200,000 years, a mutant form of this gene appeared in human race. Genes are set of developmental switches, a unit of selection in sense of Dawkin's selfish gene. Wilson explained human behaviour as product of 'scheming' gene, and thus, explains such complex issues as homosexuality, ethics and social science. Harris believes that environment and genome both have a preponderant influence on personality of a child mainly though child's peer group. The challenges of behavioural genetics research include the difficulty in defining and quantifying behaviour, the tremendous variation in behaviour within and between individuals, the involvement of many genes, and different gene functioning in different tissues at different times during the ontogeny of an organism. All of which combine to influence a single pattern of behaviour (Sokolowski 2001). Our current knowledge about gene functioning in regard to behaviour is highlighted in Table 3.2.

3.2 NEUROBIOLOGY AND BEHAVIOUR

Neuroethology, the integration of animal behaviour with neuroscience, provides frameworks for hypothesizing neural mechanisms. Careful behavioural data narrow the scope of studies and to focus on relevant input stimuli and to attend to relevant responses. The use of species specific natural stimuli has led to new insights about neural structure and function. Work in animal behaviour has shown a downward influence of behaviour and social organization on physiological and cellular processes. Differences in social environment inhibit or stimulate ovulation, produce menstrual synchrony, and induce miscarriages. Quality of social and behavioural environment have a direct effect on immune system functioning. Nervous system serves as the interface between molecular and cellular functions and the whole organism. Direct neural connections between sensory cells and muscles are found in Sea jellies and their swimming motion can change as needed. Central nervous systems and brain integrates various sensory inputs in complex animals. The coordinating parts of nervous and some sensory systems are found in anterior part of animal's body. Various regions of brain regulates learning, memory, coordination of movement, and physiological functions. Neural connections within brain allow transfer of information in such regions. Acetylcholine, serotonin, and dopamine transmit information in brain cells. Overall levels of neurotransmitters in brain affect general behaviour. Manipulation of dopamine affects wakefulness.

Nervous system, the most crucial system in the elicitation of behaviour, is formed during development by networks of interacting genes. Similar networks assemble the physiological structures necessary to generate behaviour pattern. Brain (Figure 3.6) is the master control centre of every activity. Human brain, product of millions of year of evolution, contains 100 billions of nerve cells. Billions more neurons are found in other parts of nervous systems. A single neuron holds many secrets of behaviour and mental activity. Two hemispheres of brain are similar, though not identical, work together. Human brain demonstrates extraordinary plasticity, ability to adapt to new environmental conditions. Brain also responds to feedback from senses and surrounding environment. Neurons vary widely in shape and size. They are specialized to receive and transmit information. Nervous system contains a vast number of glial cells, which hold neurons in place, provide nourishment, remove waste products, and prevent harmful substances from passing

TABLE 3.2 Genes and their Function (*Source:* Emmons and Lipton 2003)

Gene	*Gene Product*	*Behavioural or Nervous system function*
C. elegans		
cat-2	Tyrosine hydroxylase	Dopamine biosynthesis
egl-1	General cell death gene	Kill HSN neurons in male
egl-5	Hox family transcription factor	Ray differentiation
egl-19	L-type voltage-gated calcium channel	Copulation (prolonged spicule protraction)
lin-32	bHLH transcription factor	Ray development
lov-1	Putative transmembrane protein w/seq. similarity to PKD1	Copulation (response, location of vulva)
mab-23	DM-domain family transcription factor	Multiple aspects of male differentiation, copulation
mab-3	DM-domain family transcription factor	Multiple aspects of male differentiation, copulation
mab-5	Hox family transcription factor	Ray development
mab-9	T-box family transcription factor	Development of spicules, neurons, etc.
mod-5	Serotonin reuptake transporter	Sex drive (mate searching)
pkd-2	Ca++ channel with similarity to PKD2	Copulation (response, location of vulva)
tra-1	Zn–finger transcription factor	Sex determination (hermaphrodite—development)
unc-29	Nicotinic acetylcholine receptor	Copulation (rapid spicule prodding)
unc-38	Nicotinic acetylcholine receptor	Copulation (rapid spicule prodding, prolonged spicule protraction)
unc-68	Sarcoplasmic calcium channel	Copulation (rapid spicule prodding)
unc-77	Unknown	Drive (mate searching)
Drosophila		
amnesiac	Neuropeptide	Experience dependent modification of drive
courtless	Ubiquitin-conjugating enzyme	Drive
dissatisfaction	Nuclear hormone receptor family transcription factors	Aspects of courtship, copulation
don Giovanni	Unknown	Induction of female rejection signal, response
doublesex	DM-domain family transcription factor minor role in behaviour	Many aspects of male development,
dunce	cAMP-specific phsophodiesterase	Experience-dependent modification of drive
fruitless	BTB Zn finger family transcription factor	Male differentiation of CNS
he's not-interested	Unknown	Drive
lingerer	Unknown	Copulation
period	Component of fly circadian oscillator	Courtship (singing)
rutabaga	Adenylate cyclase	Experience-dependent modification of drive
stuck	Unknown	Copulation
technical knockout	Mitochondrial ribosomal protein	Courtship
tra	Splicing factor	Sex determination (female development)
Mouse		
DMRTI	DM-domain family transcription factor	Sex determination (testes development)
TRP2	TRP family ion channel	Discrimination of the sexes
αER, βER	Estrogen α and β receptor types	Male sexual behaviour
Prairie vole		
V1aR	Arginine vascopressin (AVP) V1a receptor	Affiliative sexual behaviour (olfactory investigation and grooming)

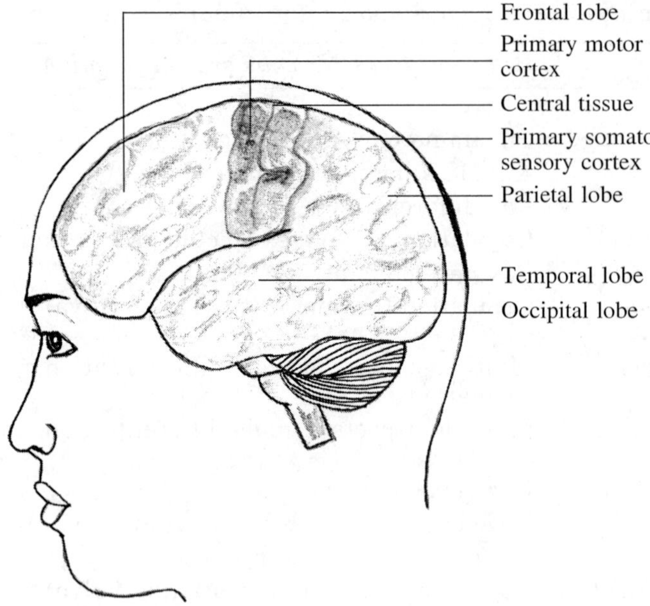

FIGURE 3.6 Human brain showing four lobes of the cerebral cortex.

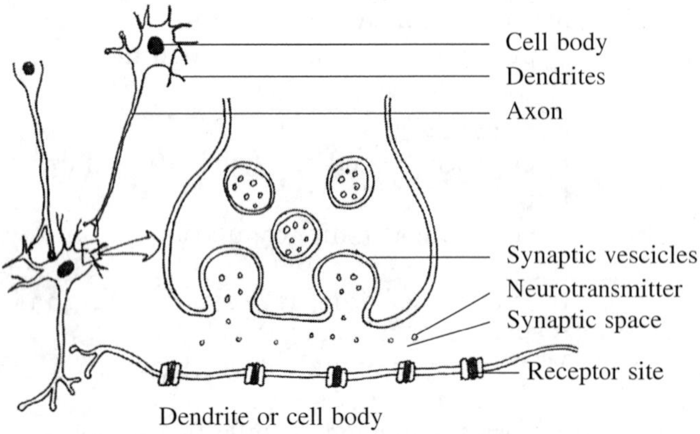

FIGURE 3.7 Synaptic transmission between neurons.

from blood stream into brain. Glial cells may play an important role in learning and memory, and thereby affect brain's response to new experiences. Axons with myelin sheaths conduct impulses very rapidly, because impulses "leapfrog" along string of pinched nodes of sheaths. Neurons without myelin sheaths tend to conduct impulses slowly. A single neuron may have many hundreds of dendrites and its axon may branch out in numerous directions. Thus, it is in touch with hundreds or thousands of other cells at both its dendrite end and axon end. A neuron may receive excitatory and inhibitory messages at any given moment. Propagation of nerve impulse requires crossing the synaptic space (Figure 3.7). Most axon terminals contain numbers of minute oval sacs called **synaptic vesicles**. When nerve impulse reaches the end of terminals, vesicles release neurotransmitter into synaptic space, neurotransmitter fit into their corresponding receptor sites ensuring that neuro-transmitter do not randomly stimulate other neurons, but follow orderly pathways. Neuro-transmitters detach from receptor site after completion of their job. They are then either reabsorbed, broken drown, and recycled to make new neurotransmitters or disposed of by body. Hundreds of neurotransmitters have been identified. Their exact functions are still being studied. Major neurotransmitters are acetylcholine, dopamine, serotonin, norepinephrine, endorphin, and glutamate.

3.2.1 Nervous System

Basic functional unit of nervous system are neurons, the cells process and to communicate information. The cell body (or soma) of neurons is the bulbous end of neuron, containing cell nucleus and much of cells metabolic machinery. Axon carries information away from cell body and may range in size from several microns to several meters in whales and giraffes. Axons may branch into terminal buttons at its end. Dendrites receive neurotransmitter secreted by axon of other neurons. Synapse is a specialized junction between cells of nervous system which permits signalling between them. Neurons conduct and process signals that involve small changes in electrical voltage generated across the neuronal membrane. At rest, inside the neuron is negative relative to cell surround which is called the *resting potential*. Special types of voltage changes feature a rapid spike in voltage resulting from changes in ion conductances. Such signals are mainly transmitted along axonal lines where neurons interact with other cellular entities over long distances.

Ohm's Law

Basic familiarity with key concepts in electricity helps to understand neuronal mechanisms. Ohm's law holds that rate at which electric charges flow (Current or I) depends on force and its direction exerted onto charged particle (Potential, V) and ease with which the flow can occur (Conductance, g) which is expressed as reciprocal of conductance, namely the obstructs in the flow of charges (Resistance, R). Flow of electricity is flowing from negative to positive.

Properties of Neurons

Voltage present across cell's membrane can be measured between a reference electrode and a conducting glass capillary tube drawn out to a fine point, and inserted into cell. At rest the neuronal interior is negative relative to outside at between –60 to –80 mV, while most non-neuronal cells feature a potential of about –30 mV. Ions carried the current moving across membrane at ion conductances when specific channels allow passing them. The axon can in electrical terms is an insulated cable. An electric potential can spread passively along any stretch of membrane. In the process its strength decays and slopes of on- and offsets become less steep with distance including resistance along its length and both the resistance and a capacitance component across it. Signals spread fastest when longitudinal resistance is low (via increased axonal diameter) and they spread furthest when resistance across the membrane is high.

Resting Potential

A neuron is at rest when it has not been begun for sometime and exhibits a negatively charged inside relative to surrounding environment. Resting Potential emerges from an unequal distribution of ions across the cell's membrane. An excess of K^+ and of various anions- exists on inside while Na+ and Cl^- exist in higher concentrations on the outside. Membrane is dotted with proteins that permit the flow of ions (ion pores) or control their flow (ion channels) across it. In a cell at rest only potassium ions (K^+) flow freely between the 2 compartments. Initially there is a net flow of positively charged potassium ions from the inside where they are more numerous to outside and where their concentration is much lower. Due to the flow of positively charged ions from the inside across the membrane, the inside becomes negative increasingly. Resting potential settles into equilibrium when many potassium ions are pushed out of cell which can be calculated using the Nernst potential for known inside and outside concentrations of any given ion

Nernst Potential (E) =

$$2.303 * RT/zF * \log([\text{ion}]\text{out}/[\text{ion}]\text{in})$$

Nernst Potential (E) =

$$2.303 * RT/zF * \log([\text{ion}]\text{out}/[\text{ion}]\text{in})$$

where R = gas constant (8.3143 $\text{Jk}^{-1}\text{mole}^{-1}$); T = absolute temperature in Kelvin (310); z = ionic valence; F = Faraday's constant (96,487 C/mol).

Depending on their respective inside and outside concentrations, different ions produce different equilibrium potentials (reversal potentials) at particular cell. Ion concentrations are similar to those listed:

Ion	[ion]in	[ion]out	E
K^+	400	20	–75 mV
Na^+	50	440	+55 mV
Ca^{2+}	0.0001	125	+155 mV
Cl^-	9	100	–65 mV

Cell's resting membrane potential combines all relevant ion currents with K^+ ions figuring most prominently due to their high resting conductance. Small number of Na^+ ions leak into cell. Resulting resting membrane potential is thus slightly lower compared to the EK^+ at around –65 mV to maintain the concentration gradient across the resting neurons membrane. ATP supplies energy continually to the Na^+/K^+ ATPase (ion pump) as ions leak across the axomembrane.

Action potential

Action potential represents a transient inward current of cations (Na+) and is generated when the graded membrane potential rises above threshold either spontaneously or because of depolarizing input. Voltage-gated sodium channels are opened when potential rises (depolarizes) from resting potential (–65 mV) to cells threshold (–55 mV). These channels close again when potential further rises to +10 mV. As sodium ions spread to neighbouring areas of axonal membrane, sodium channels open via depolarization and signal runs entire length of cell (all-or-none-principle). A refractory period follows action potential during which time neuron reestablishes normal resting potential. At the beginning of this period it is impossible for another signal to be transmitted. This is the absolute refractory phase which is followed by relative refractory phase where it is possible to send another signal but more excitation than normal is needed. Duration of an action potential is 1–2 ms in vertebrates and 1–100 ms in invertebrates. Frequency of firing ranges from <1 to about 100/s (100Hz). Amplitude ranges between 70–80 mV when recorded intracellularly and 5–200 μV when recorded extracellularly.

Synapse

Two electrically excitable cells, like neurons or muscle cells may be electrically coupled where an action potential in one cell moves directly into other via arrays of gap junctions. Electrical synapses are fast and mostly used in neuronal circuits for escape behaviours where speed of conduction is essential. Electrical synapses allow current to flow between separate neurons when ions pass through gap junctions. Connexons (where 6 connexin proteins form a hemi-channel) are the actual pores that allow ions to flow past the 2 membranes. Connexon in presynaptic membrane lines up with its respective equivalent in postsynaptic membrane, with continuous channel from one neuron to another. Many small molecules pass through efficiently with pore diameter of about 1.5 m^{-9}. Intracellular Ca^{2+} concentration, pH alters easily the rate with which ions and proteins may pass through pore. As there is no synaptic delay in transmission of current from cell to another, conduction of potential changes is faster than through chemical synapses. Electrical synapses are often bi-directional. Some synapses carry current better in one direction than other (rectifying synapse). Electrical synapses are used in time-critical processe (escape behaviour), that requires rapid synchronization of many cells.

Chemical Junctions

Cells communicate chemically across a gap (synapse), which forms a directional connection from one neuron to another cell. Neuro-transmitter is emitted from terminal endings of one cell's axon onto a dendrite or body of a second cell. Arrival of a signal leads to the release of a neurotransmitter from presynaptic terminal, diffusion across synaptic cleft, and binding to receptors in postsynaptic membrane. Metabotropic or Ionotropic receptors (binding proteins) alter ion conductances at postsynaptic cell membrane (increase in Na^+ conductance depolarizes and is excitatory; CL^- hyperpolarizes and is thus inhibitory). When excitatory, postsynaptic cell is depolarized with an Excitatory Post-Synaptic Potential (EPSP) and an action potential is elicited if threshold is reached; in inhibitory connections, postsynaptic cell is hyperpolarized with an Inhibitory Post-Synaptic Potential (IPSP). A single input rarely lead to an action potential in post-synaptic cell. Multiple EPSPs add and attain threshold during occurrence of action potentials at high rate. Chemical synapses are able to integrate a complex scenario of inputs.

Neurotransmitter is released at a synapse and diffuses across synaptic cleft to act on a receptor located on membrane of a postsynaptic cell, which may be another neuron, a muscle cell. Various chemicals are used as neurotransmitters and are released on arrival of an action potential producing changes in excitability of postsynaptic membrane. Ca^{2+} influx at axon terminus is required for synaptic release. Neuromodulator, a compound that is released in localized region of CNS, the receptor of which is not necessarily sited on anatomically apposed postsynaptic cell. Neuromodulator affect several postsynaptic cells with specificity conferred mainly by distribution of receptors. Main action affects second messenger systems. cAMP or inositole triphosphate, perhaps affect protein phosphorylation. Sometimes, same neurochemical may have rapid transmitter type effects with long modulatory influences suggesting that neurotransmitter and neuromodulator effects may be differentiated at receptor level. Activation of receptor on a protein structure involving an ion channel is called the neurotransmission. Activation of receptors coupled to ion channels is called the *neuro-modulation*.

Axons, the basic cellular units of the nervous system, transmit electrical or chemical signals from one location of the body to another. In the central nervous system, interconnecting axons store memory and other integrative functions. Vertebrate axons are covered with a myelin sheath for rapid transmission of electrical impulses from one end of the axon to the other. Invertebrate axons lack a myelin sheath and the transmission speed is proportional to the axon diameter. Invertebrate axons with critical functions in fast escape responses have large diameters and are called *giant axons*. Messages are carried between axons by neurotransmitters. The most common neurotransmitter is acetyl-choline. When an action potential or nervous impulse reaches the end of the axon, the transmitter is released causing depolarization of the membrane of the next axon or of the muscle cells. An enzyme acetylcholinesterase breaks down the acetylcholine once the message is transmitted. Other neurotransmitters are octopamine, serotonin and dopamine, which function in the central nervous system.

Axons are organized into a nervous system. Most animals have a brain which receives inputs from sensory nerves and produces behavioural responses. The brain also receives information about the animal's physiological state and serves as a centre for translating physiological needs into behavioural responses. Neurotransmitters modulate general behavioural state. In humans, serotonin and norepinephrine levels in the brain are associated with depression, while dopamine is related to pychosis. For learning and memory to function appropriately, acetylcholine must be present in the brain at the optimum level. Neuropeptides provide a link between the brain and other physiological systems. Acting as hormones, neuropeptides regulate many functions, including some behaviours. Communication between nerve cells helps explain our sensations, feelings, thoughts, learning, and memory. Neurons collect and process information as nerve impulse, work as per situation and the individual's need. Nerve impulse is generated in response to a given stimulus and passes from one cell to another, creating a chain of information within a network of neurons. Neurotransmitters attach themselves to chemical receptors in the membrane of the following neuron and promote excitatory or inhibitory changes in its membrane.

Neurotransmitters of one cell influence the nerve impulses. Synaptic transmission is the key to understand the operation of the nervous system. The nervous system controls and co-ordinates the body function. Neurotransmitters are small and simple molecules. Different types of cells secrete different neurotransmitters. Each brain chemical works in widely spread but specific brain locations and may have a different effect depending on the place where it is activated. Some 60 neurotransmitters have been identified which belong to one of four classes: cholines exemplified by acetylcholine; biogenic amines exemplified by serotonin, histamine and the catecholamine; amino acids exemplified by glutamate and aspartate as excitatory transmitters, while gamma-amino butyric acid (GABA), glycine and taurine as inhibitory neuro-transmitters. Neuropeptides are formed by longer chains of amino acids. Over 50 of them occur in the brain, and many have been implied in the modulation or transmission of neural information.

Cellular continuity between one neuron and the next is interrupted by a gap called *synapse*. The membranes of the sending and receiving cells are separated from each other by the synaptic gap. The signal cannot leap across the gap electrically. Neurotransmitters are released by presynaptic sending membrane and seep across the gap to receptors on the receiving neuron's postsynaptic membrane. The binding of neuro-transmitters to these receptors allows ions to pass in and out of the receiving cell. The normal direction of information flow is from the axon terminal to the target neuron. Thus, the axon terminal is said to be presynaptic and the target neuron is said to be postsynaptic.

There are electrical and the chemical synapses. Most mammalian synapses are the chemical synapses. Electrical synapse allows the direct transfer of ionic current from one cell to the next. Such synapses occur at specialized sites called gap junctions, and form channels that allow ions to pass directly from the cytoplasm of one cell to the cytoplasm of the other. Transmission at electrical synapses is very fast. An action potential in the presynaptic neuron can produce instantaneous action potential in the postsynaptic neuron. In the chemical synapse, the incoming signal is transmitted when one neuron releases a neurotransmitter into the synaptic cleft which is detected by the second neuron through the activation of receptors placed opposite to the release site.

The chemical binding of the neurotransmitter to the receptors causes physiological changes in the second neuron which constitutes the signal. Usually the release from the first neuron is caused by the depolarization of its membrane, and almost invariably when an action potential occurs. As an electrical impulse travels down the axon and arrives at its end, it triggers vesicles containing a neurotransmitter to move towards the terminal membrane. The vesicles fuse with the terminal membrane to release their contents. Once inside the synaptic cleft, the neurotransmitter can bind to receptors on the membrane of a neighbouring neuron.

The action potential stimulates the influx of Ca^{2+} causing synaptic vesicles to attach to the release sites, fuse with the plasma membrane and expel their supply of transmitter. The transmitter diffuses to the target cell, and binds to a receptor protein on the external surface of the cell membrane. After a brief period, the transmitter dissociates from the receptor and the response ends. To prevent the transmitter from rebinding to the receptor and repeating the cycle, either the transmitter is destroyed by the enzyme or it is taken up into the presynaptic ending. Each neuron produces only one kind of transmitter. There are two types of chemical synapses. An impulse after arriving in the presynaptic terminal releases neurotransmitter. In one type, the molecules bind to the transmitter-gated ion channels in the postsynaptic membrane. If Na^+ enters the postsynaptic cell through the open channels, the membrane becomes depolarized. In the other type, the molecules bind to the transmitter-gated ion channels in the postsynaptic membrane. If Cl^- enters the postsynaptic cell through the open channels, the membrane becomes hyperpolarized. Excitatory synapses cause an excitatory electrical change in the post-synaptic potential (EPSP) which occurs when the net effect of transmitter release is to depolarize the membrane, bringing it nearer to the electrical threshold for firing an action potential. Such effect is mediated by the opening of membrane channels for sodium and calcium ions.

Inhibitory synapses cause an inhibitory postsynaptic potential (IPSP), because the net effect of transmitter release is to hyperpolarize the membrane, making it difficult to reach the electrical threshold potential. Such synapse opens different ion channels in the membrane, typically the chloride (Cl^-) or potassium (K^+) channels. The exterior face of the membrane is negative in relation to the interior. The resting potential of the postsynaptic membrane is around –70 mV. Depolarization decreases this value causing an upward deflection. The recording of the membrane potential for an IPSP shows a hyperpolarization. If the net depolarization reaches the threshold level, the postsynaptic cell fires action potential. Different synapses may be distinguished by which part of the neuron is postsynaptic to the axon terminal. If the post-synaptic membrane is on a dendrite, the synapse is axodendritic. If the postsynaptic membrane is on the cell body, the synapse is axosomatic. Sometimes the postsynaptic membrane is on another axon, and the synapse is called axoaxonic. In certain neurons, dendrites form synapses with one another. These are called dendrodendritic synapses.

3.2.2 The Chemical Environment of Brain

The brain has no lymphatic system. The dura mater is a tough, protective and connective tissue. Under the dura mater is the subarachnoid space containing CSF, arteries and web-like strands of connective/supportive tissue called the arachnoid mater. The pia mater is a permeable membrane of collagen, and elastin fibres. Fibro-blasts on the floor of the subarachnoid space allow diffusion between the CSF and the interstitial fluid of the brain tissue. The pia mater lies on a membrane that is infiltrated with astrocyte processes. The dura mater, the arachnoid mater and the pia mater are collectively called the *meninges*.

Fifty per cent of dry brain weight is due to lipid, most of which is structural in contrast to the triglycerides and free fatty acids constituting the fat of other organs. The blood-brain barrier creates a protected environment for the brains wherein certain molecules perform functions independent of the functions those molecules perform in the rest body. This is important for the neurotransmitter's serotonin and norepine-phrine. All the known amino-acid neurotransmitters are non-essential amino acids. The major neurotransmitters of the brain are glutamic acid and GABA. There are many neurotransmitters in the central nervous system; the peripheral nervous system has only acetylcholine and nore-pinephrine.

Rats raised in enriched environment, had larger neurons, with more synaptic connections, than those raised in impoverished environments (Rosenzweig 1984). Rats raised in stimulating environments are reported to perform better on a variety of cognitive tests and develop more synapses, when required to perform complex tasks. These results suggest that neural plasticity is a feedback loop. Experience leads to changes in brain which facilitate new learning and ultimately leads to further neural change. Studies in 1990s showed that adult brains are capable of neurogenesis which has widespread implications, for treating neurological disorders. Neuroethology was born out of classical ethology as an effort to understand behaviour, based on detailed knowledge about anatomy and physiology of nervous systems. Studies on sensory processing and decision making in frog retire and brain and motor control of swimming in lamprey and pheromone searching in moths are studied. A new study on striking-looking crested auklet adds to evidence that smell is important to birds, like other animals. Sea bird found in Alaska's Aleutian Islands produces a strong tangerine-like smell, which are linked to courtship displays.

Memory is a critical function underlying behaviour. Organisms must have confidence in outcomes of action in response to various stimuli and in various contexts. Memory manifests in several forms, each associated with different parts of brain. Prevailing view of brain function involves "modularity" idea for specific parts of CNS modules, which is strongly associated with specific function. Generally, many individual components of a response rely on a separate site to orchestrate their effective expression. Limbic system is principally a subcortical circuit involving several distinct structures, but also involve cortical structure (cingulate gyrus), and hypothalamus. Subcortical structures are septum, amygdala, and hippocampus. Limbic system is responsible for most homeostatic, vegetative, visceral functions, and integrates these functions that results in emotional behaviour.

Animal obtains information through perceived stimulus from environment. A signal is coding of information capable of transformation through environment. Sensation is neuronal activity resulting from transduction of stimulus into electrical energy. A cell endowed with ability to absorb a specific stimulus. Receptors respond to a narrow cocktail of characteristics. Input signals are conveyed to nervous system. Information transfer is based on rate of action potentials of up to 500 AP/S for intense stimuli. Processing and summation of multiple graded responses influences frequency of action potentials. Tonic receptors respond with a constant rate of firing as long as stimulus is applied (pain). Phasic receptors produce a burst of activity during the onset of stimulus, but quickly reduce their firing rate if stimulus is retained (odour, touch, temperature). With sensory adaptations, organism ceases to pay attention to constant stimuli. Sensory receptors are mechanoreceptors, stretch receptors, electromagnetic receptors, electroreceptors, and thermoreceptors.

Neurons and connection amongst neurons, serve particular function (Figure 3.8). Neurons consist typically of three major parts, dendrites, cell-body, and axon (Figures 3.9 and 3.10). The neuron is an extraordinary sensitive cell. Concept of a "distributed network" now describes nervous system. Protein synthesis and subsequent genetic effects on behaviour is mediated by a distributed network. Genes affecting mental conditions code for neurotransmitter receptors, each of which is used by a number of neurons. Different sensory systems are segregated within the spinal cord, so that there are somewhat separate columns mediating pain, temperature sense, muscle sense, and touch. The descending or motor column consists of neurons going all the way down from the brain and crossing to the motor nerve cells in the ventral parts of the spinal cord (Figure 3.11). Neurons, that communicate using one neuro-transmitter, are often interspersed with neurons that use others instead. Several neurotransmitter receptors exist for a given neurotransmitter. Each type confers distinct electrical properties on neurons that house them. A single type of receptor for a single neurotransmitter is distributed over much of nervous systems in a complex pattern. Emergent properties of network are not evident in effects of most single gene or single protein. In such cases, same gene may have different effects on a behaviour depending on context. Others may have effects on multiple behaviour or cognitive function. Thus, if a genuine genetic link to manic depression is found, it might turn out that gene exacerbate symptoms of manic depression.

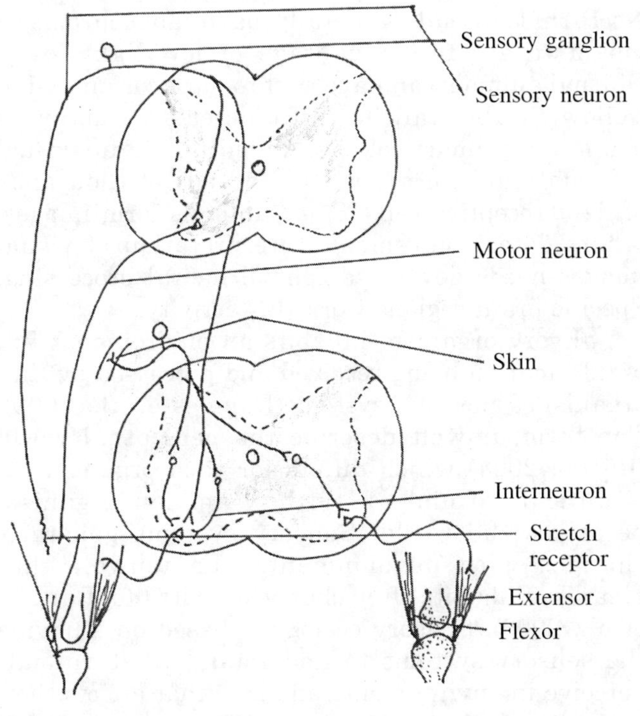

FIGURE 3.8 Connection of sensory neurons, interneurons and motor neurons.

Sensory structures like eyes and ears sense information from the environment and convert it into internal signals that animal uses to shape behaviour. Transduction transforms external energy like light, electrical fields, vibrations in air into nervous signal, or action potential. Action potential carries information from sensory organ to brain. Weakly electric fish inhabit murky water in which vision is not effective. They navigate and communicate using magneto receptors. In visual perception receptive cells for vision is mediated through pigment and the pigment is mostly composed of retinal and opsin. Retinal and opsin jointly form rhodopsin. When light energy hits rhodopsin, molecule alters its shape triggering metabolic changes leading to action potential for transmitting information to

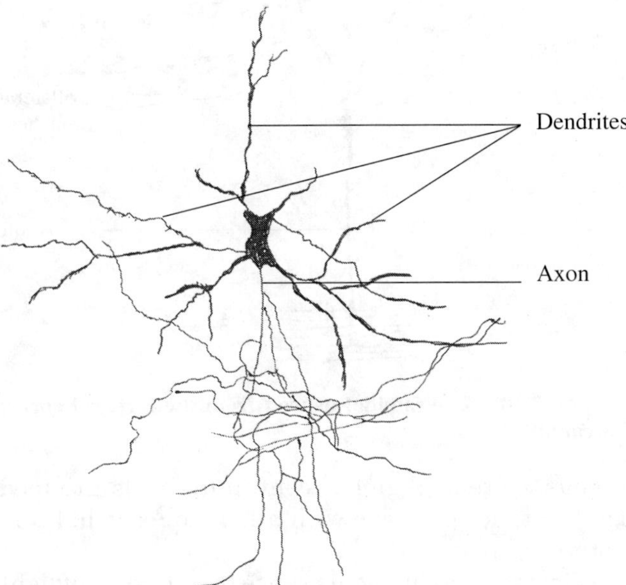

FIGURE 3.9 Neuron from the cerebral cartex of a cat.

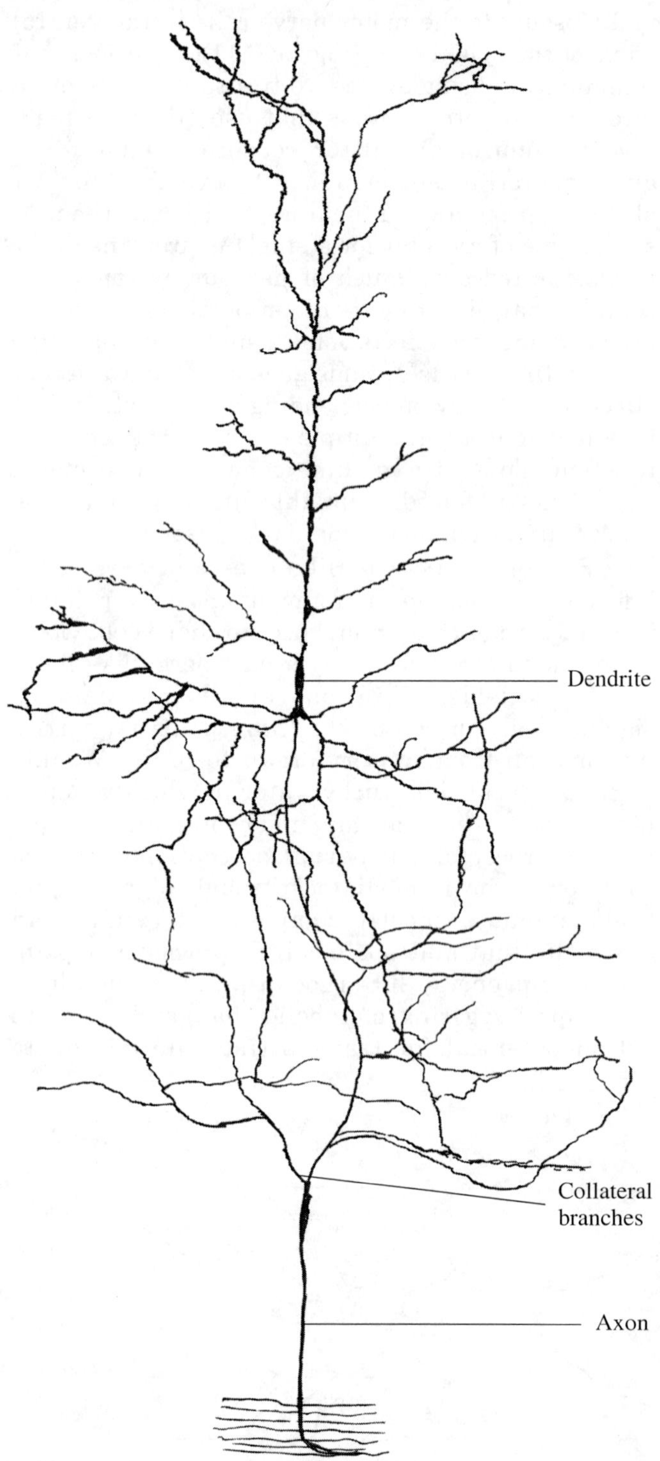

FIGURE 3.10 A pyramidal neuron from the cerebral cortex of a rabbit.

nervous system. Pigment containing cells grouped together to form an eye with a lens to focus light on photoreceptive cells.

Some visual pigments sense broad range of light frequencies from ultraviolet to infrared, producing

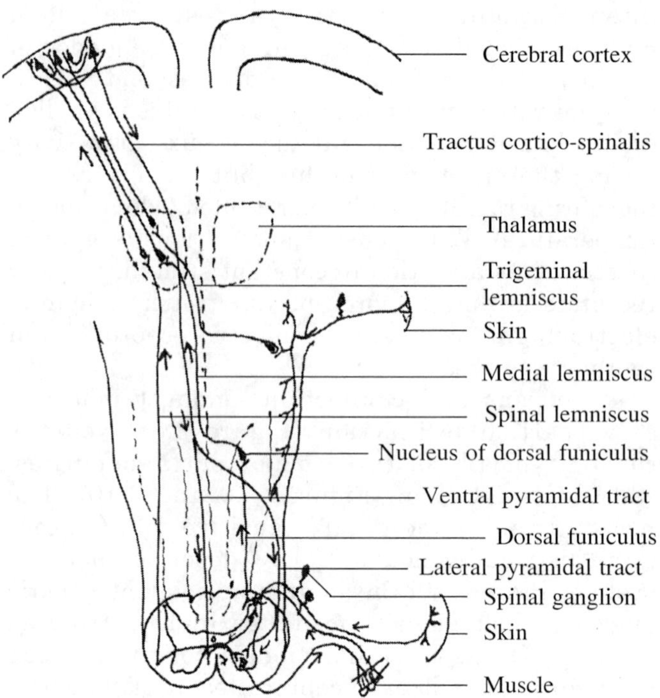

FIGURE 3.11 The ascending and descending pathways between the spinal cord and the brain, showing the crossing of pathways from one side of the nervous system to the other.

high sensitivity to light and monochromatic (black and white) vision. Some visual pigments detect narrow range light frequency and provide low light sensitivity and possibility for colour vision. Nocturnal animals generally have monochromatic vision to get full advantage of low light level. Diurnal animals can have trichromatic vision. Asian Yellow Swallowtail butterfly detects as many as 5 different primary colours. Animals integrate visual information from a few to hundreds of thousands of light receptive cells. Some animals form images, others detect movement. Interpretation of visual images needs developed central neural processing. Specific brain regions work this activity.

Every organism inhabits an environment full of information being received and processed by that organism's nervous system (Dangles et al., 2009). The term umwelt describe the perceptual world (Rüting 2004) which differs for each organism. It is difficult to understand how another organism perceives the world. Sensory systems generally changes with development. The umwelt that organism inhabits often change over its life (Dangles et al., 2009). Sensory ecology is based on studying the sensory systems to understand what animals perceive in environments and how that affects their interactions with environment (Dangles et al., 2009). Such perceptual world is dependant on senses that a particular organism possesses, although it is affected by internal workings of nervous system of an animal at any given time. An animal possesses various senses which can be categorized based on type of information that animal receives. Relative importance of particular sensory modalities varies between species to species. Ecology and evolution of species could be learned by examining different sensory modalities it possesses (Dangles et al., 2009). We examine each ability and define the stimuli they receive based on how we respond to it. We detect a small portion of total range of electromagnetic energy that exists in visible spectrum. Many animals detect stimuli that are undetectable with human senses. The prefixes "ultra" is used for stimuli that are above the range that human detect, and the term "infra" for stimuli below the range of human detection. Animal sensory ecology gives us a better insight of animal behaviour, new ways of "looking", especially during analyzing stimuli that are in the ultra or infra regions, like canine ability to detect odours, which is at least 1,000,000 times more sensitive than humans (Olender et al., 2004), or stimuli that are impossible for us to detect like ability to detect magnetic fields (Holland et al., 2010) and electric fields (Camperi et al., 2007).

Sense of vision depends on the ability of photoreceptors to detect the presence of specific electromagnetic energy. Electromagnetic energy is described based on its wavelength that we see form familiar colours of rainbow correspond to wavelengths of roughly 700 nm (red) down to about 400 nm (violet). For wavelengths beyond the range that we can see, the terms infrared (IR; wavelengths up to ~1000 nm from 1 mm) and ultraviolet (UV; 100–400 nm) (Ryer 1997) are used. IR and UV spectra are well within the range that many animals see. While we cannot detect it, infrared radiation is familiar to us as the type of warmth we feel when we put our hand near source of heat. Any object that is warmer than its environment emits infrared radiation. Some animals have receptors to detect this as a type of "heat vision" (Sichert et al., 2006) which is highly beneficial to predators that feed on birds and mammals, as their prey is almost always warmer than their environment. Pit vipers, pythons, and boa constrictors, have special pits on heads that contain IR-sensitive receptors (Pappas et al., 2004) which allow this predator to hide in secluded location and wait until an IR-emitting prey strays within range of strike (Ebert et al., 2007). These receptors enable snake to detect and avoid mammalian predators and to find locations with optimum temperature for sleeping or basking (van Dyke and Grace 2010).

Many flowers reflect light in both visible and UV range (Arnold et al., 2010). When such flowers are examined in UV light, patterns of light and dark are visible which guide bees to be part of flower where they find nectars. These patterned calling benefit plant by directing a bee to portions of flower where it picks up pollen that can then be spread to other flowers. The bees see well in visible spectrum (Kevan et al., 2001), addition of UV light gives access to information that would be lacking if they use eyes with only human sensitivity. In locations where bees do not pollinate flowers regularly, number of plant species with UV-reflecting flowers is significantly lower than in areas where bees are important pollinators (Kevan et al., 2001).

Animals detect vibrations around them using vibration-detecting structures including vibrations spread through substrate like that produced by an animal walking on a plant (Caldwell et al., 2010), or vibrations in air or water around them (Stebbins 1983). Detection of vibrations in air is performed through the hearing. Vibrations of air causing series of events inside the ear to produce electrical signals. Our brain interprets such nerve signals as sound. Vibrations are measured through frequency, the number of vibrations per second. We hear best between 2000–4000 Hz, but our audible range varies between 20–20,000 Hz (Gescheider 1997). Vibrations beyond audible range is not detected by us, but can be sensed by other animals. Elephants communicate with other members of herd with sounds below 20 Hz (Langbauer et al., 1991). Such low-frequency sounds travel long distances through air, ground, and enabling elephant groups to remain in contact over a range of tens of kilometers (McComb et al., 2003). Ranges are measured using different methodologies. Many animals produce and detect ultrasound. Cetaceans (Au 2004), rodents (Moles & D'Amato 2000), bats (Fenton 1984), frogs (Shen et al. 2008), and insects (Rodríguez and Greenfield 2003), use ultrasound for purposes like communication, predator avoidance, and echolocation. The best-known example is the echolocation in bats. Frequency range of bat echolocation calls varies between 10,000 to 200,000 Hz, depending on species (Altringham 1996). A single bat call may last few milliseconds while covering a frequency range of 50,000 Hz or more. Such high frequency sounds do not travel very far, but reflect off solid objects including small ones (Surlykke and Kalko 2008) allowing bat to detect small targets from echoes reflected from the object. Some echo locating species use vision for navigation (Bell and Fenton 1986). They emit echolocation signals during moving, showing their dependence on echolocation abilities (Masters and Harley 2004).

Their high-frequency calls travel short distances. A bat's perception is affected by distances over which its sounds travel (Surlykke and Kalko 2008). Most animals have receptors to detect stimuli over broad range. Sensory ecology deals with the sensory systems affecting animal's perception (Kevan et al., 2001) is important. An animal is likely receiving input from multiple sensory modalities (Dötterl and Vereecken 2010). Practical implications of different sensory systems have evolved. Type of sensory stimuli that an organism detects appropriate food (Siemers and Swift 2006), compete for resources, attract mate, or avoid predators (Mason and Parker 2010). Interaction of sensory systems with species' environment may drive the evolution of new species (Seehausen et al., 2008).

3.3 BEHAVIOURAL ENDOCRINOLOGY

Hormones regulate behaviour, seasonal changes in behaviour, parental care and mating. Estrogen and testosterone influence expression of sexual behaviour, development of reproductive system, aggression and territoriality. Estrogen is derived from testosterone, and both hormones shape female behaviour. Castrated males are less aggressive and manageable. Administration of testosterone increases aggression and territoriality. Oxytocin and vasopressin regulate pair-bonding, and influences parental care. Neurons produce and release neurohormones. These are produced in hypothalamus and secreted from the pituitary. Prolactin prepares females for nursing and stimulates parental care. Juvenile hormone regulates egg production in almost all insects and mating behaviour in some insects. In honeybee, juvenile hormone influence activities a worker performs in colony. Ecdysone or molting hormone may shape behaviour of insects and crustacea. Bag cell hormones control egg-laying behaviour in mollusks. Corticosteroids elevate in individual during stress, while reproductive and territorial behaviours are suppressed and escape behaviours are promoted (Wingfield et al., 1998). Corticosteroids affect learning and memory acquisition (Thaker et al., 2010). Inhibition of corticosterone elevation in lizards (Sceloporus undulatus) during an encounter with attacker impairs escape responses by lizards and limits learning and recall in future encounters. Increased corticosteroids are necessary for antipredatory acts and aversive learning in prey. Resource availability in early development in bonobos (Pan paniscus) and chimpanzees (P. troglodytes), influences whether adults are prone to sharing food (Wobber et al., 2010).

Frank Beach (Figure 3.12) is generally considered to be the founding father of the study of the effects of hormones on behaviour—an area known today as behavioural endocrinology or Psycho-endocrinology. The estrus cycle lasts four or five days (Figure 3.13) in the rat, and is associated with variations in the level of the hormone estrogen and progesterone released by ovaries. Around mid-cycle, there is a 12–15 hour period of estrus or heat during which the female is receptive to the male. Hormones have at least three effects on female sexual behaviour namely receptivity, proceptivity, and attractiveness.

FIGURE 3.12 Frank Beach.

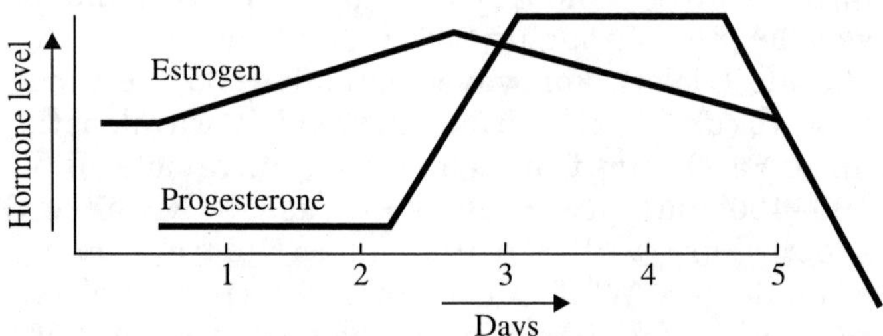

FIGURE 3.13 The estrus cycle in rat.

Social behaviour is linked to steroid hormones, prolactin, and neurohormones. Hormonal mechanisms are more taxon specific, and serve various functions in various animals. Hormones are chemical messengers which are synthesized by ductless glands, and are carried by circulating blood to another parts of the body where they evoke systemic adjustments by acting on specific cells, tissues and organs. They are contrasted for example with neurotransmitters, such as dopamine and acetylcholine, which are the chemicals for rapid communication across tiny gaps between neurons, and between neurons and muscles. An example would be testosterone which is carried by circulation form testes to target tissues. Discovery that hormones can be produced in brain itself upset this neat dichotomy. Boundaries between hormones, neurohormones, neuromodulators, and neuro-transmitters are now fuzzy. Both nervous and endocrine systems originally evolved from a system of cell-cell chemical messengers, and that is why they overlap so much (Figure 3.14). Hormones coordinate behavioural and physiological sequence over time, establish duration of events, and sequence by regulating onset and offset, and modify nervous system appropriately.

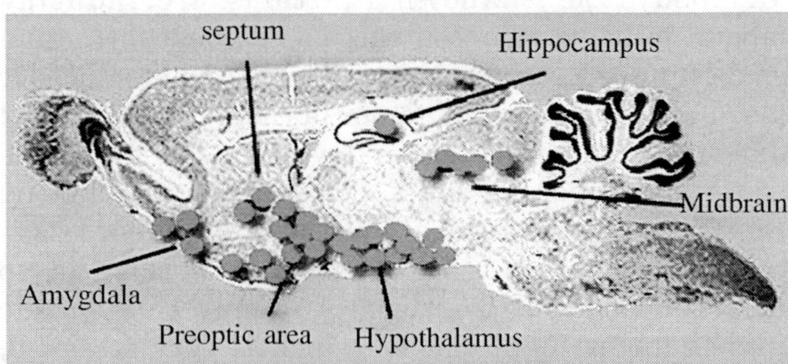

FIGURE 3.14 Overlapping in nervous and endocrine system.

Hormones affect animal behaviour in two ways; one by influencing the surrounding organs directly, and another by affecting behaviour through nervous system. During the estrous cycle of female, Follicle Stimulating Hormone (FSH) promotes follicular development, together with Luteinizing Hormone (LH) causes the secretion of ovarian estrogen, and progesterone. By the hypothalamuspituitary axis, ovarian estrogen and progesterone maintain low levels of FSH and LH in a feedback way which results in the fluctuation of LH, and induction of ovulation. Before ovulation, an increase of ovarian estrogen leads to sexual behaviour. In male, FSH stimulates the generation of sperm and LH causes the secretion of testosterone, which controls masculine sexual behaviour as androgen.

Secretion of hormone is controlled by genes. Research emphasizes the synergistic effect of gene, hormone, and nerve, and suggests that hormones act only as a medium in more cases. Specific gene controls the expression of specific behaviour. So, mutation of specific gene can lead to the change or lack of specific action. Prolactin Receptor (PRLR), whose gene has mapped to the sixteenth chromosome found in breast tissue. One of the ligands of PRLR is prolactin which is secreted by hypophysis, with its function of mammary gland development and lactation, as well as, enhancement of maternal behaviour. A nonsense mutation in this gene could result in the lack of maternal behaviour.

Hormone influences animal behaviour by impacting animal development including the secondary sexual characteristics and even nervous system. Five levels of sex determination are: chromosomal sex determination, gonadal sex determination, hormonal sex determination, morphological sex determination, and behavioural sex determination. Sex determination build on the foundation of chromosomes, which determine the gonad—testis or ovary. Hormones secreted by gonad, morphology (such as the secondary sexual charac-teristics) and behaviour display the sex sufficiently.

Behavioural sex determination
↑
Morphological sex determination
↑
Hormonal sex determination
↑
Gonadal sex determination
↑
Chromosomal sex determination

Recent studies have shown that singing behaviour of male finch is controlled by chromosomes. Brain of finch also displays different sex structures. Right part of brain has a larger vocal centre than the left one. So, it is obvious that genetic material controls the structure and feature of central nervous system and affects the response to hormones. Studies on bees also show that hormone combined with gene affects social behaviour of bee colony. Specific gene controls the expression of specific behaviour. The contributions of gene show in each period of animal development and adulthood. Therefore, gene should be taken into account, when studying the hormone dependent behaviour. cAMP as a key secondary messenger also participate in learning memory.

Several steroids show distinct daily rhythms. Glucocorticoids in mammals and in some birds, peak around the time of onset of daily period of activity (awakening from sleep), while in other birds they peak during night, when birds are inactive. Testosterone in male mammals peaks at awakening. Neuronal cell groups that produce the neuropeptides linked to social behaviour are located in characteristic brain region. In several mammals, Arginine-Vasopressin (AVP) is produced by cells in suproioptic nuclei of hypothalamus, Medial nucleus of Amygdaloid complex (MCA), Bed Nucleus of Striaterminalis (BNST), and lateral septal region.

Neuropeptides have different functions, depending on where they are produced, where the projections from those neurons go, and whether source of projection is steroid regulated. Chemical messenger regulator systems interact with each other at several levels, such as when neuro-transmitters and neuromodulators are regulated by steroids in some brain regions, and peptide and protein hormone actions require prior steroid priming. These modulatory chemicals provide flexibility, temporal, and spatial coordination necessary for adaptive behaviour. It is unlikely that any steroid or peptide hormones have a specific (one to one) relationship, to a given behaviour. Any neurotransmitter, neuromodulator, or hormone functions in more than one type of behaviour and a single type of behaviour is based on multiple chemical messengers, not on one chemical regulator.

Many aspects of social behaviour are subject to modification by experience, such as by social learning or Pavlovian conditioning. Even hormone levels themselves are subject to Pavlovian conditioning. When a male mouse smells a female, his levels of lutenizing hormone and testosterone rise within minutes. Pavlovian conditioning enables animals to anticipate what is to come, and enhances reproductive success (Adkins–Regan and Mackillop, 2003).

Castrated men have lower metabolic rates and live longer, and it is concluded that "a price is paid for a beard". Both testosterone and estradiol increase metabolic rate. Folstad and Carter (1992) proposed that testosterone is costly in part, because it suppresses immune systems leaving animals more vulnerable to pathogens and this cost keep testosterone dependent male signals honest. Testosterone levels in men vary as a function of age (Figure 3.15). There are many different neurochemicals, which are often hormones, each with ability to deliver various messages. Brain and body have their own intelligence. They constantly balance these chemical

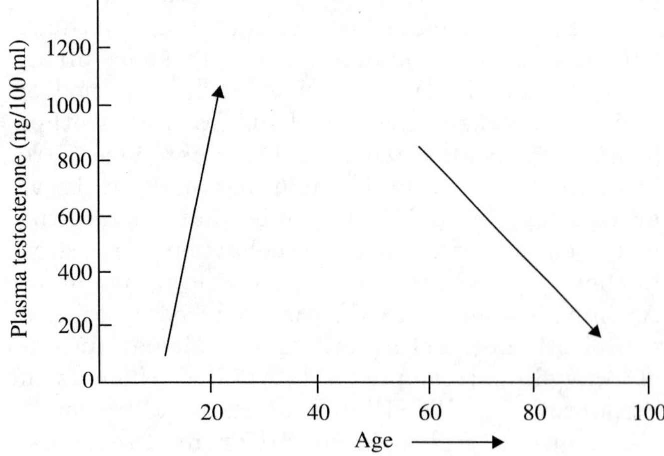

FIGURE 3.15 Testosterone level in man at various age(s).

messengers to control all bodily, and brain activities, maintaining precise, incomprehensible, interrelated control. Same neuro-chemicals can have different effects depending upon other factors, such as availability of receptors on nerve cells receiving the chemical messages, and other neurochemical messages being sent within body. Many early studies measured female sexual behaviour in terms of lordosis, which refers to a characteristic posture in the female rodent that arches her back and moves her tail to permit penetration by the male.

Oxytocin has multiple functions in both brain and body. Some known functions are inducing labour, milk ejection, contracting vagina, contracting smooth muscles in penis; intestines; and stomach. It also induces changes in major hormones that control digestion, and raise sugar levels in blood, affects water and mineral balance, blood pressure and heart rate. Adrenaline, at ideal level is an antidepressant and body increases it as needed to respond to emergencies. But overdose of it may thrown one into "fight or flight" mode, which is very stressful, and unhealthy. All inter-related neuro-chemicals tend to adjust themselves, create unknown side effects, or even counter desired effects. Hormone replacement therapy prevents osteoporosis, but long-term studies reveal an increase in strokes, and cancer among patients using this therapy. One hormone can mimic effects of another hormone. Soaking the brain artificially with oxytocin activates brain centres designed to be turned-on by another hormone, vasopressin. Nature prevents such mix-ups with pinpointed release.

Specific neurochemicals and certain behaviours are circular. It is also intriguing that oxytocin levels can correlate with levels of human Nerve Growth Factor (NGF). NGF is present in brain in large quantities during early romance, when couples experience "in love" feeling. Freudian analysis convinced, that people with repressed sexual desires, make most of the humanity ill. Most people naturally concluded that untrammeled expression of sexual energy was vital to their wellbeing. This general "hard and fast rule" has not changed since, despite its unfulfilled promise.

A male's job from biology's perspective is to impregnate as many fertile females as possible. This function is widely observed in animals, but underlying neurochemical mechanism has not been studied much. Hypothalamus, part of reward circuitry of brain integrates body and mind. As a command centre for endocrine and autonomic nervous system, it is seat of desires and emotions, concerned with every thought and impulse in controlling all sex hormones. Hypothalamus determines hunger and satiation, in both sexes. Certain behaviours and substances stimulate production of dopamine; the craving neurochemical. Over time, an addiction creates a chronic lowering of dopamine levels. Sense that "something is missing" is the basis of addictive cravings. In short, there is reward circuitry in brain which can become activated, depending upon a person's learned behaviours.

In both men and women at orgasm, oxytocin, "the bonding hormone" rise suddenly. Within 5 minutes of orgasm it returns to baseline levels in both sexes. Climax in both sexes is marked by a rise in prolactin. After orgasm, prolactin appears to act as a sexual satiation mechanism. High prolactin may be associated with symptoms, like declining libido, mood changes, weight gain, and hostility. Prolactin is also associated with long-term stress and despair. Women, whose relationships are not going well show high levels of oxytocin, which do not counter their stress. Oxytocin and prolactin interact a lot in pregnancy, birth and child rearing. Stressful environmental stimuli trigger release of Corticotrophin Releasing Factor (CRF) from nerve cells in hypothalamus into a blood system that travels to anterior pituitary. In response to CRF, anterior pituitary releases Adrenocorticotrophic Hormone (ACTH) into blood system.

ACTH is transported to remotely located adrenal gland. ACTH stimulates production of cortisol in adrenal cortex. Cortisol is released into blood stream. Cortisol affects body's reaction to stress. Pituitary-adrenal axis is a complex feedback mechanism. Cortisol feeds back to hypothalamus to control release of CRF. High levels of cortisol in blood tend to inhibit release of CRF by hypothalamus. Therefore, less ACTH is released by anterior pituitary. Consequently, amount of cortisol circulating in blood stream is reduced. This is called a **negative feedback system**. Effect, of stressors on hormones pituitary adrenal system is very sensitive to changes in an animal's environment. Not surprisingly, immobilizing an animal causes a massive release of hormones from both the adrenal medulla (epinephrine and norepinephrine) and adrenal cortex (corticosterone/cortisol). But opening the animal cage or transferring the cage to another room, elicits hormone secretion. Stress response refers to these physiological reactions. Stressors are stimuli that elicit stress response. Stress is often used to refer to either stress response, or stressor. Humans are also sensitive to stressors. For example, examinations, hospital visits, oral presentations and participating in sporting events all activate pituitaryadrenal system.

Amount of cortisol in blood stream is not just controlled by amount of stressor we experience. There is a circadian (daily) rhythm in ACTH secretion and cortisol level which is higher in morning. As well as reacting to stress there is a circadian pattern of plasma cortisol level in humans. ACTH levels exhibit same circadian pattern as cortisol, except that ACTH changes precede change in cortisol level. Cortisol is secreted in several 'pulses' during the day.

Animals gather information about the environment using organs which act as transducers, converting various kinds of energy from the environment into action potentials. The rods and cones in eyes transduce photons into nervous impulses. This information is then carried to the central nervous system, processed and used in determining the animal's next behaviours.

The hypothalamus in vertebrates coordinates the production of many hormones, and has a central role in the control of behaviour. Neuro-peptides from the hypothalamus induce the pituitary gland, which is suspended from the hypothalamus to produce a number of hormones. Perhaps the most important hormones which affect behaviour are the luteinizing hormone (LH) and follicle stimulating hormone (FSH), which activate the ovaries and testes of vertebrates. Behaviourally, the important vertebrate hormones are the estrogens and testo-sterone which affect the expression of mating behaviour, parental behaviour, aggression, and territoriality. Migratory impulses are at least partially under hormonal control.

For invertebrates, the best known hormone with behavioural effects is juvenile hormone, a product of the corpora allata in insects. Juvenile hormone affects mating behaviour, pheromone secretion, parental behaviour in adult insects, as well as worker behaviour in eusocial insects. The new born animal's nervous system is likely to be incompletely formed. It coordinates the activities to sustain the young animal's life like seeking food or its mother, hiding from potential predators and finding the suitable environmental conditions. Vertebrate species are termed either *altricial* or *precocial*. Baby sea turtles are programmed to seek the water, where they can maintain themselves without parental assistance. Mouse pups have incompletely developed eyes and lack the ability to feed themselves.

Behavioural endocrinology is an exciting field which is increasingly being applied to solve many problems. Excellent reviews in the field include the works of Naz (1998), Weintraub (1964), Buckle (1983), and Walker et al., (2005). GnRHI has shown to regulate the release of pituitary gonadotropins. Recently, GnRHII has been discovered. In the brain, GnRHI and II apparently modulate mammalian reproductive behaviours in different but complementary ways (Kauffman 2004).

3.3.1 Organization of Active Avoidance Learning

Pituitary-Adrenal Axis (PAA) may be involved in learning about stressful situations. Experiments have examined the effects of ACTH and corticosterone on avoidance learning using rats. Some of these experiments use discriminated two-way active avoidance learning in a shuttle-box. This task is an example of 'discriminated' avoidance, because a warning signal (light or buzzer) warns that shock is about to be delivered to side of shuttle-box currently occupied by rat. The task is called 'avoidance', because animal can avoid shock by moving to other side of apparatus. The task is 'active', because the rat must emit a response (move to other compartment) to avoid shock. It is a 'two-way' task, because rat must shuttle between two sides of apparatus on successive trials.

'Pole-jump' task is also used in some experiments; this is a discriminated one-way active avoidance task. The term 'one-way' refers to the response which the animal is trained to emit. Rat has to climb one-way up a pole to avoid shock delivered through floor of apparatus. But there is a crucial difference between shuttle box and pole-jump task—the role of experimenter. In pole-jump apparatus, experimenter may sometimes need to remove rat from pole, after it has made an avoidance response before next trial begins. In contrast, shuttle box task is completely automated—experimenter does not handle rat during training. Now, the problem raises is that different experimenters may handle rats in different ways and this may have some subtle effects on animal's behaviour. For example, a confident experimenter may handle rats more gently than someone who is afraid of being bitten. Rough handling may increase a rat's level of fear. One theory that accounts for avoidance learning is called the 'two factor' theory of avoidance behaviour. Briefly, it invokes a mixture of classical and operant/instrumental conditioning (the two factors), to explain avoidance learning. Through classical conditioning, rat learns to fear warning signal. In a shuttle box, the light/buzzer is CS (Conditioned Stimulus), and shock is UCS (Unconditioned Stimulus). After several CS/UCS, pairings presentation of CS (light/buzzer) elicits CR. Fear is CR.

Reduction in fear is reinforcing. Reinforced responses are learned. Therefore, a response that reduces fear by removing CS will be learned. In this way, rat learns to emit an operant response, shuttling or pole-jumping that turns off CS. This response is reinforced because it reduces classically conditioned fear. Experiments have studied stress hormone effects during acquisition (initial learning) of avoidance response, and during extinction (unlearning of a previously learnt response), when rat does not receive shock if it fails to emit avoidance signal.

Levine (Readings in *Physiological Psychology*, *Scientific American*, and *Freeman*) reviews a number of studies that have examined effects of ACTH on avoidance learning. He argues that ACTH increases fear, and thereby, improves avoidance learning. Hypophysectomy is the operation for removal of whole of pituitary gland. One consequence of this operation is loss of ACTH from anterior pituitary, which may be responsible for profound disruption of avoidance learning in rats. Injection of ACTH restores ability of hypophysectomised rats; to acquire an avoidance response. An alternative explanation is that it is not the loss of ACTH per se that disrupts avoidance learning. It could be the decline in adrenal hormones that follow hypophysectomy that leads to the avoidance impairment. Adrenalectomy (removing the adrenal glands) leads to a loss of corticosterone and an increase in ACTH due to removing feedback effect of corticosterone. De Weid has found that adrenalectomised rats make more avoidance responses under extinction conditions than intact rats.

Vasopressin V1a receptor (V1aR) modulates social behaviour in a wide variety of species. Variation in a repetitive microsatellite element in 5-flanking region of V1aR gene (avpr1a) is associated with variation in social behaviour in rodents. Evolution of these regions may have contributed to variation in social behaviour in primates. Melanin-Concentrating Hormone (MCH) levels were also investigated, but found to be more variable. Interestingly, increases in Hcrt were associated with sleep-wake transitions, and increases in MCH were associated with wake-sleep transitions. In addition, high levels of Hcrt were associated with social interaction. This pattern of Hcrt release is consistent with the sleepiness and cataplexy seen, when the Hcrt system is nonfunctional, as in narcolepsy.

Oxytocin is released peripherally (within the bloodstream) during the orgasmic phase of human sexual response in both men and women. It stimulates sperm transport in males and females by stimulating contraction of smooth muscle. Oxytocin may also be released within brain. Several effects have been associated with release of oxytocin centrally. It stimulates sex behaviour in males and females. In males, oxytocin levels increase with each ejaculation. High levels of oxytocin have been shown to inhibit sex behaviour in males.

3.4 BEHAVIOURAL ECOLOGY

In many temperate zone birds, increasing day lengths of spring stimulate massive growth of gonads, and a corresponding increase in sex steroid (Wingfield and Farner 1978). A stressful stimulus causes release of corticosterone, which interrupts mating in a male newt. Exposure to certain EDCS has contributed to adverse effects in some wildlife populations. These effects vary from subtle changes in physiology, and sexual behaviour to permanently altered sexual differentiation. Migration, hibernation, aestivation and sleep are linked to environment.

Behavioural ecology determines natural selection service in shaping genetically based behaviours. Not all behavioural changes are adaptive, and therefore, not all of the behaviours are shaped by natural selection. Genetic drift, selection of unrelated traits, and gene flow influence changes in inheritable traits that affect behaviour. Behaviours increase fitness by attracting fitter mates, decreasing vulnerability to predators, increasing energy intake for producing more offspring. Behaviours that increase fitness have a greater likelihood of being passed on to later generations. Tinbergen observed that gulls removed broken eggshells, from their nests, after eggs hatched. He designed an experiment, in which he left broken eggshells in a gull nest and observed that white of egg-shells attracted crows which prey on gull hatchlings to nest, he concluded that gull behaviour is adaptive, since it increases chance of survival of gull hatchlings.

Some behaviours may increase one aspect of an animal fitness and may be detrimental to another aspect of fitness as occur frequently in foraging and territorial behaviour, where an increase in energy intake or protection of territory comes at expense of exposure to predators. Natural selection generally favours behaviours that maximize energy gain. Other factors that increase fitness are predator avoidance and mating opportunities. Shore crabs benefit by gaining energy from feeding on larger mussels. Increased exertion required to open shell of larger mussels would cost shore crab in energy expenditure. A male bird defending a territory containing many breeding females benefit from increase in mating opportunities, and may assume greater risk of injury from fighting with competing males.

Cost of reproduction, sexual selection, and a species mating system influence reproductive behaviours. Reproduction requires use of resources at the expense of parent. Cost of reproduction refers to amount of resources an animal invests into reproducing offspring. Species and individuals exhibit either low or high investment in reproduction cost. Low investment, such as small gametes or no parental care after mating results in behaviour particularly in male of a species that increases frequency of mating opportunities. High investment, like large gametes or parental care after mating results in behaviour that encourages greater selectivity, when choosing a mate. High investment is most often required by females and results in mate choice, a behaviour in which female evaluates several potential males and chooses one, she determines to be most fit. Roles of males and females are reversed in some species where males demonstrate high investment, and display mate choice. Males and females may demonstrate equal levels of investment leading to similar mating behaviour in both sexes.

Sexual selection refers to competition for mating opportunities. Limit to reproductive opportunities for both the sexes in any animal population influence development of behaviours to enhance an individual change for reproductive success. Traits of successful individuals are passed on to offspring and have a higher likelihood of influencing gene pool of the population. Individuals in a population compete through two methods of sexual selection, intrasexual and intersexual selection. In the former, individuals of the same sex generally male, compete with each other for mating opportunities. Traits that allow one male to outcompete another, are favoured. Deer, with large antlers have advantage in fights and is likely outcompete deer with smaller antlers. Trait for large antlers is passed on, and over time, deers with large antlers become more common in population. In inter-sexual selection individuals of one sex attempt to entice individuals of opposite sex to mate with them. Sometimes benefits associated with a certain mate are direct or obvious. In other cases, benefits that lead an individual to choose a certain mate are indirect and not obvious. A female that requires protection for her and her offspring is likely to mate with the largest male. Here benefit is direct, since a larger male provides better protection.

Many female birds demonstrate indirect mate selection by choosing mate with bright colour.

Bright colour does not provide an obvious or direct benefit, but may be an indication that the mate is healthy. This selection provides indirect benefit of passing on genes for good health to the offspring. Sexual selection often leads to sexual dimorphism, which refers to differences between sexes. In many species, where intrasexual selection occurs, males tend to be larger than females. Over time, size of the males in a population increase, as larger-size males outcompete smaller males. Females in same population will remain with the same size. Three mating systems, monogamy, polygamy, and polyandry, have evolved based on their interactions among individuals and environment. In monogamy, both parents tend to provide care for altricial young, which are youngs that require long and extensive care, reducing tendency for one individual to leave and mate with other individuals. Many animals, such as birds that produce altricial young are monogamous. In polygamy, in which male mates with more than one female, the precocial young, which require little parental care are for by only one parent allowing other parent, to leave and mate with other individuals in population. Many animals that produce precocial young are polygamous. Polygyny is particularly beneficial to females in a population, where a male defends a territory and females mate with male, as found in elephant seal populations. In polyandry, one female mate with more than one male. Animals that produce precocial young may also be polyandrous. Polyandrous females, such as spotted sandpiper, mate with more than one male leaving several males to provide parental care for the offspring. Many birds, thought to be monogamous actually display polygamous and polyandrous behaviour. Scientists have concluded that females must have been mating with more than one male. Scientists call this Extra-Pair Copulation (EPC). Males perform EPC to increase number of offsprings, while females perform EPC, to either enhance genes of their offspring with those of genetically superior males, or to enlist more helpers to aid in chick-rearing.

Altruism characterizes behaviour of one individual that is beneficial to others in a population, but detrimental to individual. An individual sounding a warning call after sighting a predator allows others to hide, but draws the predator's attention. Evolution of altruism seems to contradict the expected behaviour of individual survival. If an altruistic behaviour is detrimental to an organism, it would not be favoured by natural selection and the behaviour would become diminished over time. Reciprocal altruism suggests that partnerships exist in which an individual provides a benefit to others by performing an altruistic behaviour, but benefits from others who also perform that behaviour. This also suggests no reciprocators are cut off from these benefits. Vampire bats, that find a source of blood give a small amount of it to other vampire bats in group. This behaviour prevents other vampire bats from starving. And is reciprocated when another bats find a source of blood. Vampire bats do not give blood to individuals that have failed to reciprocate in the past.

Kin selection asserts that reproductive success of a relative is beneficial to an individual who shares some of the same genetic material. A percentage of relative genes will be present in the offspring, ensuring that at least some of relative genetic material is passed on to next generation, even if they never reproduce themselves. Individual ground squirrels that have relatives nearby are more likely to sound an alarm to alert others of a predator. False altruistic acts appear to be altruistic, and are actually performed for the benefit of individual. An animal that issues a warning call to others appear to be sacrificing itself. Reaction of other individuals in population may actually direct attention away from caller. A group of animals living together in a cooperative manner form a society. Social systems present several distinct benefits to individuals living in them. Related individuals benefit from additional help provided by others in group.

Presence of more individuals decrease any individual chance of being singlled out by a predator. Individuals may learn of food sources from other individuals in their group. Social behaviour in societies is often based on altruism. Insects, such as ants, bees, wasps, and termites tend to form very structured, altruistic societies, based on caste systems. In caste systems, individuals of same species perform different tasks, such as foraging or protecting, based on characteristics, such as size and division of labour. Many vertebrates also form social groups in which individuals differ based on characteristics and tasks performed. However, vertebrate groups tend to be less organized and less altruistic than those observed among species of insects.

Notable marine predators are great white sharks, tiger sharks, killer whales, giant squid, and barracudas. They use amazing adaptations to avoid predation, and learning the same help in gaining insight into their lives. Fishes including silversides, mackerels, anchovies, herrings, tunas, jacks, and snappers live in large schools. There is the concept of safety in numbers, meaning a chance of a single fish getting captured by a hungry predator is less when a fish schools, than if that same fish swims alone in a sea. A position nearest centre of a school maximizes protective schooling benefit. A school might be seen by predators as bigger than them. Attacking a single creature that appears to be as big as a school, would mean taking too big of a risk. All fishes do not live in schools.

Solitary species use other ways to avoid predators. When danger arises, thin-bodied angelfishes and butterflyfishes escape into narrow cracks and crevices of reefs, places that their thicker-bodied predators cannot enter. When feeling threatened, razor fishes disappear into sand. Jewfishes, tilefishes and blennies take shelter in burrows. Flatfishes, like flounders, halibut, turbots and sand dabs camouflage themselves to avoid predators. They use cells in their skin (chromatophores) to match the hue and pattern of their surroundings with remarkable accuracy. Sculpins and stone fishes use their colouration, patterning, and facial appendages to blend with surroundings. Frogfishes and leaf-fishes resemble their surroundings and sit almost motionless for long periods for becoming undetected. Flattened shape of rays, skates, and sharks like angel sharks, helps them blend with substrate. Many flattened animals bury themselves in substrate. Balloon and pufferfishes swallow water and inflate their bodies, increase their size, and erect sharp spines that deter many potential threats. Smaller fishes like horn sharks, and Port Jackson sharks possess sharp spines next to their dorsal fins that are raised to puncture mouths of the predator. Spines of juveniles are sharper than of adults. Juveniles are often spit out; for avoiding predators. Triggerfishes use a modified trigger like dorsal fin to lodge into crevices in face of danger. Their dorsal fin contains 3 spines with largest forward one. When first spine is erected during fear, second spine moves forward to "lock" first spine into position. Fish positions to erect spine that prevents it from being pulled out of its hiding place.

Spines in dorsal fins of scorpion and stonefishes cause a painful and toxic puncture wound to predators. Beautiful species, like lion or turkeyfishes too are armed with a venomous defense system inflict potent poisonous stings with their long dorsal spines. Parrotfishes at night enter a sleep-like state known as torpor, zoning out on sea floor, sometimes squeeze into cracks or crevices. Members of several species routinely "sleep" out in open inside of self-spun cocoons created from combination of sand and mucus. Cocoons are believed to mask their odor, preventing predators that rely upon their olfactory system to find prey. Butterflyfishes (Figure 3.16) use a false eyespot, to fool the predator, which stands out prominently near tail, while their eyes are often masked by facial stripes and other markings. Seeing false eyespot, a predator may be deceived about which end of butterflyfish is the head. Anemonefishes hide among potent, stinging tentacles of sea anemones, able to protect themselves from powerful stings of host anemones. Bright colours and attention-getting antics of anemone fishes lure other fishes into a fatal trap that is believed to provide food for host anemone and for the anemonefish (Figure 3.17). Shark suckers, remoras and many small fishes, either attach themselves, or swim close to a larger animal like shark, turtle, grouper or barracuda, so that no other predator is likely to come close to their large predatory host, so these relatively small fishes only have to keep a watchful eye out for a single predator.

FIGURE 3.16 A butterflyfish.

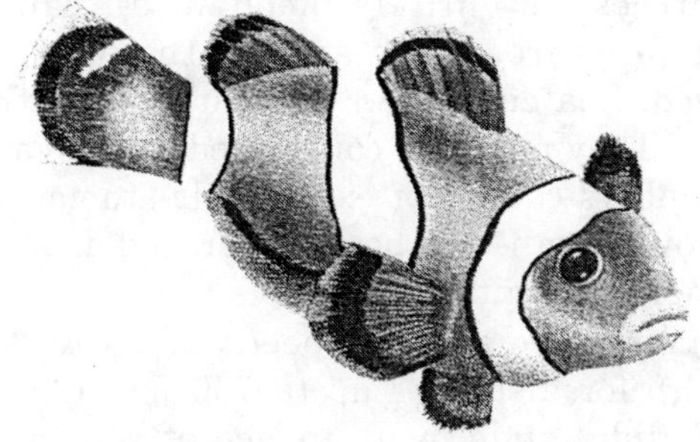

FIGURE 3.17 A clownfish.

Octopuses (Figure 3.18) and squids emit clouds of ink to evade predators. While ink obscures predator's vision, anesthetizing olfactory systems of predators, such as moray eels is thought to be ink's primary purpose. Many nudibranchs are colourful and stand out in their surroundings, secrete noxious chemicals that repel predators. Other species transfer unfired stinging cells from soft corals and anemones into cerata on their backs and nudibranchs use stinging cells as part of their own defense. Sponges

FIGURE 3.18 An *octopus.*

might appear to be an easy meal for a predator, but their tough fibrous bodies are laced with sharp, silica-based spicules. Some sponges harbour zoanthids, which help to deter fishes, like French angelfishes which are adapted to feed on sponges. In other instances, best defense is a quick retreat and few species are as adept at retreat as tubeworms. When relaxed and not feeling threatened, tubeworms extend their featherlike gill plumes to trap food and to extract oxygen from water. But instant a tubeworm senses danger, gill plumes are drawn into their tube in acts that occur in blink of an eye.

Legendary stings of corals, anemones, hydroids, and jellyfishes are both potent and fast-acting. It is believed, that firing of their stinging cells is one of the fastest cellular responses in all of nature. Firing process is entirely mechanical, caused by touch or pressure waves and no thought process is involved. Sea cucumbers are rather unattractive creatures. They might be considered an easy meal, if sea cucumbers could not expel their stomachs. The sticky, foul organ can be regenerated in a short-term, when regurgitated it repels.

Flyingfishes soar at speeds of up to 35 mph (56 kmph), for distances up to 750 feet (227 m) in a single glide, glide over surface of sea to escape predators, such as billfishes, dorado, tunas and sharks. Flyingfishes use powerful side-to-side motion of their tails, to propel themselves forward through water. Once they break free of water, these fishes extend their enlarged pectoral and pelvic fins, using them to help fishes glide over surface. When gliding across surface, flying fishes certainly seems apropos. Sea urchins use sharp spines to defend themselves against predators. Barbed spines of most sea urchins lack venom, but create a formidable defense nonetheless. Known for eating corals, crown of thorns sea star is armed with sharp protective spines. Over time, many sea stars can regenerate their lost body parts, and "lost parts" can regenerate a new body in some instances, as long as a portion of animal's central disc remains attached. So when attacked, sea stars often sacrifice one arm to escape, regenerate and live another day.

Nowicki studied mating response of female song sparrows, to songs of captive-raised males and analyzed males' songs to determine degree of accuracy with which males copied songs. Females preferred those songs closest to wild-type songs they heard when young and presumably learned as models. It was believed that female songbirds pay attention to male song as an indicator of fitness. Nowicki and others have developed experimental evidence, that there is a link between early stress, male brain development and song-learning.

Army ant (*Eciton burichelli*) colony castes consist of queen, soldier, and multi-purpose worker ants. Queen ant lays eggs. Soldiers pay attention, to defense of colony. Worker ants are split between being foragers, or tending to queens brood. Army ant is abundantly found in humid lowland forests or heavily forested areas. Temperature and precipitation changes have no affect on life cycles or migration of these ants. Social structure and life cycles of ants have two distinct phases. Colonies go through a stationary phase and a migratory (nomadic) phase, based on ability of a queen to produce many eggs in a very short period of time (up to 300,000 eggs in 5 to 10 days). Queen overpopulates a given area and then triggers migration of colony. In stationary phase, queen's abdomen swells to hold 66,000 eggs. As eggs develop into mature ants, tens of thousands of adult workers appear triggering nomadic phase. Colony activity increases exponentially. "Swarm raids" increase in size and intensity. With a colony of this size, colony must emigrate daily.

Migration continues until "larval pupation" begins. Once this occurs, colony reenters stationary phase and the cycle repeats. Army ants are called "swarm feeders." Foraging workers form a fan-shaped swarm, with a broad front. They also have a unique way of forming nests. Nest is made up of army ants themselves. Ants form walls and fasten onto each other by using their mandibles. Colony forms these walls almost anywhere in order to enclose queen and her brood. Nest is very structured, with corridors for transporting food and eggs throughout many areas of nest. Army ant colonies march and nest in different phases as well. In nomadic phase, ants march at night and stop to camp during day. Colony begins a nomadic phase when available food has decreased, during which colony makes temporary nests that are changed everyday. Each of these nomadic rampages or marches lasts for about 17 days. Army ants are notorious for eating anything that gets in their path. They normally consume lizards, snakes, chickens, pigs, goats, scorpions, tarantulas, beetles, and other ants. Army ants can also climb trees, feed on animals within its canopy, and communicate through chemical messaging, and trail pheromones. Such communication act as stimuli for changing behaviour patterns. Unlike other ants, army ants do not have compound eyes, but instead have single eyes. Army ants are blind and use their antennae, to sense smell and touch which are used during nesting and raiding.

Fighting in juvenile American lobsters *(Homarus americanus)* begins with a series of threat displays. If opponents are evenly matched, encounter progressively escalates through ritualized components of fighting, restrained forms of physical combat and finally brief periods of unrestrained fighting where opponents may even inflict injuries, on each other (Huber and Kravitz 1995). War of attrition scenarios refer to situation, where fighters attempt to grind down opponent's defenses. There is no fixed cost associated with losing or contesting, but as encounter wears on, each player accumulates incremental costs. A decision to give up arises, when one individual backs down, relinquishing access to the contested resource, rather than continuing to sustain further insults.

It is a daunting task to study behaviour, by simultaneously considering all significant factors, and interactions that impact it. Method that attempts to reduce the associated complexities by wrapping ones expectations into a set of mathematical abstractions based on a reduced number of significant variables. Resulting quantitative predictions are necessarily incomplete, often less ambiguous and easier to test. Goal is to model animal's rational decisions of what action to take and given some other information.

3.5 CONCLUSION

James contention was that humans have more instincts than animals, such as sociability, shyness, secretiveness, cleanliness, modesty, shame, jealousy. For Ridley, the most important of them is sexual attraction, which under control of hormones control activities of certain genes and specify various instincts. Ingel's experiments of injecting oxytocin and vasopressin into brain provoked different behavioural and sexual responses in male and female animals. Since oxytocin receptor genes in different animals have promoters of different lengths, Ridley speculates that this may due to variety of mating and bonding preference in different species.

That animal behaviour is controlled by specific genes was beautifully demonstrated by Benzer with his mutational analysis of fruit-fly behavioural patterns. According to Mooney, humans at birth are psychosexually neutral. Gender orientation, as male or female is not innate, but is acquired later though learning and experience. Diamond opposed the view that sexual behaviour was environmentally determined. Expression of certain genes is important in determining behavioural patterns. In some egg laying species, sex determination is controlled by environmental factors, such as temperature in incubation of eggs. Hormonal imbalances during gestation in mammals exert irreversible influences on behavioural patterns during adult life. Thus, both genes and environment play important, often complex and interdependent roles in determining behaviour.

The rapid movements which are reversible, and repeatable an indefinite number of times have been largely exploited by the animal. Muscles use energy which is stored as chemical fuel. The basic unit of such action is the neuron. Neurons communicate with each other in a similar way to the pulse codes of digital computers. A single neuron may have tens of thousands of connections with other components. The neuron has evolved in the direction of miniaturization. Neurons are cells with a nucleus and chromosomes with cell walls drawn out in long, thin, wire-like projections. Often a neuron has one long axon, the width of which is microscopic. A single axon runs the whole length of a giraffe's neck. The axons are usually bundled together in nerves and lead from one part of the body to another. Other neurons have short axons, and are confined to ganglia or brains. Brains contribute to the success of survival. Natural selection favoured animals that became equipped with sense organs. The brain is connected to the sense organs.

Behaviour possesses a definite purpose. Purposiveness has evolved the property we call *consciousness*. A gene for altruistic behaviour means the development of nervous systems in such a way as to make the organism behave altruistically. The genes are master programmers, which program their lives and are judged according to the success of their programs in coping with the hazards.

Some behaviours can be broadly labelled communication. An animal may be said to have communicated with another when it influences its behaviour or the state of its nervous system.

Examples of communication are numerous: song in birds, frogs and crickets; tail-wagging and hackle-raising in dogs; human gestures and language. Bees dance in the dark to give other bees accurate information about the direction and distance of food, a feat of communication rivalled only by human language itself. Communication signals probably evolve for the mutual benefit of both sender and recipient.

Many edible insects, like the butterflies, derive protection by mimicking the external appearance of other distasteful or stinging insects. Some beemimicking flies are even more perfect in their deception. Predators too tell lies. Angler fish wait patiently on sea bottom, blending in with the background. The conspicuous part, a worm-like piece of flesh, projects from the top of the head. When a small prey fish comes near, the angler dances its worm-like bait in front of the fish, and lures it down to the region of the angler's own concealed mouth. Suddenly it opens its jaws, and the little fish is sucked in and eaten.

Failure to link specific genes with heritable mental diseases means that environmental factors or earlier experiences should not be ignored. These include viral infections, stress and diet and so on, thus, emphasizing role of nurture. For Ridley, genes are means by which nurture expresses itself, and "ultimate explanation for schizophrenia will include both nature and nurture, neither of which will be able to claim primacy". No connection is made for development of behaviour, by specific genes. Experiments on C. elegans, suggest that development of behaviour can be environmentally plastic.

A gene expressed in brain is thought to facilitate imprinting through visual stimuli during development by its ability to control maturation of Gamma Amino Butyric Acid (GABA) system. This observation reflects one of the consequences, but says nothing about causes of imprinting. Foetal expression of SRY gene, located on Y chromosome sets off a chain reaction of downstream genes coming into play, which eventually set up male anatomical and behavioural characteristics. In mammals, but not in egg laying animals, female is "default" sex. If SRY gene is incorrectly expressed or not expressed or missing, foetus develops into female. Some have extrapolated that homosexuality may result from this gene failing to cause prenatal masculinisation in brain during gestation. Others consider mutation or mis-expression of other genes, leading to an autoimmune reaction by mother which would mark the baby for life.

Review Questions

Short Answer Questions

1. What are QTLs?
2. What is neurotransmitter?
3. Name four major neurotransmitters.
4. Define hormone.
5. What is neuroethology?
6. Mention the function of limbic system.
7. What is called bonding hormone?
8. What is extra-par copulation?
9. Name the major factors which control behaviour.
10. How does gene influence behaviour?
11. How does environment influence behaviour?
12. How does hormone influence behaviour?
13. How does nervous system influence behaviour?
14. What is the basis of "Benzerian" approach?
15. What is the basis of "Hirschian" approach?
16. Comment on the hygienic behaviour of honeybee.
17. What is quantitative genetics?
18. What is qualitative genetics?
19. What is androgen insensitivity syndrome?
20. Who wrote the book, "Hereditary genius"?
21. Name four human instincts.
22. Mention two important characteristics of human brain.
23. Name two steroids associated with rhythmic activity of human.
24. Comment on two-factor theory of avoidance learning.
25. What do you understand by the term hypophysectomy?
26. Name the factors that influence reproductive behaviour.
27. Comment on intersexual selection.
28. Comment on intrasexual selection.
29. Define polygamy.
30. Define polyandry.

31. Define meme.
32. Define gene.
33. What is the significance of genetic diversity?
34. What is muton?
35. What do you understand by the term heterozygosity?
36. What do you understand by GP1?
37. What do you understand by GP2?
38. What do you understand by GP3?
39. Define mutation.
40. What is gene mutation?
41. What is paracentric inversion?
42. What is pericentric inversion?
43. What do you understand by monosomy?
44. Name three neurotransmitters.
45. What are giant axons?
46. What is presynaptic neuron?
47. What is postsynaptic neuron?
48. What are inhibitory synapses?
49. What is action potential?
50. What do you understand by "electrical threshold potential"?

Long Answer Questions

1. Explain behavioural genetics in light of modern study.
2. Describe the hormonal basis of behaviour.
3. Describe the neuronal control of behaviour.
4. Explain with examples how environment influences behaviour.
5. Write in brief about the biological basis of behaviour.
6. Mention the role of hypothalamus in behaviour.
7. Write a note on active avoidance learning.
8. Both gene and environment play important roles in determining behaviour. Discuss.
9. Write an account of behavioural ecology in the light of our present understanding.
10. Write how hormones of pituitary and adrenal influence vertebrate behaviour.
11. Discuss the chemical environment of brain.

Chapter 4

Innate and Learned Behaviour

Learning refers to an animal's ability to change behaviour based on past experience and is divided into nonassociative learning and associative learning. Nonassociative learning refers to an animal's ability to develop a behaviour without developing a connection between two stimuli or between a stimulus and a response. Nonassociative learning is divided into habituation and sensitization. Associative learning refers to an animal's ability to change its behaviour by forming a connection between two stimuli or between a stimulus and a response. Associative learning is divided into classical conditioning and operant conditioning. Interactions between nature and nurture mould an animal behaviour during development. Behaviour can be either instinctive or a combination of instinct and learning. Imprinting refers to the process where social attachments developed as an animal matures and it influences behaviour later in life. Imprinting is divided into filial imprinting and sexual imprinting. Cognitive behaviour refers to a response that is determined through the processes of thought. The extent to which animals are capable of cognition is a subject of debate, although problem-solving behaviour observed in some animals suggests that they do think. Movement of animals including migration can be a combination of instinctual and learned behaviours.

The learning ability of insects varies greatly from one species to the next. Bees can be trained to associate food with a given colour or shape. Ants exhibit a considerable proficiency in learning a maze (Figure 4.1). Birds display a wide range of innate and learned behaviours. Innate behaviours are responses that are highly resistant to modification, stereotyped, genetically determined, and inherited. Learned behaviours are adaptive modifications of behaviours through experiences.

Instinct is inherent disposition of a living organism towards a particular behaviour. Instincts are unlearned, inherited Fixed Action Patterns (FAP) of responses to certain kinds of stimuli. Instinctive or hardwired behaviour drew the interest of Charles Darwin and Niko Tinbergen. Instinct is performed without thought and cannot be modified by learning. Instincts are simple behaviour which are exhibited in response to particular stimulus. A dog circles on it's bedding several times. A rattlesnake will strike at a moving, mouse sized, warm object. Genetic information determines behaviour when a species' environment varies little, or when unambiguous messages are sent and received. Information for prey selection, processing prey after striking and place for forage is likely innate. Many signals used in communication are innate. The constancy that comes from having signal and its genetically encoded interpretation makes the message unambiguous. Facial expressions, hair erection, and tail posture in combinations give dog (about other dogs) specific messages. Others use combined genetic and learned information in forming signals. Some birds could

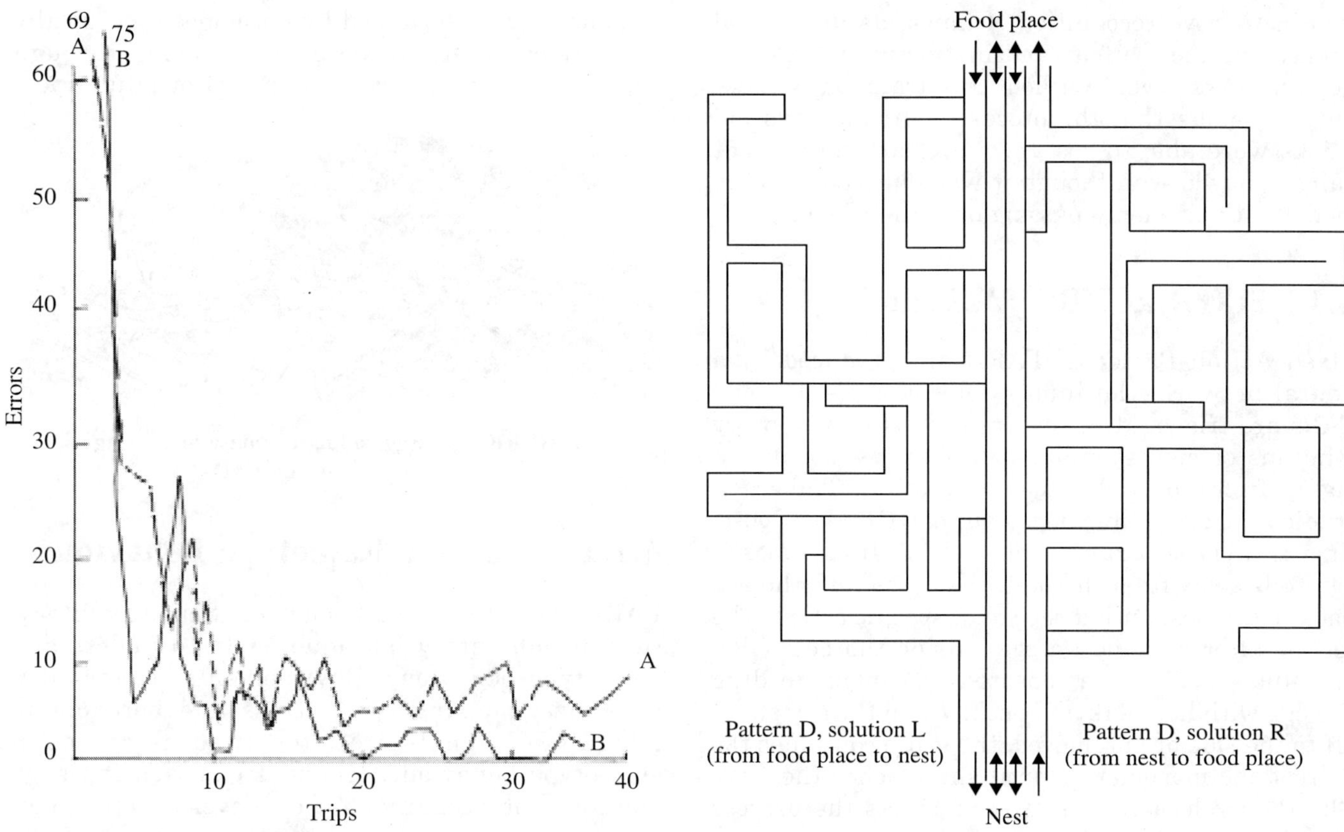

Learning curve for a fairly successful performance of *Formica incerta* in the maze shown at the right. A is outward trip to food place. B is return trip to nest.

Maze used by Schneirla (1933) in the investigation of ant learning.

FIGURE 4.1 An ants proficiency in learning a maze.

create song's elements by not hearing song of another bird, but they require to hear songs during development to repeat song of their species correctly.

Instincts like reproduction, feeding, fighting, courtship, escape function, nest building are not based upon prior experience. They are hard wired, ready to use behaviours. "Our brain is a tool for thinking, but not its cause. Our brain has not invented thinking, the thinking invented the brain". "Mock up/decoy experiments were used, to study inborn behaviour." Innate behaviour offers a unique opportunity to use genetic analysis to dissect and characterize the neural substrates of complex behavioural programmes. Courtship, in *Drosophila*, involves a complex series of stereotyped behaviour that include numerous exchanges of multi modal sensory information over time.

Tinbergen used a flock of newly hatched baby turkeys, not more than a few days old. They were kept in a circular pen, about 20 feet in diameter, with walls not more, than a foot in height. There was a vertical pole of 3 metres in height, at centre of the pen with a horizontal arm extending out from top of central pole, so that arm could sweep horizontally over the pen in a rotary motion. Then, a wooden cross was attached to the end of the arm. Little chicks were peacefully feeding in the pen. When arm with cross was slowly moved in one direction, little chicks run for cover. When same cross moved in opposite direction, birds ignore it. If cross moved in direction of its longer arm, little chicks completely undisturbed would go on pecking at their food. If cross moved in direction of its shorter arm, they immediately scream with fright and ran for cover in a hutch in centre pen. Tinbergen (1951) showed that when young turkeys see a silhouette model pulled in direction that makes it look like a hawk, they were terrified and ran for cover. When a silhouette model resembled outlines of a flying goose, little chicks ignore. When a silhouette model resembled outline of a flying hawk, chicks were terrified and ran for cover despite the fact that there was not even a mother to warn them or related her experience to little chicks. Inborn, in those little chicks' brain was instinctive recognition of hawk, as its natural enemy, instinctive fear and instinctive reaction of flee and take cover. Without any training, without any conscious thought processes, a few days old chicks were able to recognize a clear and present danger, hawk even though it was only a silhouette model of their enemy passing over their heads.

4.1 FIXED ACTION PATTERNS

Fixed Action Patterns (FAPs) are produced by a neural network, the innate releasing mechanism. FAPs are triggered in response to external sensory stimulus or sign stimulus or releaser when it is a signal from one individual to another. Tinbergen studied the egg rolling behaviour of a Graylag Goose. This is a classic example of an FAP. If one goose's egg rolls away from the nest, the goose rolls the egg back to the nest with a repeated, specific action. The female after noticing an egg outside the nest (sign stimulus), begins the repeated movement to drag the egg with her beak and neck. When the egg slides off to the side or is removed, the goose continues the stereotypic movements, until she reaches the nest. She then relocates the egg and starts the process all over again. Many birds exhibit specific sequence of elaborate movements. Sometimes both male and female birds dance together, but in many cases, ornately feathered males dance for the females. The key stimulus for this FAP is the presence of the female. The female selects her mate based on elaborate plumage and dance rate. FAPs are common in animals with simple cognitive abilities. We also exhibit FAPs. Yawning is a great example. We yawn when we see another person yawning. Once yawn starts, this instinctive response runs its course from the beginning to the end. Some species take advantage of the FAPs of other species by mimicking their sign stimulus. The term used to describe the duplication of another species' releasing mechanism is known as *code-breaking.*

Any event that initiates an instinct, is a Key Stimulus (KS). KS lead to Innate Releasing Mechanisms (IRM) which produce FAP. More then one KS may be needed to trigger an FAP. Sensory receptor cells are critical in determining FAP type. Egg-rolling response of graylag goose shows FAP (Figure 4.2). Pheromone reception, through nasal sensory receptor cells may trigger a sexual response. "Frightening sound", through auditory sensory receptor cells may trigger a fight or flight response. Neural networks of these cells assist in integrating signal from many receptors to determine degree of KS and producing appropriate response. Several responses are determined by hormones. Specifically in vertebrates, this system is funneled through hypothalamus to anterior and posterior pituitary.

FIGURE 4.2 Egg-rolling response of graylag goose showing FAP.

4.1.1 FAP and Stereotype Behaviour

FAPs show stereotype behaviour. Scientific literature on laboratory macaques help to understand stereotype behaviour. Environmental factors like cage size, housing type, stress and boredom in laboratory animals are often cited as causes of stereotype behaviour. Monkeys experiencing such environmental conditions do not develop stereotypes. Animals with stereotype tend to accentuate the behaviour. Environmental factors appear to influence the frequency of established stereotype but do not cause it. Intrinsic factors influence the onset of stereotype behaviour. Rearing history, breed, genotype, and exposure influence psychology of animal. Animals separated from their mothers in early life are prone to develop stereotype behaviour. Different breeds of macaques respond differently to stress and situations including maternal separation. Personality contributes to occurrence of stereotype which signifies poor psychological condition and constitutes a welfare problem in animals. This behaviour is complex, not completely understood but an abnormal repetitive, invariant behaviour without clear goal or function (Mason 1991) and includes crib-biting and wind-sucking in horses. Stereotypes either oral or bizarre postures or prolonged locomotion is best seen in animal walk in distinct, unchanging pattern in cage called *pacing.* Walking range in speed from slow and deliberate to quick trotting involves few circuits or prolonged. Locomotion may be combined with head toss at corners of cage, or animal rearing onto its hind feet at some point in circuit. Stereotype often go through stages of development which differs from animal to animal. In early stages, this may be easily interrupted by loud noise or other stimulus. At later stage, such interruption is difficult as stimuli escalate performance of behaviour. Animal appear to be in a trance-like state, disconnected from its surroundings. Difference between stereotype and non-stereotype behaviour is seen in eating behaviour. An animal may eat in manner, sitting in same place, using a distinct motor pattern for prolonged time but this would not be stereotype as it has obvious function. Severe aggression or inadequate maternal behaviour is not stereotype as they are not repetitive and ritualized. Stereotype is evaluated in terms of frequency and duration done by examining the animal's time budget. There may less concern for an animal that spends up to 5% of its time in stereotype than one that spends 75% of its time in stereotype which is a matterof concern and is an indication of poor welfare.

Stereotype behaviour in non-human primates

Stereotypes of non-human primates are divided into deprivation—and cage stereotypes. Deprivation stereotypes are called *self-directed stereotypes* behaviour as behaviours are performed on individual's own body. Self orality means non-nutritive sucking on one or more of body parts (fingers, tail, or genitalia). Huddle, rocking, and crouch are postures of abnormal type. Scratching, biting, and head banging are self injurious. Deprivation stereotypes found in monkeys are separated from their mothers at birth or within the first year of life and raised in social isolation (Sackett et al., 1981). Such behaviours are analogous to normal behaviour seen in infant and juvenile monkeys and become self-directed in the absence of con-specific. Self orality in isolates, particularly self-sucking of tail or digits or tail, is analogous to infant's nipple sucking (http://www.awionline.org/lab_animals/biblio/at-phil.htm). Self clasp is related to clasping of the mother. Self abuse is the outlet of frustration in such animals. Repetitive locomotion stereotypes are the pacing, jumping in place, and somer-saulting. Cage stereotypes are perhaps the result of living in stressful environment. The stereotypes are active involving dynamicand whole body movements. Dancing the back and forth quadrupedal movement is distinct from pacing or spinning. Some behaviour is not included in either category. Paulk and others (1977) used the term *walk in circles* to describe an animal pacing completely around its cage and to denote an animal pacing up and down in one side of the cage. Berkson refers to pacing with no such distinction. The term *twirling* is defined differently by Paulk et al., (1977) as partially bipedal walking, than by Capitanio (1986) who refers it as stationary quadrupedal movement.

Stereotype behaviour in Laboratory primates

Locomotion stereotypes are found in many laboratory primates which are kept in laboratory. The effects of isolation rearing in a germ-free environment in 2 marmosets was described. Marmoseto, in contrast to macaques do not show deprivation stereotypes in such rearing (Berkson et al., 1966). Squirrel monkey shows deprivation and repetitive locomotion stereotypes when get separated from their mothers at birth and nursery reared in individual cages. Animals showing self-orality, rocking, huddling, and pacing-like stereotype which are termed looping and locomotion stereotype called rolling-in-a-ball. Macaques species namely, *M. fascicularis*, *M. mulatty*, *M. arctoides* and *M. fuscata* exhibit repetitive locomotion stereotypes (Capitanio's 1986). In such studies, animals were reared in part or total isolation.

4.1.2 Animal Welfare and Stereotypic Behaviour

Stereotypic behaviour is used as a marker of welfare. The UK Code of Practice for Housing and Care of Animals in Scientific Procedures (Home Office/ British Government, 1989) states that animals must be maintained in good health and physical state, behaving in a manner normal for species and strain. Stereotypic behaviour in the UK and US is considered as a symptom of poor welfare. The US Animal Welfare Act amendment of 1985 (US Department of Agriculture 1991) aimed to promote the psychological well-being of captive primates. The useful measure of psychological well-being remains elusive. Debate and controversy has surrounded what does and does not constitute psychological well-being of primates. The International Primatological Society in the year 1933 proposed that stereotypes should be used as a behavioural indicator of poor psychological well-being and considered stereotypes as serious enough to be a welfare problem. Mason (1991) concludes that stereotypic behaviour is indicative of inadequate and possibly aversive environment. Reduced responsiveness seen in sows exhibiting stereotypes is the indication of poor welfare (Broom 1987). Such comments are reflected in voices of campaigners of animal rights.

The development of stereotypic behaviour reflects an impairment of animal's capacity to interact with environment" (Wemelsfelder 1993). This may be the direct evidence of chronic suffering. In laboratory macaques stereotyped pacing may be the sign of severe stress (Fouts, R. et al., 1997). Stereotypes are used as markers of animal welfare within the scientific community. Factors underlying stereotype behaviour focussed on environmental causes. Stereotypic behaviours may develop in deficient environment (Paulk et al., 1977). Several aspects like housing cage size, stress, and environmental simplicity in captivity of laboratory primates have been investigated. In the past, need to sterilize laboratory caging has led to the use of small cages which were cleaned with conventional cage wash machines (http://www.awionline.org/lab_animals/biblio/at-phil.htm). Use of small cages has been implicated as cause of stereotype. Specification of minimum cage sizes per body weight of animal aims to promote species-specific behaviour (including vertical flight reaction) and to reduce abnormal behaviour. Size of enclosure is considered as one of 4 significant physical factors concerning the housing of non-human primates (International Primatological Society 1993). Type of housing is linked to stereotype of laboratory primates. The macaques in laboratory were housed individually to reduce the risk of wounds due to fighting in the past. Social housing in pairs or groups whenever possible was favoured over housing for individual. Primates generally live in wild in large groups. Individual housing, even with olfactory, visual and auditory contact with con-specific is akin to solitary confinement. Thus, individual housing may influence the development of stereotype. Stress in laboratory might contribute to the development of abnormal behaviours. Laboratory primates experience many stressors, chemical and physical restraint, blood draw, cage change, and potentially including injection. Inability to escape from such experiences might lead to adoption of stereotype to cope with an aversive environment. Boredom may cause stereotype. Primates are active, intelligent with demand for complex, stimulating environments. Primates in laboratory are at risk of being bored. Development of enrichment devices for primates in laboratories and effects of these interventions on stereotypic behaviour reflect this belief (Schapiro, and Bloomsmith 1994). Research into the causes of stereotypic behaviour has focused on cage size, as it is the most easily manipulated variable. Small cage size seems intuitively to be a likely candidate for causing harmful effects in laboratory primates. The oldest and most often cited studies on cage size are the work (Draper and Bernstein 1963), in which behaviour of 12 wild-born 3 year old rhesus monkeys was evaluated in 3 cage sizes. The largest cage size was more akin to a housing pen. Any stereotype behaviour seen in the smallest cage was significantly reduced in medium sized cage, and was not observed in largest cage. A female monkey exhibited continuous backward somersaults in small cage stereotyped pacing in medium cage, occasionally throwing up her forelegs and tossing her head back. This was interpreted as beginning of a somersault which was not completed. In the largest cage, this animal did not engaged in stereotype behaviour. Housing monkeys in small cages leads to the development of stereotype locomotion as a substitute for locomotion. When the spatial restriction was removed (http://www.awionline.org/lab_animals/biblio/at-phil.htm) normal locomotion resumed.

Paulk et al., (1977) found similar links between cage size and stereotype behaviour. In this experiment 24 wild and captive bred rhesus monkeys, were studied in small and large cage. Results indicate that in 8 monkeys showing stereotype, stereotypic locomotion increased in small cage while normal locomotion decreased, probably as a result of prolonged housing in small cage. This conclusion is drawn presumably on the length of time each monkey had spent in a small cage. Animals which exhibited stereotypes were older and had spent more time in small individual cages.

Crockett and others (1993, 1995) investigated the effect of cage size on urinary cortisol levels and behaviour in long-tail macaques. All the animals were wild-born, captured and imported as adults. Five cage sizes were used. After 48 hours, the most restrictive cage (Size 0) was slightly enlarged and the cage size was also larger for males than for females to allow animals to move freely. All cages used were substantially smaller than ones used in either of 2 experiments. All the cages are smaller than legal minimum cage size for animals of this weight in UK (89 cm x 89 cm x 110 cm for a 4–6 kg animal). After a 2-week preparation phase for habituatation to the experimental room, monkeys were moved from one cage size to another every 2 weeks until each monkey experienced each cage size. The first cage size was assigned at random (http://www.awionline.org/lab_animals/biblio/at-phil.htm), with the order which was the same for each monkey (0-1-4-2-3). Each monkey experienced both increased and reduced cage sizes. The experiment phase was continued for 10 weeks. Two and one half years later the monkeys were evaluated in a follow-up phase.

Urinary cortisol responses and behavioural data were collected. Elevation in urinary cortisol accepted as indicator of stress were unrelated to cage size and no significant elevation in cortisol levels were found in any cage size. Abnormal behaviour did not differ significantly across cage sizes. Variations in cage size have no stress effect as judged by urinary cortisol. Cage size was least important of all independent variables. Some other variables examined were also the phase of the experiment, sex and days in captivity. These results do not stand alone and agree with results found for pig-tail macaques (*M. nemestrina*) in a similar experiment (Crockett et al. 1993) and by Goosen (1988) in his exploration of cage size and stereotyped locomotion. Investigation on the differences in heart rate and behavioural patterns in relation to cage size in female rhesus monkeys showed that increased cage size had no significant effect on behaviour, heart rate, or activity levels (Line et al., 1990). Long-tail macaques taken from the wild as adults develop substantial levels of abnormal behaviour (Crockett et al., (1995), even when such animals were kept individually for more than 3 years in laboratory caging with minimum provision of social contact and enrichment devices. This indicates that causes of stereotypic behaviour are not as straightforward as previously described. There may be other complicating factors contributing to the development of stereotypic behaviour in laboratory macaques. Few areas of interest namely, rearing history, breed and genotype of animals, and differences in individual disposition are to be explored (Crockett et al., 1995).

Animals separated from their mothers at birth and raised in partial or total social isolation develop deprivations stereotypes. The animals used by Crockett et al., (1993, 1995), were wild-born animals which were captured and imported as adults failed to develop significant levels of stereotypic behaviour. In the heart rate study, female rhesus monkeys (Line et al., (1990) were reared in social housing about for 20 months of age. Of the 8 animals that showed stereotype (Paulk et al., 1977) in cage size experiment, six had been raised in a laboratory and separated from their mothers at 3 months of age. The animals in Draper and Bernstein (1963) cage size study were 3 years old at the time of experiment and had been in laboratory for years. They had been captured and separated from their mothers when they were less than one year old. The nursery-reared squirrel monkeys (Roy 1981) were separated from their mothers at one day after birth. The maternally reared controls in the experiment did not develop stereotypic behaviour. Thus, possibly the rearing histories of animals contribute to the development of stereotypic behaviour. The factors influencing stereotypic behaviour in primates in a zoo highlights effect of rearing (Marriner and Drickamer 1994). The animals which showed the highest level of stereotype were those that had been hand-reared by humans. Thus stereotypic behaviour is probably influenced by rearing history of the animal. Goosen in the year 1988 suggestss that infants be weaned only when they are 3 to 6 months old. He could not observed stereotypic locomotion in nursing infants but observed "prestereotypic" locomotion in weaned juveniles as agitated and non-functional but lacking the fixed appearance of true stereotypic pacing. Goosen observed this "prestereotypic" shortly after weaning in such animals. The author concludes that stereotypic locomotion is a response to early maternal separation in such animals. Wesseling (1988) at the Primate Centre TNO in Netherlands examines faeces smearing by laboratory macaques. This behaviour is considered abnormal. The behaviour of both wild-born and laboratory-born animals was examined. The captive-born animals had been weaned at 3 months and housed together in groups until 2 years of age at which time they were moved to individual housing. Most of the wild-born animals were older and had been captive and in individual housing longer than laboratory-born animals. Over half of laboratory-born animals showed this abnormal behaviour while none of wild-born animals did. The ten monkeys that had been born in a safari park and moved to centre when they were older than one year did not show abnormal behaviour. Rearing history influence whether or not that animal will show abnormal behaviour. When an animal stays long with his/her mother, (http://www.awionline.org/lab_animals/biblio/at-phil.htm) the animal less likely develop abnormal or stereotyped behaviour.

4.2 PHENOTYPIC PLASTICITY

When two instincts contradict each other, animals may resort to a displacement activity. A favourable trait like an instinct is selected through competition and improve survival of life forms those possess it. Baldwin offered "a new factor in evolution" termed phenotypic plasticity—ability of an organism to adjust to the environment during lifetime. Learning is obvious example of phenotypic plasticity. Baldwin pointed out that new factor explains punctuated equilibrium. Over time, this theory became Baldwin effect, which functions in two steps. First, phenotypic plasticity allows an individual to adjust to a partially successful mutation, which might otherwise be useless to an individual. If this mutation adds to inclusive fitness; it succeeds and proliferates in population. Phenotypic plasticity is very costly for an individual as learning requires time, energy and on occasion involves dangerous mistakes. In second step, evolution may find an inexorable mechanism to replace plastic mechanism. Thus, a behaviour that was once learned in time becomes instinctive. Criteria for instinctual behaviour are to (a) be automatic, (b) be irresistible, (c) occur at some point in development, (d) be triggered by some event in the environment, (e) occur in every member of species, (f) be unmodifiable, and (g) govern behaviour for which organism needs no training. Behaviours like hibernation, migration, nest building, mating are instincts. Psycho-analysts refer instinct to human motivational forces. Use of term motivational forces is replaced by instinctual drives. Instincts in humans are seen in instinctive reflexes. Babinski reflex, fanning of toes when foot is stroked is seen in babies and is indicative of development stages. A sense of fairness could be considered instinctual. Innate behaviour may be classified into seven types—irritability, tropism, kinesis, taxis, reflexes, instincts, and motivation. Taxis and kinesis belong to orientation behaviour.

4.3 ORIENTATION

Understanding the position and movement of a body in space require understanding of three main axes—longitudinal, transversal, and dorsoventral. Body can rotate about any of these three axes. Rotation about dorsoventral axis is rolling, transversal is pitching, and dorsoventral is yawing. Body can also move linearly; forward-backward on longitudinal axis, left-right on transversal axis, and up-down on longitudinal axis. Orientation accounts a number of factors under headings:

1. Resources including all factors for promoting well being.
2. Stress sources that are detrimental to well being.

Orientation is spatial adjustment in response to various stimuli. Orienting components are quite variable in comparison to stereotyped form of Fixed Action Pattern (FAP). Orientation in space depends upon external stimuli, which triggers and directs it during performance. Difference between orientation and FAP is best explained in retrieving egg back into the nest by rolling movement in graylag goose involving two movements, i.e. stretching of neck, placing underside of beak over egg and pulling back the neck and performing lateral balancing movements of neck, to prevent egg from slipping away. If one removes egg (external stimuli) while the bird is still pulling its back, animal continues retrieving movements, but lateral balancing movements cease. Egg rolling, then is a FAP that once initiated continues, whereas balancing movements of body are orienting responses, whose presence and course depends on external stimulus. Here, FAP and orienting components are going on simultaneously. Sometimes, both components occur one after other. After noticing a fly, a frog first turns towards it and line it up with its body axis, and only after this movement comes FAP (flicking out of tongue and catching the pray). All kinds of orientation are broadly put under kineses and taxes.

4.3.1 Kineses

The simplest form of spatial orientation is unlearned and movement of animal in response to undirected stimuli is called **kinesis**. Animal response is proportional to stimulus intensity, but it is independent of stimuli spatiality. Kineses can be of two types.

Klinokinesis

Here, rate of change of direction increases in proportion to increase in light intensity. *Dendrocoelum lacteum*, found in wet damp conditions, prefers darker places. If it is kept in diffused or dim light, it turns occasionally. When light intensity is increased, rate of turning of *Dendrocoelum* (Figure 4.3) also increases. This increase, helps to locate a darker place of animal preference. Protozoans usually swim in a cork-screw manner. In favourable location, random deviation from line of movement is increased keeping animal in favourable environment longer, as also seen, in higher crustacean and grazing mammals. *Paramoecium* exhibits klinokinetic response towards local concentration of CO_2 (Figure 4.4).

Orthokinesis

Here, speed of locomotion is related to stimulation intensity. Ammocoetes larvae of lamprey are found buried in bottom of lakes and ponds with their head pointing below. Increase in light intensity and their exposure to it causes an active swimming, with their heads pointing below, which helps them burrow again and move away from light. When wood lice are allowed to move about on gause, a portion with high humidity and another portion with low

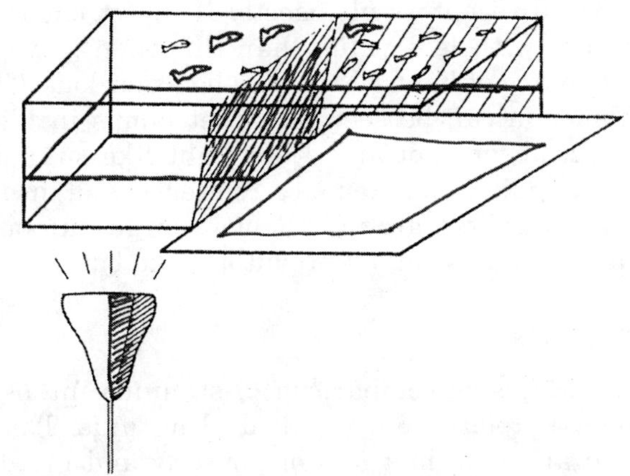

FIGURE 4.3 *Dendrocoelum* showing klinokinesis.

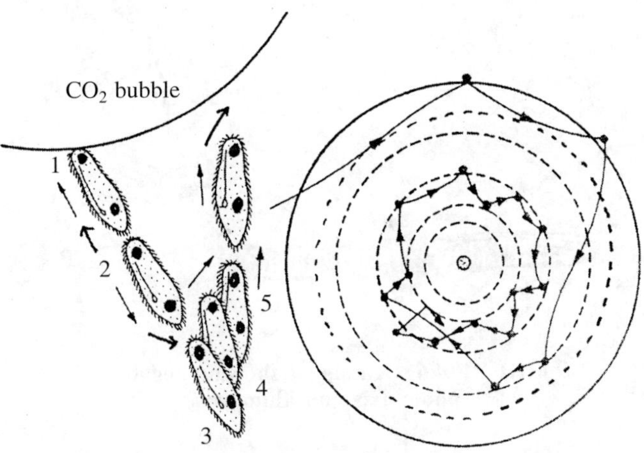

FIGURE 4.4 Klinokinesis of *Paramecium*. *Paramecium* with-draws sensing high CO_2 concentration, turns through a certain angle and advances again, but its new direction is not related to direction of stimulus.

humidity, ultimately a cluster of animals builds up on the damp part of gause (Figure 4.5).

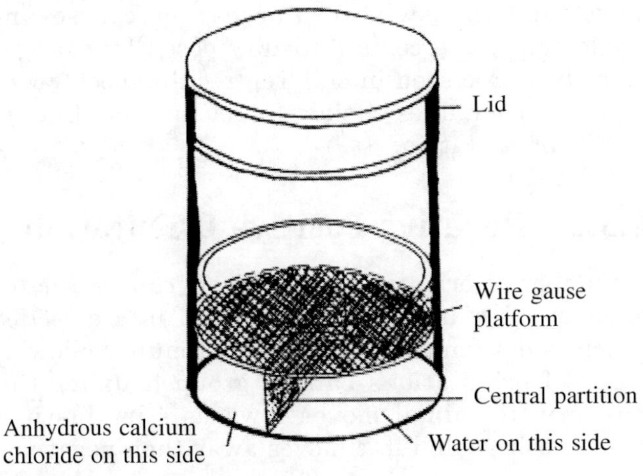

FIGURE 4.5 Experiment on orthokinesis of wood lice.

4.3.2 Taxes

This occurs with reference to direction of stimulus where spatiality is of significant importance. In Beach fleas (Amphipodia), if fleas are carried away from sea to a dry spot, they head straight back to water. Fleas, when placed in closed glass tubes and kept away from sea on dry beach, they always gathered in that part of glass tube which is closest to the sea. Along with moisture, position of sun is utilized by fleas to orient them towards sea. There are a variety of taxes depending upon relationship of stimuli and receptors.

Tropotaxis

When a stimulus is received simultaneously by two or more receptors, it can then compare and steer a course towards or away from stimulus. *Planaria* reacts with positive tropotaxis response to water current carrying scent of food. Animal aligns itself in water current, in such a way that both sides of its three cornered heads are simultaneously and equally stimulated, and then, it moves upstream. Tropotaxis is found in pill wood louse which has an active and inactive phase in its life. When it has fed enough, it moves away from light (negative phototaxis) under stones and fallen trees, where it leads an inactive life. When need of food arises it comes out in open (towards light: positive phototaxis) and enters into active phase of life. Photoreceptors play a role in controlling phases. Animals use paired receptors to achieve a balance in stimulus reception. If one receptor is destroyed, then animal is unable to reach this balance and keep on trying to achieve balance resulting in typical movements, like moving in circles, e.g. graywing butterfly escapes from enemy by flying towards sun. If it is blinded in one eye, it would keep on flying in circles. Door beetles use their right eye for orientation by solar compass in forenoon, and their left eye in afternoon. Biological significance of this type of orientation is still not known.

Telotaxis

Mechanisms keeping certain stimulus at a fixed place on a receptor, mostly eye, are called **telotaxis**. When a sprinter does a 100 metre dash at an athletic meet, we turn our eyes to get image of athelete on fovia centralis area of retina, where vision is sharpest. From then onwards, even though we turn our eyes, head or body, to follow movement, the image is kept locked on that specific location on retina. So in spite of image being fixed at a place, movement of sprinter is perceived and motion appreciated by our brain. This appreciation comes through motor organization involved in movement of eye, head or body. All messages come from different regions are integrated into continuous perception of movement. Simultaneous comparison of stimuli is not involved. It can be called a goal directed orientation. Hermit crabs with well developed eyes, when presented in dim and bright light sources, always move towards brighter one. Male graywing butterflies, after being blinded in one eye, would still fly towards passing females, showing goal directed orientation.

Menotaxis

Mechanisms for maintaining a constant direction by steering a course at constant angle to incidence of stimulus are called **menotaxis** (meno = I remain). Here, principle of negative feedback is involved, and output of regulating cycle acts as input. In light compass reaction, insects use sun for guidance in homing. Ants, *Lasius niger* use sun as a reference for homing when kept confined to a dark box for few hours while in middle of their homeward journey. After their release, they would maintain same angle to sun as one when they were kept in the box. They would not compensate for movement of sun during these hours which changed its angle. This would then lead them to a different path then one taking them home. An ant taken far away from its hill return along a straight line, that is when ant is displaced in open and is allowed to see sun, but if it is put in a box and then released, it will make a slight directional error. Error in angle of direction occurs corresponding to change in position of sun during time when the insect was in box. Ant, bees and locusts may be fooled with a mirror that reflects sun's rays (while actual sun is cut off by a shield of some kind). Then, insect go off in a direction opposite to which they would normally take. Ants can also compensate for this movement of sun and continue in original direction. Animals, like *Velia curens*, beetle—*Geotrupes sylvaticus* and honeybee—*Apis mellifera* show this ability. Central nervous system may at times, make it difficult to interpret regulating cycles in terms of negative feedback because of the choice of reference value which may be directed by central nervous system anytime. Importance of negative feedback can be demonstrated well, when requirements for its operation are lacking. Insects may face dangers, if they get close to a light. Requirement would be that light be at a great distance, so no harm is caused to insect. If light is at a great distance, then its rays will be parallel, and would subtend an angle of 90° or more to receptors of insects. If insect choose a reference, with less value than this then it would end at source of light which may be hazardous. This phenomenon can be seen at street lamps in night where in absence of any other light like sunlight, street lamp light is taken as a reference, and insects ultimately collect around them in large numbers, and are preyed upon, by predators, like bats.

Klinotaxis

When successive comparison of stimulus intensity is taken as reference it is called **klinotaxis**. This is seen in animals (maggots of housefly and blowfly; Figure 4.6), when they are confronted with gradients of chemical stimulation.

FIGURE 4.6 A maggot showing negative phototaxis and klinotaxis.

Mnemotaxis

Mnemotaxis is based on memory. Water shrew (*Neomys fodiens*) exhibits this orientation, when it follows the same path every time. If its memory map is disturbed by changing a part of its path, it will stop, and start exploring surroundings, it will then retrace its steps till it locates a known land mark, and then will resume earlier course and would try to overcome difficulty posed by changed path. It is also seen in children reciting poetry who repeat earlier lines if they forget any one line, in middle of recitation.

4.3.3 Negative-Positive Orientation

A number of orienting mechanisms can be related to their effect in turning an animal in a direction which is determined by external stimuli. Following types of orientations can be grouped under this category. If animal moves towards stimulus it is called positive and if it moves away then negative.

1. **Hydrotaxis:** Response to water,
2. **Thermotaxis:** Response to temperature,

3. **Phototaxis:** Response to light,
4. **Thig-motaxis:** Response to touch,
5. **Chemotaxis:** Response to chemicals,
6. **Galvanotaxis:** Response to electrical field,
7. **Geotaxis:** Response to gravity,
8. **Rheotaxis:** Response to water currents, and
9. **Geomagnetotaxis:** Response to earth's magnetic filed.

Many migratory fishes orient and navigate with the help of odour. Silver Salmon of Far East were caught in North America. In half of fishes, nose cavity was plugged with cotton, they were then released. Fishes without cotton plugs could follow correct path. Plugged nose fishes were lost in way. Bees and many other insects are very sensitive to odours and utilize them in orientation. Male gypsy moth responds to aromatic substances, secreted by glands of females in a concentration of a few molecules per cubicmetre. This enables males to locate females at distances up to 6–8 kilometres. We favour darkness to have a good sleep, and larva of house fly looks for a dark place to pupate. These orienting responses are negative phototaxes, as the stimulus (light) is avoided here. A grain beetle calendar granary has tendency to climb in a vertical plane, showing negative geotaxis. *Planaria*, when kept in a test tube of water at constant temperature and light, it aggregates at bottom of tube showing positive geotaxis. Orientation mechanisms have also been grouped according to their function.

4.3.4 Orientation Types Based on Function

Orientation in space

Stimulus is used as a reference to maintain normal position of an animal. An animal do not seek such stimuli goals. Examples are (1) maintaining normal body position against gravitational attraction, directional control of aquatic animals, control of body by optical and tactile perception, orientation related to light, compass and sun compass orientation, orientation with land marks, electrical and magnetic orientation, and orientation in air and water currents. (2) Stability of posture and movement by mechanical sense organs, like semicircular canal and with eyes. (3) Objects orientation, where the stimulus is the goal are optical target orientation (predation), echolocation, tactile location, and chemical orientation.

Orientation is a form of adaptive behaviour to patterns of existence of factors, like food, predation and physiographic conditions. Ability of organisms to minimize their distance from resources and maximize it from stress-sources is called **orientation fitness**. Orientation requires spatial information and sources have to be there, to cater to it. These information are grouped into four classes: (i) immediate environment, (ii) learned, (iii) innate, and (iv) random information. Immediate environment information is about direction and distances through sensory system. A predator would turn in direction of its prey after seeing it, and then approach it to capture or kill it. Learned information comes into play along with immediate information. After sensory information is received, animal has many options for orientation response that he can perform and selects one from this. This selection of proper response for given situation is directed by information already available to animal through memory or instinct. This information is either learned or inherited. A kangaroo, immediately after its birth, crawls only in one direction that would take it to its mother's pouch and not in any other direction. A young calf directs its head between horizontal and vertical axis, where it would find udders. Random information may result from ignorance. An organism would locate a resource or stress source through trial and error or random movements and would then plan its next orientational response.

Orientation is very relevant, to spatial nature of environment, and orientation behaviour in this context can be categorized into six types: (1) Positional-, (2) Object-, (3) Strato-, (4) Zonal-, (5) Topographic-, and (6) Geographic orientation.

Positional orientation

Most mobile animals maintain their position in space through this orientation with reference to gravity and substrate. Large animals maintain an upright position, whereas lighter animals rely more on the position of the substrate than gravity, like a fly walking on roof. In pigeons, less specialized gravity sense organs are located in intestine, along with main sense organs in head to maintain position in flight. Many insects, like locusts and dragon fly use dorsal light orientation to maintain flight position. Flies have special structure, halters, which are stick like organs that oscillate and any aberration is detected by small sensory organs at base of these halters. Many aquatic animals rely on direction of light to maintain their position along with the gravity. Larger aquatic animals prefer to keep their longitudinal body axis horizontal, while microscopic animals like *Daphnia*, and Nauplius larva keep their longitudinal axis vertical. Positional orientation also helps to ward-off displacement, especially for swimming and flying animals, which can be displaced by currents of water or air. Positional orientation also provides a means of ensuring placement of locomotor organs on the substrate. A kind of camouflage is also provided by such orientation. For example, aquatic animals have dark backs and light bellies, and hence, are difficult to locate against background, when observed from below or above.

Object orientation

Ability to find and approach resources and avoid stress sources is object orientation. Typically, this is two phased; search and approach. Initially searching begins in a larger area, which is called ranging, largely a straight movement, with occasional turns. If some sources are detected then searching occurs in smaller area and is called **local search**, which involves mainly turns with little straight movement. If it leads to a resource, then it is approached and if to a stress source, then it is avoided.

Strato orientation

Movement across vertical layers of a habitat is strato orientation. Importance of this movement is more for smaller forms, as it is easier for them with their small weight and also because they can not moves fast. It becomes taxing to move great distances horizontally to encounter variation in physical factors, whereas less vertical movement produces same changes. Sensory input for strato orientation is very easy because of the earth's gravity as reference, and can be sensed by all animals. Light may also be used as reference. Best example is vertical migration of aquatic zooplanktons. Every day scores of these animals move up for night and down for day, guided by direction and intensity of light and gravity. Internal biological clock also play active role in this, with apparent advantage of avoidance of predation in day and utilization of abundance of food in upper layer of water in night.

Zonal orientation

Factors like variations in ground level and degree of forestation lead to this orientation. Donax lies on marine beaches, where wave action provides plenty of organic material as food. Tides shift this habitat back and forth. In order to remain there, clam digs themselves out of sand and moves with outgoing tide, and dig in again at proper place to get optimum supply of food. *Talitrus saltator*, a crustacean prefers beaches with certain level of humidity. It goes down slope on dry sand and up slope on wet sand to find such a place. Such simple zonal orientation response may be supplemented by compass orientation, where some directional or celestial clues are utilized. Zonal orientation is also important to avoid stress sources. A frog would jump into water if some predator approaches. A rodent would jump into its burrow in such a situation.

Topographic orientation

When an animal retraces its path to point where it started from, is a form of topographical orientation. This complex orientation requires components, like distance orientation, directional and orientation by land marks as well as learned information. In some spiders and crabs, it is the last component which is involved in topographical orientation. They somehow seem to know the length of its outward journey, and at end of it turn 180° and travel same distance. Bees, ants, and wasps use all components, in their topographical orientation. Bee measures distance with reference to energy spent to make movement. They use compass orientation to maintain proper course to their nest or feeding sites, and use landmark and this comes handy, especially when the sky is overcast.

Geographic orientation

Furnarius rufus is the most common bird species in Brazil. The relationship between nesting behaviour and environmental characteristics is often observed in bird species from regions with stressing seasonal climatic changes, such as temperate species. This may not be the case for tropical birds, where climate is usually less severe. According to anecdotal information for Argentinean populations of *F. rufus*, nest opening has a south-southwestern orientation and such pattern is in accordance with local wind direction. Usually, local climate conditions, mainly wind and rainfall, lead a species to build their nests in a specific direction. In species with wide geographical distribution, and experience of climatic gradients throughout their range, a geographical component in nesting behaviour is expected as could be found throughout *F. rufus* geographical distribution.

4.3.5 Irritability

Nervous system is concerned with universal property of life called **irritability**. Capacity of cells and whole organisms to respond in a characteristic fashion to changes in the environment are called **stimuli**. Specific reaction elicited by an external or internal stimulus is termed as response. Response generally yields an adjustment that promotes well being of entity. Stimulus response reactions are usually rapid and afford a continuous mechanism for maintaining internal constancy in face of environmental change. Complex repertoire of often subtle voluntary responses are associated with irritability in primates that affords a far greater range of adjustments to environment than is the case with tropisms, which are stereotyped and limited types of behaviour.

4.3.6 Motivation

The term motivation refers to, generally reversible and often short-term changes in the behaviour. At the same time, an animal may be hungry and hot, or need to sleep, and a mate. Focusing on one activity at a time results in success, than to achieve simultaneously multiple, possibly conflicting goals. In analyses of behavioural choices, some outcomes are obvious. Grooming is often given a low priority than other behaviours, and occur during inactivity. Making decisions for compelling drives like foraging and mating is less understood, although mating and parenting trump other activities. During mating season, or nurturing young, an adult deplete their nutritional reserves. Research on the neuroscience of competing behavioural needs would provide insight into mechanisms animals use to rank their activities. By controlling mating, herds and flocks of useful animals were domesticated which behave differently than their wild progenitors. Selective breeding was importantly in human history. That both genes and environment influence behaviour, and studying behaviour focusses on interaction between these 2 factors. Genes influence morphology and physiology, build a framework within which the environment acts to shape the behaviour. Environment could affect morphological and physiological development. Thus behaviour develops as a result of that animal's shape and internal workings. Genes create the scaffold for learning, memory, and cognition that allow acquiring and storing information about animal's environment for shaping their behaviour.

As an animal changes from feeding to drinking and then back again to feeding, all within the course of a few hours that we refer to as being, due to changes in motivation. Motivational changes are often not just animal's response to one specific stimulus that fluctuates, but a whole range of responses that are functionally related to one another. An animal's threshold of response to all stimuli connected with food and feeding behaviour rise and fall together along with those connected with sexual behaviour, and so on. Because of this association of effects with different sets of stimuli, workers have talked about "specific motivational states". Existence of a set of internal causal factors implies to affect not just the behaviour, but the whole functional groups. Many patterns of instinctive behaviour can be interpreted in terms of drive towards a goal. Attainment of goal results in reduction of drive and this phase is known as **satiation**. A specific motivation is called **drive**. For example, a dog deprived of food (goal) for a long period may exhibit a high feeding drive. Dog moves about restlessly and is highly responsive to smell or sight of food. Thus, motivated behaviour is a drive that leads to goal directed behaviour and satiation. Lorenz's hydraulic model (Figure 4.7) of motivation and Tinbergen's hierarchial model (Figure 4.8) are mentionable in this connection. In mammals, excitatory mechanisms present in the hypo-thalamus contribute to arousal of motivated behaviour. Inhibitory mechanism reduces motivated

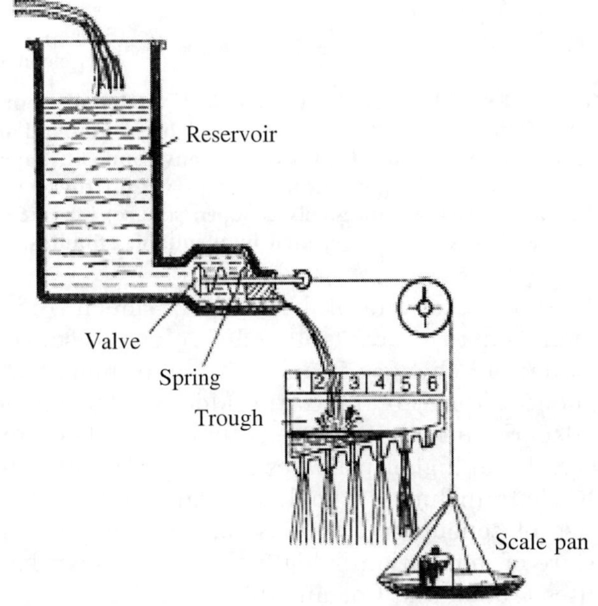

FIGURE 4.7 Lorenz's hydraulic model of motivation. Water represents the action specific energy accumulates, when behaviour is not expressed. Behaviour occurs, when water passes out of the reservoir (representing animal's drive level). Higher threshold aspects of behaviour are numbered as 4, 5 and 6.

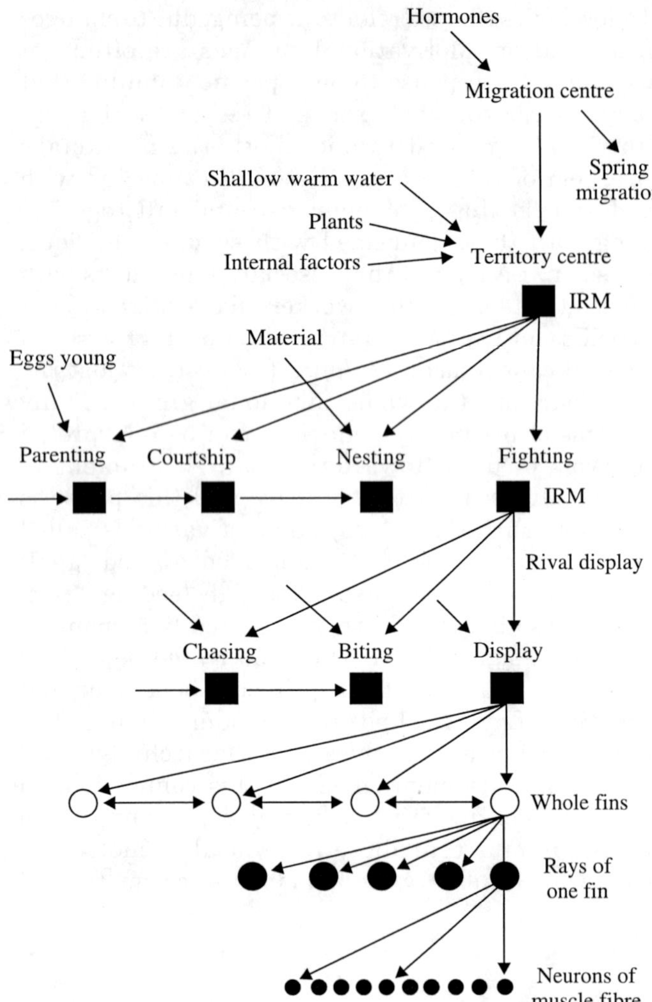

FIGURE 4.8 Tinbergen's hierarchical model of reproductive instinct in male three-spined stickleback. Motivational impulses (arrows) come from external environment, superordinate centres or spontaneously from within the centre itself. Innate releasing mechanisms (open square) inhabit the discharge of centres until released by a suitable stimulus.

behaviour. Action, of both these mechanisms is believed to be controlled by relevant sensory stimulation, changes in internal environment and influences from cerebral cortex. Motivated behaviour is also regulated to a great extent by learning, especially in higher primates and man. A motivated behaviour includes a goal searching phase and a phase of consummating acts—a quiescent period. Nature of motivation is highlighted in the works of Toates (1986), and Colgan (1989).

4.3.7 Tropism

Loeb developed a new theory of animal behaviour called tropism or involuntary forced movement. Animals are structured to react to certain kind of energy—mechanical and chemical. Loeb found that an animal's response is a direct and autonomic function of a reaction to a stimulus. In other words, behaviour is said to be forced by a stimulus. It does not require any explanation in terms of an animal consciousness. Loeb's work was viewed as an objective and mechanistic approach to animal psychology. He argued that consciousness among the animals was due to association of memory, which is when an animal learns to act to a desired stimuli in a certain or desired way. He used examples of jellyfish, starfish and worms, and then he extended the results to higher forms of life.

4.3.8 Reflex

Neuron or nerve cells are considered as anatomical unit of nervous system. Reflexes are regarded as functional unit of this system. A reflex is an unvarying and automatic response to a specific stimulus. Probably the simplest innate behaviour among animals is possession of a nervous system along with its ability to exhibit reflex action. Reflexes are of two types—tonic reflexes and phasic reflexes. Tonic reflexes are relatively slow, long lasting adjustment that maintains muscular tone, posture and equilibrium. This is also known as **stretch reflex**. Phasic reflexes are rapid, short lived adjustments, such as that seen in sudden withdrawal of hand form a hot object. This is also called **flexion reflex**. Reflexes can be classified in several ways. Generally, they are classified according to the level of nervous systems involved. Accordingly reflexes are: spinal, medullary and cerebellar types. Reflexes which involve skin, cornea, etc. are superficial reflexes, those which depend on proprioceptive input for muscles are called myotatic reflexes. Visceral reflexes are known to concern with contraction and dilation of pupil, speeding and slowing beating of heart. Other types of reflexes are: exteroreceptor reflex which is due to stimulation of receptor on outer surface of body, entero-receptor reflex is due to stimulation of receptors of internal organs, conditional reflexes and unconditional reflexes. Sherrington (1906) considered the way in which reflexes operate, and how the central nervous system integrates them into adaptive behaviour.

Classical conditioning

It requires an unconditional reflex, where an unconditional stimulus brings about an automatic unlearned response. If a neutral stimulus tends to precede it, an association is made, and conditional response transfer into conditional stimulus, and a conditional reflex is learned. Pavlov (1941) presented food to dogs and measured their salivary response. Then, he began ringing a bell just before presenting food. At first, dogs did not salivate until food was presented. After a while, dogs salivate when sound of bell was presented (Figures 4.9 and 4.10). Sound of bell became equivalent to presentation of food. Stimuli, on which animals react without training are called **unconditional stimulus**. Physiological basis of conditional response lies in neuronal activity, in auditory areas of brain, to motor neurons controlling salivation transfer by appropriate neurons. This involves development and/or strengthening of neural circuits which are characteristics of all forms of learning. Conditional reflex is an excellent tool for determining sensory capabilities of animals. Honeybees can be conditioned to seek food on a piece of blue cardboard. By offering other colours to a blue conditioned bee, Frisch found that honeybees can discriminate between yellow-green, blue-green, blue-violet and ultraviolet.

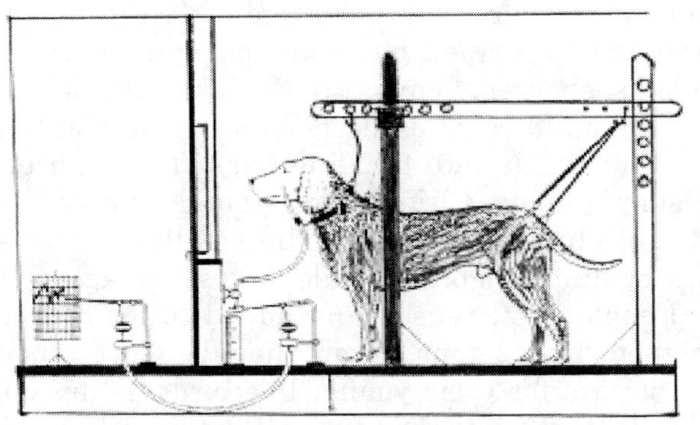

FIGURE 4.9 Salivation experiment in dog.

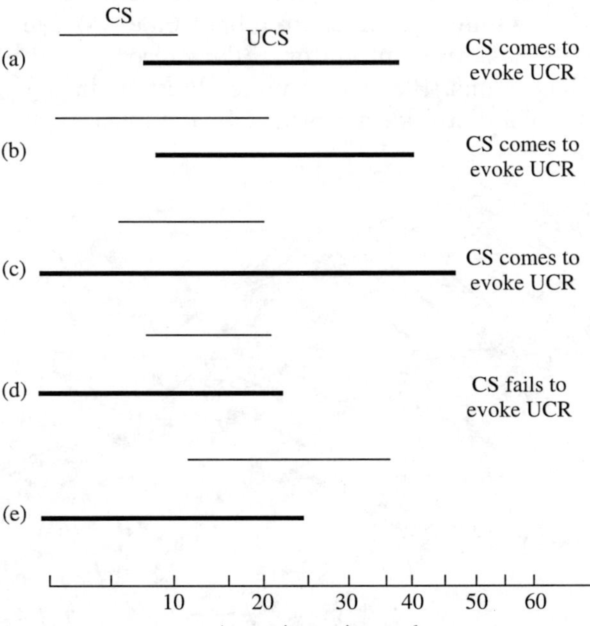

FIGURE 4.10 Conditioned reflex. Thin line denotes the duration of CS, thick line denotes UCS. The results are given on the right. CS must not end with or persist beyond the UCS if a positive conditioned reflex is to be established.

Operant conditioning

Operant conditioning forms an association between behaviour and a consequence, also called response stimulus conditioning because it forms an association between animal response and stimulus that follows consequences. This conditioning also called instrumental conditioning was first extensively studied by Thorndike (1898, 1911). He investigated the behaviour of a cat trying to escape from various home-made puzzle boxes. When first constrained in boxes, cat took a long time to escape, with experience, ineffective response occurred less frequently and successful responses more quickly enabling the cat to escape in less time, over successive trials. Skinner extended theories proposed by Thorndike. His ideas of animals operating on environment leads him to analyze behaviour change and its consequences. Skinner box consisted of a bar on wall that when pressed triggered release of food pellet (Figure 4.11). Skinner believed that by rewarding an animal when an appropriate action occurs, it increases likelihood of repetition of that behaviour. When rats, placed in a box accidentally tapped on the bar on wall, a pellet was released, Skinner observed amount of time the rat took to find the pellet. As rat learned that a pellet released each time, the rat happened to step on bar. Rat learned to press bar, and to immediately find food. This training is operant conditioning. Where classical conditioning illustrates $S \rightarrow R$ learning, operant conditioning is often viewed as $R \rightarrow S$ learning, since there are consequences that follow response that influences, whether response is likely or unlikely to occur again. Through operant

FIGURE 4.11 Instrumental conditioning.

conditioning, voluntary responses are learned. Five basic factors in operant conditioning are as follows:

1. **Positive reinforcement:** Reinforcement always indicates a process that strengthens behaviour. In this case, a positive or pleasant stimulus is used in process and reinforcement is added. In such reinforcement, a positive reinforce is added after a response and increases frequency of response.
2. **Negative reinforcement:** A negative stimulus is used in the process and the reinforcer is subtracted. After the response, removal of negative reinforcer increases frequencies of response. Two types of negative reinforcements are escape and avoidance. A learner must first learn to escape before he or she learns to avoid.
3. **Response cost:** If positive reinforcement strengthens a response by adding a positive stimulus, then response cost has to weaken behaviour by subtracting a positive stimulus. After response, removal of positive reinforcer weakens frequency of response.
4. **Punishment:** If negative reinforcement strengthens a behaviour by subtracting a negative stimulus, then punishment has to weaken behaviour by adding a negative stimulus. After a response, an added negative stimulus weakens frequency of response.
5. **Extinction:** No longer reinforce a previously reinforced response results in weakening of frequency of response. If reinforcement fails to occur after a behaviour that has been reinforcing in past, the behaviour might extinguish. This process is called **extinction**. A variable ratio, schedule of reinforcement makes the behaviour less vulnerable to extinction.

Most of reflexes that maintain posture are two neuron reflexes. Those that are involved in withdrawal of a body part from a painful or noxious situation usually require interpolation of an interneuron in spinal cord. They are three neuron reflexes. Simple reflex arc is named for physical path actually taken by impulse as it moves from receptor of afferent neuron to effectors of motor neuron. Reflexes are mechanisms for maintaining appropriate posture, regulating blood pressure, and orienting body to environmental conditions that threaten organism. Loss of certain extensor or flexor reflexes may be used clinically to assess damage to central nervous system.

4.3.9 Nest Building: An Innate Behaviour

Many people believe that nests are animal houses, but most nests are only temporary shelters built to hold young. Nests are constructed out of everything, form bubbles to spider webs. They provide a safe area where eggs incubate and youngs are guarded from predators. Birds are most well known nest builders (Figure 4.12). Their nests can be just a scraped location on ground like those made by auks, or they can be tall stacks of sticks like those built by rufousfronted thorn birds. Most birds build familiar cup shaped nests out of plant materials, such as grasses, leaves and sticks. Orioles and weavers build complex nests by weaving plant and grass materials into hanging bags. They can even tie simple knots. Cliff swallows build mud nests on sides of cliffs or dwelling. The smallest nests are built by hummingbirds, while largest by eagles. One bald eagle nest, reused and added to for 30 years weighed over 2 tons. Even though, most reptiles do not care for their young, like birds, many build nests to protect their eggs. Alligators pile rotting grass, vegetation and mud into mounds as large as 6 feet across. They bury between 20–70 eggs inside mound, while eggs incubate (about 65 days), female guards nest from predators, like raccoons, skunks and opossums. Black rat snakes bury their eggs in nests of mulch-like materials. Sea turtles excavate

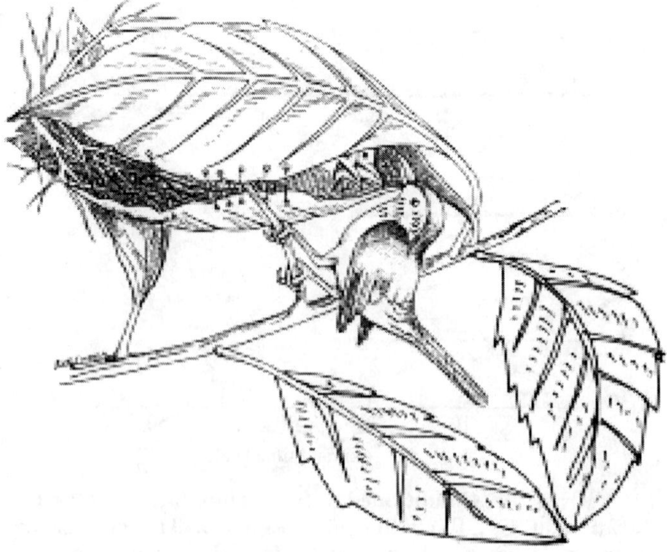

FIGURE 4.12 Nest building in birds.

pits in beach sand and bury their eggs. Some fish are also nest builders. Siamese fighting fish males make bubble nests, when female lays her eggs. Male catches them in his mouth and spits them into bubbles. Sunfish males clean out a dinner plate sized area for females to lay their eggs. They then protect eggs until they hatch.

Insects also create nests. In spring, paper wasp queens emerge from their winter burrows. They chew wood and plant fibers with special mouthparts to make a paper like substance. They mold paper into hollow tubes and hang them from trees, buildings or other surfaces. After several tubes are constructed, they lay their eggs inside. Queen feeds young in larval stage. When they emerge as adults, they become workers and help the queen to build a large nest, and raise youngs. Yellow jackets build spectacular round nests that sized of a basketball over summer. Majority of birds builds some sort of nest or at least chooses a nest site to improve reproductive success. Nests and their location function as protection against sun, rain, wind, heat, cold, and predators. It also protects the offspring. In most birds, even if sexes differ in some behavioural or morphological traits, both sexes together engage in nest building, incubation and feeding of nestlings. Although, a nest and raising offspring is crucial for offspring survival, it is difficult to determine the contribution of each sex to offspring survival with regard to nest building. In many weaver-birds, males build nest frame without the participation of females (Emlen 1973). These nest frames get accepted by females, who line them, and pad inside with soft material. Contribution to offspring survival is clearly divided between sexes. In polygynous weaverbird species, with described division of labour between sexes, male reproductive success is determined by plumage or courtship behaviour, and ability to build many high quality nests. Nests can influence male reproductive success in two ways. First, males with more nests built during a season can attract more females. Second, females could choose males based on their nests, and evaluate the quality of nests.

Red Bishops males built nest frame, which get accepted by females, whereas females incubate eggs and raise offspring without male assistance as is typical for polygynous weaverbird species. Male reproductive success is mainly determined by total number of nests built within a breeding season, which in turn depends on duration of territory tenure, and number of nests built per week. Nuclei from all major forebrain subdivisions are perhaps involved in processing sensor and motor information of many complex behavioural pattern. Nest-building behaviour in weaverbird is a complex sequential motor pattern and needs to be learned in first two years of life. It is assumed that neural control of nest building is as complex as neural control of singing, involving several different regions in forebrain. Regions involved in control of nest-building behaviours are identified. Wood ducks typically nest in tree cavities created by wood-peckers or by limb breakage (Sherman 2001).

Study of nest building behaviour in higher apes is based on roughly 50 years of field research and about 200 years of observation. In 1929, chimpanzee, gorillas, and orangutans were systematically studied. In them, nesting behaviour illustrates appearance, and phylogenetic development of dependence of self-adjustment to increasing dependence of manipulation, or modification of environments as a method of behavioural adaptation. Nest building is to a great extent learned behaviour. Earlier zoologists considered it merely motor programmed instinctive behaviour.

Construction, form and nest types

Higher apes are nomads. In search of food, animals daily wander over a more or less clearly defined home range, and rest at "customary used nesting places". Every night is spent in different places. This is source of a routine for nest building behaviour. All three species of higher apes from a certain ages (3–4), every night build themselves at least one new nest. On average, this means about 10–15 thousand nests over the life of an ape, a virtual tower of 11 to 16 times the height of Eiffel tower in Paris. Construction process is rather stereotyped. On the trees, the animal stands or squats on two legs, with its arms it pulls about three thick branches towards its body, bends them, presses them down under his feet and weaves them into stable round platform of about 60–80 cm in diameter. Then thinner branches and twigs are woven into a wreath. Finally, the platform is cushioned, with twigs broken off and with plucked leaves. Uneven spots are levelled by knocking with back of hand. Whole procedure lasts about one to five minutes, depending on animal abilities. At the end, constructor lies comfortably down in finished nest, and falls asleep. High up in crown of tree, it safely passes the night, until the first light of next morning.

There are ground nests, which may take the form of grass, foliage or twig nests close to ground or other construction in underwood bushes or bamboo groves, reaching heights of 2–4 m. Former in general are simply made by heaping up materials in a circular form, whereas later is stable construction standing vertically in space. In bamboo groves, heavy animal hangs on bamboo stalks by his arms, bends them down and weaves them essentially, in standing position into a stable framework. After checking his work, animals climb up and lie down for sleep or rest.

Distinction of tree and ground nests

Differentiation between tree and ground nests is extremely important because these two environments are entirely different in regard to animal's movement. In crowns of trees, apes move vertically by climbing and horizontally by swinging (arboreal locomotion), in contrast with terrestrial locomotion. Animals use stable surface to move on four to two legs in the latter case, like man and have a quite different access to plant, etc. than in vertically structured environment. The hands freed from use in locomotion. In case of tree nest, some strong horizontal branches are enough to support platform. Stability is provided by physical condition of trees. Variation is limited by a limited situation. In contrast to this, materials offered on ground are of much greater variety and this variety is accessible. Ground nests are, therefore, more differentiated in terms of construction and form. They vary from heaped types on ground to woven structure on bushed, and to stable and vertically standing structures. This typology can be interpreted in terms of an elementary evolution. Bamboo tower certainly resembles human structure. Its stalks are joined by weaving and tying knot, like slings and it shows tectonic qualities, e.g. standing vertically stable in space. Their foundations are naturally provided by rooted stalks.

Protection: Nest primarily gives security to a protected and stable place during nocturnal phase of apes' daily life. In this phase, animals are not fit for locomotion. In dark, three dimensional visions of higher apes are practically inoperative. With their platform, animals not only protect themselves against predators, perhaps, more importantly, they bridge darker half of their existence during which they are not adapted to their environment. They also satisfy the physical need to their great bodies for recreation in a horizontal position.

Home: Van Lawick-Goodall has described nests of chimpanzees as "sickbeds". Animals shot by hunters; use their last energies to build their last nest, which prevents them from falling. Nest is the last place of refuge.

Mother-child nest: Nest of mother and child develop mother and child relation. Apes spend about two to three years in nests of their mother. During this time, the child gradually learns to build a nest. Long close contact to mother's bodily warmth contributes to a sense of safety in the nest.

Settlement: Nests of a group form an elementary settlement in which individual characters and social relationships are expressed by the relative distance and location of the nests.

Instinctive behaviours may be of two types—closed instinct and open instinct. Closed instincts are programmed fixed motor patterns that are functional from the moment the neural circuitry is in place and are not modified by the environment. Closed instincts are quite common in courtship signals, where there is a premium on clear cut messages and benefit associated with modification of the message. For example, a mature male spider that locates a receptive female, mechanically performs a series of behaviour pattern in response to sign stimuli. If he is fortunate, copulation follows (Figures 4.13 and 4.14). Open instinct is seen in behaviour of many species that is functional, when first performance is capable of modification as a result of interaction with the environment.

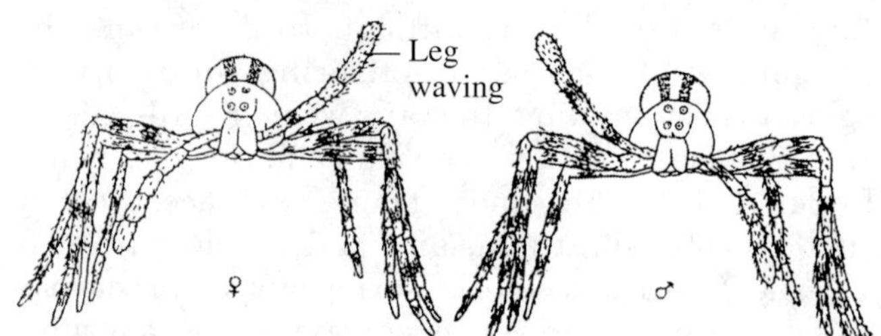

FIGURE 4.13 Courtship signal in a male spider (*Lycosa*) before approaching the female.

Orientation: Durability of nest artifacts is remarkably longer than factual use, which in general is only one night. Nests, left behind by animals last two months on average. Unfortunately, their semantic function in system of ape's orientation has not yet been studied.

Problems with instinctive behaviour

Genes are inherited, but behavioural patterns per se are not inherited. Genes do affect behaviour, but all animals develop within some sort of environment (e.g. an animal's mother provides an environment before birth). 'Sameness' of behaviour between members of same species does not exclude possibility that all members of a particular species

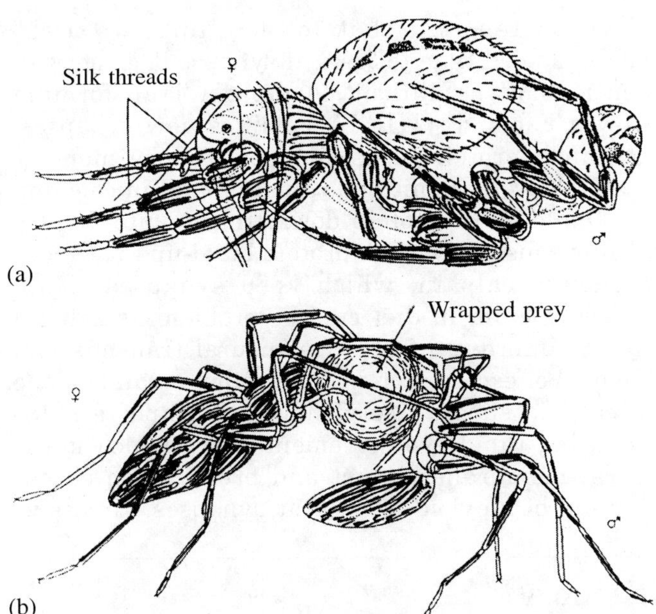

FIGURE 4.14 (a) Male *Xysticus* fastened his male to the grounds with silk threads for depositing sperm. (b) Male *Pisaura* presenting wrapped prey to the female prior to mating.

share common learning experiences. Deprivation experiments (in which animals are raised in social isolation, to remove environmental influences) have been criticized because "deprived" environment is still an environment and learning can still take place in the deprived environment. It is logically impossible to test, whether behaviour would develop the same in all environments. Learning is a process that changes pre-existing behaviour.

4.4 LEARNED BEHAVIOUR

Behaviour that changes in response to input from an animal's social and physical environment is called learned. Learning is especially adaptive in changing environment and unpredictable. It can modify innate behaviour for making it more appropriate. Learning is characterized by persistent and measurable changes in behaviour, which are not associated with fatigue, altered motivation or maturation. Genetic make up and physical structure of an animal body determines what kinds of behaviours are possible. An animal can learn to do only that is physically capable of doing. A dolphin can not lean to ride a bicycle because it has no legs to work pedals and no fingers to grip handle bars. Learning is thus, acquisition of information that makes this possible, and memory is retention and storage of that information. These two are closely related and should be considered together. Learning in insects is well-documented (Quinn et al., 1974, Lee and Bernays 1990, Turlings et al., 1993, Dukas 1999). Dukas and Bernays (2000) also examined the potential fitness-related benefits in grasshopper, *Schistocerca americana.*

Memory: From a physiologic view point, memory is divided into explicit and implicit forms. Explicit memory, also called **declarative** or **recognition memory** is associated with consciousness or at least awareness, and is dependent for its retention on hippocampus and other parts of medial and temporal lobes of brain. It is divided into episodic and semantic memory. Episodic memory refers to the ability to explicitly recall information about a specific event occurred at a specific time and place. If something significant occurs at any particular moment, such as life threatening event; it may be encapsulated as episodic memory. Semantic memory refers to the ability to consciously recall knowledge of facts that are independent of a specific time and place. Such knowledge is usually learned on many occasions, and thus, not tied to any specific time in one's life. Implicit memory does not involve awareness in act of recollection and is also called **non-declaration memory**. Explicit memory can be non-associative, associative, skills and habits, priming. In associative learning, the organism learns about relation of one stimulus to another. In non-associative learning, an organism learns about a single stimulus. Skill and habits once acquired became unconscious and automatic. It also includes priming, which is facilitation of recognition of words or objects by prior exposure to them. Explicit memory and many forms of implicit memory involve short-term and long-term memories. Various aspects of memory have been worked out by Gazzinga (2000), Squire et al., (1987, 1972).

Short-term memory: This allows recalling within several seconds to a minute. Its strength depends primarily on attention and not rehearsal. It is highly vulnerable to disruption, when attention shifts elsewhere. Amount of information held is on order of 4–6 numbers, which can be boosted by grouping then into distinct chunks. Ability to recall such information is contingent on transient patterns of neuronal activity in regions of frontal and parietal lobe.

Long-term memory: Such memory lasts hours to months, depend on a transfer of information form short-term, using repeated rehearsal. Hippocampus is an essential structure in such routine. Sleep improved consolidation of information possible by hippocampal replaying of activity form previous wake period.

Memory mechanism: Gazzaniga et al., (1998) reported that memory occurs through nervous system. Every thought is "felt", throughout the body as receptors for chemicals in brain are found throughout the body. When chemicals are activated across synapses, message is communicated to every part by chemotaxis that allows cells to communicate using blood and cerebrospinal fluid. Body sometimes buries painful memories in muscle tissues. Upon recurring deep tissue massage, muscles are stimulated and memories reactivate causing a person to experience repressed emotions. Electro-chemical messages come down to brain cells. An electrical impulse travels down axon or "outgoing branch" ultimately, release neurotransmitters. Dendrites or "incoming branches" or other neurons pick these up.

It is reported that for learning to "stick" synapses need time to "gel". If synapse does not gel, then memory is difficult. A protein Transforming Growth Factor-B (TGF-B) solidifies new synapses. Too much protein clogs synapse, reduce memory recall. Neurotransmitters, calpain found in calcium, keep build-up or protein down. Inadequate dietary calcium causes too much protein build-up in absence of enough calpain, which keep synapses clean. Excess calcium in diet creates problem as calpain start to interfere with proper neural transmission. Removal of excess protein from synapse can be done by electric shock. Acetylcholine is important for activating rapid eye movement sleep, which keeps neural membranes intact and break down excess build-up of amyloid protein at synapses. Excessive

Anatomy of Memory

FATIK B. MANDAL

Memory is a biological storehouse of information, events, and images. It functions through the nervous system, especially the hippocampus, thalamus, and amygdale. Biomolecules like proteins, hormones, and neurotransmitters are associated with the memory.

Thalamus, a structure situated in the brain, receives sensory inputs, and sends these inputs to cerebral hemispheres and other parts of the brain. Thalamus is connected with almost all the sensory pathways engraved with memories. The amygdala, a part of the limbic system, is also associated with the functioning of emotional memory.

Memory is categorized into short-term and long-term memory. Short-term memory lasts for 60 seconds, or more depending on attention, and not rehersal, and is vulnerable to disruption during shifting of attention.

Long-term memory lasts up to months, requires long term potentiation (LTP) with the involvement of N-methyl-D-aspartate (NMDA). LTP is the tendency of nerve cells, with exposure to a quickly repeated stimulus, to respond strongly long after the intense exposure. An excited neuron releases a neurotransmitter called glutamate which alters the electrical signal of the axon terminals of the neuron. Glutamate is taken up by the NMDA receptor of the dendrite of another neuron. This results in the production of protein called kinases, which increase the passage rate of electrical signals. Neuron which receives the message probably produces nitric oxide. Nitric oxide stimulates farther production of glutamate by the first neuron. Through such mechanism along with the formatting of new synapse, the memory remains preserved. The protein, Transforming Growth Factor-b, is produced under the influence of calcium that solidifies new synapse.

Memory is also classified as explicit, or declarative memory, and implicit, or non-declaration memory. Declarative memory is believed to be stored in the cerebral cortex, and is again divided into episodic and semantic memory. Episodic memory is the ability to recall explicitly information about a definite event at a definite time and place. Semantic memory is the ability to consciously recall information that is independent of a definite time and place. Implicit memory does not involve consciousness during the process of recollection.

Cortisol, a hormone released by adrenal cortex, interferes with verbal declarative memory. A recent study published in the *Journal of Neuroscience* has reported a "memory molecule" in brain called "CamKII".

Source: http://www.associatedcontent.com/pop_print.html?content_type=article&content_type_id=6

stress and obesity produce an over-production of glucocorticoids. Over exposure to glucocorticoids, damage and destroys neurons in brain hippocampus. Exercise burn-off excess cortisol. Failing to store information properly or lack of enough emotion or personal importance connected to information or for transmatic information forgetfulness occurs.

Effect of cortisol on memory: It has been suspected for some time that glucocorticoids released from adrenal cortex during stress have adverse effects on cognitive functions, like learning and memory. High level of stress hormone, cortisol is known to interfere with verbal declarative memory. Results show that high steroid levels disrupt this memory task. Effect is not permanent. Probably the effect of cortisol on memory is due to reduced ACTH, caused by feedback of cortisol to hypothalamus. An experiment by de Quervain et al., (1998) makes this explanation unlikely. This experiment involved effects of stress on rats' ability to remember position of a submerged platform in a water maze.

Recent findings: An interesting article describes that there appears to be a "replay" mechanism, where higher brain (neocortex) can query the part of brain responsible for short-term memory (hippocampus). In other words, there is a "dialogue between hippocampus; where initial memories of day's events are formed, and neocortex; the sheet or neurons on outer surface of brain that mediates conscious thought and contains long-term memories." A previously unknown process of DNA methylation can occur in neurons of adult brain, in response to life experiences allowing memory formation based on learned behaviour. Recent research has shown that our memories are not holistically spread across the brain, but may be stored in very specific locations. A new study, published in the *Journal of Neuroscience*, has identified a "memory molecule" in brain called "**CamKll**". It turns out in area of brain known as hippocampus. One can artificially create and then biochemically erase short-term memories, using CamKll protein molecules. Learning methods are habituation, sensitization, imprinting, insight, latent earning, and reasoning.

4.4.1 Habituation

This is a common form of simple learning and may be defined as a decline in response, to a continuous repetition of stimulus. Animal learns not to respond to stimuli without significance. For example, young birds learn that a butterfly is something to eat and not to be afraid of. Various birds are preyed upon by hawks. Tinbergen has shown that these birds will flee, if a hawk silhouette is displayed overhead. Silhouettes of other shapes do not evoke escape behaviour (Figure 4.15). Characteristics that are important in the bird's identification of hawks are the wings shape, the long tail and short neck. Models without these characters do not elicit escape. The same model may be interpreted in different ways, depending upon its direction of movement. The model pictured in Figure 4.16 will stimulate escape behaviour, if it is moving in the direction of the short end of the body, it does not do so if it moves in a reverse direction. Ability to habituate prevents an animal, from wasting its energy and attention, on irrelevant stimulus. However, behaviour is permanent and no response is ever shown, to stimulate, after period of habituation. *Stentor* retracts, when touched, but gradually steps retracting, if touching is repeated continuously. Sea-anaemone, which lacks a brain, exhibits similar

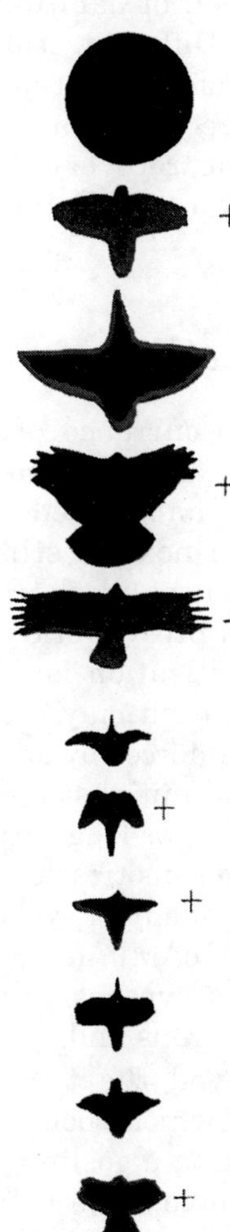

FIGURE 4.15 Habituation: Models used to test bird's response to silhouettes of birds of prey. Models are marked with a + stimulated escape response.

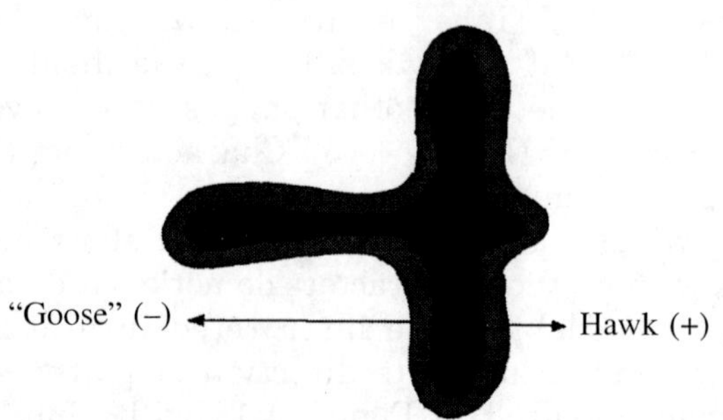

FIGURE 4.16 Habituation: Model used to test bird's response to direction of a silhouette, when the model is moved to right, it evokes an escape response (+), when to the left, does not stimulate escape (–).

response. In *Nereis*, habituation occurs more rapidly when stimuli are given close together. For instance, when animals were exposed, to a bright flash of light at half minute intervals, it took less than 40 trials, whereas it took about no exposure, when interval was 5 minutes. Speed of habituation depended on nature of stimulus. Different stimuli produce their characteristic rates of habituation. Animals recover form habituation effect after a lapse of some time. Clark detected some recovery within an hour in *Neries*, and animal was completely recovered within 24 hours.

4.4.2 Sensitization

This is prolonged occurrence of augmented post-synaptic responses, after a stimulus to which an animal has become habituated is paired once, or several times with a noxious stimulus. In *Aplysia*, noxious stimulus causes discharge of serotonergic neurons that end on presynaptic endings of sensory neurons. Thus sensitization is due to presynaptic facilitation. Sensitization may occur as a transient response or if it is reinforced by additional pairings of noxious stimulus and initial stimulus, it can exhibit features of short-term, or long-term memory. Short-term prolongation of sensitization is due to a Ca++-mediated change in adenylyl cyclase that leads to a greater production of cAMP. Long-term potentiation also involves protein synthesis and growth or pre- and post-synapticneurons and their connections.

Lorenz employed the term "imprinting" to describe process by which social bond was formed. He implied that during a gosling, or duckling's first encounter with a moving object, the image of object is somehow stamped irreversibly on nervous system, and for many years. This was accepted conception of process. Now, it has come to be realized that traditional view of imprinting is incorrect, and imprinting is neither rapid nor irreversible, as was claimed by Lorenz and his followers. Latest finding conclude that imprinting occurs in many species including man, and that it entails much more plastic and forgiving mechanisms than were claimed by Lorenz. The work also proves compelling evidence of the social bond that develops, through imprinting, entails an addictive process that is mediated, by release of endorphins. This surprising insight helps us to better understand a number of otherwise puzzling issues with respect to how to deal with each other and without children.

Imprinting involves learning of specific piece of information at right stage of development. Learning through imprinting is restricted to short time span called *critical period*. Imprinting shows how genes largely shape behaviour. Evolution creates window for learning important information about variation in environment. Imprinting helps to learn key variable components of environment retaining largely innate behavioural patterns. Flexibility may be shown in the development of food preferences, as food availability varies from habitat to habitat. Insects imprint on chemistry of leaves they eat as caterpillars. In adult stage they choose to lay their eggs on plants with a chemistry that matches the leaves they ate when young. Young birds and mammals prefer food based on food shared by adults, and by observing feeding preferences of adults, and sampling the possible food items.

Imprinting works as follows: to be an appropriate target for social bonding of an object (it could be a person or an animal) has to provide stimulation that is pleasurable and in this sense comforting. This will happen, when some aspect of object (its shape or texture or motion) has capacity to innately stimulate production of endorphins. When a young duckling or gosling or human body is exposed, to such an object, it is immediately comforted, and if exposure is extensive, initially neutral features of object, gradually acquire capacity to themselves stimulate production of endorphins. This happens to be thoroughly understood learning process. Once this learning has occurred, object will have been rendered familiar. As a result, it will continue to be comforting, when development has proceeded to point, when any unfamiliar, but otherwise appropriate object will elicit a competing fear reaction. When viewed in this fashion it becomes clear that so called critical period is merely period in development prior to onset of fear of novelty (at about day three) in ducklings and goslings and fear of strangers (at about 8 months in our own babies). Subjects that are beyond critical period, an appropriate but unfamiliar object must eventually become familiar, provided that subject have sufficient exposure to it. Once this happens, object will no longer elicit a competing fear response, and since object already has capacity to stimulate production of endorphins, it will now serve as a potential target, for social bonding.

4.4.3 Filial Imprinting

The best known form of imprinting in which a young animal learns characteristic of its parent, as found in birds, which imprint on their parents and then follows them around. It was first reported in domestic chickens by Douglas Spalding, and rediscovered and popularize by his disciple, Lorenz, when working with graylag geese. Lorenz (Figure 4.17) demonstrated how incubator-hatched geese would imprint on first suitable moving stimulus, they saw within what he called a "critical period" of about 36 hours, shortly after hatching. Most famously goslings would imprint on Lorenz himself (more specifically on his wading boots), and he is often depicted being followed, by a gaggle of geese that had imprinted on him. Filial imprinting is not restricted to animals that are able to follow their parents in child development. The term used to refer to process by which a body learns, who its mother and father are. Process is recognized as beginning in the womb, when an unborn baby starts to recognize its parents voice.

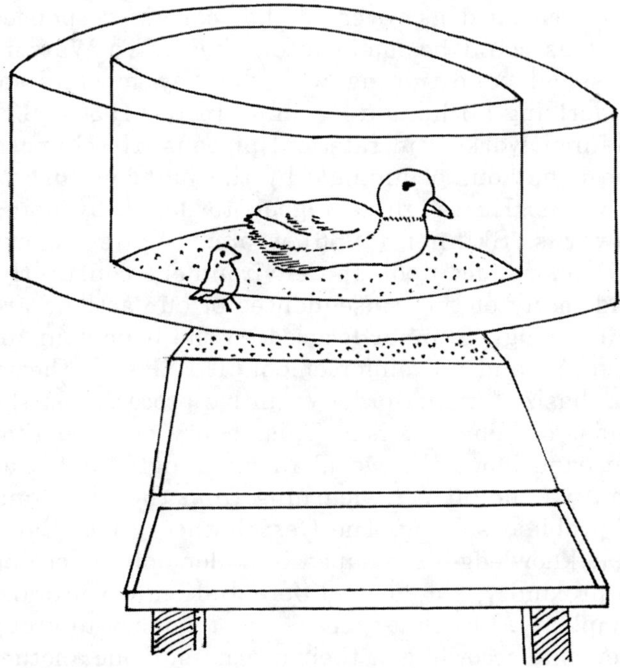

FIGURE 4.17 Imprinting: A decoy duck is moved around the centre of the ring, which a duckling flows along a circular runway of the apparatus used.

4.4.4 Sexual Imprinting

Sexual imprinting is a process by which a young animal learns characteristics of a desirable mate. For example, male zebra finches appear to prefer mates, with appearance of female bird that rears them rather than mates of their own type (Immelmann 1972). John Mooney called it love map. Sexual imprinting on inanimate objects is a popular theory, concerning development of sexual fetishism. According to this theory, imprinting on shoes or boots (as with Lorenz geese) would be cause of shoe fetishism.

Westermarck effect

Reverse sexual imprinting is also seen when two people live in close domestic proximity during first few years in life. Either one or both people are desensitized to later close attraction known as **Westermarck effect**, first described by Westermarck. Westermarck effect has been observed in many places and cultures, including in Israeli Kibbutz system. In the Chinese shim pua marriage customs, as well as in biological related families in case of Israeli Kibbutzim (collective farms), children were reared somewhat communally in peer groups, groups based on age, not the biological relation. A study a marriage patterns of these children later in life revealed that out of nearly 3000 marriages that occurred across Kibbutz system, only 14 were between children from same peer group. Of those 14, none had been reared together during first 6 years of life. This report provides evidence not only that Westermarck effect is demonstrable, but that it operates during critical period from birth to the age of six. When close proximity during this critical period does not occur, for example where a brother and sister are brought up separately, never meeting one another, they may find one another highly sexually attractive, when they meet as adults. This phenomenon is known as **genetic sexual attraction**. This observation is consistent with theory that Westermarck effect evolve, because it suppresses inbreeding. This attraction may also be seen with cousin couples.

4.4.5 Trial and Error Learning

Through such learning animals acquire suitable responses to stimulus, through experience. A hungry toad seldom captures a bee. Having its tongue stung results in trial and error learning with a signal experience. The toad then excludes bees and similar insect as food. Through such learning, a child learns the taste of good or bad food, that a store can be hot, not pulls the cat's tail.

4.4.6 Latent Learning

Not all behavioural activities are apparently directed to satisfying a need or obtaining a reward. Animals explore new surrounding and learn information that may be used in later stage. Knowledge of immediate environment of its burrow may help mice to escape from a predator. During acquiring this knowledge, it had no apparent value.

4.4.7 Insight

In some circumstances, animals solve problems suddenly without the benefit of previous experience. Kohler observerd that a hungry chimpanzee would stack boxes to reach a banana suspended from the celling, without any training. Epstein and others observed that pigeons under controlled conditions exhibit similar insight. Pigeons moved a box beneath a suspended banana and then stood on the box to reach the banana.

4.4.8 Reasoning

This is a mental process of taking inferences from two or more than two statements or happenings. Reasoning contributes to behaviour in primates in such a way that clearly marks off them from the sub-primate animals. Several types of tests are devised, to test an animal's reasoning ability. Deutor problem test consist of placing the animal in an environment, where it must follow a circuitous route to a food source or an escape path. A direct pathway to the food is blocked, and the animal must go away from the food to succeed. Only higher primates are good at this type of problem.

4.5 LEARNING THEORIES

Learning theories explain how learning takes place and why learning occurs. Such theories interpret learning and consider variables that are crucial in achieving the desired goals. Learning theories trigger the organizational improvement. Different learning theories overlap, and learning theory strategies are concentrated along various points of a continuum. Cognitive strategies are useful in solving tactics where defined facts and rules are applied in unfamiliar situations. Constructivist strategies are suited to deal with ill-defined problems.

4.5.1 History

In the 19th century, Darwin published, *The Origin of Species by Means of Natural Selection*. Scientists soon realized that humans share many similarities with animals. Studying biological processes in animals could therefore shed some light on the same processes in humans. Around the turn of the 20th century, Thorndike developed an objective experimental method to study the behaviour of cats and dogs by designing "puzzle box" with a lever that would release the door lock if the lever was pressed. The animal had to learn to press the lever to open the box. Thorndike was interested in discovering whether animals could learn through imitation or observation. He noticed that when an animal with an experience of a problem situation is more likely to perform the same action that had earlier brought the desired reward. The reward of being freed from the box strengthened the association between the stimulus and a suitable action. Thorndike concluded that rewards act to strengthen stimulus-response associations. J.B. Watson was familiar with the classical conditioning work of Ivan Pavlov. Pavlov's research on dogs revealed that certain responses in dogs could be made into a habit. To Watson, classical conditioning was the key mechanism underlying all human learning. In the 1930s, B.F. Skinner worked on rats and pigeons. He changed the behaviour of animals by the judicious use of rewards. He taught a pigeon to dance by using rewards. Skinner (1966) wrote: "While we are awake, we act upon the environment constantly, and many of the consequences of our actions are reinforcing." Skinner focused on the behaviour (or operant) and its reinforcement. The Gestalt theory emphasized higher-order cognitive processes in the midst of behaviourism. This theory claimed that we experience the world in meaningful patterns and we should view learning from the viewpoint of problem solving. The Gestalt theory postulates that knowledge is grouped into elements according to proximity, similarity/differentiation, closure and simplicity. Proximity means our tendency to group elements according to their nearness to one another and the patterns that they form. Similarity implies our tendency to group together similar items in some respect. Closure indicates our tendency to first look for a single, recognizable pattern. When we see the incomplete figure, we attempt to fill in missing information to form a complete figure.

Köhler deviated from the experiential approaches used by behavioural psychologists. He designed a series of problem situations for the chimpanzees and in each case all the elements were available to the animal to solve the problems. From his experiment, he concluded that learning took place through an act of insight. Wertheimer provides a gestalt interpretation of problem-solving episodes of reputed scientists. Successful problem-solving behaviour according to him is being able to see the overall structure of the problem. Jean Piaget believed that the human capacity to think and learn was an adaptive feature that enabled humans to deal effectively with the environment. According to Piaget, children shape their conceptions of reality through interaction with their environment. Cognitive development occurs as children adapt to their environment. Piaget considered knowledge growth as a continual sequential process consisting of logically embedded structures. This is divided into stages of development, and children move from one stage to the next by maturation and exploration. Piaget identified the developmental stages like the sensorimotor stage, the preoperational stage, the concrete operational stage, and the formal operational stage. He outlined several principles for building cognitive structures. Lev Vygotsky was convinced that social interaction plays a fundamental role in the development of cognition, and culture was a determinant of individual development. Humans have cultures, and every human child develops in the context of a culture. Human cognitive development is affected to some extent by the culture, including family environments. Children acquire much of the content of their thinking (cognition) from culture and they acquire the processes of their thinking from culture. At any given time in a child's development, he/she will be more susceptible to new knowledge which he called the "zone of proximal development". The Bruner's theory linked to child development research. Bruner identified the three stages of development as the enactive stage, the iconic stage and the symbolic stage. Bruner believed that learning situations should be structured to enable the learner to learn.

4.5.2 Connectionism

Numerous views regarding learning exist today. Such views are classified into four paradigms—behaviourism, cognitivism, constructivism, and social learning theories. The Thorndike's theory of learning called **connectionism** describes learning as the association between sense impressions and impulses to action. Earlier associationism linked one idea with another. Functional analysis of responses (R) in the background of stimuli (S) led Thorndike to consider neural nature of S-R connection. He trained cats to come out from a puzzle box which is made up of pole or a chain hanging from the top. The cat could push the pole or pull the chain to come out. This is an example of trial-and-error learning. The puzzle box experiment suggests that the time to solve the problem decreases as a function of trials.

In "puzzle box the door pivoted on a screw and could be pushed aside to the right or left." Fifteen boxes were made mainly of wooden slats and hardware cloth each containing a door that the cat could open by manipulating a device. A cat could open the door, for example, by pulling on a wire loop suspended six inches above the box floor or by pushing it aside or by pressing a lever. One box required the performance of three distinct responses: The cat had to depress a treadle, pull on a string, and push a bar before opening the door. The cat placed in such a box tries to squeeze through any opening without paying much attention to the outside food, but strives instinctively to come out from confinement. For eight or ten minutes, it will claw and bite and squeeze. When Thorndike put a cat into a given box repeatedly, the whole demeanor of the animal modified. The cat's behaviour appeared at first to be almost random. Gradually, it became orderly and efficient. Thorndike recorded the time required for the cat to come out and prepared a "time curve". The graph depicts about 64 trials over four days. Such a curve shows much variability in performance and slower learning than when escape resulted from manipulation of some part of a box. Slope of the curve showed the rate of learning. Whether the cat climbed up to the wire netting of its pen after each of the statements was recorded. Of the error data in two frequency graphs, one shows the failure to climb to the netting at the first signal, and the other shows climbing the netting at the second signal. Thorndike proposed that animal learning occurs as a result of "trial and accidental" success with no connection to reasoning or the association of ideas and theorized that a cat is innately equipped with various "action impulses" when placed in a box. It has the impulse to bite the bars and to scratch at objects within the box. If pulling a loop results in escaping from a box, the impulse to pull on loops is strengthened; if the impulse to scratch at bars does not result in escape, that impulse is weakened. Only the association of sensations, such as the sight of a loop, and the impulse to perform certain acts, such as clawing at things, were suggested by Thorndike.

Born in Williamsburg, Massachusetts, on August 31, 1874, Thorndike studied animal behaviour and the learning process to led to the theory of connectionism. He specified three conditions that maximize learning. The Law of Effect states that the likely recurrence of a response is generally governed by its consequence or effect generally in the form of reward or punishment. The Law of Recency states that the most recent response is likely to govern the recurrence. The Law of Exercise states that stimulus-response associations are strengthened through repetition.

Thorndike created 13 basic rules which are listed as follows:

1. Trial and error learning is the most basic form of learning.
2. Learning is incremental.
3. Learning is not governed by ideas.
4. All mammals learn in the same way.
5. Interference with goal-directed behaviour causes frustration and causing someone to do something they do not want to do is also frustrating.
   (a) When one is ready to do some act, it is satisfying to do so.
   (b) When one is ready to do an act, it is annoying not to do so.
   (c) When one is not ready to do an act and is forced to do so, it is annoying.
6. We learn by doing, and forget by not doing, although to a small extent.
   (a) Connections between a stimulus and a response are strengthened when they are in use.
   (b) Connections between a stimulus and a response are weakened when they are not in use.
7. If the response in a connection is followed by satisfying affairs, the connection strength is increased, and if followed by annoying affairs, then the connection strength is marginally decreased.
8. Before it is actually solved, a learner should keep trying multiple responses to solve a problem.
9. What the learner already possesses is the present state of the learner while it begins learning a new task.
10. Different responses to the same environment would be evoked by various perceptions from the environment. Different perceptions would be subject to the prepotency of various elements for different perceivers.
11. Using solution techniques employed to solve analogous problems new problems are solved.
12. Let stimulus S be paired with response R. If stimulus Q is given simultaneously with stimulus S repeatedly, then stimulus Q is likely to get paired with response R.
13. Learning is more effective than if the relationship is unnatural.

4.5.3 Behavioural Theories

J.B. Watson, the father of behaviourism, studied response to conditioning based on the experiments of Pavlov, and concluded learning as a sequence of stimulus and response actions in observable cause and effect relationship. Classical conditioning and operant conditioning are universal learning processes. Classical conditioning is the major cornerstone of behavioural theories. Operant conditioning occurs when a response to a stimulus is reinforced. When a behaviour is rewarded, this is repeated. Reward or reinforcement strengthens the response and results in changes in behaviour. Pavlov showed that the application of neutral stimuli could be used to elicit a response in animal. Watson and Skinner showed that these principles are applicable to humans with the addition of a reinforcement element and showed that responses related to more complex behaviour could be achieved, which they termed "operant responses." The behavioural theory maintains a focus on the change in observable behaviours as the expression of learning. Changes in behaviours occur due to the influence of the external environment, rather than by the internal thought process. People learn desired behaviours due to stimuli from their external environment that recognize and reinforce the behaviour in a positive manner. Undesired behaviours can be eliminated by an absence of attention to such behaviours. Behaviourism comprises several individual theories with a common theme. The first common assumption is the emphasis on observable behaviour rather than internal thought processes that create learning. Second, it is ultimately the environment that enables to learn and it determines what is learned. Lastly, it is the ability to understand the overall process, and the ability to repeat that process that is a common thread (Merriam and Caffarella 1999). The behavioural theory and training are key components of animal training and skill training in humans. Pattison (1999) suggested that American adult education originated in liberal arts education and then continued in progressive education by quoting others (Elias and Merriam 1995). Behaviourism coupled with progressive education would help "control human behaviour, and viewed education as a tool for bringing about societal change". The behavioural theory presents learning in short manageable blocks that build on previously learned behaviours. Kearsley (1994) identified three fundamental principles common in behavioural learning:

1. *Positive reinforcement* of the desired behaviour will most likely prompt the same behaviour.
2. Learning should be presented in small manageable blocks.
3. Stimulus generalization of learning can produce *secondary conditioning*.

Since the late 1800s, psychologists using behavioural principles have established hundreds of tests to identify both how learning and memory occur in varying complexities of brain structures. When the reinforcing agent is "painless," then learning occurs in the cerebellum in many species. However, if there is an emotional connection to the reinforcer, then learning and memory occur in the amygdala (Kolb and Whishaw 2005). The environment of an individual reinforces behaviours either positively or negatively and all of learning takes place through environmental influences. The behavioural theory was founded in the early decades of the 20th century, but there still exist many examples of support for the theory.

4.5.4 Cognitive Theories

Cognitive theories of learning view knowledge as absolute. These theories view learning as involving the acquisition of the cognitive structures through which human beings process and store information (Good and Brophy 1990). The classical Gestalt theory and the Tolman's sign learning theory, otherwise known as purposive behaviourism, are the important cognitive theories. The gestalt psychologists explain that learning is a matter of neither adding new traces nor subtracting old ones but of modifying one gestalt into another. They treat learning as a purposive, exploitative, imaginative, and creative method of developing new insights (Hill 1963). Motivation is a crucial aspect in learning process (Hill 2002) and is related to arousal, attention, anxiety and reinforcement. The behavioural theories focus on rewards while the cognitive theories deal with goals (Weiner 1990). Three characteristics of the Tolman's theory (Hillgard and Bower 1975) state that it is concerned with goal-directed behaviour, and explains learning in terms of the effects of external stimuli on behaviour, and considers that behaviour is changed through an organism's experience.

4.5.5 Constructivism

The constructivist view assumes various forms of knowledge and view knowledge as a constructed entity. The constructivist approach is based on the fact that by reflecting on our experiences, we construct our understanding of the world. Individuals use their own mental constructs to make sense of their experiences. Learning is a search for meaning, and learning must be based on the issues that require personal inter-pretation. Construction of meaning needs an understanding of "wholes" as well as parts. Parts must be understood in the context of wholes. The learning process focuses on primary concepts. Emphasis is given on the application of knowledge rather than on mere acquisition of decontextualised facts. Social aspects of learning form a crucial part of the constructivist view. Constructivism as a unique theory promotes a more open-ended learning experience where the methods and results of learning are not easily measured and may not be the same for each learner. All humans can construct knowledge in their own minds through discovery and problem solving. Piaget (1970) observed human development as a progressive stage of cognitive development. His four stages characterize the cognitive abilities necessary at each stage of development to construct meaning of one's environment.

Constructivism is a synthesis of multiple theories. It is the assimilation of behaviourialist and cognitive ideals. This is a combination effect of using a person's cognitive abilities and insight to understand his environment. This concept is easily translated into a self-directed learning style, where the individual has the ability to take in all the information and the environment of a problem and learn. Various constructivist theories agree "that learning is a process of constructing meaning; it is how people make sense of their experience"

(Merriam and Caffaerall 1999). Two viewpoints of constructivist theories include the individual constructivist view and the social constructivist view. The individualist constructivist view understands learning to be an intrinsically personal process whereby "meaning is made by the individual and is dependent upon the individual's previous and current knowledge structure", and as a result can be considered an "internal cognitive activity." The social constructivist view premises that learning is constructed through social interaction and discourse, and is considered, according to Driver et al., (1994), to be a process in which meaning is made dialogically (Merriam and Caffaerall, 1999).

4.5.6 Social Learning Theories

Bandura's social learning theory is widely accepted because of its complete and parsimonious interpretation of social learning (Manz and Sims, 1981). This theory views human behaviour in terms of a continuous reciprocal interaction between cognitive, behavioural, and environmental determinants. Learning takes place as a result of experienced responses and through observing the effects on the social environment of other people's behaviour. Bandura (1977) considers attention, retention, motor reproduction, and motivational processes. These processes explain the acquisition and maintenance of observational learning (Davis and Luthans, 1980). Social learning theory provides a general framework for many aspects of management education.

Remark: Another type of learning involves aversions, which develop at any point in animal's life. Birds and mammals learn lifelong aversions to specific poisonous foods (monarch butterflies). Some preferences and aversions may be innate, or at least to be driven by physiological needs for nutrients like salt. Many animals learn key information for survival often very specific to a specific context. A species may be very adept at learning facts that are required for its survival. Native birds in Guam were failed to learn how to evade predation by brown tree snakes, which were introduced into Guam. Some animals cache food from a central location and centralized caches require strong defense against thieves, a notable ability of honeybees. Cached food can be scattered through habitat. Tree squirrels and gray jays are notable for scatter caching (Steele et al., 2008) that stand out as a challenging context for learning complex information about locations. Mammals and Birds that cache food display impressive abilities to recall cache locations. Location of a nest or burrow is unlikely to remain constant across many generations. Ability to return home requires ability to incorporate much environmental information. The desert ant, Cataglyphis cursor, use learning into navigation by using path integration, the ability to remember the distances and directions travelled, to sum them, and determine their return path (Müller and Wehner 1988). Learning and calculation abilities integrate navigational path. Other animals use landmarks, like position of the sun for knowing the outward path, which they use in opposite for returning home. Evolution has given innate mechanisms for incorporating learned environmental information in cache retrieval and homing.

Review Questions

Short Answer Questions

1. Define learning.
2. Define non-associative learning.
3. Define associative learning.
4. Define imprinting.
5. Define innate behaviour.
6. What is instinct?
7. State the difference between instinct and learned behaviour.
8. Give two examples of instinct.
9. Give two examples of learned behaviour.
10. What is key stimulus?
11. What is FAP?
12. What is phenotypic plasticity?
13. Define orientation.
14. What is klinokinesis?
15. What is orthokinesis?
16. Define tropotaxis.
17. Define telotaxis.
18. What do you understand by menotaxis?
19. What is positional orientation?
20. What is strato orientation?

21. What is zonal orientation?
22. Define reflex.
23. Give an example of reflex.
24. Define tonic reflex.
25. Define phasic reflex.
26. Define classical conditioning.
27. Give an example of classical conditioning.
28. Define operant conditioning.
29. Give an example of operant conditioning.
30. State the difference between classical and operant conditioning.
31. Comment on behavioural extinction.
32. Define habituation.
33. Give an example of habituation.
34. Define sensitization.
35. Give an example of sensitization.
36. What is Westermarck effect?
37. What do you understand by the term "insight"?
38. What is "trial and error" learning?
39. What is latent learning?
40. Define motivation.
41. Give an example of latent learning.
42. Give an example of insight.
43. Define irritability.
44. Comment on tropism.
45. Name two types of nest studied by you.

Long Answer Questions

1. Describe the experiment conducted by Pavlov in conditioning.
2. Describe various types of learning with examples in brief.
3. Discuss various biological aspects of memory.
4. Write a note on sensitization with suitable examples.
5. Write a note on habituation with suitable example.

Chapter 5

Animal Communication

Communication is the production of a signal by one organism that changes behaviour of another organism in a way which is beneficial to one or both. It mostly occurs between members of same species, but may occur between members of different species. When one animal responds to the signal sent out by other animals, communication occurs. Members of same species compete with one another for food, space, and mates. Communication often resolves such conflicts with minimal damage. Ability to communicate effectively with other individuals plays critical role. For example, when moths attract mate, ground squirrels inform about nearby predators,and chimpanzees maintain positions in dominance hierarchy communication is said to occur. A primer about types of communication signals used and variety of functions they serve are known. "... action of or cue given by one organism (sender) is perceived by, and thus alters the probability pattern of behaviour in another organism (receiver) in a fashion adaptive to either one or both of the participants" (Wilson 1975) through communication. While both sender and receiver must be involved for communication to occur, sometimes, one player get benefits from the interaction. Female Photuris manipulate small male Photinus by mimicking the flash signals. When males investigate signal, they are voraciously consumed by larger (Lloyd 1975). This is a case where sender benefits and receiver do not. In fringe-lipped bats, *Trachops cirrhosus*, and tungara frogs, *Physalaemus pustulosus*, receiver is the only player that benefits from interaction. Male tungara frogs produce calls to attract females to their location; while signal is designed to be received by females, eavesdropping fringe-lipped bats detect calls and use that information to locate and capture frogs (Ryan et al., 1982). In many cases sender and receiver both were benefitted from the exchange of the information. Greater sage grouse shows such "true communication". Males exhibit expensive displays, and females use this information about male quality to choose the mating partner (Vehrencamp et al., 1989) during mating period. In Vervet monkeys, *Chlorocebus pygerythrus*, adults give alarm calls to warn colony members about the presence of specific type of the predator. This is valuable as it conveys the information needed to act appropriately given the characteristics of predator. Emitting a cough call shows presence of an aerial predator like an eagle. The colony members respond by seeking cover amongst vegetation on ground (Stewart and Cheney 1980). Such an evasive reaction would not be appropriate if a terrestrial predator, like leopard, were approaching.

Most animals use "body language", sound and smell to communicate with one another. While communication among species is common throughout the nature, communication between species is much less frequent. Animals including insects, wolves, deers, and humans communicate by smell. They release pheromones, the airborne chemicals which play an important part in reproduction, and other social behaviour. Even an *Amoeba* communicates with other amoebas chemically. One *Amoeba* attracts others to it for reproduction. Bees dance, when they found nectar. Chimpanzees greet each other by touching hands. Male fiddler crabs wave their giant claw to attract female fiddler crabs. White-tailed deer show alarm by flicking up tails. Dogs stretch their front legs out in front of them and lower their bodies when they want to play. These forms of communication are affected and influenced by the genetic make up of a species, their own environment and their experiences. Communication is used for resolving conflict, courtship and pair-maintenance.

Animals communicate, although correspondence is generally limited to instinctual signals expressing danger, territory, and courtship. A dog growls at a perceived threat, a frog croaks at dusk, a male bird of paradise clings to a branch upside down while preening its feathers. Signalling as acquired behaviour is much less common. Blue whales rumble at volumes louder than a commercial jet at takeoff, to broadcast their location and probably their heading to other blues. They vocalize directly at a thermocline using it as an underwater PA system, to amplify and echo calls over distances measured in oceans. Most complex examples of animal communication occur within species that stand relatively low on evolutionary scale. The best known example is probably the honey bee's waggle dance, first observed by Karl von Frisch (Figure 5.1) in 1940.

FIGURE 5.1 Karl von Frisch.

In the 1920s, South African naturalist, Eugene Marais, concluded that termite colonies are best understood as a composite organism possessed of a group purpose and even a group mind. He observed that the tasks of the group were communicated through some unknown mechanism comprising elements of syntax, and vocabulary, now known to be chemical fragrances called **pheromones**. A few animals use sound to communicate a vocabulary. In 1980, biologists Struhsaker and others discovered that vervet monkeys in Kenya possess a vocabulary based on predators in their life. A certain grunt is the actual word for eagle. When it is vocalized, all the vervets in earshot scan sky. A bark means leopard, prompting the monkeys to scamper to top of a tree. Other sounds express territoriality, kinship and social standing. When a vervet infant screams out, the word signifying a certain predator, only its mother responds directly. The other monkeys react to mother, recognizing whose baby is in distress. Alarm call of a robin attracts robins, but also blue jays, orioles, and catbirds, which help drive off the predator. The birds arrive not out of curiosity or because of some universal sensibility to pain and suffering, but because certain alarm calls mean the same thing to several bird species. The timbre and frequency of that particular call overrides territorial protocol.

Horses and their riders communicate in subtle ways, combining vocals, touch and body language to disclose direction, pain, fear and joy. Rudyard Kipling told about "what, where, who, when, how and why of learning and intellect". Dogs, parrots, elephants, and even pigeons have been documented in communicating what, where, who, and arguably, how. But they all lack the other two, when and why. Science of interspecies communication is in its infancy and saddled with more controversy than it probably deserves.

5.1 THE EVOLUTION OF COMMUNICATION

Consider the evolutionary costs and benefits of the signal, along with the morphological and physiological features that influence its production. when we expect communication to be honest and when we expect it to be unreliable. Then we must focus on two hypotheses about the evolutionary origin of signals. communication is the transmission of information from one animal to another animal. Researchers through the 1970s emphasized a cooperative view of communication. Here both sender and receiver enjoy the accurate transfer of information (Dawkins and Krebs, 1978). In this condition, selection must make signals unambiguous, efficient, and reliable. Dishonest, or inaccurate, signalling was thought to be unlikely in animals; after all, even in humans, lying is risky and hard to do (Smith, 1977). Communication benefits both partners, which seems reasonable at first glance. the complex role of communication in coordinate signals, even if it is sometimes deceived. After all, not all signals are dishonest. Often, the signaller benefits from conveying accurate information, such as in the case of the female moth alerting males that she is ready to mate. Thus, signals are reliable but deceptive. Many species perform this via species-specific pheromones. Females broadcast their availability. males detect chemical signals of prospective mates. After potential partners find each other, they may engage in a complex give-and-take of displays that culminate in mating. Both partners receive fitness benefits by communicating their intentions to each other. Many forms of communication uphold the view that their function is to share information.

5.2 CHANNELS FOR COMMUNICATION

Communication involves many sensory channels. These are chemical, touch, vision, auditory, and electrical fields. The channel used for a specific signal depends on the signal's habitat, biology, and function.

5.3 VISUAL COMMUNICATION

Visual systems provide various signals. Because most animals perceive various stimuli. These are the animals' colour, brightness, and changing spatial and temporal patterns. Environmental conditions influence the importance of one type of visual stimulus or another. Agonistic displays aimed at conspecifics; many shark species depress their pectoral fins. They hold them for long periods in a symmetrical downward position (Martin 2007). Many sharks inhabit clear water, such as nearshore and reef habitats. They have conspicuous markings on their pectoral fins. They have black or white tips and margins to increase the visibility of postural displays. Sharks inhabit areas where light is scarce dependent on their posture. Visual signals have disadvantages if the sender is not visible. Then such signals are useless, as obstructions block the vision. Sometimes, the visual displays allow animals to avoid obstructions. As the spawning season approaches, razorback suckers move from deep water to shallow water. The males rest within their territories along the river bottom. When a roving male approaches a territorial male, the latter rolls his eyes downward. They exposed his eyes' whites to light from the water's surface. Two quick flashes of reflected light signal the territorial male's presence to the interloper, who then retreats. when sediment increases in the river, making transmitting visual signals difficult.

At night or in dark places, light-producing species use visual signals. Most animals cannot produce light. Many animals are active at dawn and dusk, in the presence of some light. They are nocturnal in low light it is difficult to differentiate colours. Thus, visual signals focus on contrast rather than colour and use white at dawn and dusk, as observed in eagle owls (Bubo bubo). Males and females of this owl have a white badge of feathers on the throat. This badge is visible when the throat is inflated and deflated, which peaks at dawn and dusk (Delgado and Penteriani 2007). Breeding eagle owls use white material to mark the area around their nest. They accumulate white faeces at defecation sites. They display the white feathers of prey at conspicuous plucking sites (Penteriani and Delgado 2008). In such ways, nocturnal animals use visual signals.

The size of visual signals (their conspicuousness) reduces with distance. Animals adjust their visual signals based on *receiver* distance. Male fiddler crabs exhibit conspicuous claw-waving displays. A male waved his single, enlarged claw. In the absence of a receiver, males of *Uca perplexa,* a fiddler crab, exhibit their courtship displays. A courting male turn and faces the female and directs his display toward her. Males do more than face the female at showtime. Males adjust their claw-waving display to measure the distance to females (How et al., 2008). As the distance to the female decreases, the interval between claw waves decreases along with the duration of each wave. These changes translate into an intense display when the females are close. The claw-waving display is a signal of male quality. The mode of its display changes as the distance to the female decreases. The horizontal distance swept by the tip of the enlarged claw decreases when the distance to the female decreases.

Animals that rely on vision for finding their way about and detecting prey tend to communicate through visual signals. Such signals can be active in which a specific movement or posture conveys a message. It may be passive, in which the size, or colours of the animals conveys important information, like its sex, and reproductive state. Visual signals are influenced by the fact that some wavelengths of light are scattered, or absorbed more than others in their journey form sender to receiver. This effect is striking in seawater; where blue light of about 475 nm travels well, but red light (longer wavelength) or ultraviolet (shorter) are filtered out by the water. This is the reasons why so many reef fises are blue that is visible in seawater (Lythgoe 1979).

Birds like *Rupicola rupicola* prefers to display in places in forest, and at times of day when their signals maximize their colour and brightness (Endler 1993). It is also striking that visual signals that operate over long distances, such as white tails of rabbits and some deer have a broad range of frequencies. White is very conspicuous (Manning and Dawkin 2002) at all distances. Visual displays are important types of communication. Mating display is mentionable in this connection. Visual communication helps in species recognition. Visual communication helps in concealment (cryptic colouration) or colouration (bright and bold colour). Sexual presentation in the chacma baboon is noteworthy.

Advantages

They are instantaneous and can be rapidly modified to convey a number of messages over a short period. This communication is quite and generally do not alert distant predators, although signaller makes itself conspicuous to those nearby.

Disadvantages

Generally ineffective in darkness and in dense vegetation, though female fireflies signal potential mates by species-specific pattern of flashes. In fish living in deep water, where light is scarce, they make their own light.

5.4 COMMUNICATION BY SOUND

Sound signals send over long distances in water. Sound transmits more than light. As a rapid means of sending a signal at close range, the transitory nature of the sound makes for rapid exchange and quick modification. But it does not allow the signal to linger. After sending the message, the complete signal disappears completely. In times of low visibility, like in night sound signals get a benefit to convey a message. Sound signals can be complex. The temporal variation of its frequency (pitch) and amplitude (loudness) creates a variety of sounds. Depending on the species, animals vary in either or both aspects of sound. They may like a drummer who alters the pattern of presentation. How the animal produces them determines the types of sounds used by a particular species. The respiratory structures may generate sounds, striking objects in the environment, or rubbing appendages together. Many structures specialized for sound production have evolved along with respiratory structures. Mammals and birds use larynx and syrinx respectively to create sound. The anatomical structure and location of these organs are different. Both allow the production of complex sounds. The environment is in use to produce auditory signals. Humans often tap their toes when listening to music. For some other animals, such as rabbits and deer, foot stamping itself is the signal. Beavers slap the water with their tails. Woodpeckers drum on trees with their bills. Insects produce auditory signals by rubbing together parts of their exoskeleton. By opening and closing wings crickets create sound. Each wing has a thickened edge called a scraper. It rubs against a row of ridges, the file, on the underside of the other wing cover. The scraper moves across the file, and a pulse of sound is produced. This method of generating sounds is called stridulation.

Stridulation in insects, males of the club-winged manakin, and a small bird of the Neotropics are on record (Bostwick and Prum 2005). In courtship displays, a perching male creates a tick-tick-ting sound. Movements of the wings and modification of secondary feathers create the sound. The male rotates his wings forward and flips the feathers above his back. Then brings the wings together, causing the secondary feathers to collide to create the sound. The ting sound is produced in the same manner except after the secondary feathers collide. The male shivers his wings, causing oscillations of the secondary feathers. The result of these movements is that one modified feather rubs back and forth against the ribbed surface of the adjacent feather. This excites resonance in the enlarged rachis and puts the finishing touches on this auditory signal.

Some animals make sounds that humans cannot hear. Bats, rodents, and cetaceans produce and detect ultrasounds through their echolocation. Ultrasounds with frequencies more than about 20 kHz are inaudible to us. Ultrasonic communication in mammals and amphibians is on record. The concave-eared torrent frog hat lives along noisy streams and waterfalls. Unlike other frogs, they lack ear canals. their eardrums are located at the skin surface. The males in this species have ear canals and recessed eardrums, like mammals. The concave-eared torrent frogs suggest that ultrasonic communication allows this unusual species to avoid the masking effects of the constant low-frequency background noise of the local streams and waterfalls in their habitat. The inaudible and ultrasonic parts of male calls evoke vocal responses in male conspecifics (Feng et al., 2006). An auditory processing centre of the midbrain, when frogs are exposed to bursts of ultrasound, confirms this ultrasonic hearing capacity. The females, before ovulation, produce calls with ultrasonic components (Shen et al., 2008). When female calls play, they evoke calls from nearby males, who then approach the loudspeaker. This is called positive phototaxis. We cannot hear sounds with frequencies above 20 kHz. These sounds are ultrasounds. We cannot hear sounds whose frequencies are less than about 20 Hz. The sounds below this limit are infrasound. The elephant's voice is the trumpet blast. In fact, most elephant calls are infrasonic having frequencies of 15 to 25 Hz (Poole et al., 1988). Elephants drive air from the lungs to set the vocal folds of the larynx in motion (Garstang 2004). Elephants are very social animals. They live in matrilineal family groups. In such groups, daughters remain with mothers. The sons left to live in a bachelor group or as lone bulls. Family groups range over large areas for food and water. Long-distance communication between family members and between different family groups is critical. Because such communication occurs between sexually receptive females and males. In comparison with high-frequency sounds, low-frequency sounds are less degraded by processes like absorption, refraction, and reflection. Infrasound works especially well for long-distance communication (Garstang 2004).

Sound can be transmitted through darkness, dense forest, and water. Low resonant sound of humpback whale can be heard by other whales up to hundreds of miles away. Animals signal over long distances by elevating themselves above ground. Crickets that sing from trees can spread their signal over 14 times the area of those that sing from the ground, and consequently attract more females. Motivation changes can be signalled by a change in loudness or pitch of sound. While studying vervet monkeys, Ethologist T. Struhsaker found that they produce different calls in response to threats from each of their major predators—snakes, leopards and eagles. Response of other vervet monkeys to each of these calls is appropriate to the particular predator.

Animal sounds have a potential to provide extensive information about internal state, sex, subspecies, reproductive state, social status, stress and welfare of animals, and there are now many positive results in the area. This approach is based on the vocal behaviour of animals and on the structure of their calls (Volodina and Volodin 1999). It has been shown that bioacoustics methods may be useful for determination of subspecies status in black gibbons. Phokin (1983) reported that the subspecies may be recognized in one-day old chicks by their vocalizations. In squirrel monkeys, Jurgens (1979) found a relationship between the vocalizations evoked by electrical brain stimulation and the degree of aversion in the animals' emotional state. Sex determination in birds which lack sexual dimorphism was initially developed by Tichonoff and colleagues (1988). Carlson and Trost (1992) elaborated a method of sex determination in the whooping crane by analysis of guard calls.

5.5 VIBRATIONS

Vibration has a history of only the last 30 years in biology. Mechanism of vibration production and nature of wave produced are not always understood. Recently, technical advances have been made how animals send and receive signals through substrate. Use of vibration in communication is much more widespread than previously thought. Vibration provides information used in predatorprey interactions, recruitment to food, mate choice, intrasexual competition and maternal/brood social interactions in animals ranging from insects to elephants.

5.5.1 Detection and Use of Vibration

Perception of vibration as a sensory channel predates that of vertebrate ear mechanism. Extinct amphibians were able to detect vibrations through their jaw in contact with ground, and conduction through quadrate bone to inner ear, via bony tissue. A reduced hyomandibula associated with quadrate in a fish's visceral skull evolved to form stapes, or columella. Caecilians, urodeles and some anurans, snakes, amphisbaenians, and some lizards have a stapes which may be attached to shoulder girdle or skin, and are well suited to detecting low frequency vibrations from substrate. Massive ear ossicles are found in large mammals that use acoustic information, through bone-conducted vibration at the expense of auditory acuity at higher frequencies. Modern-day golden moles have specialized structures for hearing low-frequency sounds emitted by their prey, including a complex hyoid apparatus in contact with their tympanic bulla and many have massive ear ossicles. Frogs are exceptionally sensitive to seismic stimuli and use saccule of ear to detect these vibrations. Sandfish lizards use vibration to locate prey also has a very large saccule.

Somatosensory mechanisms involving mechano-receptors in sensillar hairs and sub-genual organs of insects (Bell 1980), basitarsal compound-slit sensilla of scorpions, and trico-bothria and metatarsal lyriform organs of spiders process vibrational signals. Simple nerve endings in snakes perceive vibration perception. Herbst corpuscles in birds and Pacinian corpuscles in Eutherian mammals perceive vibrations. Lamel-lated corpuscles in legs of macropod marsupials detect ground-to-bone vibration produced by approaching predators. Initiation of substrate vibrations during courtship is reported in vertebrates. *Chamaeleo calyptratus* produces plant-borne vibrations that may be used in communication. Chameleons might also detect substrate vibrations. Two species of frogs belong to genus *Leptodactylus* produce substrate-borne vibrations. Allopatric species of genus *Dipodomys* produces foot drumming in species-specific patterns. Banner-tailed kangaroo rats, *Dipodomys spectabilis* produce individually distinct signatures in their foot drumming that remain constant over time allowing discrimination between neighbours and strangers. Airborne signals are best for communication with distant neighbours, while outside burrow on windless nights. Substrate-borne vibration provides a channel for communication from inside burrow to near neighbours on windy nights. Species-specific head drumming of *Spalax ehrenbergi*, may function as an isolating mechanism between species and a long-range communication system within species.

"Spiders produce vibration by drumming with palps and abdomen by stridulating or by plucking threads of their own or other spiders' webs". Information from vibration in their environments affects many aspects of spider ecology: prey-catching, courtship, territorial behaviour, and social interactions. In species, sharing a common web, spiders locate prey through vibrations of web, but *Cupiennius salei*, a spider that does not live on a web is also capable of reciprocal signalling between mates at least one metre apart on banana plants.

Workers of the leaf-cutter ant of genus *Atta* produce both airborne and vibration components of a stridulation sound using a file-and-scraper mechanism. When they are buried by a cave-in nest-mates, they do not respond to airborne sound, but rescue their buried relatives in response to substrate vibration. Low frequency component of stridulation output is emphasized underground, where radiation conditions are much better than in air for an animal of this size. Since vibrations also mechanically aid the cutting process, communication in this instance of foraging is a secondary effect for a process that first increased efficiency in food handling.

Caterpillars produce secretions rich in amino acids and sugars that feed the ants, but they appear to attract the ants with vibration signals. Common communication pathway known to ants is exploited by butterfly caterpillars and, appears to have evolved at least three separate times in related lineages. Vibrations serve termites in pathogen alarm behaviour. Termites hang their heads and produce substrate-borne vibrations as an alarm signal in response to a disturbance of nest by predators. Young damp wood termites, *Zootermopsis angusticollis*, also produce a vibratory alarm in response to exposure to spores of a fungal pathogen. Unlike disturbance alarm signal, pathogen alarm induces nest mates to flee rather than move toward stimulus source.

In prairie mole crickets, *Gryllotalpa major*, vibrations are used to attract flying females and may represent bimodal communication, important in male-male spacing. Male bush crickets, *Tettigonia cantans* produce both sound and vibration. Vibration, mainly in the form of bending waves, seems to play a role in mate location rather than competition with other males and reinforce acoustic signals. Male *Balamara gydia*, an Australian cricket, has a complete stridulatory apparatus, but communicate with females by tapping their abdomens on vegetation. Both sexes tap, but neither has a tympanal organ. Female meadow katydids, *Conocephalus nigropleurum* discriminate among tremulation signals of males to choose a larger male, even in absence of a signalling male. Female wandering spiders, *Cupiennius getazi* uses vibration signals for mate recognition, but not female choice. But female wolf spiders, *Hygrolycosa rubrofasciata* actively choose males, based on their drumming rate. Drumming activity does predict male survival, and so females use drumming rate as an indicator of male fitness, rather than as an index of male body mass. Drumming has both acoustic and vibratory components in wolf spiders, and relative importance of either of these is yet to be determined. In addition to communication during courtship and reproduction, vibrations can transfer information among social groups, including sibling groups. Nymphs of treehopper, *Umbonia crassicornis*, display a sibling-group alarm signalling, when one is attacked by a predator. Their mother responds to the plant-borne vibration to defend her offspring, but she only responds to the group signal. Female in turn uses vibration to signal to her offspring at a low rate throughout day. These signals appear to play a major role in maternal care. By signalling through plant tissue, mother and offspring can communicate outside perception range of their most common predators.

Male fiddler crab, *Uca pugilator*, drums the ground with his large chela during courtship, especially at night, when waving chela is ineffective as a signal. Vibration can be vehicle of predator-prey interaction. Wandering spider, *Cupiennius salei* uses vibration for responding appropriately to the context of vibration signal. Scorpion, *Paruroctonus mesaensis*, can interpret vibrations in sand to determine both direction and distance of prey species, while burrowing cockroach, *Arenivaga investigata,* uses vibration to detect and avoid approach of scorpion. Jumping spider, *Portia fimbriata*, a predator of spiders, uses aggressive mimicry by generating a context-specific repertoire of simulated vibration signals on prey's web.

Certainly animals send and receive signals through substrate in environments, where this mode is most efficient and most economical. Animals in closed burrows have limited options, for receiving input, through airborne signals, especially, if cospecifics are also underground. Those with open burrows may more efficiently transmit information on location and sex of individual through substrate vibrations to near neighbours, especially in a windy or otherwise noisy environment. Fiddler crabs that wave to attract females, during day use vibration to communicate at night (Hill 2001).

5.6 CHEMICAL SIGNALLING

The senses of smell and taste are channels for communication based on the movement of odour molecules from signaller to receiver (Wyatt, 2003). Information may be carried by chemicals over long distances, especially when assisted by currents of air or water. The rates of transmission and fade-out time are slower than for visual or auditory signals. Based on the nature of the signal's function this is a benefit. In the demarcation of territorial boundaries, a durable signal is efficient. Because it remains after the signaller has gone. Some mammals increase the signal life of chemicals used to mark territories by secreting them with oily carrier substances, such as those from sebaceous glands, or by associating them with urinary proteins that slow the release of the signals (Wyatt ,2003). Another benefit is that these long-lasting chemical signals do not need continued energy expenditure by the sender. Chemical signals can be used where visibility is limited. The ease with which the sender of a chemical signal can be located varies with the chemical emitted. But it is usually more difficult to locate a signaller that uses chemicals than one that uses visual or auditory signals. Some signals are complex blends of chemicals. Thus, the proportions of different chemicals affect the sender. The presence of specific chemicals in the mixture does not affect the sender. the full chemical "image" of the signal, called the odour mosaic creates effects (Johnston 2000). Scent marking is the act of placing a chemical mark in the environment. The chemical image is left by common marmosets (*Callithrix jacchus*), small primates endemic to the forests of Brazil and recognized by their white ear tufts and forehead blaze. Females of this species mark their territory using a mixture of faeces, urine, and secretions of the reproductive tract. Female common marmosets are trained to deposit a scent mark on a special collection device. A detailed analysis of the scent marks revealed the presence of 162 distinct chemicals. Females could differentiate between the scent marks of familiar and unfamiliar conspecifics. Based on the ratios of chemicals, especially the volatile ones, in the scent mark, each female has a unique scent signature (Smith, 2006). The meaning of a particular signal sometimes varies with the context. A case with a chemical signal sent by a queen honeybee may be cited here. The chemical trans-9-keto-2-decenoic acid is picked up from the queen as the workers groom her and are distributed throughout the hive, along with the food that is shared by the workers. When attained in this manner, the chemical prevents the rearing of any more queens. The queen exudes the chemical as she soars skyward on her nuptial flight. In this context, the same chemical causes males to gather around her. It serves as a queen inhibitor or a sex attractant, depending on the context (Robinson, 1996). The detection of chemical cues may occur at a distance. when volatile chemical cues become airborne and reach mammal noses or insect antennae; this is remote chemical chemoreception. Detection of chemical cues through direct contact with the chemical is contact chemoreception. One ant rapidly and repeatedly touches the body of another ant with its antennae. Contact chemoreception has an associated with non-volatile chemical cues.

A vomeronasal (or Jacobson's) organ plays a role in chemical communication between mates, and parents. Some species of mammals, reptiles and amphibians have the organ. It is separate from other chemosensory structures. Its neural wiring goes to brain regions different from those associated with the main olfactory system (Halpern and Martinez-Marcos, 2003). The vomeronasal organ is found between the nasal cavity and the mouth or on the roof of the mouth. So communicative chemicals must reach it through the nose, mouth, or both. This special organ detects non-volatile chemicals, which must be brought to the organ. In a snake, the roof of the mouth contains the vomeronasal organ. The tongue delivers the chemicals. A mammal licks or touches its nose to the chemicals found in urine or special body secretions of conspecifics. Following this contact, many mammals make a facial grimace called a flehmen. This helps transfer the chemicals to the vomeronasal organ. The head is raised, and the lips are curled back in flehmen. Chemicals convey messages to other members of the same species. These are pheromones. Releaser pheromones immediately affect the recipient's behaviour. The sexual attractant is a releaser pheromone. The female *Bombyx mori* has a unique sex attractant called bombykol. She emits a minuscule amount of the sex-attractant from a sac at the tip of her abdomen. The wind carries this pheromone. The males in the vicinity detect it and the pheromone binds to the receptor hairs on the antennae of males. About 200 bombykol molecules immediately affect the male's behaviour. He flies upwind for the emitting female (Schneider, 1974). The specific olfactory receptors are found on the antennae of male silk moths (Sakurai et al., 2004. Releaser pheromones in insects include trail pheromones. This influences the foraging efforts of others. The alarm substances of the pheromone warn about danger. Vertebrates produce releaser pheromones. Lactating rabbits emit mammary pheromones. This causes their pups to find and grasp onto a nipple (Moncomble et al., 2005). Quick attachment by pups to a nipple is critical. Because mother rabbits return to their nest to nurse pups only once a day for about three to five minutes. Primer pheromones slowly change the physiology and behaviour of the receiver. In insects, queens use primer pheromones to control the reproductive activities of nest mates. The mandibular gland of a queen honeybee releases compounds. These ensure the existence of the queen as the only reproductive individual in the colony (Robinson, 1996). This pheromone coats the queen's entire body surface but is most concentrated on her head and feet. Most of the pheromone spreads through the colony through the activities of the workers that are attending to the queen. But some spread through the wax of the comb (Naumann et al., 1991). The pheromone prevents the workers from feeding the larvae the special diet that causes them to develop into rival queens. When the queen dies, the inhibiting substance is no longer produced, and new queens can be reared (Wilson, 1968). Vertebrates also produce primer pheromones that influence reproductive activity. They may regulate reproductive activities so that reproduction occurs in the best social or physical setting.

The vomeronasal organ needs attention for pheromones, and the main olfactory system is for smelling odorants. No distinct functional differences between the vomeronasal organ and the main olfactory system are known (Baxi et al., 2006). The vomeronasal organ gets stimulated by substances other than pheromones. In a hunting snake, the vomeronasal organ responds to chemical cues of prey by the flicking tongue. Chemicals from prey species do not fit the definition of a pheromone, and the behaviour of interest is foraging, not communication. Some chemical signals such as the rabbit mammary pheromone function via the main olfactory system. The scent marks left by female hamsters around their territories (Swann et al., 2001). These marks contain vaginal fluid pheromones that a male detects via his main olfactory system and that prompts him to locate the female. Once the male found the female, another component of the vaginal secretion was perceived through his vomeronasal organ, prompting him to investigate and mount her. In this way, it works for sexually inexperienced males. Sexually experienced males have learned the odour cues of receptive females. They no longer need input via the vomeronasal organ to stimulate mounting. Thus, either the vomeronasal organ or the main olfactory system perceives the pheromones. In some situations, the two systems work together, although this relationship changes after experience.

5.7 COMMUNICATION BY TOUCH

Animals also communicate by touch. Tactile messages can be sent rapidly. It is easy to locate the sender, even in the dark. Honeybee scouts inform nestmates of the location of a food source by dancing. The recruits cannot see the choreography because the hive is so dark. but they follow the dancers' movements by touching them with their antennae. Tactile signals are only effective over short distances and are not effective around barriers. A message sent via touch varied in several ways. For example, how the recipient is touched, where it is touched, and touch duration (Hartenstein et al., 2006a).

5.8 COMMUNICATION BY ELECTRICAL FIELDS

Two distantly related groups of tropical freshwater fish produce electrical signals used in both orientation and communication. These groups are the knife fishes (gymnotiforms) and the elephant-nose fishes (mormyriforms). Gymnotiforms and mormyriforms are "weakly electric," in contrast to torpedo rays or electric eels. The latter generates very strong electric discharges to stun prey or predators. Electric organs derived from muscle generate electrical signals generated by electric organs derived from muscle. When a normal muscle cell contracts, it generates a weak electrical current. The modified muscle cells in an electric organ also generate a weak electrical current. Because they are arranged in stacks, their currents are added. This results in a stronger current. When the electric organ of a gymnotiform or mormyriform discharges, the tail end of the fish, where the electric organ is located, becomes negative on the head. Thus, an electrical field is created around the fish. This electrical field is the basis of the signal. Diverse signals can be created by varying the shape of the electrical field, the discharge frequency, and the timing patterns between signals from the sender and receiver, as well as by stopping the electrical discharge. Electric organ discharges are detected by special sensory receptors in the skin called electroreceptors (Bullock et al., 2005).

Two general patterns of electric organ discharge occur in electric fish: wave-type and pulse (Zakon et al., 2008). These two distinct patterns have evolved in both gymnotiforms and mormyriforms. Those species whose discharges are classified as wave-type patterns produce signals, and the waveform of the signal is monophasic. Species whose electric organ discharges are emitted as pulses produce discharges at higher rates when active and lower rates when resting. The waveform of the signal has a complex, multiphasic structure. Electrical signals are suited for transmitting information that fluctuates, such as aggressive tendencies (Hagedorn, 1995). An electrical signal does not propagate away from the sender but instead exists as an electrical field around the sender. Because an electrical signal is not propagated, its waveform is not distorted during transmission. As a result, the waveform of the electrical signal may be a reliable indicator of the sender's identity (Hopkins, 1986a). Although the waveform is constant for an individual, it is different in males and females and among different species (Stoddard et al., 2006). (Hopkins, 1986a).

Electrical signals are well suited for communication in the environments in which electric fish live. Both gymnotiforms and mormyriforms are active at night. Generally. They live in muddy tropical rivers and streams or at depths in tropical rivers with poor visibility. Electrical signals move around obstacles and are undisturbed by the suspended matter. They are effective over short distances of 1 to 2 m, based on the depth of the water and the relative positions of the sender and receiver (Hopkins, 1999). The shortness of their effective distance may be an advantage. Different electric species may coexist in an area. So, the short effective distance of the signal may reduce electrical "noise" when many individuals signal at once. Electric fish use electrical signals to convey the same messages that other organisms send through other channels. Males of some species not only advertise their sex and species by electrical signals. They also court females by "singing" an electrical courtship song (Hagedorn, 1986). In some species, a courting male and female engage in electrical duetting. This is a coordinated pattern of communication in which the signals of one individual alternate with those of the other (Wong and Hopkins, 2007). Electrical signals are used during agonistic encounters, where certain patterns of discharge have been associated with aggression, dominance, and submission (Hupé and Lewis, 2008). Parents and offspring communicate via electrical signals to maintain proximity to one another (Crampton and Hopkins, 2005).

5.9 MULTIMODAL COMMUNICATION

Animals do not always use a single channel for communicating. The displays of animals have signals that come from two or many sensory modalities. This type of communication is multimodal. Signalling in various channels can occur (Partan and Marler, 2005). The production of multimodal signals is interesting. The courtship display of a male bird contains visual and auditory signals. During grooming, a monkey touches the skin of its companion. At the same time making a certain facial expression and vocalization. When humans communicate through speech, signals are passed through the visual channel. It is easy to understand a spoken message if we have visual and auditory signals from the sender. An elephant's vocalization could have seismic and auditory components. In multimodal communication, the messages conveyed using signalling channels are either enough or inefficient. Measuring whether the messages are redundant requires presenting each component of the display to a recipient. It also needs to know how the recipient responds (Partan and Marler, 2005). In redundant messages, the recipient responds in the same way to each component presented. In non-redundant messages, the recipient should respond to the different components. Let's consider the courtship display of male brush-legged wolf spiders (Gibson and Uetz, 2008). Males court females using a complex "jerky-tapping" display using both visual and seismic signals. The visual part of the display consists of the male raising and lowering his first pair of legs. Each of these legs is adorned with a conspicuous tuft of bristles. The seismic components of the male courtship display include:

(1) stridulation produced by organs on the pedipalps,
(2) up and down bouncing of the entire body, and
(3) striking the substrate with chelicerae.

The signals in both channels elicit receptivity in females. The information content of visual and seismic signals appears similar. Both allow a female to assess the quality of the displaying male. Visual and seismic signals contain information in the male wolf spider's display. If females respond to individual presentations of visual and seismic signals, nonredundant messages result.

Multimodal communication has benefits for senders and receivers (Partan and Marler, 2005). For nonredundant multimodal signals, more information can be sent per unit of time. For redundant multimodal signals, the message will be received and recognized even if one sensory channel is noisy. In the case of the brush-legged wolf spiders, the leaf litter in which they live make communication difficult. Under such conditions, redundant signals prevent miscommunication between males and females. This led to the male becoming the female's next meal rather than her mate.

Miscommunication leading to mating with the wrong species would also be costly (Gibson and Uetz, 2008). Multimodal communication has costs for senders and receivers (Partan and Marler, 2005) .Signalling in many sensory modalities may need more of the sender's energy. Recipients may need more energy to receive and process signals. Signalling in many channels makes senders susceptible to predation. Even receivers are susceptible to predation. Because they have fewer senses to detect signals from conspecifics and predators.

The courtship display of a male bird contains visual and auditory signals. During allogrooming, a monkey touches the skin of its companion. At the same time, the monkey makes certain facial expressions and vocalizations. When humans communicate through speech, information is often passed through the visual channel. It is easy to understand a spoken message if we receive visual and auditory signals from the sender. An elephant's vocalization could have seismic as well as auditory components. During multimodal communication, the messages conveyed through various signalling channels can be redundant or nonredundant. This can be determined by presenting each component of the display to a recipient and seeing how the recipient responds (Partan and Marler, 2005). In the case of redundant messages, the recipient should respond in the same way to each component presented. In the case of non-redundant messages, the recipient should respond to the different components.

5.10 OTHER CUES USED IN FORAGING

As alluded to earlier, whether waggle dances convey information has been the subject of controversy (Munz, 2005). Wenner (2002) thought that bees rely on odour cues. The dance language is not used by recruits to gain information. Although most researchers accept the waggle dance as a communicatory system, Wenner's work helped focus on odour cues used by bees. The waggle dance directs bees to a particular area. The location of the flower is indicated by odour cues. Dance followers can detect food scents on the dancers and even contact the dancers in the appropriate place. hive-mates foraged for pollen and contacted the dancers' legs. Dancers that had collected sugar solutions had head-to-head contact (Daz et al., 2007). Besides the chemicals they pick up from food, dancing foragers also produce other chemicals from their abdomens. Four chemicals were released by dancers by comparing air samples taken from dancing and nondancing bees and analyzing them (Thom et al., 2007). When the researchers injected these chemicals into the air inside the hive, more foragers exited the hive. Thus, it seems that these chemicals cause the bees to become primed to look for food.

5.11 THE EVOLUTIONARY ORIGINS OF SIGNALS

The signal has meaning for both the sender and the receiver. These signals did not spring into being. How did they originate? How does an incipient signal get meaning? Two different approaches need consideration of the evolutionary origins of signals. One focuses to identify the behaviours of the senders which is the raw material for signals. Ritualization was the focus of some observers. How signals exploit the receiver's sensory biases is the second focus. These two evolutionary pathways may have played a role in each system.

Communication involves many sensory channels. These are chemical, touch, vision, auditory, and electrical fields. The channel used for a specific signal depends on the signal's habitat, biology, and function.

A communication system involves the following seven components:

- **Sender:** An individual, which release a signal.
- **Receiver:** An individual, which receives a signal when behaviour is changed by the signal.
- **Signal:** Conspicuous behaviour patterns released by the sender. Signals include odours, postures, colours, sounds, shapes and motion.
- **Channel:** A pathway through which the signal travels, e.g. vocal-auditory channel.
- **Noise:** Background activity in channel that is unrelated to signal.
- **Context:** Setting, in which a signal is released and received.
- **Code:** Complete set of possible signal and contexts.

Signals

Signals serve as sexual advertisement to attract mate. Successful reproduction needs suitable mating partners, or mate quality. Male satin bowerbird, *Ptilonorhynchus violaceus* use brightly coloured elaborate bowers for attracting female. When a female approaches the bower, male dance elaborately, which may not allow male to mate with female (Borgia 1985). In the absence of visual signals males find less chance to mate. Due to more reproductive investment females are the choosy sex. In species with reversal sexual roles females signal to attract males. Females during mating period produce temporary striped pattern which are attractive to males than unornamented females in *Syngnathus typhle* (Berglund et al., 1997). Signals resolve conflict including territory defense. When males compete for females, costs for engaging in physical combat is high. Hence natural selection evolves communication systems to allow males to assess the fighting ability of their opponents without combat. Red deer, *Cervus elaphus* exhibit a complex signalling system. Males strongly defend a female group during mating season, yet fighting among males is uncommon. Male's signals of fighting power include roaring and parallel walks. Altercation between 2 males accelerates to physical fight when individuals are closely matched in size and exchange of visual and acoustic signals are insufficient to know which animal is likely to win a fight (Clutton-Brock et al., 1979). Communication signals help animals to relocate and identify their own young. In species producing altricial young, adults leave their offspring at nest, to forage and gather resources. On returning, adults identify their own offspring, which is difficult in highly colonial species.

Sophisticated communication signals help in integration of individuals into group and maintaining group cohesion. In group-living species forming dominance hierarchies, communication maintains ameliorative relationships between subordinates and dominants. The lower-ranking individuals exhibit submissive displays toward higher-ranking individuals like releasing "pant-grunt" vocalizations in chimpanzees. Reconciliatory signals of dominants show low aggression. Communication coordinates group movements. Contact calls used by various birds and mammals inform individuals about location of group mates which are not in visual range. Communication gives insight into inner worlds of animals and helps to better understand evolutionary questions. Divergence over time in structure of signals to attract mate may result in 2 reproductively isolated populations. Even if populations converge again, clear differences in critical communication signals cause individuals to select mates from their own population. Three species of closely related and identical lacewings are reproductively isolated due to differences in low-frequency songs produced by males. Females show response to songs readily from their species compared to the songs from the other species (Martinez, Wells and Henry 1992). Animal communication systems are important in taking effective decisions for conserving threatened and endangered species. How human-created noise affects communication in other animals is clear from the research study (Rabin et al., 2003).

Signal Modalities

Animals use various sensory channels, or signal modalities to communicate. Visual signals are effective in daytime. Some visual signals are permanent advertisements. Bright red epaulets of male red-winged blackbirds, *Agelaius phoeniceus* are always displayed to defense territory. In blackened epaulets, males were subject to high rates of intrusion by other males (Smith 1972). Some visual signals are produced by individual in suitable conditions. Male *Anolis carolinensis* bob their head and extend coloured throat fan (dewlap) during signalling the ownership of the territory. Acoustic communication is abundant in nature as sound is used in various environmental and behavioural conditions. Sounds vary in duration, frequency structure and amplitude, which affect the distance the sound travel and the receiver localize position of the sender. Many passerine birds release pure-tone alarm calls making localization difficult, while same species produce broadband, more complex, mate attraction songs to allow conspecifics to find easily the sender (Marler 1955). Specialized acoustic communication is found in micro chiropteran bats and cetaceans that use high-frequency sounds to localize prey. After the release of sound, returning echo is detected and processed, allowing animal to build a picture of their surrounding and make accurate assessments of prey location.

Compared to visual and acoustic modalities, chemical signals travel slowly since they diffuse from point source of production. They transmit over long distances and fade slowly once produced. In many moths, females produce chemical cues and males follow trail to female's location. Researchers worked to know the role of the chemical and visual signalling in *Bombyx mori*, by giving males the choice between a female in transparent airtight box and a piece of filter paper soaked in chemicals produced by sexually receptive female. Males were drawn to source of chemical signal and did not respond to the sight of isolated female (Schneider 1974). Some animals have specialized vomeronasal organ to detect chemical cues. Male Elaphus maximus use vomeronasal organ to process chemical cues in female's urine and detect if she is sexually receptive (Rasmussen, et al., 1982). Tactile signals, in which physical contact occurs, can transmit over short distances. This is important in building and maintaining relationship among social animals. Chimpanzees groom other individuals regularly and get great cooperation and food sharing (de Waal 1989). Eectrical signalling is ideal way for communicationin animals inhabiting murky water. Several mormyrid fish species produce species-specific electrical pulses, for locating prey via electro location, but allow individuals searching for mating partner to mark the conspecifics from the heterospecific individuals. Foraging sharks detect electrical signals using specialized electroreceptor cells in head region, which are used for eavesdropping on weak bioelectric fields of prey (von der Emde 1998).

Signals are perceivable indicators of qualities those are not directly observable. Signalling occurs in competitive environment. Interests of the sender and receiver seldom align exactly, and often they are quite at odds with each other. Sometimes, the competition is fierce and overt, as with prey and predators. Potential prey may signal to predators that they are poisonous. Potential competitors may signal their strength to each other. Cues are "any feature of the world, animate or inanimate that can be used as a guide to future action" (Maynard Smith and Harper 2003). Cues need not be intentional and the information gleaned from a cue may not be beneficial to person or animal producing the cues. In fact, signals are a subset of cues. They can be used as a guide to future action, but not cues are signals. The requirement that signals must be intentional means that for the most parts, producing the signal must be beneficial to the signals. If it is not, the signaller will cease to produce it.

Cues as signals to the reproductive status

In many mammalian species, mating activity is restricted to the 'estrus' period that coincides with the fertile phase of the female ovulatory cycle. Apart from the behaviour, the patterns of olfactory or morphological cues displayed by a female may also change during 'estrus'. For males, who are not able to detect ovulation itself, these cues can serve as indirect signals of the fertile phase. As sexual behaviour in monkeys and apes is less strictly controlled by hormones than in other mammals, the term 'estrus' has often been considered inappropriate for these species. Although in general, signalling their readiness to mate lasts much longer than the fertile phase and is more variable in Anthropoid primates than in other mammals, the variety of female cues that can stimulate male mating activity remains the same (Reichert et al., 2002).

Most obvious morphological cues of impending ovulation are sexual swelling of many old world primates. Here, sexual skin around the perineum begins to swell or redden during follicular phase of the ovulatory cycle showing a conspicuous maximum around the presumed time of ovulation (Girolami and Bielert 1987). These cyclic changes in appearance of sexual skin usually reflect cyclic fluctuations in secretion of ovarian hormones during female cycle (Dixon 1998) and markedly influence the attractivity of females to males. However, moderate changes in external genitalia can be observed in many non-primate mammals and prosimians during periovulatory phase (Dixon 1998). Occurrence of exaggerated sexual swelling is mainly restricted to old world primate species with multi-male mating systems (Hrdy and Whitten 1986). Duration of tumescence and of maximum tume-scence of swelling varies considerably between species, the latter ranging from 3 to more than 20 days (Aujard et al., 1998).

Functional significance of sexual swelling has been the subject of much discussions and several hypotheses have been suggested. Sexual swellings have evolved as honest signals of ovulation that increase paternity confidence for a high-ranking male, and thus, allocate paternal care (Hamilton 1984). They increase male-male mating competition, and therefore, female's chances to mate with a superior male. Consequently she gains profit from the male's superior quality, either for herself or for her offspring. Swelling enables females to mate with many males, and therefore, confuse paternity and probably minimize the risk of infanticide. They are a graded signal that allows females to follow a mixed strategy of biasing and confusing paternity by mating with dominant male at peak swelling and with multiple males outside peak swelling. Sexual swelling conceals ovulation, force males into long lasting consortships and informs males about female quality and serve as a social passport during inter-group transfer.

Analyzed endocrine, morphological and behavioural changes across the ovarian cycle of Tonkean macaques suggest that their sexual swellings are a reliable indicator of periovulatory phase (Aujard et al., 1998). Compared with this species and also to chimpanzees, sexual swelling in bonobos (*Pan paniscus*) are characterized by a longer and more variable duration (duration of maximum tumescence in Tonkean macaques, 6–15 days; in chimpanzees, 7–17 days; in bonobos, 6–24 days). In bonobos, mating activity was found to peak at maximum tumescence, but females also mate, when tumescence is below maximum. Thus, it has been concluded that sexual swelling does not allow male bonobos to assess the time of ovulation reliably (Furuichi, 1987). Heistermann et al., (1991) investigated the timing of ovulation with respect to maximum swelling and concluded that sexual swelling is a poor predictor of ovulation in bonobos.

Evolution of chemical cues into signals

Cues used for social recognition of kins, clans, and colony members are complex, greatly varied mixtures of many compounds. Differences between odour mixtures are message. The sex pheromone, each elephant produces is highly individual-specific odour mixtures and can be used by other elephants for recognition of kin, clan or social group, and perhaps individuals. Elephants spend much time sniffing each other. While pheromones are detected by 'sniffing' air or water after travelling some distance from signaller, many chemical cues are detected by contact chemoreception, as in case of an ant tapping its antennae on a fellow ant to detect complex mixtures of chemicals on its cuticle that differ between colonies and allow distinction of nest mates from strangers. Pheromones may be transferred directly from signaller to receiver. For example, male Queen butterflies (*Danaus gilippus*) deposit crystals of pheromone, danaidone from their hair pencils directly onto antennae of female. In this same continuum included molecules passed together with sperm to female during mating in many species.

Chemical senses are shared by all organisms including bacteria. So animals are pre-adapted to detect chemical signals in the environment. Chemical information is used to locate potential food sources, to detect predators and to receive the chemical signals in social interactions. Signals are derived from movements, body parts, or molecules already in use and are changed in the course of evolution to enhance their signal function. Thus, pheromones evolve from compounds like hormones, host plant odours, chemicals released on injury, or waste products originally having other uses.

Evolution also occurs in senses and response of receiver. Ubiquity and extraordinary diversity of pheromones are evolutionary consequence of olfactory system. Most animal olfactory systems have a large range of relatively non-specific olfactory receptors. This means that almost any chemical in rich chemical world of animals, stimulate some olfactory sensory neurons and can potentially evolve into a pheromone. If detection of a particular chemical cue leads to greater reproductive success or survival, there can be selection for that chemical. In some cases, animals evolve a finely tuned system including specialized sensory organs and brain circuits, as found in male moths, which is used to detect and respond to female pheromones. Any pheromone signal that overlaps receiver's pre-existing sensory sensitivities, for example for food odours, is likely to be selected over others. This is the phenomenon of sensory drive. For example, female moths use plant odours to find host plants when egg laying, so their olfactory system is already tuned to these odours, and male pheromones have evolved to exploit the sensory bias of females.

Alarm calling

Many organisms, particularly birds and mammals use vocalizations to communicate danger to other members of a group or serve as a warning of a predator. Alarm calling seems to be an altruistic event because caller appears to put itself in danger by attracting the predator's attention, while learning time for other group members to seek refuge. These calls may be species-specific, age-specific, sex-specific or even predator-specific. Sound spectograms of typical alarm calls given by adult male and female vervet monkeys to leopards, martial eagles and pythons are recorded.

5.12 FUNCTIONS OF COMMUNICATION

Communication is integral to animal behaviour. Some signals bring individuals together, and others help keep them apart. Some signals settle conflicts, whereas others incite them. Some cause alarm, and others are pacifying. Some signals attract suitors, and others repel them. Descriptions of communication are woven. Here we will highlight a few of the messages that signals convey.

Species Recognition

Many species of animals are attuned due to their conspecifics, which are likely to be competitors. Conspecifics are likely to be competitors. They want the same shelter, food, and mates. Other conspecifics might be friendly members of a social group. They may be potential mates that they persuade to support. The term heterospecific is not applicable to conspecifics. "Malechickadeese respond a little to the song of a grey-crowned rosy finch. When creases then a speaker broadcasts the song in another sturdy, its call rate (Charrier and Sturdy, 2005) species use all sensory channels for species recognition in one species or another acoustic feature, Bradbury and Vehrencamp, 1998) a specific species. The chickadees use many acoustic features whether the produced call belongs to a stranger or a conspecific (Charrier and Sturdy, 2005). Birds use song features such as frequency and syntax. Crickets make sounds by scraping together a file and rasping on their wings. They are incapable of modulating the notes of their songs. They rely on their rhythmic pattern differences. Some insects use olfactory cues. Some species use specific pheromones to attract mates. Others rely on visual cues, such as display and colour patterns.

Digitization and manipulation of bird songs to detect the omission, addition, or playing and the speed of components in another order. Playing back the modified calls can detect the response of free-living birds. The chickadees use many acoustic features, whether the produced call belongs to a stranger or a conspecific (Charrier and Sturdy, 2005). Crabs (Uca) signal by waving their large chelipeds. Different Uca species wave in different patterns. U. rhizopharae waves up and down, and *U. annulipes* forms large circles. *U. pugilator* waves in small circles (Crane 1941). These display patterns are important for species recognition. In *U. mjoebergi* and *U. capricornis*, species recognition depends not on display behaviour but on the colour of the cheliped. Female *U. mjoebergi* prefer a male with a bright yellow cheliped, even if it is a male of another species with a painted yellow claw (Detto et al., 2006).

In species selection, different signals will be strongest where species occur together in the same place, as in frogs. Male frogs and toads usually attract their mates by calling at night. the response of nearby females, these calls seem to be the amphibian equal. The calls of different species range from high-pitched peeps to trills to bass-drum booms. Males of several species often serenade together in a chorus. A female chooses one of her own kind from the various callers at the local pond. Female's ability to differentiate is important. Males are unselective and grab any female (or male) of about the right size (Gerhardt 2001). Thus, selection has favoured clear species differences in calls. A skilled person can identify the species of many frogs by their calls alone. Selection has shaped the ability of female green tree frogs to differentiate conspecific calls from other calls. In parts of its range, Hyla cinerea shares its breeding ponds with the related *H. gratiosa*. In other regions, *H. gratiosa* is not present, and H. cinerea lives alone. Female *H. cinerea* that shares the same area with H. gratiosa prefer calls that distinguish them from *H. gratiosa* calls (Höbel and Gerhardt 2003).

Animals cannot always distinguish conspecifics from others. Males of many species court females. Australian Julodimorpha Bakewell beetles try to collect the beer bottles. As the bottles look like females, shiny brown and bumpy, they overcome their inability to use behavioural signals (Gwynne and Rentz, 1983). Even more selective females sometimes make erroneous choices (Gröning and Hochkirch, 2008).

Mate Attraction

Animals that spend most of their time alone face the challenge of locating each other during breeding periods. Besides to being species-specific, the signals that attract a mate must be easy to locate and effective over long distances. Males and females find each other, even if the species are distributed. Chemical and auditory signals are used for this cause but not for attraction. For example, the sex-attractant pheromone of the female silk moth (*Bombyx mori)* has already been described. The pheromone may attract males from 100 meters away. The pheromone of *Actias selene* is also potent. In one study, males displaced 46 km away could move hatched females at the original site (Immelmann ,1980). Auditory signals carry well when amplified by communal display or by special anatomical or environmental features. The courtship songs of birds, frogs, and crickets attract mates from long distances.

Courtship and Mating

Once the individuals are close enough to interact, they court before mating. Communication serves to identify a partner of the appropriate species and sex, assess mate quality, coordinate mate behaviour, and coordinate mate physiology. In some species, it maintains pair bonds after mating. Animals also communicate their sex. presence of antlers and secondary sexual features often differentiate between males and females. Some species have displays that showcase aspects of their bodies and their sex. A female stickleback enters a male's territory. It reduces the probability of attack in a head-up position that displays her egg-swollen abdomen. It distinguishes her from an intruding male (Tinbergen, 1952). In other species, differences are more subtle. Male blue-ring octopuses cannot distinguish males from females until late in courtship. Octopuses insert their modified third right arm into the mantle cavity of the female and release a spermatophore. Male blue-ring octopuses insert their arms into both males and females. But only release spermatophores when inserted into females (Cheng and Caldwell, 2000). Male-male interactions are not aggressive. The fitness costs of making an insertion into a male are low.

Mate Assessment

Courtship may allow a female to judge the qualities of her suitor and choose the most likely one to enhance reproductive success. Courtship displays provide a means for evaluating a suitor's qualities. This includes his physical prowess and ability to provide food for the offspring. Male common terns catch fish and offer them to the female. She compares the quantity of fish provided by her various suitors and chooses the best fisherman. The number of fish a male provides during courtship correlates with the quantity he later provides to his chicks. The quality of the courtship offering is a reliable indicator of the male's ability to provide for the pair's offspring (Wiggins and Morris, 1986). Males of a small bird from Spain show their ability to their mates by collecting stones with their beaks and carrying them to potential nest sites. In a study, males carried 277 stones, on average weighing 1.8 kg, over the course of a single week (Morena et al., 1994). Females watch the males carry stones (Soler et al., 1996, 1999). Male wheatears that carried heavier stones scored better on a test of immunocompetence. This is an indicator of male health (Soler et al., 1999).

Coordination of Behaviour and Physiology

Males and females have different reproductive systems that are not always synchronized. Courtship displays a function to coordinate the couple's behaviour and physiology. Ring doves (*Streptopelia risoria)* are a well-studied example. The behaviour and hormonal state of each partner in the mating sequence, from courtship through nest construction, copulation, egg laying, and feeding nestlings, is programmed (Fusani, 2008). Displays used for coordinating receptivity are generally between partners near each other. Many of them are visual or tactile. Some rely on pheromones at close range. A male mountain dusky salamander applies a courtship pheromone to the female by pulling his lower jaw across the female's back. Then he angled his snout so that his premaxillary teeth scraped the female's skin. then injecting the pheromone into the female's circulatory system. The female then indicates her receptivity by assuming a tail-straddling position. The male deposits a packet of sperm called a spermatophore. The male courtship pheromone makes the female more receptive (Houck and Reagan, 1990). They staged a total of 200 courtship encounters between 50 pairs of salamanders on each of the four nights. Removing the mental gland producing the pheromone from males prevents the males to deliver the pheromone during courtship. These glands were then used to create an elixir containing the courtship pheromone. Thirty minutes before some encounters, each female was treated with this pheromone-containing elixir. Before other encounters, the same female was treated with a saline solution. After a female received a pheromone treatment, she then assumed a tail-straddle position, showing receptivity. She mated 43 minutes and 59 minutes sooner than she did after receiving a saline injection.

Maintenance of Pair Bonds

A final function of communication is the formation of bonds between monogamous pairs of animals. As with displays that coordinate mating, pair-bond displays often occur at close range. Thus, they are often visual or tactile, such as dusky titi monkeys that sit with their tails intertwined. Let's examine a particular pair bond in detail (Sogabe and Yanagisawa, 2007; 2008). Pipefish are long, skinny fish related to seahorses. Like seahorses, males have a brood pouch into which females deposit eggs. The males care for the offspring for one to eight weeks. Pairs are monogamous and only form bonds with other individuals when their mate disappears. Male and female pipefish conduct a greeting ceremony every morning. They approach one another, swim in parallel side-by-side, cross over each other's backs, arch their bodies, or rise into a vertical position. Although the home ranges of pairs overlap with those of other individuals, greeting ceremonies are performed by members of a pair. Greetings are even carried out during the non-reproductive season. Members of a pair meet at a particular site every morning and exchange greetings for several minutes. Then become disconnected from contact for the rest of the day. Because no other benefits seem obvious to this off-season greeting, the pairs do not cooperate in any other behaviour. It serves to maintain the bond with the partner in preparation for the next breeding season.

In some species, social group members, not mated pairs, use communication to maintain their bonds. These communicatory signals are based on contact: resting together, nuzzling. Touching in general tends to form firm social bonds (Eibl-Eibesfeldt, 1975). Many greeting signals exchanged by animals as they encounter one another serve as an assurance of non-aggression. Chimpanzees often greet by touching hands or sometimes by placing a hand on the companion's thigh (Goodall, 1965). Sea lions rub noses, and lions rub cheeks. African wild dogs greet one another by pushing their muzzles into the corners of each other's mouths (Schaller, 1972). You may have been head-bumped by your cat—you should take this as a compliment. In many mammals, social bonds appear to be built and maintained through social grooming, also called allogrooming. Maria Boccia compared several aspects of social grooming and self-grooming in rhesus monkeys. She reasoned that if the primary function of both social grooming and self-grooming is hygiene, these physical aspects of grooming may be the same in both. Social grooming is different from self-grooming in each of these respects. She concluded that skin care is not the most important factor in moulding the form of social grooming (Boccia, 1983). She showed that the message of the tactile signal varies according to the body site being groomed (Boccia, 1986). The recipient monkey's response depended on which part of its body was groomed. The animal being groomed was likely to move away from the groomer when the posterior part of the body was groomed. The responses to the five body regions groomed may reflect a continuum, from a tendency to maintain an affiliation at one extreme to a tendency to end the interaction at the other. In primates, grooming can smooth over tension and restore relationships after conflicts (Watts, 2006).

5.13 ALARM

Alarm signals warn another animal of danger. Danger presents itself in the form of predators. but individuals may guard against other members of their species bent on infanticide or some other form of aggression. Alarm signals have different functions. Their characteristics depend on their function (Bradbury and Vehrencamp, 1998). Many alarms signals cause those who hear them to flee or take cover. We can predict that "flee" signals should share several characteristics. These signals should be easy to make before it is too late. From the signaller's point of view, it's also best if the signal is difficult to locate. Alarm signals from several sensory channels fit the bill. Examples include rapid visual signals like the flash of a deer's tail; volatile pheromones; and high-pitched sounds. these are hard to localize. Due to similar selective pressures, species often share similar alarm signals. Some passerine birds have features that make them difficult to find.

They begin and end and use only a few wave predators. They emit staccato grunts that are loud and low-pitched. These features allow the grunts to be located and transmitted over long distances. Thereby broadcasting the position of the predator. Other monkeys run into thickets, where the dense brush makes it difficult for a swooping eagle to catch them (Struhsaker, 1967). The responses described are also typical of responses to playback tapes of these three types of calls. The responses, then, are specific to the nature of the alarm call and not to the appearance of the predator (Seyfarth et al., 1980).

Other calls vary according to context rather than the species of the predator. California ground squirrels whistle when a predator arrives. Then there is little time to escape and give a chatter-chat call when predators are at a distance (Seyfarth and

Cheney, 2003). We'll talk more about the evolution lengths centred on 8 kHz. Some species even respond to the signals of other species. Eurasian red squirrels flee or increase their vigilance when they hear the alarm calls of Eurasian jays (Randler, 2006).

Most species use the same signal to show any source of danger. although some use specific calls to name the type of threat. Vervet monkeys classify their most common predators into three groups: snakes, mammals, and birds. Considering the hunting strategy of each type of predator. the features of the alarm call and the response of conspecifics within hearing distance seem to be adaptive. The low-amplitude alarm call is emitted when a snake seeks the attention of individuals near the caller. The caller might be in danger from the slow-moving reptile without attracting other predators in the area. Other monkeys respond by looking at the ground, the most likely place to find a snake. When they see a large mammalian predator such as a leopard, monkeys emit very loud, low-pitched, and abrupt chirps. Such qualities make the call audible from a distance as well as easy for conspecifics to find the caller. The common response is to scatter in the chirp and run for hide in the trees to escape the attack of a leopard. When monkeys spot an avian, they congregate into a group to defend a resource or drive off predators (Bradbury and Vehrencamp, 1998). Compared to signals that cause others to flee, assembly signals generally need to be easier to localize and longer lasting. Often, they are repetitive. You may have heard a flock of crows cawing as they mobbed a hawk. Responses to assembly alarms can be complex. Let us consider the pattern of response pattern of ants to alarm pheromones (Yamagata et al., 2007). First, they freeze. Then they raise their heads while waving their antennae. Then they next move toward the pheromone source and emit the pheromone themselves. Finally, they begin biting the potential enemy. Many neurons complicate this response because they are sensitive to alarm pheromone components.

5.14 AGGREGATION

Besides assembling in response to alarm signals, animals often aggregate for other reasons. to hibernate, to share a resting place or a roost, or to prepare for migration. To select a particularly unpleasant example, consider bedbugs. Bedbugs are household pests that come out of hiding places, such as cracks in a bedframe. in the middle of the night and bite sleeping humans for a blood meal. They have recently been in the news because there has been a resurgence in bedbugs, even in expensive hotels. Bugs enjoy aggregation for many reasons. These include decreased sensitivity to desiccation, protection from predators, and ease of finding mates. To find one another, bedbugs release an aggregation pheromone (Siljander et al., 2008). Humans will be able to exploit this pheromone to control bedbugs.

Agonistic Encounters

Animals conflict with conspecifics for territory, mates, food, and their dominance hierarchy. Agonistic behaviours are the actions involved in the conflict. These are aggressive behaviours, such as threats and attacks. Submissive behaviours are appeasement or avoidance. Familiar examples are bighorn sheep butting heads, cats hissing, and dogs rolling on their backs to show their vulnerable bellies to a strong rival.

5.15 BIRD'S SONG

Melodic songs of thrushes indicates the arrival of spring. Birds attract mates and defend their territories against nonspecific rivals with their elaborate songs. Bird sounds can be calls or songs. Call is a short and simple vocalization. It gives signals for flight or danger throughout the year (Catchpole and Slater 2008). *Song* is the long and complex vocalization during breeding season and are organized into several phrases (or motifs) consisting of series of syllables which are in turn made up of collection of single notes (or elements). Each bird's own song repertoire consisting of different versions of song called *song type*. Variation exists in repertoire size between species. About a third of all songbird species have single song type in their repertoire. In about 20% of all species, repertoire consists of more than 5 songs (MacDougall-Shackleton 1997). In brown thrashers (Toxostoma rufum), song types exceed 2,000. White-crowned sparrows (Zonotrichia leucophrys) after a period of few months after hatching learn songs (called close-ended learners). Open-ended learners like European canaries (*Serinus canaria*) and starlings (Sturnus vulgaris) learn and add new songs to their repertoire throughout lives. Only male sings in some species. Males and females sing equally in duets.

Birds communicate information about danger, food, sex, group movements and other purposes via acoustic signals. A subset of these is termed song which features melodic characteristics. Zebra finch's song includes several introductory notes. Such notes are followed by string of syllables in an extended melodious pattern. Sonograms as a primary tool are used for comparing bird songs. Muscles force air streams from large air sacs through bronchi. Membranes in syrinx vibrate when the air expelled from the bronchi and passes over them. Sound producing structures, the syrinx muscles in left and right act independently and many birds sing in harmonies with themselves. Song helps in species recognition. Babbling birds practice coordinated movements of sound producing organs and related structures. During sensory-motor phase, young birds produce plastic song that consists of vocalizations with clear syllables and recognizable elements. Such work include elements from song of tutors and elaborate them into various syllables and phrases that even exceed which is used in adult song (Wada 2010).

Song complexity frequently coincides with presence of ornate plumage. It also stimulates and synchronizes courtship behaviour, reproductive readiness in females and pair bond maintenance. Local song dialects are present in many species. Song in most adult male song birds depends on memorizing calls of a conspecific tutor during an earlier, sensitive phase in life. Appropriate song repertoire is acquired in a series of distinct stages. Young birds listen to a conspecific tutor; obtain information about characteristics of its own song. Very specific subset of surrounding songs is actually accepted as suitable, suggesting presence of an in-built song template. Then young birds vocalize them. Their sub-song is an atonal, noisy, meaningless repetition of sounds lacking recognizable syllables. Young birds spontaneously produce plastic song consisting of vocalizations with distinct syllables and recognizable elements. Ability to hear its own vocalization is critical for normal development. In transition to mature phase, birds adopt a crystallized song with syllables and syntax structure that are the characteristics of species. These song patterns remain fixed in many species, and are presented intact during subsequent breeding season. Open-ended learners retain capacity to adjust their song throughout life. Song production is under control of multiple hormonal systems from embryonic gonads. Injections of testosterone induce adult males to sing, even out of season, while similar injections in females have no such effect. Presence of estrogen during male development appears to be essential. When estrogen is blocked in developing males, testosterone injection fails to elicit song. When estrogen is delivered to developing females, injection of testosterone elicits song in them.

Higher vocal centre (Figure 5.2) is a group of neurons in forebrain that is larger in (singing) males than in (nonsinging) females. Damage to it blocks song production in adults. Nucleus of archistriatum in males is larger in females and its neurons increase in size. Denritic arborization occurs during song learning. Damage to this area blocks song production in adults. Lateral magno-cellular nucleus of anterior neostriatum is neither sexually dimorphic nor shows seasonal change in neuron size or number. Its ablation in young birds interferes with song acquisition. Area X of paraolfactory lobe is sexually dimorphic and new neurons are added in song learning. Damage to it interferes with song acquisition in young birds, but not in adults. According to some scientists, songs are of two types—primary song and secondary song. Primary songs are used in territorial defense and mate attraction. Secondary songs may be whisper songs or duetting. Song frequencies tend to be associated with vegetation. Frequencies of song are low in dense vegetation while there are high in sparse vegetation.

Song of Zebra finch

Song production requires flow of air through semi-independent vibrators in syrinx and vocal tract. Singing is largely restricted to males and is under control of androgens. Emergence of song in adult illustrates interactions of genetic and environmental factors in behavioural development. After listening to song of tutors, it starts its own partial vocalizations, rehearsing and adapting its own song, species-specific adult version slowly emerges. Song circuits exihibit extensive plasticity even in adults with ongoing neurogenesis and seasonal changes in neuronal morphology.

In natural world, there is a cost to make noise of any type. Among other things, it can alert potential predators. So, it makes sense that if there is a risk to vocal communication, there must also be some important benefits. Most migratory songbirds are territorial and fight for places for next year, but one of them might be vacant, just from some birds that died over winter. With little energy, the songbird has found a suitable place to rear its young, just by listening to other birds sing about their parental success.

Development of birdsong

Research on song learning in late 1950s was pioneered by the noted scientist William Thorpe. Thorpe in 1958 observed that nestlings of chaffinch (*Fringilla coelebs*), reared in laboratory condition without getting chance of exposure to adult males of same species produced abnormal song. When youngs was birds exposed to the recordings of wild chaffinch tutor song, they sang species-specific songs like the adults. It shows that the song learning happens early in the life of the birds. According to Peter Marler and others, song dialect is learned during sensitive period (Wada 2010). They also added that birds possess innate predisposition to learn the songs of conspecifics (Marler 1970).

Auditory and Motor Phase

Song learning is a 2-stage process. Birds first memorize tutor song to form a "template," in their brain (sensory phase). They then translate the inner template into the motor activity, and refine the songs (sensorimotor phase).

In the ***Sensory phase***, with few exceptions like Dumetella carolinensis, Acrocephalus schoenobaenus, Molothrus ater, birds when raised in acoustic isolation produce typical songs like the adults. This indicates that most young birds learn species-specific song in the first year. The wild birds grow up listening the songs of various other species. Even without previous experience of hearing of own species' songs, young birds are able to response to conspecific songs (Wada 2010). This again shows their innate capability to identify their species-specific song (Brainard and Doupe 2002). Young birds preferentially learn conspecific song than the heterospecific song. birds raised in acoustic isolation sing abnormal songs with species-specific elements. Timing of sensory phase varies among the species.

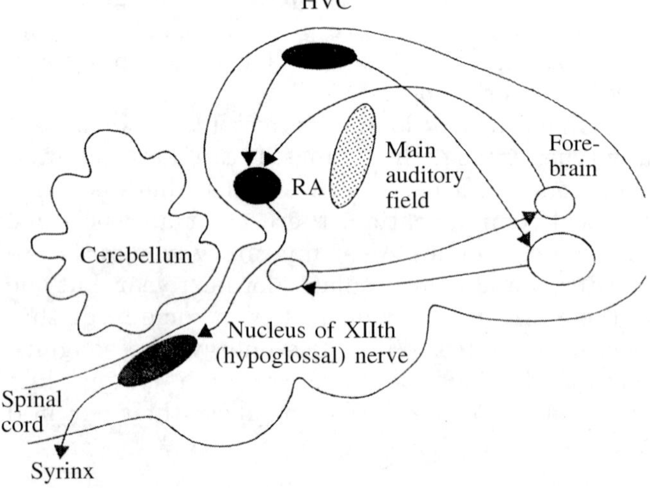

FIGURE 5.2 A simplified diagram of the main neural centres concerned with the control of song in a bird. The brain is depicted in longitudinal section, anterior to the right. The dark-shaded nuclei, control the descending motor pathway for song production. The unshaded nuclei, especially those of the forebrain, are known to be involved with song learning. Main nerve tracts linking the nuclei are indicated by arrows.

End of sensory phase varies among the species depending to some extent on experience. Young birds raised with heterospecific song only, learn conspecific song later than those raised hearing conspecific song (Brainard and Doupe 2002). In the birds are raised in acoustic isolation, sensory phase are extended even into adulthood in some bird species (see Wada, 2010).

At the onset of **sensorimotor phase**, young birds first produce variable and quiet vocalizations called the sub song (Brenowitz et al., 1997). They then produce variable, structured and louder songs called the plastic songs containing the elements of tutor song. The Songs then crystallize to stable stereotyped songs. At this phase, birds hear their own vocalization to produce normal songs. When juveniles are deafened after sensory phase but before sensorimotor phase, they produce aberrant songs (Konishi 1965). Certain avian species produce more sounds in sensorimotor phase than they produce in the adulthood (Wada, 2010). White-crowned sparrows integrate the songs that match dialects of neighbouring males into crystallized song (Nelson and Marler 1994). This may be advantageous as males that sing local dialects have higher reproductive success than those that sing foreign dialects (MacDougall-Shackleton et al., 2002). Cowbirds attain songs that are effective in triggering mating-like behaviours in females (West and King 1988). In both case, song selection during crystallization is based on functional significance to maximize reproductive output (Wada, 2010).

5.16 ECHOLOCATION IN BATS

Echolocation is an auditory imaging system, used by various animals for navigation and perceiving location of prey, when vision is ineffective. This involves emission of vocalizations, detection of echoes and their uses to produce three-dimensional information about environment. Organisms using echolocation are toothed whales, small mammals, such as rats and shrews. Two groups of birds also use acoustic reflections. Echolocation requires specialized neural mechanisms and complex computations. Neural circuitry underlying echolocation allows for perceptual organization of auditory information, which guides complex spatially-guided behaviours.

Auditory signals used can vary a great deal. Narrow band signals, cover a narrow range of frequencies, have a long duration, and allow for detecting targets over long distances. Broadband signals, cover a large range of frequencies, typically lasts for less than 5 milliseconds and best adapted for vocalization. Many dolphin species produce ultrasonic echolocation signals. Sperm whale and dolphin produce signals that are audible to humans. Because sound travels faster in water than in air, signals produced by toothed whales are of a much shorter duration than those produced by bats.

Lazzaro Spallenzani in Italy, and Louis Jurine, a Swiss biologist advocated that bats use their ears, but neither of them had a clue as to how they did it. Throughout the next 150 years at least 20 authorities addressed the problem and proposed theories ranging from wings sensitive to air motion, through the idea that all five senses are involved to some extent, on to postulating a mysterious sixth sense. H. Hartridge actually suggested in 1920 that "bats during flight emit a short wave-length note and this sound is reflected from objects in the vicinity". Two near-novice graduate students solved the old problem using three independent methods:

- They demonstrated that bat ears respond to sound frequencies several octaves above upper limit of human hearing, which means bats are equipped to hear echoes of any very high frequency cry they might emit.
- They compared the obstacle avoidance performance of normal animals before and after being deprived reversibly of vision, hearing, and the ability to utter cries. Only deaf and silenced bats are helpless when flying around obstacles.
- They used a device that converts ultrasonic into audible sounds, to demonstrate that flying bats emit cries, we cannot hear when they avoid obstacles, and that successful avoidance is highly correlated with where and when the emission rate is highest.

Most research into echolocation has been carried out in bats, in which it has attained sophistication. Bats have been echolocating for over 50 million years. Within Microchiroptera suborder are about 800 bat species, most of which are insectivorous and use echolocation to catch food. Echolocating bats have enlarged ears and they can detect faint echoes. Chinese species *Rhinolophus paradoxolophus*, send sound waves through their nostrils, and have evolved large nose.

Mechanism of echolocation

A few bat species are able to compensate for changes in frequency of sound waves produced by their own movement and that of prey. Frequency of vocalization echoes that occur while a bat is accelerating towards a target is higher than frequency of emitted sounds because acceleration 'squeezes' the peaks of sound waves closer together. Doppler shift compensation involves an adjustment of frequency of emitted vocalizations, so that frequency of echoes fall within the range that can be perceived by its auditory system.

Echolocation provides detailed information about a target. It enable bats to discriminate acoustic image of small flying insects in the environment, containing dense vegetation with thousands of surfaces, which themselves reflect sound waves. Shape of a target can be perceived by spectrum of echoes from different parts of the target. Other data obtained from echo spectrum include relative velocity of a target, flutter of target, and target's size and elevation. The time delay between production of click and reception of echo provides information about distance of objects. In air, sound travels at 340 metres per second, so a delay of 2 milliseconds between production and reception corresponds to a target that is 34 centimetres away. In sperm whales, vocalization increase in frequency as they approached their prey, so do bats produce 'feeding buzzes' as they come-in on insects.

In bats, auditory cortex carries out computations, which transform echo spectrum into estimates of distance. In bats, audition is primary sensory modality. Sound waves are directed by ears, and sometimes the nose, to inner ear where they fall on basilar membrane of cochlea. Basilar membrane vibrates in response to sound waves which stimulates hair cells to produce nervous impulses, which are sent to auditory cortex via brainstem. Cochlea of bat contains hair cells, each of which is adapted to respond to sound waves of a specific frequency. Frequencies of sound waves produced and perceived by bats (14–100 kHz) are well outside the hearing range of humans. In auditory cortex, neurons in specialized ganglia are also adapted to respond to sounds of specific frequencies.

Specialized neuroanatomical structures of bat are enlarged brainstem nuclei of ascending auditory pathway. Neurons in some nuclei of lateral lemniscus are arranged in columns which receive tonotopically arranged inputs via conspicuously large calyx synapses. Neurotransmission across these synapses occurs particularly quickly, and neurons in these nuclei function as time markers. These cells process frequency-modulated components of echolocating signals, and are specialized for detecting interneural time differences.

Bat emits a sonar sound that travels outward to impinge on insect at some target range. Reflections return to bat from various insect body parts are separated by a small difference in range. These sources of reflections are called **glints**. Insect is depicted here as a 2-glint target. Sound-pressure waveforms of bat have transmitted sonar sound and reflections from insect's glints. Transmitted sound is heard directly by bat to initiate processing by establishing a zero-time origin for reception of subsequent echoes. Echoes return to bat's ears, after a delay related to target range at the rate of 5.8 milliseconds/metre. Reflections from 1st and 2nd glints add together at a small time separation to interfere with each other, when they form a combined echo arriving at bat's ears. For each transmitted sound, acoustic stimulus received is not just echo, but a pair of sounds—outgoing broadcast received directly at moment of emission followed by returning. Combined echo comprised of many reflections are returned by target's glints.

Broadcast signal contains two sweeps, a 1st harmonic from 45 to 22 kHz and a 2nd harmonic from 80 to 45 kHz that appear as sloping ridges one over the other. Echo also contains these two harmonics, but sloping ridges for combined echo spectrogram are not as smooth as those in broadcast spectrogram, because interference between overlapping reflections from 1st and 2nd glints create alternating peaks and notches that appear as ripples or undulations. Notches are locations where energy in echo has been cancelled by interference; their locations are principal defining characteristics of ripples, because their frequencies are related to delayed separation of reflections from glints. Bat's inner ear transforms spectrograms of broadcasts and echoes into neural spectrograms composed of spikes that register the FM sweeps in terms of tuned frequencies of different neurons and times or latencies of spikes.

5.17 ECHOLOCATION IN MARINE MAMMALS

Dolphins (Figure 5.3) make a wide array of sounds, including clicks, moans, chirps, creaks, barks, squeaks, yaps, mews, and whistles. They make sounds resembling the engine of a motorboat, and laugh of their trainer. They use clicking noises in echolocation, which allows them to navigate, identify prey and friends, avoid obstacles and predators, and bounce off objects underwater. They use whistles to maintain contact, within their pods or, with other pods of dolphins. Whistles may signal danger, a call for help, or simply identification, may also help dolphins to hunt cooperatively and to coordinate migratory movements. Each dolphin has its own signature whistle. Sounds are carried a long distance underwater. Sounds of captive dolphins make it impossible, to decipher which animals were making which sounds, since marine mammals do not move their mouths, when they vocalize.

FIGURE 5.3 A dolphin.

Whales and dolphins echolocate, emit clicking sounds that bounce off prey or objects in order to locate and identify them. Captive whales and dolphins probably do this much more than wild populations. Although all whales chatter up a storm to keep in contact with each other to communicate, some whales may not use echolocation at all, but find their prey by listening. Sperm whales and right whales are struck by boats. This suggests they may not be using echolocation, to avoid these large objects. Dolphins may turn off their echolocation while travelling, which would explain why they become entangled in drift nets instead of leaping over them. Whales stay in touch with each other using a variety of vocalizations. Killer whales make plaintive wailing noises. Fin whale pods, typically, have one whale that dominates the conversation, while others take turns answering. Blue whales call for 20 seconds, pause for 20 seconds, and then repeat the 20-second call. As underwater noise pollution increases from boat engines, submarines, and oceanographic experiments, it is critical to learn more about the way whales use sound in wild, in order to prevent human activities from disrupting their lives.

Water is a more effective and efficient conveyer of sound. Echolocation may be thus more effective, for detecting objects underwater than light-based vision on land. Sound with a broad frequency range has a more complex interaction with objects that reflect it than does light. For this reason, sound can convey more information than light. Whales, porpoises and dolphins emit pulses of sounds and listen echo. These mammals use sounds of many frequencies and a highly direction-sensitive sense of hearing, to navigate and feed. Echolocation provides these mammals with a highly detailed, three-dimensional image of their environment. Whales, dolphins, and porpoises have a weak sense of vision and of smell. They first emit a frequency-modulated sound pulse. A large fatty deposit, sometimes called **melon**, found in its head helps them to focus sound.

Echoes are received at a part of lower jaw, sometimes called **acoustic window**. Echo's vibration is then transmitted through a fatty organ in the middle ear, where it is converted to neural impulses and delivered to brain. Brains of these sea mammals are at least as large relative to their body size as is a human brain relative to size of human body.

They can locate tiny objects and thin wires and distinguish between objects made of different metals and of different sizes as an object's material, structure, and texture all affect the nature of echo returning to porpoise. Toothed whales have specially adapted structures in their head, for using echolocation. Some species of toothed whales have a bony structure in head that insulates back of skull, where sounds are received from front of skull where sounds are produced. Middle ear cavity is divided into a complex sinus that may help to acoustically separate right and left ears. This would enable whale to more easily glean information from echoes it receives. Other structures help to reduce confusion of transmitted and received sound throughout skull.

At the present time only toothed whales, including beluga whales, sperm whales, dolphins, and porpoises, have been shown to use echolocation during feeding, but it is thought that other groups of marine mammals may have potential to use echolocation during feeding too. Echolocation is typically used by toothed whales to capture single prey items such as fishes or squid. Toothed whales that use echolocation, send high frequency click sounds into environment. Sounds then bounce off distant objects, and echoes are received by animal that produced them. Returning echoes sound differ from original click produced by the animal. Differences between sound of original click and returning echo provide echolocating animal with information about size, shape, orientation, direction, speed, and even composition of object. Dolphins have an amazing ability to detect and identify a target of the size of a golf ball at a distance of 100 metres. Beam of echolocation clicks is also very directional and can be moved with a slight turn of animal's head.

Young dolphins typically stay with their mothers for several years, three to six years or even longer. Even after they leave their mother's pod, they may return for a visit, and daughters often return, when they have their first child. Generally these pods consist of young dolphins between the ages of three and thirteen years. Here, two males may often form strong bonds that may last 10 to 15 years or longer. These male dolphins travel from one pod to another, staying for short periods of time and then moving on. There are pods of coastal dolphins that have lived year-round in the same area for many generations. Norris of the University of California at Santa Cruz examined dolphin skulls, to find evidence that sounds are conducted through lower jaw to dolphin's ears. Barklow found similar structures in hippos that he believes may make it possible for hippos to listen sounds above the surface with their ears, while simultaneously monitoring underwater sounds through its submerged jaw.

Toothed whales and baleen whales produce other sounds to increase their chances of success during feeding. Humpback whales have developed a feeding technique, called bubble. Bubble feeding involves one or a few whales' blowing air from their blowhole, while underwater. This produces sound as bubbles form a cloud, curtain, or column that rises toward surface. Bubbles trap the prey between surface and whales mouth. A bubble net is formed, when bubbles emitted by whales, form a ring and concentrate prey inside. Both sound and bubbles work, to concentrate prey, so humpback can capture more food per mouthful. Bottlenose dolphins also make use of sound and bubbles. Dolphins foraging in sea grassbeds in Australia and Florida use a technique called **kerplunking** to drive fishes from protection of sea grasses. A dolphin lifts its tail, lower body out of water and crashes it down on water surface. This causes a loud splash and creates a trail of bubbles under water. This startles fishes hiding in sea grass and flushes them from their hiding places, making it easier for dolphin to detect them.

Some marine mammal species may use this technique to find prey. Fishes make a variety of sounds that cetaceans and pinnipeds may detect. Transient killer whales sit and quietly listen for sounds of other marine mammals in the area before they make their attack. It benefits killer whale to be as quiet as possible while listening, so it does not scare away its prey. Listening for prey sounds is a mechanism that is also used by seals and sea lions. Once the seal or sea lion detects a fish by sounds, it waits for it to get close enough to use its sensitive whiskers to track path of the fish as it tries to escape. Dolphins produce two main types of echolocation signals—short duration broadband signals and narrowband signals of longer duration in organs called nasal sacs, located on top of the brain. A structure called **melon** acts as a lens, focusing sound waves into a narrow beam which is projected forward. Echoes are received by panbone in dolphin's lower jaw. Fatty tissue behind panbone transmits sound waves to middle ear and then to brain. Teeth are also believed to be involved in transmitting echoed sounds to dolphin's brain.

Sperm whales emit a series of clicks during their dives. These clicks enable whales to find prey by echolocation. As whales descend into depths, frequency of clicks increases enabling whales to obtain detailed information regarding position and movements of the prey. Regular clicks function as long-range biological sonar. Clicking frequency continues to increase as whales approach their prey, until clicks are so close together that they sound like a continuous buzz. Whales descended an average of 392 metres between start of regular clicking and first buzz, which corresponds to point at which they are about to catch a squid.

5.18 PHEROMONES

Word pheromone comes from Greek word 'pherein' which means to carry or transfer, and 'hormone' means to excite or stimulate. Action of pheromones between individuals is contrasted, with action of hormones as internal signals within an individual organism. Pheromones are molecules used for communication, between animals. A broader term for chemicals involved in animal communication is semiochemical. Pheromones are a subclass of semiochemicals, used for communication within species. Semio-chemicals acting between individuals from different species are allelochemicals and are further divided depending on costs and benefits to signaller and receiver. Pheromones were originally defined as 'substances secreted to outside by an individual and received by a second individual of the same species, in which they release a specific reaction, for instance a definite behaviour (releaser pheromone) or developmental process (primer pheromone). Pheromones are by far the most important signal used by organisms of all kinds. The way that a flock of birds fly or a school of fish swim may involve more than individuals' simply judging the distance between themselves in the group.

Communication happens when one animal's behaviour influences behaviour of another. Signals are means through which these effects are achieved. Signals may often be ritualized, that is made conspicuous and exaggerated. Reutilization could be evolution of pre-existing chemicals as a pheromone, for example in the way that sex pheromones of goldfish have evolved from hormones leaking out across gills. However, not all signals evolve to be conspicuous. Pheromone signals, like recognition cues in social insects and mammals, may be subtle and complex. Pheromones can be used as honest signals, which provide reliable information because they accurately reflect signaller's ability or resources. Female tiger moths *(Utetheisa ornatrix)* choose a male with most pheromones. His pheromone is derived from same plant poisons used to protect eggs, which he will pass to female at mating. Pheromone load is correlated with male's gift that is given. Male garter snakes court larger snakes which have more pheromones. In mammals, production of pheromone is directly related to hormone levels and so scent marks tend to be honest. Mammals and lizards scent mark their territories leaving signals that are inherently reliable.

Scientists have presented five models describing how animals may receive communication signals and discuss how signal reception affects formation of different patterns, both moving and stationary. Model not only explains five known group patterns, but also reveals five previously unknown patterns. The five animal communication models include attractive, repulsive and align forces. Scientists explain that attractive forces work at long ranges, repulsive forces at short ranges, and alignment forces at intermediate ranges. Five models differ because, for example not all animals can receive signals from neighbours in front, behind, or moving away from themselves, resulting in different individual reactions, and thus, different patterns.

In model M3, animals can only receive signals from individuals in front of them. This means that individuals at the edge of the group that are facing away can leave the group and will not return, since they do not receive information from behind. This is called the feathers pattern, and is one of at least three patterns that M3 communication can result in. In model M4, individuals can receive signals from other animals ahead or behind, but only if those animals are moving towards the reference individual. If the individuals in front are moving faster than those behind them, they can lose the group. But later, the group often picks them back up. This can result in expanding and contracting groups called **breathers** as well as travelling breathers. Scientists suggest that schools of fishes and flocks of birds may use breather behaviour as an antipredatory technique. In model M5, individuals can only receive signals from animals that are ahead and moving toward them. This behaviour seen in Myxobacteria can manifest in a ripple pattern, when two groups approach each other. While a few individuals will continue moving in the same direction, most will turn around due to repulsive forces. This behaviour gives impression that two waves pass through each other. Because in different models, different amounts of information are received by an individual, scientists could also infer that amount of information received seems to influence strength of alignment force required to produce a group pattern. In other words, animals that can receive signals from more directions can align more easily.

Elephants and moths are unlikely mates. It was surprising, when it was discovered that Asian elephant (*Elephas maximus*), shares its female sex pheromone with some 140 species of moth. Compound is a small, volatile molecule. Elephants and moths-convergent pheromones discovery explains that elephants and some moths share the sex pheromone (Z)-7-dodecen-1-yl acetate. This particularly interesting because it illustrates important points emerging about pheromones in mammals and insects, and animals in general. It illustrates the ubiquity of pheromones across the animal kingdom, and more interactions are mediated by pheromones than by any other kind of signal.

Second, shared use of a compound as a signal illustrates a relatively common phenomenon of independent evolution of particular molecules, as signals by species that are not closely related. Such coincidences are a consequence of the common origin of life: basic enzyme pathways are common to all multi-cellular organisms, and most classes of molecules are found throughout the animal kingdom. However, despite sharing an attraction to (Z)-7-dodecen-1-yl acetate, male moths and elephants are unlikely to be confused. Apart from mating difficulties, male moths are unlikely to be attracted by the pheromones in female elephant urine because moth pheromones contain multi-components. The (Z)-7-dodecen-1-yl acetate would be only one of perhaps five or six other similar compounds making up a precise blend for each moth species. Male elephants are unlikely to be attracted to a female moth because she releases such small quantities (picograms per hour) that would not be noticed by a male elephant. In some cases, same compound is used for similar functions in different species. More commonly, the arbitrary nature of signals is revealed by different uses for the same compound. Pheromones are often divided by functions, for example into sex pheromones and aggregation pheromones. Individuals from other species can perceive signals broadcast to wider world. Predators are using bark beetle pheromones as kairomones. Animals of one species can emit signals that benefit themselves at cost of receiving species. Chemical signals used in such deceit are termed allomones as bolas spiders synthesize particular moth pheromones to lure male moths of those species into range for capture. Semiochemicals benefiting both signaller and receiver in mutualism, as between sea anemones and anemone fishes are termed synomones.

5.18.1 Types of Pheromones

Alarm pheromone

Many alarm pheromones provoke fight or flight in receivers and appear to be evolved from compounds released during fighting or injured co-specifics. Over evolutionary time, defensive compounds may gain a signal function. For example, most ant species use same chemicals, for defense and alarm, to repel enemies to alert, and recruit nest mates. This pattern is found in arthropods. In other animals, alarm pheromones may derive from compounds which to make the flesh unpalatable or toxic to predators. These compounds released by injured animal, for example, anthopleurine in sea anemone and bufotoxins and larval skin extract which elicit an alarm response in toad tadpoles. Alarm pheromone of fish is not an antifeedant, but may have evolved with a primary function, such as control of skin pathogens. In response to danger or stress; ants release a chemical signal which is sensed by other ants nearby. In response, these nearby ants travel towards the scent and work quickly to help their endangered nest mate. Honeybees have alarm signals that are released via pheromones.

Trail pheromone

Ants release a scent when they are returning to their nest with food. This pheromone signals to other ants that there is food at nest. As more ants follow this scent, it is renewed to tell more ants. As food quantities decrease, ants release a different scent that tells others that there is no more food.

Sexual attractant

Females, mostly in insects most often release pheromones that attract males.

Releaser pheromone

Releaser pheromones bring forth specific behaviour. Many mammals release pheromones that signal to others that they are in their "territory." This is found in dog's urine and also in many other mammals. Female rabbits and some other mammals release a scent that tells their young that they are ready to nurse. This scent then causes their young to immediately begin nursing.

Primer pheromone

Primer pheromones cause a shift in endocrine system of the animal receiving it. Many mammals, including rats and mice, release pheromones that cause sexual behaviour in other sex. When at peak fertility, females release a scent that attracts a male and causes him to become aroused and exhibit mounting behaviour on the female who is also exhibiting lordosis. Pheromones are sensed by Vomeronasal Organ (VNO) in brain. Female rats mature faster when exposed to scent of an adult male rat. Scent of a male rat of one strain can induce miscarriage in a female rat pregnant by a male of a different strain.

Information pheromone

Information pheromones provide information about an animal's identity. By smelling another animal of the same species, they are able to tell what the animal ate last, if they are in heat, their level of dominance, and their level of health.

5.18.2 Evolution of Pheromones

Molecules in coelomic fluids normally released with the sperm by sexually mature adults may become pheromones. Marine polychaetes release sex pheromones with their gametes, which prompt other sex to release its gametes. Hormones or other molecules associated with reproductive cycles have evolved into pheromones by eaves-dropping in many animals. In elephants and mice, some pheromones are excreted in urine. In fish and lobsters, pheromones may have evolved from molecules excreted in urine or leaking into water across the gills. Hormone-based sex pheromones in goldfish, *Carassius auratus* provides a good model system. Male goldfishes are extraordinarily sensitive to steroid and prostaglandin hormones and their metabolites are released into the water by females. Released molecules reflect blood concentrations of hormones in female and are a reliable indicator of her biological state.

5.19 BEE DANCE

Honeybees exhibit excellent communication abilities. A foraging bee, returning from a good source of food performs a waggle dance on vertical sheets of honey comb. Dance specifies to other bees the direction of food. Dance takes the form of a flattened 'figure 8'; during crucial part of manoeuvre the forager vibrates her body. Angle of this part of run specifies direction of food, points up that source is in direction of the sun. Whereas, if it is aimed, for example 700 left or vertical, the food is located 700 left to the sun. Number of waging motions specifics distance to food. Complex of this dance language has paved the way of studies of higher animals. *Apis mellifera* forages around the hive in search of resources. Usually, this resource constitutes food in form of pollinating flowers. Several landmark studies were performed in middle of 20th century to analyze how these social organisms communicated exact distances and directions to one another to effectively locate these resources. In an experiment, food dishes were offered to test the accuracy of foragers recruited by the dance about distance and direction of food source (Figure 5.4). Karl von Frisch determined that honeybees perform two distinct dance routines that coincide with two different distance approximations made by foraging bee. These two dances, round dance and waggle dance, communicate to other the approximate distance from hive to new resource. Only the waggle dance communicates direction (von Frisch 1967). Round dance is performed by returning

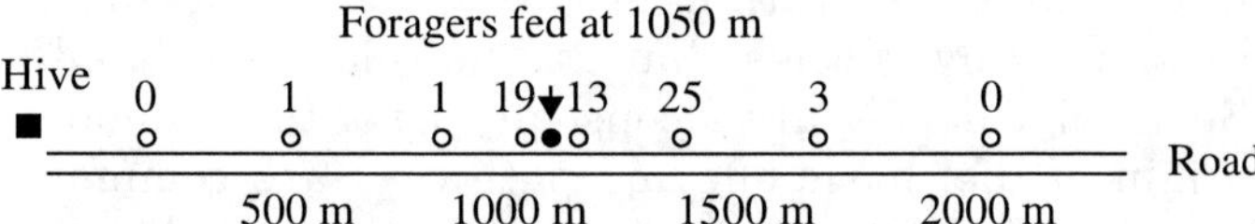

(a) Foragers were trained to a dilute scented food, 1050 m east from hive. Then a series of scented plates without food were kept within 2000 m in the same direction. Dancing was induced by suddenly increasing the sugar concentration at the feeding dish. *a* Recruits were counted but not captured, as they approached various scent plates. The numbers above them record the number of units made to each sent plate. Most recruits appeared at plates close to the feeding station.

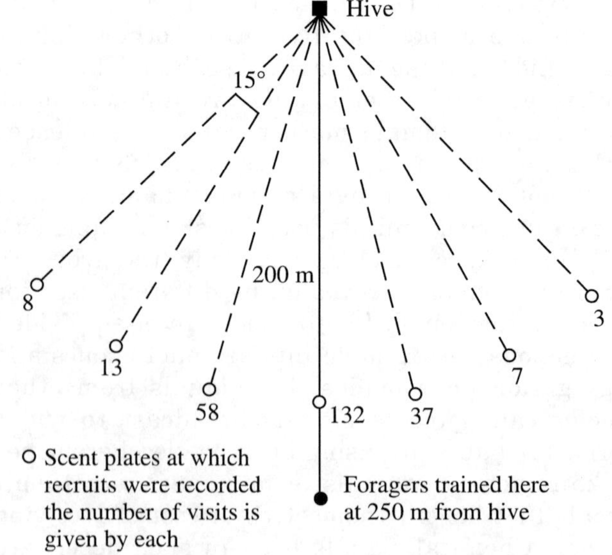

(b) Direction test: Scent plates are put in an array at the same distance but in different directions from the hive. Majority of visits were made to dishes close to the bearing of training station.

FIGURE 5.4 Experiments by Von Frisch and others. (After von Frisch 1967)

bee, usually in complete darkness, vertically on a honeycomb. Circuitous motion attracts other foragers, which then learn that resource is within approximately 50 metres from the hive. No direction is given by this routine. As a result, the newest foragers leave, to search in all directions surrounding the hive. Behaviourally, this dance is energetically favourable, due to short distances travelled. In contrast, waggle dance is energetically unfavourable to individual, but beneficial to hive. Waggle dance is performed, primarily, when resource is far than 50 metres. Returning forager either performs the dance on a vertical surface or a horizontal one. To determine the distance and direction in which the bee orientates itself is relative to the sun. Any deviation from this point gives the angle, the new foragers should pursue. If vertical, bee orientates itself to gravity. Perpendicular to ground becomes the reference point (the sun). Deviations from such relay direction accordingly. Distance is communicated by the length of abdomen shake that forms middle of a 'figure 8' dance. Kirchner et al., determined that round dance does not convey direction. By tipping several hives onto their horizontal axis and placed a resource 10 metres away, the reference heading was experimentally altered. Regardless of hive orientation, number of successful returning foragers was constant. Once the distance crossed 50 metres, success rate decreased for hives orientated horizontally, indicating direction is dependent on hive orientation, for waggle dance to be effective.

Honeybees foraging behaviour can be interpreted from a historical or adaptationist view. Lindauer performed historical analysis that reveals that distantly related bees might have evolved through 3 stages of development: First stage analysis reveals that the genus *Trigona* conveyed direction by buzzing to gain their hive mates attention. Odours trapped on bee trigger others to forage in search of odour. Secondary stage involves marking the path from resource to hive with the mandibular pheromones. She then buzzes, and bees follow the scent markers. The third stage involves an in-flight "waggle dance" in direction of resource (Lindauer 1961). Seely provides an adaptationist theory on bee dance evolution. Foragers seek new resources to provide increased hive fitness. By dancing, cost/benefit ratio is reduced to individual level more, so than a group of foragers expending energy searching in all directions all the time (Seely 1992). The study demonstrates that dancing has become an evolutionarily desirable trait because less time foraging allows more time for collecting the resource. Honeybee dancing provides an evolutionarily advantageous behaviour, which optimizes the hive's fitness. Karl von Frisch in his major book, "*The Dance Language and Orientation of Bees*" (1967) provided a full survey of the whole dance system. Experiments carried out by von Frisch and his co-workers has illustrated in Figure 5.5.

(a) Workers cluster around a bee (returned of sun from a foraging trip).

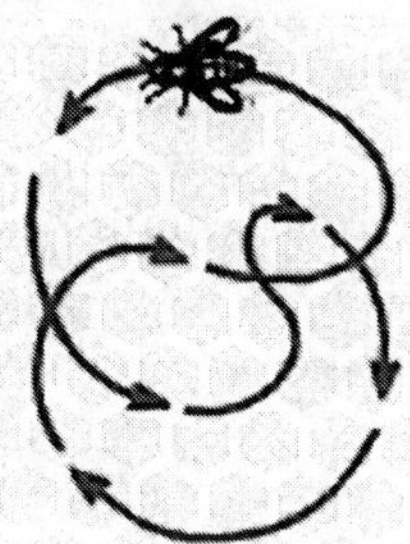

(b) Round dance indicating food's presence nearby.

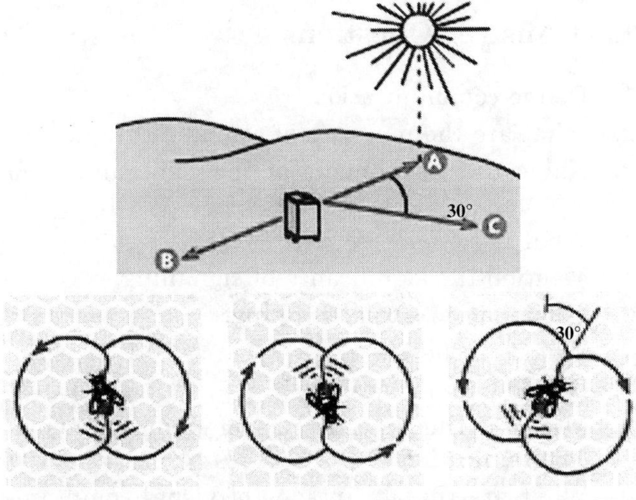

Location A: Food source is in same direction as sun.

Location B: Food source is in direction opposite to sun.

Location C: Food source in 30° right of sun.

(c) Waggle dance performed when food is distantly located resembles a 'figure 8'.

FIGURE 5.5 Honeybee dance language.

5.19.1 Round Dance

It stimulates other workers to leave the hive and search nearby. It conveys the information of searching, within 50 metres and about some olfactory cue. When food source is scented, the dancer carries this scent along with her body. When source is not scented, forager marks it by opening Nasanoff scent gland on her abdomen during drinking. Such dance forms a roughly circular path, just over her body's length in diameter. The bee moves in circles alternately to left and right, stays approximately in the same place on the comb and may dance for up to 30 seconds, before moving on. Other foragers face dancer, often with their antennae in contact with her body and follow her movements closely.

5.19.2 Waggle Dance

This dance occurs at about 100 m–5 km distance. Bee moves in a straight line, shaking abdomen back and forth (waggling) and buzzing the wings. Direction of waggling indicates direction of food source. Richness of food source is communicated by the vigour of dance and how long it continues. Rate at which the bee circles communicate the distance of nectar source form hive; the higher the rate, the closer the food source. About 9–10 complete cycles per 15s occur, when food is located at 100 m distance, but only two, when food is at 6 km. Foraging bee uses the sun as a compass and records position of food source, with respect to it. When tail-wagging run is directed upward, food is located towards the sun. If tail-wagging run is directed downward, food is located away from the sun.

Inclination of run to right or left of vertical, indicates angle of sun to right or left of food source. An internal clock compensates for passage of time between discovery of food and start of dance. So the information is correct even though the sun has moved during interval.

(a) When dance is performed on a wall inside hive, angle that straight run deviates from vertical represents the angle between the sun and flowers, (b) On a horizontal surface outside, waggling run is aimed directly towards the flowers.

Waggle dance is shaped in a 'figure 8' shape, looping first to left and then to right, with the waggling of abdomen. Typically, it consists of a straight run, while abdomen is waggled vigorously. Then bee makes a semicircular turn and waggles in a straight line again. This is followed by another semicircular turn in opposite direction and another straight waggle run. Dance is repeated many times. Direction of straight run indicates direction of food, and tempo of dance signals distance.

Review Questions

Short Answer Questions

1. Define communication.
2. What are the purposes of communication?
3. Name the components of the communication system.
4. What is signal?
5. Mention the significance of signalling.
6. Comment on sexual swelling.
7. What are cues?
8. Define pheromones.
9. Mention the significance of pheromones in communication.
10. What are the advantages of visual communication?
11. What are the disadvantages of visual communication?
12. Define bird's song.
13. Mention two functions of bird's song.
14. What is higher vocal centre?
15. Define echolocation.
16. Mention the significance of echolocation.
17. What is queen substance?
18. What is aggregation pheromone?
19. Define alarm pheromone.
20. Define releaser pheromone.
21. Comment on the primer pheromone.
22. State the significance of vibration in the animal world.
23. Name two animals that use vibration.
24. What is information pheromone?
25. What do you understand by waggle dance?
26. What is round dance?

27. What is the significance of bee dance?
28. Name the scientist famous for studying bee dance.
29. When is round dance performed?
30. When is waggle dance performed?

Long Answer Questions

1. Write an essay on animal communication.
2. Discuss the interesting features of bird's song.
3. Discuss the interesting features of echolocation in bat.
4. Discuss the phenomenon of echolocation in marine mammals.
5. Discuss various biological aspects of pheromone.
6. Write a note on vibration as mediator of communication.
7. Describe bee dance in details with suitable diagrams.

Chapter 6

Defensive Behaviour and Biomimicry

Various kinds of organisms exhibit a number of defensive behaviours. Such behaviours help them to escape from the predators, and increase their survival chance. Such behaviours are very much interesting from the evolutionary point of view. Animal colouration often aids in defense of animals.

6.1 CAMOUFLAGE

In nature, every advantage increases an animal's chances of survival, and therefore its chances of reproducing. This has caused species to evolve adaptations, which help them to find food and keep them of from becoming food. One of the most widespread and varied adaptation is camouflage, an animal's ability to hide itself from predator and prey. Most animal species have developed some sort of natural camouflage. Specific nature of camouflage varies considerably from species to species. Basic form of camouflage is colouration which matches an animal's surroundings. An animal's surroundings may change from time to time. Many animals have developed adaptations to change their colour as their surroundings change. Shifts in animal's surroundings occur with change of seasons. Many birds and mammals deal with this by producing different colours of fur or feathers, depending on the time of the year. In most of the cases, either changing amounts of daylight or shifts in temperature, trigger a hormonal reaction in animal that causes it to produce different biochromes. Feathers and fur in animals are like human hair and fingernails. They are actually dead tissues attached to an animal, but since they are not alive, animal can not alter their composition. Consequently, a bird or mammal has to produce a whole new coat of fur or feathers, in order to change the colour. In many reptiles, amphibians, and fish, colouration is determined by biochromes in living cells. Biochromes may be in cells at skin's surface or in cells at deeper levels. These deeper-level cells are called **chromatophores**.

Cuttlefish manipulate their chromatophores to change their overall skin colour. They have a collection of chromatophores, each of which contains a single pigment. An individual chromatophore is surrounded by a circular muscle, which can constrict and expand. When muscle constricts, all of the pigment is squeezed to the top of chromatophore. At the top, cell is flattened out into a wide disc. When muscle relaxes, cell returns to its natural shape of a relatively small blob. Blob is harder to see than the wide disc of constricted cell. By constricting all chromatophores with a certain pigment, and relaxing all with other pigments, animal changes overall body colour. Cuttlefish can generate a wide range of colours, and many interesting patterns. By perceiving colour of a backdrop and constricting right combination of chromatophores, animal blend in with all sorts of surroundings. They may also use this ability to communicate with others. Famous colour-changer chameleon, alters its skin colour using a similar mechanism, but not usually for camouflaging purposes. Chameleons tend to change their skin colour when their mood changes, not when they move into different surroundings.

Nudibranchs change their colour by altering their diet. When a nudibranch feeds from a particular type of coral, its body deposits pigments from that coral in skin and outer extensions of intestines, and animal gets same colour as of coral. Since coral is not only creature's food, but also its habitat, colouration is perfect camouflage. When creature moves on to a differently coloured piece of coral, its body colour changes with new food source. Some parasite species, such as fluke, take colour of their host, which is also their home. Many fish species, gradually produce different pigments without changing their diet. This works like seasonal molting in mammals and birds. When fish changes environments, it receives visual cues of a new surrounding. Based on this stimulus, it releases hormones that produce pigments. Over time, fish colour changes to match new surroundings. Camouflage is viewed as a result of natural selection, involves colour blending with the surroundings, through landing on areas of similar colour, and aligning body correctly. It involves costs, like finding a suitable resting spot, and benefits as seen in polymorphism which allows individuals of same species to have different camouflage, making prey detection more difficult for predators. Colour may change, during seasons, organism's life cycle, or brief intervals as in chameleon. Organisms living in same environment, may have similar colouration as a result of convergent evolution. Animals living in caves, underground, deep sea, in addition to nocturnal animals, can not even see. Their colour may have little or no adaptive value. These organisms rely primarily on olfaction and hearing, and even electroreception. Goldenrod Crab Spider (*Misumena vatia*) changes colour by secreting a liquid yellow pigment into outer cell layer of the body.

In general, resemblance consequence of colouring produces effect to blend with the environment, not requiring alteration of shape and outline. This is common in animals inhabiting uniformly coloured expanse of earth's surface, like ocean or desert. In special resemblance, equally diverse forms are defended by their sandy appearance. Effect, of a uniform appearance may be produced by a combination of tints in startling contrast. Black and white stripes of zebra (Figure 6.1) blend together at a little distance, and their proportion exactly match pale tint of arid ground in moonlight.

FIGURE 6.1 The black and white stripes of zebras have a survival value.

There are several factors that determine, what sort of camouflage a species develops. Camouflage develops differently, depending on physiology and behaviour of an animal. For example, an animal with fur, develop a different sort of camouflage than an animal with scales, and an animal that swims in large schools underwater develop different camouflage than one that swings alone through trees. Simplest camouflage for an animal is to match the "background" of its surroundings. In this case, various elements of natural habitat may be referred to as model for camouflage. Since ultimate goal of camouflage is to hide from other animals, physiology and behaviour of predators, or prey is highly significant. An animal does not develop any camouflage that does not help it to survive. So not all animals blend in with their environment in the same way. For example, there is no point in an animal replicating colour of its surroundings, if its main predator is colour-blind.

For most animals, "blending in" is the most effective approach. Deer, squirrels, hedgehogs, and many other animals have brownish "earth tone" colours, that match brown of trees and soil at forest ground level. Sharks, dolphins, and many other sea creatures have a grayish-blue colouring, which helps them blend in with underwater. Animals may also produce colours via microscopic physical structures. Essentially, these structures act like prisms refracting and scattering visible light, so that a certain combination of colours are reflected. Polar bears, actually have black skin, but appear white because they have translucent hairs. When light shines on hairs, each hair bends it a little bit. This bounces light around, so that some of it makes it to surface of skin and rest of it is deflected back out producing white colouration. In some animals, two types of colouration are combined. Reptiles, amphibians, and fish with green colouration, typically, have a layer of skin with yellow pigment and a layer of skin that scatters light to reflect a blue colour. Combined these layers of skin produce green.

Both physical and chemical colouration is determined genetically. They are passed on from parent to offspring. A species develops camouflage colouration gradually, through the process of natural selection. In wild, an individual animal that more closely matches its surroundings is more likely to be overlooked by predators and so lives longer. Animal, that matches its surroundings is more likely to produce offspring than an animal, which does not match. Camouflager's offsprings are likely to inherit same coloration, and they are likely to live long enough to pass it on. In this way, the species as a whole develops ideal colouration for survival in their environment.

Means of colouration depends on an animal's physiology. In most mammals, camouflage colouration is in fur, since this is the outermost layer of the body. In reptiles, amphibians, and fishes, it is in scales; in birds it is in feathers; and in insects, it is in part of exoskeleton. Actual structure of outer covering may also evolve to create better camouflage. In squirrels, for example the fur is fairly rough and uneven so it resembles the texture of a tree bark. Many insects have a shell that replicates smooth texture of leaves. Camouflaging colouration is very common in nature. But it is less common for an animal to be able to change its colouration to match a changing environment.

6.2 CRYPTIC COLOURATION

Cryptic colouration is found in both the predator and prey. Cryptic colouration is used for defense or attack, may be general or special. Special resemblance is seen on diversified places, as in shores, shallow water, floating masses of algae on ocean surface. Here, cryptic colouration is usually aided by modifications of shape, and instinct. Complete stillness and assumption of an attitude play an essential part in general resemblance on land. Attitude is often specialized, through which concealment is effected. Combination of colouring, shape, and attitude produce a more or less exact resemblance to particular objects in environment, such as a leaf, a twig, a patch of lichen or a flake of bark. Object, which has no interest to prey is selected. Animal do not remain hidden, but become indistinguishable from its background as in the cases of general resemblance, but it is mistaken for some well-known object. *Spongophorus guerini* (Figure 6.2) is one of many extremely cryptic membracid bugs. Another membracid bug resembles an outgrowth of the stem of its food plant (Figure 6.3).

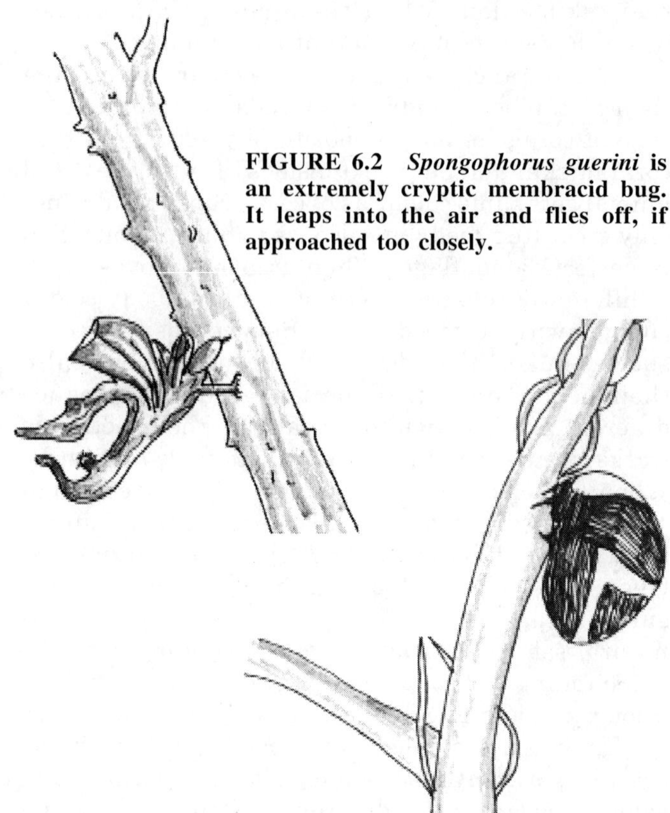

FIGURE 6.2 ***Spongophorus guerini*** **is an extremely cryptic membracid bug. It leaps into the air and flies off, if approached too closely.**

FIGURE 6.3 A membraid bug resembles an outgrowth of stem of its food plant.

Crypsis works only if the animal is resting on the appropriate background and usually only when the animal is not moving. An Indian leaf butterfly exemplifies cryptic colouration. Many vertebrates are darker on back and become gradually lighter on sides passing into white on belly. This gradation obliterates appearance of solidity due to shadow. Colour-harmony, essential to concealment is produced because the back has same tint as environment bathed in cold blue-white of sky, while belly being bathed in shadow and yellow earth reflections, produces the same effect. This method of neutralizing shadow for the purpose of concealment by increased lightness of tint was first suggested in larva and pupa.

In an analogous method, an animal in front of a background with dark shadow, may have part of its body obliterated by existence of a dark tint as found in mimicry. A common aid to concealment is adoption of two or more different appearances by different individuals; each resembles some special object to which an enemy is indifferent. Kallima present various colours and pattern on under side of the wings, each of which closely resembles dead leaf. *Trephine pronuba* is polymorphic on upper side of its wings, which are exposed as it suddenly drops among dead leaves. Caterpillars and pupae are also commonly dimorphic, green and brown. In many cases, cryptic colour changes in individual life, either seasonally as in ptarmigan or Alpine hare, or according to the environment which the animal inhabits in course of its growth. In insects, with more than one brood in a year, seasonal dimorphism is often seen. Differences are sometimes appropriate to alter condition of environment as seasonal change. In certain species, concealment is effected by use of adventitious objects, which are employed as a covering. Examples of this allocryptic defense are found in tubes of caddis fly larvae, and objects made by crabs of genera *Hyas* and *Stenorhynchus*. Such animals remain concealed in any environment. If sedentary, they are covered up with local materials. If wandering, they have instinct to reclothe. Allocryptic methods may be used for aggressive purposes as antlion larva almost buried in sand, or the large frog *Ceratophrys,* which covers its back with earth when waiting for its prey. In allocryptic defense, advantageous colour may be dissolved in blood or secreted in superficial cells of the body. Certain insects use chlorophyll of their food. Although most perfect cryptic powers are widely prevalent in fish, but such power also occurs in amphibians and reptilians. Analogous powers of certain crustaceans and cephalopodans in rapid changes of colour are due to changes in shape or position of superficial pigment cells, which in turn is controlled by nervous system, stimulated by light through eye and optic nerve. Parti-coloured surfaces do not produce parti-coloured pupae, probably because antagonistic stimuli neutralize each other in central nervous system, which then disposes superficial colours, so that a neutral or intermediate effect is produced over the whole surface.

Cryptic colour may incidentally produce superficial resemblances between animals. Thus, desert forms remain concealed in same way may gain a likeness to each other. And in the same way, special resemblances, e.g. to lichen, bark, grasses, pine-needles, may sometimes lead to a tolerably close similarity between animals, which are thus remain concealed. Such likeness may be called syncryptic or common protective resemblance. And it is to be distinguished from mimicry and common warning colours in which likeness is not incidental, but an end in itself. Syncryptic resemblances have much in common with those incidentally caused by functional adaptation, such as mole-like forms produced in burrowing Insectivora, Rodentia, and Marsupialia. Such likeness may be called syntechnic resemblance, incidentally produced by dynamic similarity, just as syncryptic resemblance is produced by static similarity.

6.3 DISRUPTIVE COLOURATION

Many animals have distinctive designs on their bodies, which might be spots, stripes or a group of patches, which help animals in a couple of ways. They may match the pattern of background of their surroundings. Animals that inhabit areas with tall, vertical grass often have long, vertical stripes. They may serve as visual disruptions. Usually, patterns are positioned "out-of-line" with body's contours. That is, pattern seems to be a separate design, superimposed on top of the animal. This makes it hard for a predator to get a clear sense of where animal begins and ends pattern on body seems to run off in every direction. This disruptive colouration is particularly effective when animals are grouped together. To a lion, a herd of zebras does not look like a whole bunch of individual animals, but more like a big, striped mass. Vertical stripes all seem to run together, making it hard for a lion to stalk, and attack one specific zebra. Stripes may also help a single zebra to hide in areas of tall grass. Since lions are colourblind, it does not matter that zebra and surrounding environment are of completely different colours. Many fish species are similarly camouflaged. Their vertical stripes may be brightly coloured, which makes them stand out to predators, but when they swim in large schools, their stripes all meld together. This confusing spectacle gives predators the impression of one big swimming blob. Generally, this sort of camouflage does not hide an animal's presence, it merely misrepresents it. A related camouflage tactic is for an animal to take on appearance of some other object. One famous example is walking stick, an insect that looks like an ordinary twig. A predator can easily distinguish a walking stick from its surroundings, but predator thinks it is only a stick and so ignores it. This sort of camouflage is also seen in some katydid species, which have evolved, so that they look just like tree leaves.

6.4 WARNING COLOURATION OR MIMICRY

Warning colouration is exact opposite of camouflage. Its function is to render animal conspicuous to its prey, so that it can be easily seen and avoided in future. Warning colours are associated with features like unpleasantness, danger, unpalatability, evil odour, sting, and fang. As objective is to warn an enemy, these colours are also called **aposematic**. Recognition markings, on the other hand are episematic, assisting individuals of same species to remain together, when their safety depends upon numbers or a place of safety. Episematic characters are less common than aposematic. Unpalatability, or even possession of a sting is not sufficient defense, unless there is enough food of another kind to be obtained at same time and place. Hence, insects with warning colours are not seen in temperate countries, except at time, when insect life as a whole is most abundant and in warmer countries, with well-marked wet and dry seasons, warning colours are proportionately less developed. The South American poison arrow frog advertises its unpleasant taste with bright and constant colour patterns (Figure 6.4).

FIGURE 6.4 An arrow poison frog.

In African butterflies belonging to genus *Junonia,* including subgenus *Precis*, wet-season broods are distinguished by more or less conspicuous under sides of wings, those of dry season, being highly cryptic. Warning colours are assisted by special adaptations of body-form and movements, which assist to render colour as conspicuous as possible. For this reason, animals with warning colours generally move or fly slowly. In butterflies, warning patterns are similar on both upper and under sides of wings. Many animals, when attacked or disturbed, imitate a dead body by falling motionless to ground. In well-concealed animals, this gives them a second chance of escape. When detected; animals with warning colours are enabled to assume a position, so their characters are displayed to full. In both cases, a definite attitude is assumed, which is not that of death.

Other warning characters include sound that is used by disturbed rattlesnake. Large birds, when attacked, often adopt a threatening attitude accompanied by a terrifying sound. Cobra warns an intruder chiefly by attitude and dilation of flattened neck, effect being heightened in some species by spectacles. In such cases, combination of cryptic and sematic methods is found. Animal being concealed until disturbed, when it instantly assumes an aposematic attitude.

In small unpalatable animals with warning colours, enemies would only first become aware of unpleasant quality by tasting and often destroying their prey. But kin of organism killed may gain experience, thus, conveyed, even though individual might suffer. Insect-eating animals do not come into world with knowledge. They learn by experience and warning colours enable this education. Great tenacity is usually possessed by animals with warning colours. Tissues of aposematic insects, generally possess great elasticity and power of resistance, so that large numbers of individuals can recover after severe treatment. Animals with warning colours are probably attacked less by ordinary enemies of their class, but they have special enemies, which keep numbers down to average. Thus, cuckoo, an insectivorous bird, which freely devours conspicuously coloured unpalatable larvae. Effect of warning colours of caterpillars is often intensified by gregarious habits. Another aposematic use of colours and structures is to divert attention from vital parts. Thus, given animal under attacks an extra chance of escape. Large, conspicuous, easily torn wings of butter-flies and moths act in this way as found by abundance of individuals, which may be captured with notches bitten symmetrically out of both wings, when they were in contact. Eye-spots and tails so common on hinder part of hind wing, and conspicuous apex so frequently seen on fore wing, probably have this meaning. Their position corresponds to parts, which are often found to be notched. In some cases, tail and eye-spot combine to suggest appearance of a head, with antennae at the posterior end of the butterfly, deception being aided by movements of hind wings. The moth *Automeris coresus* has cryptically coloured forewings, which provide camouflage when it is settled. However, if touched as by a predator, it flashes open the forewings to reveal bright eye spots (Figure 6.5). Flat-topped tussocks of hair on many caterpillars look like conspicuous fleshy projections of body and they are held prominently when larva is attacked. If seized tussock comes out and enemy is greatly inconvenienced by fine branched hairs. Tails of lizards, which easily break off are to be similarly explained. Certain crabs, similarly, throw off their claws, when attacked and claws continue to snap most actively. Tail of dormouse, which easily comes

(a) *Automeris* moth in sitting position.

(b) *Automeris* with exposed eyespots on hind wings

FIGURE 6.5 ***Automeris coresus.***

off and extremely bushy tail of squirrel is probably of use in the same manner.

6.4.1 Batesian Mimicry

Animals with warning colours often tend to resemble each other superficially, the fact first pointed out by H.W. Bates. He showed that conspicuous, presumably unpalatable, tropical American butterflies, belonging to different groups are mimicked by others and tend to resemble each other likeness being often remarkably exact. These resemblances were not explained by his theory of mimicry. Bates first proposed an explanation of mimicry based on theory of natural selection. Numerous examples of mimicry among tropical American butterflies were discussed by Bates. One of the most popular types of mimicry involves the warning colouration found on inedible or toxic organisms such as monarch butterfly (Figure 6.6). Once these toxic organisms have adapted this warning colouration, which warns predators to stay away, other organisms may start to mimic this warning colouration in an attempt to stay alive. Batesian mimics are those mimics that imitate unpalatable species even though they are palatable. Therefore, one species is harmful while other is harmless. Wasp is a great example of Batesian mimicry. Wasp is the model species in this example as it possesses a sting which enables it to escape from predators. Bright warning colouration of wasp has been mimicked by many other insects. Even though mimics are harmless, predator will avoid them due to bad experience with wasps with same colouration. Some animals have evolved a way to enjoy the benefits of warning colouration without the costs. These animals mimic colouration of poisonous animals. This type of mimicry is referred to as Batesian mimicry, named after the nineteenth-century British naturalist who first described it. The best-known example of Batesian mimicry in United

FIGURE 6.6 Monarch butterfly (the model).

States and Canada is probably Viceroy butterfly (Figure 6.7) that looks remarkably like the poisonous Monarch butterfly. The two species are unrelated and caterpillars feed on different plants and do not look anything like one another. However, the adults of both species look so similar that most people and more importantly most birds cannot tell them apart. It is important that Batesian mimic be less common than toxic model species. For example, if Viceroy were more common than Monarch, birds would end up eating a lot of Viceroys before eating a Monarch and would not "learn the lesson", colouration thus acts to teach.

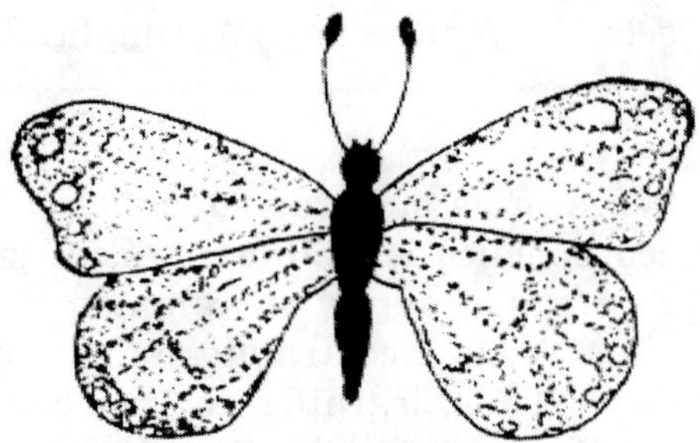

FIGURE 6.7 A viceroy butterfly (the mimic).

6.4.2 Mullerian Mimicry

With Mullerian mimicry, many unpalatable species share a similar colour pattern. Mullerian mimicry proves to be successful as predator only has to be exposed to one of species, in order to learn to stay away from all other species with same warning colour patterns. Black and yellow striped bodies of social wasps, solitary digger wasps, and caterpillars of cinnabar moths warn predators that the organism is inedible. This is a great example of Mullerian mimicry, as all of these unpalatable, unrelated species have a shared colour pattern that keeps predators away. In 1879, F. Muller suggested that life is saved by this resemblance between

MIMICRY: A Survival Strategy

FATIK B. MANDAL

Members of various species interact with one another when they search for habitats, for food and try to avoid being eaten. Predator and prey populations exert environmental pressure on one another which results in co-evolution. Mimicry is the result of predator–prey interaction in which an animal resembles some other living or non-living object.

Mimicry is classified into three types, viz. protective mimicry, aggressive mimicry and conscious mimicry. Protective mimicry affords protection to the mimic. It may be concealing or warning. In concealing mimicry, mimic tries to conceal it from its enemies by mimicking the model. The dead leaf butterfly in resting condition among leaves resembles the leaf. In warning mimicry, warning is achieved through resemblance of the organisms to other poisonous or distasteful organisms. Warning mimicry may be Batesian or Mullerian mimicry. Batesian mimicry, named after H.W. Bates, involves one or more unpalatable species that serve as models and one or more palatable species that mimic the models' colouration and behaviour. This types of mimicry is common in insects and some vertebrates. Batesian mimics are rare. Mullerian mimicry, named after F. Muller, involves two or more unpalatable and aposematic species. The natural resemblance of Mullerian mimics confers protection to all members of an association in a given area.

In aggressive mimicry, resemblance with the objects may not be for protection of the animal, but for the aggression. Such mimicry helps the mimic in catching the prey. In conscious mimicry, animals on the face of danger behave as dead. Mimicking the same group of models may result in evolutionary convergence in various groups of animals. Mimicking various groups of models may result in evolutionary divergence in the same group of animals. Co-evolutionary interactions in Batesian mimicry are reported to be similar to those occurring in host–parasite complexes. Various members of Mullerian complexes would have different effects on predators, since they pick up *poisons* from various sources. It is advantageous to the Batesian mimic to become distasteful, as it does so without sacrificing too much physiologically. Mimicry increases the survival value of a species.

www.associatedcontent.com/pop_print_html?content_type+article&content_type_i

warning colours in as much as education of young inexperienced enemies is facilitated. Each species, which falls into a group with common warning (synaposematic) colours, contributes to save lives of other members. Thus, learning and remembering, consequently of injury and loss of life involved in the process are reduced, when many species in the one place possess same aposematic colouring, instead of each exhibiting a different danger-signal. Such resemblances are often described as Mullerian mimicry as distinguished from true or Batesian mimicry (Table 6.1).

Difference between a typical aposematic character appealing to enemies, and episematic intended for other individuals of same species is seen, when we compare such examples as huge banner-like white tail, conspicuously contrasted with black, or black and white body by which slow-moving skunk warns enemies of its power of emitting an intolerably offensive odour. Small upturned white tail of rabbit is only seen when it is likely to be of use, and when the owner is moving and if pursued, very rapidly moving towards safety. Both Batesian and Mullerian mimicry are found in African butterflies (Figure 6.8).

6.4.3 Aggressive Mimicry

In cases of aggressive mimicry, an animal resembles some object which is attractive to its prey. Here, an

TABLE 6.1 Comparison between Batesian and Mullerian Mimicry

Batesian mimicry	*Mullerian mimicry*
Both mimics and models do not have warning colours	Both are warningly coloured
Models are common than mimics	Both models and mimics are common
Similarity between mimic and model is close	Similarity is not so close
Model and mimic are found together	Not found together

FIGURE 6.8 Mimicry in African butterflies. Dimorphic *Papilio* on bottom, left (male), on right (female). Three model from top (*Bemanistes epaea*, *Amauris echerina*, and *A. niavivus*. Similarity between *B. epaea* and *amauris* is an example of Mullerian mimicry, since all are distasteful.

organism will mimic a signal that is either deceptive or attractive to its prey. Examples are found in flower-like species of mantis, which attract insects on which they feed. Such cases are described as possessing alluring colours and are examples of aggressive (anti-cryptic) resemblance. One example of this involves praying mantis who will mimic flowers to attract insects that they can then capture and eat.

The term mimicry is used, to include all the superficial resemblances, between animals and any part of their environment. Wallace, however, separated cryptic resemblances, and most naturalists follow this arrangement. In mimicry, an animal resembles some other animal, which is specially disliked by its enemy, or some object, which is especially attractive to its prey, and in so doing becomes conspicuous. Essential element in mimicry is false warning (pseudoaposematic), or false recognition (pseudoepisematic) character.

Conditions under which mimicry occurs as stated by Wallace:

- Imitative species occur in same area and occupy same station as imitated.
- Imitators are always more defenseless.
- Imitators are always less numerous in individuals.
- Imitators differ from bulk of their allies.
- Imitation, however, minute is external and visible only, never extending to internal characters or to such as do not affect external appearance.

Mimicry has been explained independently of natural selection by supposition that it is common expression of direct action of common causes, such as climate, food, etc. also by supposition of independent lines of evolution, leading to same result without any selective action in consequence of advantage in struggle; also by operation of sexual selection. Many insects produced from burrowing larvae mimic those, whose larvae live in open. Mimetic resemblance is common in female than in male, a fact readily explicable by selection as suggested by Wallace for female is compelled to fly more slowly, and to expose itself while laying eggs, and hence, a resemblance to slow-flying freely exposed models is especially advantageous. Facts that mimetic species occur in same locality fly at same time of the year as their models, and are day-flying species even though they may belong to nocturnal groups, are also more or less difficult to explain, except on theory of natural selection, and so also is the fact that mimetic resemblance is produced in most varied manner.

Frightening colouration and protean displays

Sometimes the animal changes body form and attains the form of a frightening animal. For example, in the lantern bug *Fulgora lanternaria* (Figure 6.9), the anterior part has a hollow extension which is so patterned as to resemble the head of an alligator with its mouth partly opened. The behaviour of sphinx moth caterpillar, *Leucorhampha* is perhaps the most remarkable protean display in the entire animal kingdom (Figure 6.10).

6.4.4 Adaptive and Evolutionary Significance of Mimicry

1. Mimicry increases the survival value of a species.
2. When a species mimics several models, the result is polymorphism, as in *Papilio sp.*

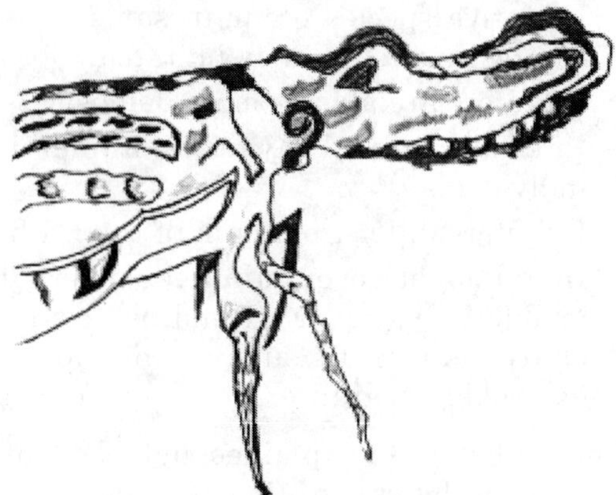

FIGURE 6.9 *Fulgora lanternaria.*

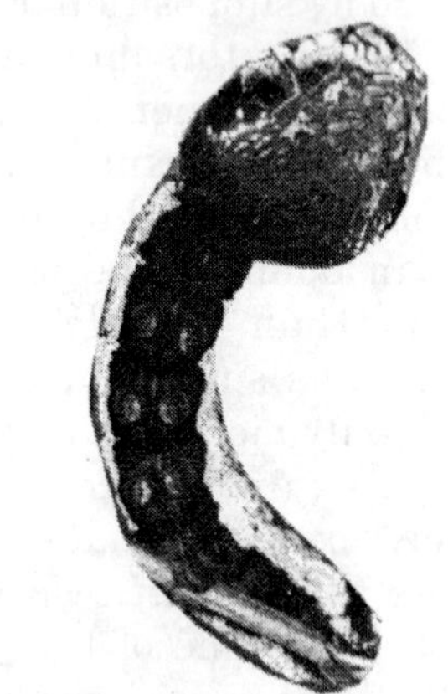

FIGURE 6.10 *Leucorhampha.*

3. Mimicking same group of models may result in evolutionary convergence among various groups of animals.
4. Coevolutionary interactions in Batesian mimicry are similar to those occurring in host parasite complexes. The mimic plays the role of the parasite and takes advantage of the model without destroying it.
5. It is advantageous to the Batesian mimic to become distasteful as it can do so, without sacrificing too much physiologically. In butterflies, it appears that the usual source of noxious elements are plant biochemicals. So the food-plant relationships must play a large role in evolutionary dynamics of any given situation.

A butterfly has several different routes to distastefulness open to it. If it eats a plant, which does not produce an appropriate compound it may switch food plants. If it is feeding on a plant, with an appropriate compound, or its precursors, the butterfly may evolve the ability to use the compound or synthesize a noxious compound from precursors. Finally, the food plant of butterfly may evolve an appropriate compound, which then may be picked up by the butterfly. As the food plant becomes more and more toxic, the butterfly must find ways of "breaking even" by avoiding poisoning or winning by turning the poison to its own advantage.

6.5 VENOM AND POISON

There is a difference between venomous and poisonous organisms. Venomous refers to animals that deliver, often inject venom into their prey, when hunting or as a defense mechanism. Poisonous organisms refer to plants or animals that are harmful when consumed or touched. In poisonous organism, poison tends to be distributed over a large part of organism's body. Venom is typically produced in specialized organs. One species of bird, the hooded pitohui, although not venomous, is poisonous, secreting a neurotoxin onto its skin and feathers. Slow Loris, a primate, blurs the boundary between poisonous and venomous. From patches inside elbows, it secretes a toxin which is smeared on young to prevent them from being eaten. However, licking these patches causes a venomous bite. Venom is secreted by animals for either defensive or offensive purposes. Venom originated from digestive enzymes that were originally located in stomach. Type of venom depends on the type of the animal. In spiders, venom is kept rather simple. Spiders (Figure 6.11) use their venom to turn their hard shelled insect meals into nice and nutritious. In insects, venom is used predominantly as a defensive weapon. Wasps (Figure 6.12), bees and ants use formic acid in their stings to cause a painful burning sensation, that either kill or injure their enemy,

FIGURE 6.11 A spider.

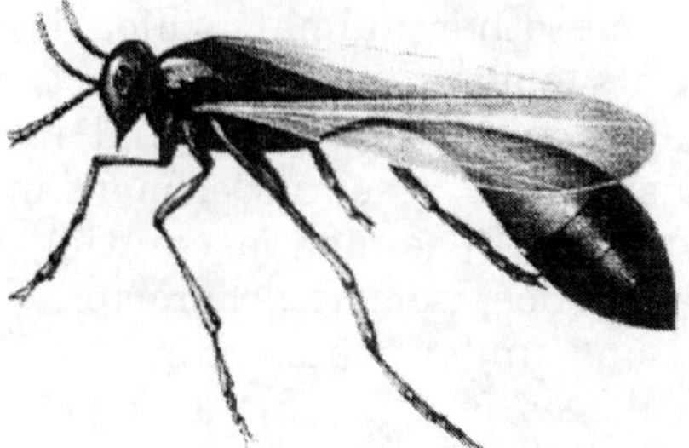

FIGURE 6.12 A wasp.

enough to make them think twice about attacking them again. Amphibians use their venom for defense. In amphibians, venom is secreted through glands in skin to make the animals unpalatable. Venom is predominantly a defensive adaptation. An adaptation that found its way into every class of vertebrates except one, the birds. Fish comprises too many venomous species. All venomous reptiles are squamates and of them, snakes make up the bulk. Two species of venomous squamates that are not snakes are lizards of the genus *Heloderma*. These lizards use their venom for defense, as well, and can deliver powerful and painful bites. In snakes, venom has found a new use for offense.

6.5.1 Venomous Fish

Venom is found in fish, such as the cartilaginous fishes—stingrays, sharks, and chimaeras, and the teleostfishes including monognathus eels, catfishes, stonefishes and waspfishes, scorpion-fishes, lionfishes, gurnard perches, rabbitfishes, surgeonfishes, scats, stargazers and weever. Any shark can inflict painful and often fatal injuries, either through bites or through abrasions from their rough skin. Rabbitfish occurs mainly on coral reefs in the Indian and Pacific oceans. They have very sharp, possibly venomous spines in their fins. The fishes are considered edible by native peoples where the fishes are found, but deaths occur from careless handling. Surgeon-fishes are often beautifully coloured and provided scalpel like spines in tail. Wounds inflicted by these spines can bring about death through infection, envenomation, and loss of blood, which may incidentally attract sharks.

Toadfish are dull coloured. They have sharp, very toxic spines along their backs. Poisonous scorpionfishes are mostly found around reefs in the tropical Indian and Pacific oceans. They have long, wavy fins and spines and their sting is intensively painful. Stonefish can inject painful venom from their dorsal spines. Weeverfish has venomous spines on the back and gills. Puffer-fishes have short spines and can inflate themselves into a ball when alarmed or agitated. Their blood, liver, and gonads are so toxic that their venom as little as 28 milligrams (1 ounce) can be fatal. Triggerfishes are characterized by their large and sharp dorsal spines.

6.5.2 Venomous Amphibians

In amphibians, the poisonous arrow frogs of Central and South America have very potent poisons. Some Indian tribes in South America use the toxins from these frogs to poison the end of their darts, which they use in blowguns for everyday hunting.

6.5.3 Venomous Reptiles

Cobras (Figure 6.13), mambas, kraits, coral snakes, sea snakes are venomous snakes and can be found in warmer parts of the world, except Europe and Madagascar. Viperidae includes venomous vipers and pit vipers. They live in most parts of world, except in Australia and Madagascar. Three types of venomous snakes are opisthoglyph, proteroglyph and solenoglyph. The first type is mostly harmless or mildly venomous snakes. Their fangs are enlarged rear teeth, with a groove for venom flows, while they are swallowing their prey. However, there are snakes of this type that are known to have killed humans before, for example the Boomslang (*Dispholidus typus*).

FIGURE 6.13 An Indian cobra.

Snake defense is rather limited. Some snakes secrete nasty substances that make them unwanted. For a venomous snake, biting a predator means wasting precious venom, a commodity that does not come without its price. Venom costs energy, to make and takes a while to refill when empty. In fact, venomous snakes have adopted so many warning strategies from warning colours, to hoods, to rattles; venomous snakes do everything in their power to avoid biting the enemy to save venom. Some snake species inject venom into their prey through hollow fangs. Venom is produced by glands below eye and delivered to victim through tubular or channeled fangs. Snake venoms contain a variety of peptide toxins (proteases), which hydrolyze protein peptide bonds, and nucleases, which hydrolyze the phosphodiester bonds of DNA. Snakes use their venom principally for hunting, though threat of being bitten serves also as a defense. Snake bites cause a variety of symptoms including pain, swelling, tissue damage, low blood pressure, convulsions, and hemorrhaging. Aristolochic acid inhibits the activity of snake venom phospholipase (PLA2) by forming a 1:1 complex with enzyme. Since phospholipase enzymes play a significant part in the cascade leading to inflammatory and pain response, their inhibition could lead to relief of problems from scorpion envenomation.Venom is also found in a few reptiles besides snakes, such as Gila monster and Mexican beaded lizard.

6.5.4 Methods of Production

Venom is made in special glands located on the head of the animal. Venom glands differentiate into false and true glands. False venom glands are made up either from mucus producing suprallabial glands that run on either side of head, extending as a continuous strip from near snout to below and well behind eye. These then lead to several ducts that lead to the bases of many maxillary teeth. Alethinophidians are known to have this type of arrangement. Most colubrids have a different arrangement. Rather than using those modified salivary glands, they use a larger gland, known as Duvernoy's gland. This gland is situated right under the skin, above and near the angle of jaw. These glands open from a duct at the base of one or more posterior usually enlarged fangs that may or may not be grooved. These glands do not have a lumen so the snakes must give off a continuous stream of venom into their prey, which means that they must continue to hold on to the animal to ensure envenomation.

True venom glands are made of thick connective tissues. They contain a lumen, a separate compressor muscle and a duct connecting them to a single fang on each side of jaw. These glands dominate all elapids and viperids. Some cobra types such as rinkhals (*Hemachatus haemachatus*) and many species of Afro-Asian cobras (*Naja* sp.) have ability to spit their venom at predators. Their fang tips have bevelled, circular apertures on anterior surface just above the tip, where venom is ejected. African spitters have spiral grooves in their fangs that force a spin on venom allowing for greater accuracy. This is only used in defense and is effective. By expending venom in these little droplets, snake is guaranteed the maximum use of venom when battling a predator. *Naja nigricollis* emptied its venom glands by spitting fifty seven times in only twenty minutes. In venomous snakes especially in viperids "dry bite" is known. Dry bites occur when a snake is cornered and forced to bite in defense. By giving off a dry bite, venomous snakes need not waste their venom supplies. In fact, over half of all rattlesnake bites are dry ones. In reptiles, there are venomous lizards living in the South-western United States and Northern Central America, all the way down to Southern Guatemala.

6.5.5 Venomous Mammals

The male platypus or duckbill (*Ornithorhyncus anatinus*) has a poisonous spur on each hind foot that can inflict intensely painful wounds. Jellyfish-related deaths are rare, but the sting they inflict is extremely painful. The Portuguese man-of-war resembles a large pink or purple balloon floating on the sea. It has poisonous tentacles hanging up to 12 metres below its body. The huge tentacles are actually colonies of stinging cells.

6.5.6 Venomous Arthropods and Other Invertebrates

Spiders and centipedes also inject venom through fangs. Scorpions and stinging insects inject venom with a sting. Many caterpillars have defensive venom glands associated with specialized bristles on body, known as urticating hairs, and can be lethal to human. Bees synthesize and employ acidic venom (apitoxin) to cause pain in those that they sting, whereas wasps use chemically different venom designed to paralyze prey. Box jellyfish is considered the most venomous creature in the world. There are poisonous species among corals. One of the most potent toxins against acetylcholine receptors, which play an important role in neuromuscular transmission, has been isolated from corals. An extremely venomous animal is cone shell, of which there are approximately 600 species. All of them are venomous, but only a dozen or so of them are life-threatening to humans. Their toxins act on nerve cells as well as muscle cells. Sea anemones have toxins inhibiting the function of sodium channels, which play an important role in action potential of nerve and muscle cells. Some sea urchins and sea stars also produce potent toxins.

Blue ringed octopus (*Hapalochlaena lunulata*) can inflict a deadly bite. Portuguese man-of-war resembles a large pink or purple balloon floating on the sea with poisonous tentacles hanging up to 12 metres below its body. The huge tentacles are actually colonies of stinging cells. Most known deaths from jellyfish are attributed to the man-of-war. Other jellyfish can inflict very painful stings as well.

Scorpions use their venom for defense and immobilizing prey. Among the many scorpion species *Leiurus quinquestriatus* is called **Death stalker**. Other venomous scorpions are *Androctonus australis, Odontobuthus doriae, Centruroides noxius, Androctonus mauritanicus, Androctonus bicolor and Androctonus crassicauda.* Venom of scorpion is primarily neurotoxic with abundant cardiotoxic components.

6.5.7 Nature's Poisonous Bounty

Toxin Tetrodotoxin (TTX), the most powerful neurotoxins with no known antidote reveals a fascinating convergent evolution. It is about 1200 times more toxic than cyanide to humans simple molecular structure of this alkaloid forming a cyclic series of carbon atoms, although active site is derived from guanine. By its guanidinium group, TTX binds to voltagegated sodium channels on nerve membranes, block nerve conduction cause motor paralysis and finally death due to suffocation. TTX was known to occur only in pufferfish (Tetraodontidae) but now found to occur in a wide variety of animals. It was named on its discovery in Spheroides rubripes in 1910. The toxin is found in a variety of animals. A species of goby (Yongeichthys criniger) possess the toxin as well

Origin of tetrodotoxin

In the pufferfish, TTX produced by endosymbiotic bacteria often passed down the food chain. Puffers reared in hatchery without access to TTX-containing food was found to be non-toxic. Such non-toxic individuals became toxic when fed TTX-containing food or TTX-producing bacteria. The large variation in pufferfish toxicity between individuals and regions supports the idea of an exogenous intoxication than production of toxin by fish themselves. Various bacteria produce TTX—mainly those in genera *Vibrio* and *Pseudomonas*, but an actinomycete, *Nocardiopsis dassonvillei* suggests that TTX synthesis may be convergent. Bacterial origin of TTX implies symbiosis between TTX-producing bacteria and animals that evolved independently numerous times. *Atelopus varius* reared from the eggs in captivity showed no detectable levels of TTX in their skin as adults. The newt *Cynops pyrrhogaster* has certain amount of TTX in egg (inherited from parents) that fades away in larval stage. In normal environment, juveniles accumulate TTX quickly, but in laboratory with no TTX-containing food, they were non-toxic as adults.

Resistance to tetrodotoxin

Resistance to TTX is centred on voltage-gated sodium channels which are disabled by toxin. These channels consist of large protein with 4 repeating domains. Each domains contains a highly conserved region that contributes to pore formation. Key substitutions of amino acids in pore region reduce affinity to toxin and allow the channels to function normally. In various species of pufferfish striking convergences at this molecular level in one or more of channel protein domains is found. Changes involve few sites, indicating that selection for normal channel function limits possibilities for resistance-conveying mutation. Tetrodotoxin resistance is an example of natural selection acting on complete gene family, repeatedly arriving at diverse but few adaptive changes within same genome. Resistance to TTX indicates that pufferfish use it as effective chemical defence and selectively feed on TTX-bearing organisms that TTX-sensitive fish have to avoid. Thamnophis sirtalis that prey on invulnerable newt *Taricha* have convergently acquired TTX resistance. Snakes that do not encounter newts in wild are poisoned when fed newts in laboratory, indicating a co-evolutionary interaction between predator and its prey.

In fish

Twenty-two pufferfish species in the family Tetraodontidae are listed as TTX-bearing, but closely related porcupinefish (Diodontidae) and boxfish (Ostraciidae) contain no TTX (Noguchi and Arakawa 2008). Distribution of TTX in bodies of pufferfish varies among species. The liver (ovary during the spawning season) has the highest toxicity in marine species. The skin appears to be the most toxic part in animals inhabiting brackish- and freshwater. Tetrodotoxin serves in defence, in combination with bizarre inflation behaviour that is typical of pufferfish and their relatives. when threatened Puffers inflate their body by swallowing water and excrete TTX from their skin. The eggs can be highly toxic, to protect them from predators. In marine species like Fugu niphobles, females use TTX as pheromone to attract males during spawning and increase chances of their eggs being fertilised.

In marine invertebrates

Many marine invertebrates contain TTX. The best examples are the blue-ringed octopuses (*Hapalochlaena*), containing TTX in salivary gland and proboscis, which indicates that it helps in prey capture. Nemerteans like Lineus fuscoviridis use it as predatory venom. TTX are also found in other marine worms of different phyla like flatworms (*Planocera*), arrow worms (chaetognaths, Parasagitta spp.) and annelids (Pseudopolamilla occelata), various genera of marine snails (*Charonia* and *Niotha*), crustaceans (*Carcinoscorpius rotundicauda* and xanthid crabs) and starfish (*Astropecten*).

In terrestrial vertebrates

In 1964, TTX was first discovered in the eggs of newt *Taricha torosa*. Presence of a toxin in these newts was reported in 1930s, but later it was identified as TTX. TTX occurs outside the oceans as bioactive natural products. Toxin is found in many amphibians of various families. The *Taricha* newts are so far known as the most toxic. About all members of the family Salamandridae contain TTX, with Salamandra salamandra which employs steroidal alkaloid samandarine, being exception. Frogs appear to be less toxic than tailed amphibians. TTX is found in about all bufonid species of the genus Atelopus. Some members of 3 other families contain the toxin. Neotropical dendrobatid or poison dart frogs are known for their venom. They employ various other alkaloids. Colostethus inguinalis contains low dose of TTX in skin. In Polypedates), the highest TTX levels documented actually approach those of Taricha. *Brachycephalus ephippium* of the family Brachycephalidae possesses relatively high dose of TTX in skin and liver.

Other remarkable poisons

Frogs-batrachotoxin: From *Phyllobates terribilis*, the name says it all, one of the range of poison arrow frogs. Acts on sodium channels.

Snails-conotoxins: A range of poisons from the venom of sea snails which use them to paralyze their prey (fish). Different forms of conotoxins act on sodium and calcium channels.

Snakes-bungarotoxin: From the krait family, bungarotoxin prevents neurotransmitter from acting on acetylcholine receptors on muscle; result in paralysis. From the mamba family, dendrotoxin blocks potassium channels. Many snake venoms also contain other chemicals that disable prey by clotting the blood and digesting flesh as well as nerve cell poisons.

Spiders-agatoxins: The agatoxins which block calcium channels and atracotoxins which cause uncontrollable opening of sodium channels are components of the venom injected through the fearsome fangs of the funnel web spider family.

Scorpions-venom: The venom components have been found which block potassium channels and modify sodium channel function.

6.6 ANIMALS AND FISHES POISONOUS TO EAT

Livers of polar bears are toxic due to their high concentrations of vitamin A. Another toxic meat is flesh of hawksbill turtle. Many fishes living in reefs near shore or in lagoons and estuaries are poisonous to eat, though some are only seasonally dangerous. Majority are tropical fishes. Some predatory fishes, such as barracuda and snapper, may become toxic if the fishes they feed on in shallow waters are poisonous. The most poisonous types appear to have parrot like beaks and hard shell-like skins with spines and often can inflate their bodies like balloons. However, at certain times of the year, indigenous populations consider the puffer a delicacy. Puffer is more tolerant to cold water. One can find them along tropical and temperate coasts worldwide, even in some of the rivers of Southeast Asia and Africa. Their blood, liver, and gonads are so toxic that as little as 28 milligrams (1 ounce) can be fatal. These fishes vary in colour and size, growing up to 75 centimetres in length.

6.7 BIOMIMICRY

It is a revolutionary interdisciplinary science. It analyzes nature's best ideas. Being a heterogeneous field of ideas and solutions drawn from ecology, physiology, molecular biology, physics, chemistry, computer science, mechanical and civil engineering, electronics, and economics, it generally comes, either from innovator's observation on a natural process and attempts to use it in existing technology, or innovator's studies on existing design problem and taking help from the nature. Thus, it is a bridge between biology and engineering. Biomimicry and synthetic biology have evolved as subfields of intentional biology. Ways through which animals and plants use sun and simple compound in producing totally biodegradable fibers are followed in biomimicry. Animals use plants for millions of years to keep them healthy, and their knowledge would be consulted to find new drugs. In her 1997 book, *Biomimicry: Innovation Inspired by Nature* Benyus writes, "Our planet-mates (plants, animals and microbes) are patiently perfecting their ways for more than 3.8 billion years, turning rock and sea into a life-friendly home". Biomimicry attempts to use incredible efficiency of nature in handling waste. Nature produces materials that are used by the organism for food, shelter, etc. Bio-inspired products can be cheaper than industrial products as they use inexpensive ingredients and made with less energy than conventional materials. Biomaterials are less toxic than traditional materials. Synthetic biology sees nature as a code, asset of instructions, guiding the fashioning of living organisms. Biomimicry research falls into 3 broad areas. Nanotechnologists, chemists, and physicists are developing new kinds of materials that imitate molecular structures, generated by plants and animals. Aerospace scientists, mechanical engineers, and roboticists are drawing inspiration from natural structures to create new designs for propulsion systems, bridges, and robots. Architects, industrial engineers and agriculturalists are trying to replicate natural processes for maintaining stable environments, manufacturing complex materials, and growing food. Eskimo cultures of North America learned from polar bears to build snow houses and to use fur for warmth. Similarly people, living in Amazon basin, observed toxic effects of dart frog on its predators, and therefore, they applied the same toxin to their arrows. Following examples of biomimicry are well documented in literature.

6.7.1 Property of Lotus

Lotus demands attraction in its ability to stay clean, even in muddy flats. Lotus cleanliness is due to tiny indentations that trap air, and prevent water, or dirt, from clinging to its surface. A transparent material has been made that is used as a coating on windows, using the mechanism of lotus. Developers have applied this lotus effect on paint. When paint dries, tiny bumps remain on surface that help water droplets to remove dirt. Similar property is seen in many plant leaves (Figure 6.14).

FIGURE 6.14 Cleanliness property of a leave.

6.7.2 Spider Silk

Spider silk, produced by special glands in a spider's body is light and flexible, about 3 times stronger than steel. Tensile strength of radial threads of spider silk and steel is 1154 Mpa, and 400 Mpa, respectively. Spider web (Figure 6.15) consists of two types of silk, major ampullate silk forming dragline and web frame, and viscid silk forming glue-covered catching spiral. Incredible properties of spider silk are attributed to its unique molecular structure. Silk is composed of long amino acid chains that form protein crystals. Majority of spider silk also contain beta-pleated sheet crystals. These amino acid sequences are composed of an 8–10 residue polyalanine block and a 24–35 residue glycine-rich block. Beta-sheet crystals crosslink fibroins into a polymer network, with great stiffness, strength and toughness. This crystalline component composed of amorphous network chains of 16–20 amino acid residues long, is embedded in a rubbery component that permits extensibility. Extensibility, tensile strength, light weight together enable webs, to prevent damage from wind, and their anchoring points, from being pulled off. Silk genes of two spider species has successfully expressed in milk of a transgenic goat. Silk can be precipitated from milk as a web-like material, called **biosteel**. Spider silk

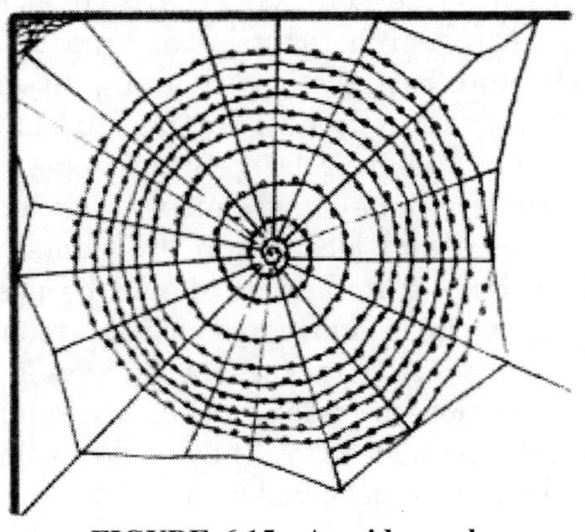

FIGURE 6.15 A spider web.

could be used in wound healing, production of super-thin, biode-gradable sutures for neurosurgery. No company has yet successfully manufactured artificial silk, but several are trying. Spider's silk is stronger than steel and has potential, for use in defense.

6.7.3 Properties of Geckos

Geckos have microscopic setae, as small as one tenth the diameter of human hair, and are themselves covered in even smaller projections, called **spatulae**, which are sometimes so tiny as to be smaller than wavelength of visible light. Each time one of these setae comes into contact with a surface, tiny hairs form a temporary atomic bond. This molecular bonding is relatively weak, but when multiplied over thousands of setae, it becomes quite potent, strong enough that every single seta was bonded to a surface. An adult gecko could support nearly 300 pounds of weight. This ability of Gecko is now being exploited to create a synthetic gecko tape that can be applied and removed easily.

6.7.4 Properties of Shellfish

Shellfish produce exceptionally durable shells from calcium carbonate and proteins. An abalone shell consists of many alternating layers of calcium and organic polymer that gives it strength and flexibility, necessary to form curved shapes. Polymers orient calcium molecules in ways that maximize their strength. Other shellfish produce strong adhesives for attaching themselves to rocks and reefs. Scientists have customized soy proteins to create an adhesive similar to that blue mussel's glue.

6.7.5 Robotics

Robot cockroaches have been developed. Their legs move in same scurrying pattern as natural cockroach. A research team is working on a robot cockroach that can infiltrate and poison nests. Robot lobster can explore bottoms of lakes and rivers. Microrobots are relatively cheap, as most use open-source or off-the-shelf sensors and actuators, microprocessors and locomotion programs, and are deployed in large numbers. It should be possible to make them swarm, to move in a coordinated manner, and work together on tasks. If a package of dozens of robotic scorpions and dragonflies are send, rather than a single complex robot, they could explore more territory and take more risks that are unthinkable with today's technology. Swarms of microrobots could be used for battlefield reconnaissance, and in search and rescue. Biomimetic technology behind robot fish is based on common carp. Robot navigates and propels itself, through water with its fins, just like a real fish.

6.7.6 Probable Use of Biomimicry

Natural resource management

Namib Desert beetle *(Onymacris unguicularis)* is now a model, for water collection. Beetle inhabits driest parts of world, where moisture comes from morning fog. Beetle wings are covered with a waxy, hydrophobic coating, punctuated by hydrophilic ridges. In morning, it raises its wings, and fog forms droplets on ridges, which then trickle down the wing into beetle's mouth. Fog collectors attempt to use this strategy of combining hydrophilic and hydrophobic surfaces to increase efficiency.

Human resource management

Today, biologists view model of self-organization and swarming behaviour. Parallels have already made between pack and swarm behaviour. Work is continuing for applying insect ventilation systems to buildings. In "biobricks" genes, proteins, and cells can be snapped together, like Tinkertoys to build living systems. Berkeley's Keasling Lab has made strides in converting bacteria into chemical factories that produce anti-malaria treatment, artemisinin at a very low cost. Costly anti-cancer drug, taxol may be produced naturally by Pacific yew tree. Anti-AIDS drug derived from *Samoan mamala* tree may be produced. Microbes are also in use that hone in on body tumors and release drugs to attack cancer cells. *Diatom* shells could be employed in filtration systems. Fabrication of more powerful computer chips, containing circuits patterned in 3 dimensions is also continuing. Biomass, specifically as a source of ethanol, is often touted as a source of renewable energy. Ethanol is produced by fermenting crops like corn or sugarcane. To met the energy need, fast growing plants are engineered, which fertilize themselves and convert carbon dioxide, sunlight, water and nutrients into biomass. Termites have evolved cellulose-conversion mechanism inside their stomachs. Synthetic biology may lead to bacteria outfitted with cellulose-converting proteins that could convert biomass and biowaste into fuel. Keasling and others began designing an organism for fermenting cellulose as a source of renewable energy.

Bacteria and virus-assembled batteries

Certain bacteria convert glucose or human waste directly into electricity. Gardner is exploring, whether bacteria can be engineered, so that their energy output is enough for larger-scale fuel-cell applications. "Some devices do not need much power or could benefit from ability to use unusual fuel sources, a medical implant that gets power from blood, for example, never needs to get charged". Robots in field could grab a plant and convert it to power. MIT researchers have reported an important advance towards building microscopic batteries. They used a virus to assemble anodes on top of electrolyte layers. Advantages of virus assembly include functioning at room temperature and precise control over size and spacing of nanomaterials, leading to uniform and easily reproducible devices. Batteries could be used in "labs on a chip" or miniature medical devices.

Engineered organisms

Venter suggested that carbon dioxide, spewing from power plant smokestacks, could be converted by microbes into natural gas to power boiler. Scientists are also designing biological systems to convert sunlight directly into hydrogen through, photosynthesis. Some organisms have natural ability to degrade environmental contaminants. Research is underway to modify *E. coli*, so that it will "swarm" around a nerve agent and degrade it. Microorganisms might be used to decontaminate hazardous waste spills and nuclear disposal sites. Engineered *E. coli* and *Pseudomonas aeruginosa* precipitate heavy metals, uranium, and plutonium on their cell wall. Once the cells accumulate metals, they settle out of solution, leaving cleaned waste-water." Recently, scientists from Duke University and Office of Naval Research devised a way to alter sensing proteins *in E. coli*, so that they bind to different molecules, such as TNT. A glucose-sensing protein could be used for monitoring diabetes, or a TNT-sensing protein to assist the U.S. Navy's underwater robots in locating explosive devices. Leaves of engineered plants could shift colour, in presence of some environmental toxins. Patches of sensor plants might be grown near chemical facilities. Engineered *E. coli* sense light by adding a light-receptor protein from blue-green algae. That sensor may then genetically wire to bacteria's pigment-making system, so that light switches off a gene that regulates pigment production. When light is projected on a petridish coated with "photosensitive" *E. coli,* a biochemical print is created. The novel genetic circuit could be used to build a cell that's switched on and off with a beam of light.

Properties of sea cucumber

A new plastic has developed that mimic sea cucumber's ability to quickly change rigidity of its skin. When dry, new plastic is hard and stiff. When wet, it softens and becomes flexible, and capable of switching quickly back and forth between two states. In new plastic, material is composed of a rubber-like base, supported by fibers of cellulose, attached to one another by hydrogen bonds, creating a rigid surface when dry. When water is introduced to system, hydrogen bonds are broken and support structure becomes loose making the material pliable. When material dries out again, hydrogen bonds quickly regenerate and rigidity is restored. This new plastic could be incorporated into the brain, during treatment of Parkinson's disease, or even spinal injuries. New plastic is hard enough, and in dry condition may be easily implanted. Other potential uses are new forms of casts that would strengthen and soften as needed in the bullet proof vests that could be turned on and off. A gene in sea cucumber blocks transmission of *Plasmodium* causing malaria. Idea is to incorporate the gene into mosquitoes, causing them to produce protein lectin, which is poisonous to malaria parasite, early in development. Genetically modified mosquitoes would be released into the wild hoping that they reproduce and spread new gene to future generations and tamp out spread of malaria. A patent has been submitted for process of centrifuging collagen from sea cucumber flesh into layers to be used in artificial corneal transplants, and connective tissue of sea cucumbers might be perfect as replacements for torn tendons in humans. Even certain cells, believed to be responsible for regenerative healing process of sea cucumber, are being isolated and tested to see, if they can help healing.

Living machines

Wetland water filter secret is amazing. Water released from this system is often clear than municipal tap water. By getting inspiration from this system, John Todd developed living machine for wastewater treatment, where microorganisms use waste input as nutrients, they digest them and in the process purify water. Water released from these systems is often cleaner than municipal tap water. This technology is not only capable of meeting tough new sewage outflow standards, but uses no chemicals and is relatively inexpensive (Ashokan 2005).

Hedgehog quills

Spiny hedgehog (Figure 6.16) has unique ability to drop from a relatively high place and bounce away

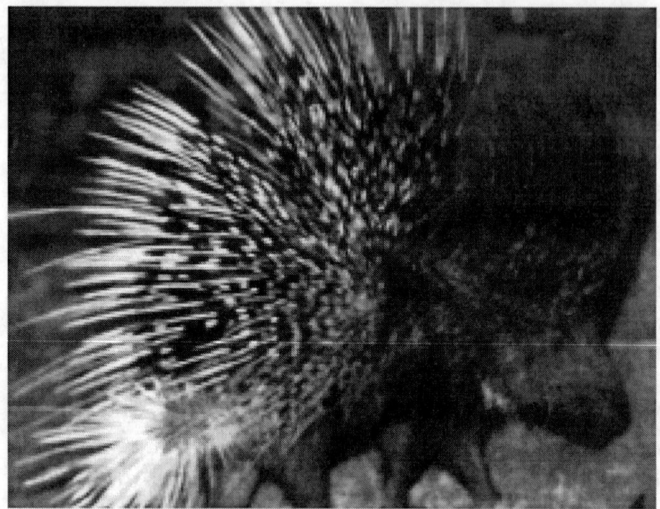

FIGURE 6.16 A hedgehog quill.

safely on its quills. Hedgehog's spines work so well because even though they show a coarse honeycomb structure, they are perfect for absorbing shocks. Besides, because of its weight-energy cost efficiency, hedgehog principle could be exploited for designing light-weight structures, such as space station. The design could also be helpful in creating better wind resistant varieties of wheat and barley.

Glass sponge fibers

Euplectella, commonly known as "venous flower basket" is capable of transmitting light as well as, if not better than industrial fiber optic cable, used for telecommunications. Researchers believe that these fibers might even be tied into a knot, without breaking or cracking, a property that lacks in industrial optic fibers. Sponge prepares fibers at cold temperature using natural materials. Scienists hope to duplicate this process in order to avoid problems created by current fiber optic manufacturing methods that require high temperature and produces relatively brittle cable. With ability to add traces of sodium to fibers, thus increasing their ability to conduct light, sponge naturally accomplishes something that science currently cannot (Ashokan 2005).

Compound eyes of insects

Surface of fly eye is covered by highly packed protuberances, which potentially increases visual efficiency through increased photon capture. Alumina replica has made by removing fly compound eye template at high temperature, which was crystallized simultaneously. Besides morphology, unique antireflection property has inherited by alumina replica. Alumina replica of a fly eye is an efficient antireflection structure of visible light, at an incident angle up to 80°. Such structure would be particularly useful on cured corneal surface, as it increases transmission of incident light through cornea compared with a smooth surface.

Properties of locusts

Locusts avoid running into each other in swarms by using highly evolved eyes that allow these insects to see in several directions simultaneously. Automobile designers mimicked the locusts' vision, when developing sensors that detect movement directly surrounding a car and warn drivers of impending crashes.

Other examples

Human attempts to create effective flying machines often involved studying bird's flight. Early humans learned hunting, shelter, and survival techniques by observing animals as they interacted with their surroundings. Whale hearts hold clues to making pacemakers, and lizard skins are showing how to cut friction in electrical appliances as companies mimic nature to develop high-tech goods, a UN-backed report said. Humpback whales pump six bathtubs of blood around their bodies and regulate the beats with electrical signals, passing through thick non-conductive blubber shielding the heart from cold. A cheap operation for humans that bridged damaged heart muscles by mimicking the tiny "wiring" could cut demand for battery-powered pacemakers in humans, based on research at the Whale Heart Satellite Tracking Program in Colombia.

Through an examination of existing biomimetic technologies, it is apparent that there are 3 levels of mimicry-organism level, behaviour level and ecosystem level. Organism level refers to a specific organism, like an animal and may involve mimicking part of, or the whole organism. Second level refers to mimicking behaviour and may include translating an aspect of how an organism behaves or relates to a larger context. Third level is the mimicking of whole ecosystem and the common principles that allow them to successfully function. Within each of these levels a further five possible dimensions to the mimicry exist. The design may be biomimetic (construction), how it works (process), or what it is able to do (function).

6.7.7 Behaviour Level

Some organisms encounter the same environmental conditions that humans do, and need to solve similar issues that human face. These organisms tend to operate within environmental carrying capacity within limits of energy and material availability. These limits and pressures that create ecological niche adaptations in eco-systems, mean not only well-adapted organisms continue to evolve, but also well adapted organism behaviour and relationship patterns between organisms or species (Reap et al. 2005). Organisms that are able to directly or indirectly control the flow of resources to other species who may cause changes in biotic or abiotic materials or systems and therefore, habitats are called **ecosystem engineers**. Ecosystem engineers alter habitat, either through their own structure or by mechanical or other means. Humans are undoubtedly effective ecosystem engineers, but may gain valuable insights by looking at how other species are able to change their environments, while creating more capacity for life in the system. Several examples demonstrate the details of organisms altering their own habitats, while facilitating the presence of other species, increasing nutrient cycling and creating mutually beneficial relationships between species. The building behaviour of other species is often termed "animal architecture" (Von Frisch and Von Frisch 1974) and may provide further examples of such ecosystem engineers.

The examples of North American beaver (*Castor canadensis*) demonstrates, how through its altering of landscape, wetlands are created and nutrient retention and plant and animal diversity is increased, helping in part to make the ecosystem more resilient to disturbance. In behaviour level biomimicry, it is not the organism itself that is mimicked, but its behaviour. It may be possible to mimic the relationship between organisms in a similar way. An architectural example of process and function of biomimicry at behaviour level is demonstrated by Mick Pearce's Eastgate Buildings in Harrare, Zimbabwe and CH_2 Building in Melbourne, Australia. Both buildings are based in parts on the techniques of passive ventilation and temperature regulation observed in termite mounds, in order to a thermally stable interior environment. Water, which is mined (and cleaned) from the sewers beneath the CH_2 building is used in a similar manner, to how certain termite species will use the proximity of aquifer water, as evaporative cooling mechanisms. Ability of termites to maintain virtually constant temperature and humidity in their SubSaharan Africa homes despite an outside temperature variation from 3°C to 42°C is interesting. Eastgate Centre stays cool without air conditioning, and uses only 10% energy of a conventional building of its size, using termite technology. Architects have already used the termites' self-cooling designs, in construction of energy-efficient buildings.

Biomimicry is often described as a tool to increase sustainability of human designed products, materials and the built environment (Baumeister 2007). However, it should be noted that a lot of biomimetic technologies or materials are not inherently more sustainable than conventional equivalents, and may not have been initially designed with such goals in mind. The built environment is increasingly held accountable for global environmental and social problems, with vast properties of waste material and energy use and green house gas emission attributed to habitats, humans have created for themselves. It is becoming increasingly clear that a shift must be made in how the build environment is created and maintained. Mimicking life, including the complex interactions between living organisms that make up ecosystems is both a readily available example for humans to learn from and an exciting prospect, for future human habitats that may be able to be entwined with the habitats of other species in a mutually beneficial way.

Biomimicry in Sustainability

—FATIK B. MANDAL

Mimicry is the somewhat adaptive resemblance, in which an animal bears resemblance to other object in form and color, to escape from the clutch of predator, or exhibits an apparent aggressiveness which is not at all real. A new discipline has emerged, which attempts to mimic the characters of organisms, their behaviour, or the whole ecosystem in which organisms live, for solving many problems of human society sustainably.

Biomimicry is a revolutionary interdisciplinary science which uses incredible efficiency of nature and imitates flora, fauna, or entire ecosystem for design. Biomimetic approach which has gained significant importance to architectural design takes idea from natural ecosystem, to create a more sustainable and even regenerative-built environment, called sustainable architecture. Three levels of biomimicry are organism level, behaviour level, and ecosystem level.

Living species have been evolving for millions of years, and have tolerated the environmental changes over time. Their evolutionary history tells us the ways to combat various natural hazards, the ways to get access to resources, and the ways to adapt with climate change.

Namibian desert beetle, *Stenocara*, inhabits desert with negligible rainfall. It absorbs moisture from swift moving fog. Droplets form on the beetle's back and wings roll down into its mouth. This fog-catching ability of the beetle has been translated into action, at the Hydrological centre for the University of Namibia. Surface of the beetle has been studied in detail, and used for other potential applications such as clearing fog from airport runways, and to improve dehumidification equipment.

Tirau's iconic dog building at New Zealand exhibits biomimicry stylistically or aesthetically and functions like a natural ecosystem. Mimicking the natural ecosystem rules has tremendous potential to increase sustainability in architectural project.

Architectural example of biomimicry is also demonstrated by Mick Pearce's Eastgate Building in Harare, Zimbabwe and the CH_2 Building in Melbourne, Australia. These buildings mimic termite mounds to produce a thermally stable interior environment. Mimicking the survival strategies of organisms would increase the sustainability and regenerative capacity for human-built environments and mimicking the whole system rather than single organism is suggested to be more appropriate.

The article is based on information of a research article, "Biomimetic approaches to architectural design for increased sustainability" published in SB07 New Zealand (Paper number: 033).

Source: http://www.associatedcontent.com/pop_print.html?content_type=article&content_type_id=6

Review Questions

Short Answer Questions

1. Define camouflage.
2. Define mimicry.
3. Define Batesian mimicry.
4. Define Mullerian mimicry.
5. State the difference between Batesian and Mullerian mimicry.
6. What is aggressive mimicry?
7. Mention two evolutionary significances of mimicry.
8. What is the difference between venomous and poisonous organism?
9. Give two examples of venomous reptiles.
10. Give two examples of venomous mammals.
11. Comment on the defensive behaviour.
12. What is cryptic colouration?
13. Name a scientist associated with Biomimicry.
14. Mention the potential of Biomimicry.
15. Mention the use of microbes in Biomimicry.

Long Answer Questions

1. Describe camouflage with examples.
2. Describe cryptic colouration with examples.
3. Give a comparative account of Batesian and Mullerian mimicry in detail.
4. Give an account of venomous vertebrates.
5. Write an essay on Biomimicry with examples.

Chapter 7
Chronobiology

F. Hallberg coined the term chronobilogy in the beginning of 20th century. "Chrono" means related with time; "biology" means science of life. Thus, chronobiology is the study of science of life in relation with time, i.e. studying the time sense of the organism (biological clock). Chronobiology is a field of science that deals with periodic (cyclic) phenomena in living organisms, and their adaptation to solar and lunar related rhythms. These cycles are known as **biological rhythms**. Related terms chronomics, and chronome have been used, to describe either molecular mechanisms involved in chronobiology, or more quantitative aspects of chronobiology, particularly, between comparisons of cycles among organisms. Daily or circadian rhythm remains deeply ingrained in physiological make-up. World is full of rhythms, and it would be incredible if these rhythms did not exert a significant effect on physiological, and performance potential. In recent years, chronobiology has confirmed that this is indeed the case. Zeitgebers, a German term meaning time giver was coined by Jurgen Aschoff (1960) for environmental agent that entrains the organisms behaviour to external environment.

Variations of timing and duration of biological activity in living organisms, occur for many essential biological processes. Most important rhythm in chronobiology is circadian rhythm, a roughly 24-hour cycle shown by physiological processes in plants and animals. Many other important cycles are also studied, including infradian rhythms, which are cycles longer than a day, annual migration or reproduction cycles found in certain animals, or menstrual cycle in human. Ultradian rhythms are cycles shorter than 24-hours, such as 90-minute REM cycle, 4-hour nasal cycle, or 3-hour cycle of growth hormone production. Tidal rhythms, commonly observed in marine life, follow roughly 12-hour transition from high to low tide and back. Different physiological rhythms, found in animal body, remain linked together to form a clock like system called **biological clock**. This clock usually independent of environmental factors, expresses the behavioural periodicity, programmes a particular phenomenon, and predicts the onset of environmental changes. Migratory birds, kept alone after hatching, migrate without any previous experience. Their gonads ripen, and body deposits sufficient fat before migration.

Animals commonly confront cyclical environmental changes, such as days, years, seasons and tides. Continuous influences of exogenous rhythms successively develop physiological rhythms, which are linked together to from a time sense endogenous clock. Endogenous rhythm gradually deviate, form exogenous cycle and become independent. Biological rhythm exhibits endogenicity, in the existing variation of natural activity period of the organism. Measurement of activity period of *Rana pipiens* under constant conditions reveals circadian rhythm of 23 hours and 10 minutes, and 24 hours and 33 minutes.

Study of biological rhythmicity began as a curiosity-driven science. Daily patterns of behaviour in organisms are known, for more than two centuries. However, reluctance to accept the existence of biological clocks, for controlling rhythmic behaviour, persisted until second half of 20th century. Nature is full of rhythmic changes. Environment changes from day to day, season to season, and perhaps from year to year, in more or less predictable manner. Behaviour and animal abundance often reflect these rhythmic changes in environment. Cycles in nature are universal in occurrence with profound importance and are induced primarily by the sun, as short-term variations in light, temperature and humidity in solar daily cycle, or long-term flux in these variables, in longer-term annual solar cycle. These changes modify lives of organisms. In fact, animal behavioural patterns are highly adapted to avoid harmful effects of environmental cycles, and utilizing beneficial ones. Although, sun is primarily responsible for changes in earth's environment, the moon also exerts important effects. Information from many organisms indicates that parts of their activities recur at regular periods, related to parts of lunar cycle. Animals synchronize their activities to a great diversity of natural, geophysical rhythms, e.g. diurnal (24 hours), tidal (12.4 hours), semilunar (14.8 days), lunar (29.5 days), annual, seasonal, or photoperiodic (365 days).

Environmental variables in complex ways stimulate and control activities of animals. How these activities are timed, maintained and vary widely are not understood. Behavioural patterns may depend on receipt of direct stimuli from environment, or repeated pattern of environmental stimuli, which results in short-term imprinting of a behavioural sequence in animal. This may persist temporarily in laboratory; sometimes indefinitely as repeated pattern of stimuli in the past has led to a recurrent and persistent pattern of behaviour in species. Thus, some animals exhibit rhythmic behaviour, dependent on environmental variables, while others are independent, but display cyclic behavioural patterns, which nevertheless correlate with cyclic fluctuation in environmental variables. Scientists used these differences to define exogenous (dependent) and endogenous (independent rhythms. The latter was called the biological clocks. Endogenous rhythms arose by long-term adaptation and synchronization, to persistent cycles in physical environment, cycles induced almost exclusively by the sun and moon. Since much of evolution of life occurred in intertidal zone, over hundreds of millions of years, ancestors of many organisms may have been subjected to rhythmic fluctuations of tide. This may explain origin of animal rhythms, and tidal nature. Some animals are found to respond to lunar, rather than tidal stimuli.

7.1 CIRCADIAN RHYTHMS

Significant progress is continuing in the field of chronobiology that studies circadian rhythms. Human body is orchestrated by internal biological clocks marching to several internal rhythms that pace themselves hourly, daily, monthly, seasonally and even yearly. Central to timekeeping mechanism of body and mind is Suprachaismatic Nuclei or SCN. SCN is master clock, actually two clusters of 50,000 neurons, one on each side of brain. SCN sits inside hypothalamus region and works with several time keeping genes. SCN and timekeeping genes together make up central clock, which governs many aspects of physiology and behaviour. They orchestrate daily rhythms and cycles that control ebb and flow of hormones, chemicals and neurotransmitters. Photosensitive proteins and circadian rhythms are believed to have originated in the earliest cells with the purpose of protecting the replicating DNA from high ultraviolet radiation during the daytime. As a result, replication was relegated to the dark. The fungus *Neurospora*, which exists today retains this clock-regulated mechanism. Rhythmicity appears to be as important in regulating cyclic bio-chemical processes within an individual as in coordinating with the environment. This is suggested by the maintenance (heritability) of circadian rhythms in fruit flies after several hundred generations in constant laboratory conditions as well as the experimental elimination of behavioural, but not physiological circadian rhythms in quail.

The simplest known circadian clock is that of the prokaryotic cyanobacteria. Recent research has demonstrated that the circadian clock of *Synechococcus elongates* can be reconstituted in vitro with just the three proteins of their central oscillator. This clock has been shown to sustain a 22-hour rhythm over several days upon the addition of ATP. Previous explanation of the prokaryotic circadian timekeeper was dependent upon a DNA transcription/translation feedback mechanism.

7.1.1 Importance in Animals

Circadian rhythms are important in determining the sleeping and feeding patterns of all animals, including human beings. There are clear patterns of core body temperature, brain wave activity, hormone production, cell regeneration, and other biological activities linked to this daily cycle. In addition, photoperiodism, the physiological reaction of organisms to the length of the day or night is vital to both plants and animals, and the circadian system plays a role in the measurement and interpretation of day length. Timely prediction of seasonal periods of weather conditions, food availability or predator activity is crucial for survival of many species. Although not the only parameter, the changing length of the photoperiod (day length) is the most predictive environmental cue for the seasonal timing of physiology and behaviour, most notable for timing of migration, hibernation and reproduction.

7.1.2 Impact of Light-dark Cycle

The rhythm is linked to the light-dark cycle. Animals including humans kept in total darkness for extended periods eventually function with a free running rhythm. Each "day" their sleep cycle is pushed back or forward, depending or whether their endogenous period is shorter or longer than 24 hours. The environmental cues that each day reset the rhythm are called **Zeitgebers** (from the German, time givers). It is interesting to note that totally blind subterranean mammals (e.g. blind mole rat, *Spalax sp.*) are able to maintain their endogenous clocks in the apparent absence of external stimuli. Free running organisms that normally have one consolidated sleep episode will still have it, when, kept in an environment shielded from external cues, but the rhythm is of course not entrained to the 24-hour light-dark cycle in nature. The sleep/wake rhythm may in these circumstances become out of phase with other circadian or ultradian rhythms. Schematics of activity/rest cycle of a diurnal animal in presence (top) and in absence (bottom) of light-dark cycles is given in the Figure 7.1.

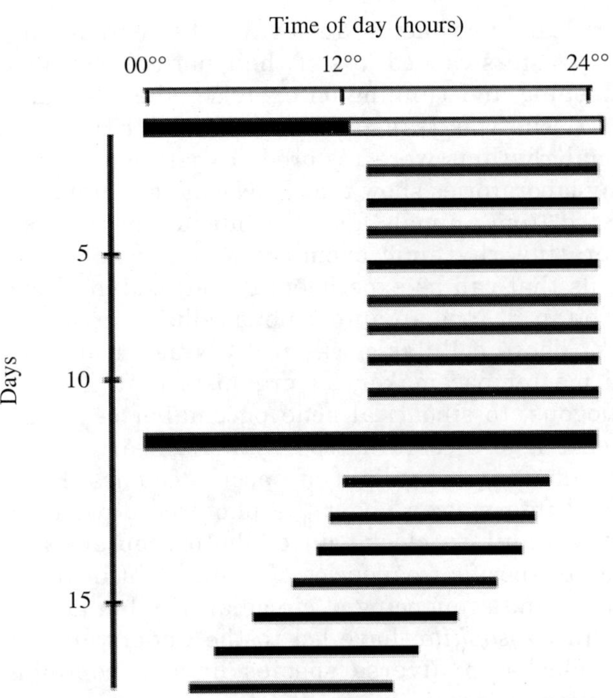

FIGURE 7.1 Diagramatic representation of light/dark cycles. Abscissa represents time of day and ordinate represents number of days. Activity/rest data is arranged one below another. Dark bar represents activity and its absence represents rest. Activity rest rhythm of the animal entrained to 24-hour light-dark cycles.

7.1.3 Brief History of Circadian Science

Study of circadian rhythms dates back to 19th century. Significant observations began in earliest 1960's. One pioneer was Curt Paul Richter, whose innovative concepts laid foundation for studying circadian rhythms in humans. The word circadian is Latin in origin; from word circa, meaning "about" or "approximately" and dian, meaning "day". The term was coined, when Franz Halberg showed that blood count varies, according to a strict rhythm of about a day. Halberg discovered that rhythms actually went somewhat longer than a 24-hour day. He was fascinated by cyclical patterns of behaviour in plants, which actually slowly move over the course of a day, to take advantage of changing light conditions. Halberg is regarded as father of chronobiology, although circadian rhythms have been observed and described since 1700 century.

In 1950s Karl von Frisch, Gustav Kramer, and Colin Pittendrigh, each independently discovered compelling evidence, for internal clocks in animals. Their investigations marked the beginning of modern field of circadian rhythms research. In 1960s Jürgen Aschoff, working in Germany, initiated studies of human circadian rhythms, under controlled laboratory conditions. He built a bunker in which volunteers lived for several weeks. For some periods, they were isolated from the outside world, and at other times the door to real world was opened. Aschoff found that even when subjects were in isolation and under constant light conditions, their sleep-wake cycles persisted. He also found that they drifted out of synchronization with the real world. It was assumed that nervous system contained clock in animals.

Hypothalamus governs many bodily functions, including sleep, hormone production, and metabolism.

In critical experiments, removal of SCN from rats and hamsters caused loss of their normal rhythms of sleeping and running on exercise wheels. When SCN tissue was transplanted back into hamsters, normal rhythms were restored. Experiments from many laboratories show that SCN can be completely isolated from animals and still function as a clock, generating rhythmic chemical and physiological signals that can be experimentally measured. Each neuron in SCN is an autonomous cellular circadian clock. These cellular clocks in SCN act in unison as a master pacemaker, for organism in a manner analogous, to sinoatrial node pacemaker of heart, which triggers coordinated rhythmic heart beats. Experiments, using light of specific colours, have yielded information about the photoreceptors that supply input to clock, since photoreceptors are tuned to specific frequencies of light. Identification, cloning, and sequencing of circadian rhythm genes, first in *Drosophila*, have led to the understanding that clocks in diverse species have a common evolutionary origin controlled by a related set of genes. Clock in brain uses specific biochemical and neural pathways for passing its timing message to other body parts. Chemical reactions and physical properties are dependent on temperature. Circadian clocks keep accurate time, in spite of temperature variation. In mammals, it resides in hypothalamus, specifically, in a small cluster of cells on each side of brain called **Suprachiasmatic Nucleus (SCN)**.

Beginning in 1980, several studies were conducted, including placing people in an isolated environment, without any external cues, or clues. These studies conclude that if left alone, healthy humans undergo changes based on an internal clock that is longer than 24 hours. These changes in biorhythms can be measured in physical strength, aerobic capacity, body temperature, blood pressure, mental alertness, production, and secretion of neurotransmitters and hormones, and many other body functions. No one knows for sure, why the cycle is of more than 24 hours, as opposed to precise 24 hours. If internal circadian clock of an individual is "free-running", it is only a matter of time, before lifestyle realities clash with free-running circadian time. This clash means hormone and neurotransmitter secretion becomes confused, and create problems, for their health. Zeitgebers, body's way of synchronizing, reconcile difference between nature's exact 24 hour cycle and internal circadian rhythms. As modern lifestyles demand more flexibility in schedules, man has lost touch with valuable zeitgeber, the sun.

7.1.4 Biorhythm in Human

A biorhythm in human is a theoretical process by which human body and mind are regulated according to set patterns. Biorhythms are usually separated into 3 distinct groups—emotional, mental, and physical cycles. Classically, lengths of these cycles are 28 days, 33 days, and 23 days, respectively. Biorhythms are thought to follow a regular wave, oscillating back and forth over length of cycle. On first day of one's cycle, one is thought to be at an optimal functional level, while at low point a person is viewed as being at his or her worst possible level of functional capability. Practice of tracking body's biorhythm dates back to the end of 19th century, when a number of doctors began observing, what they perceived as repeating cycles in a number of ailments, and immunesystem weaknesses. Lengths of biorhythm cycles date from this period, and are based on early observations of these physicians. There is a scientific evidence for a number of biological cycles—such as a woman's menstrual cycle. It is highly unlikely that a biorhythm would be same for all people. Everyone needs different amounts of sleep, but general consensus is that adults require between six and eight hours of sleep per night. Aspects of mood disorders are linked to biological rhythms with 3 different periodicities—daily, menstrual, and annual. Rhythms are related to mood disorders in several different ways. Some conditions are expressed cyclically; that is, they have a period linked to an identified internal or external periodicity, such as menstrual and annual rhythms of depression. Other conditions, like cases of bipolar illness express cycles which are not tied to any identified periodic stimulus.

7.1.5 Involvement of Circadian Rhythm

Circadian clock regulates many aspects of metabolism, physiology and behaviour in humans and many other organisms. Circadian rhythms control timing, quantity and quality of hormones and neurotransmitters. Hormones and neurotransmitters are elements that determine our feelings, sleep patterns, appetite, sex drive, and mood-related issues. Circadian rhythms create circadian balance. When out of balance, quantity, quality and timing of hormone, neurotransmitter secretion suffer, and bodies begin to suffer from a Circadian Rhythm Disorder (CRD). Nearly every organism must adapt to biorhythm, to thrive. It is advantageous, to be able to anticipate changes in environment, and most organisms, from humans to cyanobacteria, have internal mechanisms that track time. Such mechanisms have many uses. They may regulate daily activity patterns, trigger reproduction, and even aid in navigation. Circadian rhythm of wheel-running activity in a flying squirrel under conditions of constant darkness is recorded.

7.1.6 Types of Circadian Rhythm

Daily pattern of circadian rhythms can be studied experimentally by removing cues, such as light, darkness, and temperature. Removal of cues results in so-called Constant Conditions (CC). If an animal under CC performs a behaviour at approximately same time as an animal that is in its normal environment, CC animal is said to have a circadian rhythm. Such a rhythm is considered endogenous. If an animal in CC does not perform a behaviour at approximately the same time as an animal under natural conditions, or if the animal does not change its behaviour at all, CC animal clearly does not have a circadian rhythm. Such a rhythm is exogenous.

Connections between psychiatric disorders and biological rhythms were reviewed more than 40 years ago in a landmark monograph by Richter, whose innovative concepts and studies gave birth to many fields of psychobiological research, and establishing key concepts and analytic methods for studying biological rhythms. Richter described a variety of shorter- and longer-period rhythms in rodents, and a few in humans. He identified a 90- to 120-minute cycles of gastrointestinal activity, a 24-hour periodicity, 28-day menstrual cycle, and 280-day periodicity of pregnancy. But he stressed that 24-hour periodicity that formed core of his observations in rodents was poorly expressed in normal humans, and that evidence for its existence was "not very definite." Richter recognized that healthy humans showed small changes in body temperature and other functions during the day. Rhythms continue even when all external environmental conditions remain constant. They originate within organism itself, that is, they are endogenous. Shorter rhythms, such as rhythms of a beating heart are termed "**ultradian**." Cycles longer than 24 hours are called "**infradian**." If they last a year, such as migration and hibernation, they are called "**circannual**" rhythms. Biological clocks play an important role in synchronizing internal and external events of organismal physiology.

7.1.7 Mode of Action of Circadian Rhythm

Light is primary stimulus for regulating circadian rhythms, seasonal cycles, and neuro-endocrine responses in many species. Ocular photoreceptors transduce light stimuli for circadian regulation, and clinical benefits of light therapy are unknown. Retinohypothalamic tract, a distinct neural pathway that mediates circadian regulation by light, projects from retina to Suprachiasmatic Nuclei (SCN). A neural pathway extends from SCN to pineal gland. By this pathway, light and dark cycles are perceived through mammalian eyes; entrain SCN neural activity, and rhythmic secretion of melatonin from pineal gland. In virtually all species, melatonin secretion is high during night, and low during the day. In addition to entraining pineal rhythms, light exposure can acutely suppress melatonin secretion. Acute, light-induced melatonin suppression is a broadly used indicator for photic input to SCN, which is used to elucidate the ocular and neural physiology for circadian regulation. Studies using rodents with retinal degeneration, suggest that neither rods nor cones used for vision participate in light induced melatonin suppression, circadian phase shifts or photoperiodic responses. Nucleation of rodless coneless transgenic mice abolishes light-induced circadian phase shifts, and melatonin suppression. Light-induced melatonin suppression and circadian entrainment are demonstrated in humans with complete visual blindness, and with specific colour vision deficiencies. These studies on different forms of visual blindness suggest that melatonin regulation is controlled, at least in part by photoreceptors that differ from known photoreceptors for vision. That monochromatic light at 505 nm is approximately four times stronger than 555 nm in suppressing melatonin in healthy humans is known. It has been confirmed that ocular photoreceptor primarily responsible for pineal melatonin regulation in humans is not three cone systems that mediates photopic vision. Developing an action spectrum is a fundamental means for determining input physiology for circadian system. It is suggested that there is a novel photo pigment in the human eye that mediates circadian photoreception.

Recent studies with various vertebrate species have identified several new molecules that may serve as circadian photopigments. These putative photopigments include both opsin-based molecules, such as vertebrate ancient opsin, melanopsin, and peropsin, as well as non-opsin molecules, such as bilirubin, and crypto-chrome. Among these new photopigments, only melanopsin has been specifically localized to human neural retina, and cryptochrome has been localized to mouse neural retina. Cryptochromes have been studied extensively as circadian photoreceptors in plants and insects and have been proposed as circadian photoreceptors in mammals.

Recently, a study on humans showed that there are separate photoreceptors for mediating circadian entrainment versus acute suppression of melatonin. It is reasonable to hypothesize that a variety of non-visual effects of light, such as melatonin suppression, entrainment of circadian rhythms, and possibly some clinical responses to light are mediated by a shared photoreceptor system. Modern industrialized societies use light extensively in homes, schools, work places, and public facilities to support visual performance, visual comfort, and aesthetic appreciation within environment. Given that light is also a powerful regulator of human circadian system, future lighting strategies will need to provide illumination for human visual responses, as well as homeostatic responses. Hence, new approaches to architectural lighting may be needed, to optimally stimulate, both the visual and circadian systems.

7.1.8 Suprachiasmatic Nucleus (SCN)

Daily oscillations in physiology and behaviour are regulated by a brain clock located in the Suprachiasmatic Nucleus (SCN) (Antle and Silver 2005). Circadian rhythmicity is abolished by SCN lesions as recorded by (Stephan and Zuker in 1972) and restored by SCN transplants (Lehman et al., 1987). Some cells within SCN exhibit self-sustained rhythmicity driven by autoregulatory transcription-translation feedback loop that regulates expression of the period and cryptochrome genes (Reppert and Weaver 2001). Cells in the SCN core receive direct retinal intervention and express c-fos, per1 and per2 in response to phase shifting light pulse (Yan and Silver 2002). It is also reported that some cells within the SCN rhythmically express "clock" genes, whereas others exhibit induced expression of the genes, after the organism has been exposed to light pulse. The coordinated interaction of these functionally distinct cells is integrated to the coherent functioning of the brain clock.

In mammals, a master circadian pacemaker driving daily rhythms in behaviour and physiology resides in the SCN. SCN, aside the third ventricle and atop the optic chiasma in the anteroventral hypothalamus, each consist of about 8000 cells packed into an approximately 0.5 mm by 1 mm football shape. This small brain region is often described as circadian pacemaker (Klein, Moore and Reppert 1991).

The SCN is necessary for expression of most, perhaps all sustained circadian rhythms. Destruction of these cells eliminate circadian rhythms including drinking, locomotor activity, body temperature, estrous cycling and secretion of prolactin, melatonin, growth hormone and cortisone (Meyer–Bernstein et al., 1999). In 1974, Pittendrigh proposed that the mammalian circadian clock is composed of two interacting oscillators differentially affected by light, with one synchronized to dawn and the other to dusk. In order for a biological clock to be truly useful, it must be set to the local time. This occurs through the process of entrainment by which the period of the biological clock is made equal to the period of the light cycle. The classical model for entrainment (Pittendrigh 1981) suggests that synchronization occurs through periodic corrections in the angular position of the pacemaker. These phase shifts at dawn and dusk set the period to 24 hours and in doing so, establish a specific phase relationship between the pacemaker and the light cycle. Such a process, commonly referred to as non-parametric entrainment (Herzog and Tosini 2001).

7.1.9 Role of Melatonin

The pineal hormone melatonin has been hypothesized to be an internal zeitgeber for the circadian clock of vertebrates. In mammals, melatonin is synthesized and secreted by the pineal gland in a rhythmic manner and is regulated both by the circadian clocks and the L-D cycles. Melatonin level begins to rise in the evening, and remains high through night. Hence, the level of melatonin varies periodically in mammals. Besides this, in mammals the phase shifts evoked by external melatonin administrations vary periodically. It has been suggested that the SCN, the circadian clock of mammals are sensitive to melatonin. Melatonin is also ubiquitous with respect to permeation of all cells, in all tissues, in all organs and its synthesis and periodic release is directly controlled by SCN (vide Sharma and Chandrasekharan 2005).

7.1.10 Molecular Choreography

Benzer and one of his students, Konopka (Konopka and Benzer 1971) discovered the first circadian rhythm gene, and mapped it to a specific location on X chromosome of fruit flies, and named it period (*per*). *Per* was identified through random mutations. Flies, carrying mutations in *per* had abnormal circadian rhythms. Three variants of *per*, each producing a very different behavioural pattern were identified. One type of mutant fly had a clock that kept a short day period of about 19 hours. Second type of mutant had a clock with a long period of about 29 hours. Third type of mutant had no clock at all, and its activity rhythms were random. These 3 changes in clock can result from changes made to *per* gene. In 1984, using modern molecular genetics tools, Rosbash and Hall cloned and sequenced period gene. Identification of period gene and its DNA sequence raised possibility for related genes in other animals, including humans. Although *per* gene is central to function of circadian clock, no gene acts alone, and *per* gene could be used to bootstrap to other genes, with equally important roles in generation and control of circadian rhythms. Control of circadian rhythms at molecular level is very similar in all animals studied so far. New genes are identified that play important roles in functioning of circadian clock.

Fruit fly has about 100,000 neuron cells. A group of neurons near eyes control circadian clock. Flies and humans share many similarities at molecular level. Different variants or alleles of *per* gene exist those cause short-period day, long-period day, and random rhythms. Details of *Drosophila Clock* are complicated. At least, seven additional key genes are identified, and others would be found. Both mRNA and *per* protein itself oscillated on a 24-hour cycle, in normal flies. *Per* mutants, that altered period length also similarly, altered period of molecular oscillations. *Clock* mechanism, thus, was a biochemical circuit, and synthesis and degradation of *per* protein was an intimate part of circuit, and feedback of protein on its own metabolism is central to clockworks. Circadian rhythm genes—*timeless*, *double time*, and *vrillei* in *Drosophila* appear to have counterparts in mammals, as well. Rosbash's and others also cloned first photoreceptor gene involved in resetting *Drosophila Clock* by light—a cryptochrome (*cry*) gene that is related to blue-light photoreceptor genes in plants. Taghert and Hall showed that *pdf* gene, which codes for a peptide called pigment-dispersing factor, is essential for *Clock* function. *Pdf* protein is secreted by a cluster of neurons, where fly's master clock resides, which is near brain's visual centres. CLOCK and CYCLE proteins are transcription factors, positive limb of feedback loop. They combine in cell's nucleus, to drive transcription of PER and TIM proteins. Then, in cell's cytoplasm, PER and TIM bind together, to form a heterodimer—a protein complex made of two different proteins. As evening progresses, PER/TIM heterodimer enters cell's nucleus, in sufficient amount to inhibit functioning of *Clock* and *cycle*, thus, leading to shutting down of further PER and TIM production. As PER and TIM turn over during late night and early part of morning, *Clock* and *cycle* go back to work, and a new cycle begins. Double time protein (DBT) is believed to phosphorylate PER and thereby increase PER degradation. Light causes the *Drosophila Clock* to entrain, to synchronize with 24-hour day-night cycle. Photoreception by cryptochromes appears to trigger rapid degradation of TIM, and since TIM stabilizes *per*, *per* also disappears. Additional proteins may modify action of various proteins that have already been identified, and there may be additional transcription factors regulating gene activity.

Discovery of additional circadian *Clock* genes led to a more elaborate model of negative feedback loop that underlies clock mechanism. Genes and their protein products are either part of positive part of feedback loop (clock and cycle), or part of negative part of feedback loop (*per* and *timeless*). *Double time* gene makes an enzyme that phosphorylates PER, increasing PER degradation, and thus, contributing to maintain appropriate levels of PER, for a particular time of the day. Light leads to an increase in TIM degradation. A cryptochrome (*cry*) gene identified in flies probably makes a photoreceptor protein which is important for resetting the clock by light. *Pdf* gene codes for a peptide called pigment dispersing factor, which is probably responsible for communicating the *Clock's* timing message, to parts of nervous system that are directly responsible for controlling fly's movement.

The mass identification of novel circadianly regulated genes revealed insight into the diversity of processes and likely to be regulated by the *Clock*. These involve functions like vision, olfaction, mechanoreception, detoxification, learning and memory, ion channel activity and certain metabolic events (Clridge–Chang et al., 2001, McDonald and Rosbash 2001, Stempft et al., 2002, Ueda et al., 2002). Moreover, reproduction has been shown to be regulated by the clock, in terms of both mating behaviour and physiological processes underlying overall reproductive fitness (Sakai and Ishida 2001).

Circadian genetics story for mouse is similar to that for *Drosophila* at least at start. Takahashi and his colleague produced random mutations in mice by exposing them to mutation causing chemicals, screened mice for defective behaviour, mapped suspect mutant gene to a specific chromosome, and cloned first mammalian circadian gene *Clock* for "circadian locomotor output cycles." Up to that point, only three other circadian rhythm genes had been cloned, and sequenced: *per* and *tim* in *Drosophila* and frequency *(frq)* in *Neurospora*. In 1998, *Clock's* partner in mouse, *B mal1*—the counterpart to *Drosophila* cycle gene was identified. Other laboratories cloned not one, but three mammalian *per* genes in mice, now known as *per1*, *per2*, and *per3*. This provided first evidence that *Drosophila* and mammalian circadian genes are related. Two mouse Cryptochrome genes, *Cry1* and *Cry2* involved in photoreception were characterized, bringing total number of known clock genes in animals to eight. A ninth circadian rhythm gene has also made its appearance. This gene responsible for a circadian rhythm mutation in hamsters, known as *tau*. Gene, disrupted in *tau* mutants, encodes for enzyme casein kinase I epsilon and is counterpart of *double time* gene in *Drosophila*.

Although core set of circadian genes is conserved in *Drosophila* and mice, there are fundamental differences in their regulation. There are at least six core genes that are thought to play a role in circadian clock system of mice (*Clock*, *Bmal1*, *mPer1*, *mPer2*, *mCry1*, *mCry2*). *Clock* and *Bmal1* are positive elements that act upstream of *mPer* and *mCry* genes, to activate their transcription. As levels of PER and CRY proteins accumulate, they interact among themselves, and eventually translocate into nucleus, where they inhibit action of CLOCK/BMAL1 on their own transcription. With decline of *mPer* and *mCry* transcription, proteins then turn over and CLOCK and BMAL1 are released from negative feedback, leading to a new cycle of transcription. Mammalian transcriptional feed-back loop is similar to that seen in *Drosophila* at level of *per* gene regulation. Cryptochromes play a completely different role as potent negative regulators. A role for *Drosophila timeless* gene is absent in mammals, and pathways for light input are divergent. Diurnal changes in gene expression in the sheep pituitary has shown in Figure 7.2.

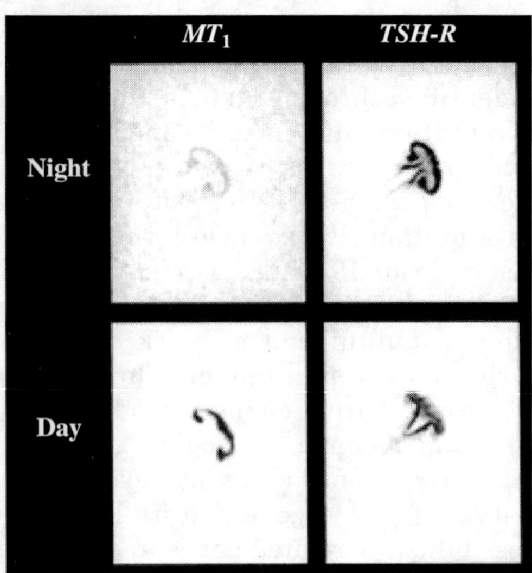

FIGURE 7.2 Diurnal changes in gene expression in the sheep pituitary.

7.1.11 Molecular and Evolutionary Analysis of Circadian Rhythms in Insect Vectors of Tropical Diseases

Many aspects of vector behaviour play an important role in the dynamics of disease transmission. One particularly relevant area in this respect is the study of circadian rhythm. Daily patterns of activity and blood feeding presented by insect vectors, for example are controlled by internal biological clock, which in turn is under genetic control. Despite medical importance of haematophagous insects, very little is known about the molecular control of their circadian clocks.

Current model for circadian pacemaker in *Drosophila* involves cyclical regulation of *period* and *timeless* gene transcription, and translation via a negative feedback loop. Clock and cycle encode proteins that have both a dimerization domain called PAS and a bHLH DNA-binding domain. These transcription factors activate cyclic expression of period and timeless. Late at night, proteins PERIOD and TIMELESS enter the nucleus of pacemaker cells, where they interact with CLOCK-CYCLE dimer, repressing the transcription of *period* and *timeless*. In the morning, CLOCK-CYCLE dimer is derepressed and is free again to promote transcription of *period* and *timeless*, closing negative feedback loop. A second loop involving two other genes, *vrille* and *Pdp1*, is responsible for cyclic expression of *Clock*. Activity rhythms and circadian expression of *period*, *timeless*, *Clock*, and *cycle* in the sand fly *Lutzomyia longipalpis*, the main vector of visceral leishmaniasis in Latin America have been carried out and analyzed. Analysis of activity rhythms in *L. longipalpis*, a crepuscular, nocturnal insect in the wild, revealed that sand fly shows more nocturnal activity, under controlled laboratory conditions, than does *D. melanogaster*. Circadian expression of four clock genes also revealed interesting differences. Although sand fly's *period* and *timeless*, two negative elements of feedback loop, show similar peaks of mRNA abundance in early evening as observed in *D. melanogaster*, expression of *L. longipalpis Clock* and *cycle* (positive transcription factors) differs markedly. *Clock* mRNA levels peak in late afternoon and early evening in sand flies, while in *Drosophila*, the peak occurs late at night and early in the morning. In addition, *cycle*, a constitutive gene in *Drosophila* also exhibits a rhythm in mRNA abundance in *L. longipalpis*. Another interesting difference between *Drosophila* and sand flies is that *L. longipalpis* CYCLE protein has a C-terminal activation domain that is missing in *Drosophila*.

Effect of blood feeding in expression levels of these four clock genes in *L. longipalpis* have carried out. While *Clock* and *cycle* sustain their levels in blood-fed female sand flies, down-regulation of period and timeless expression was observed, which is correlated with an approximately 40 per cent reduction in locomotor activity. These results suggest that circadian pacemaker and its control over the activity rhythms in this hematophagous insect are modulated by blood intake. Genes controlling circadian rhythms are good molecular markers for evolutionary studies, given that they are likely to determine species-specific differences in time of mating, differences that are potentially involved in reproductive isolation and speciation. Analysis of molecular polymorphism in period gene has shown that Brazilian populations of *L. longipalpis* do in fact, represent a species complex.

Two new components of biological clocks are conserved in flies, mice and possibly humans. These clocks conceivably could regulate every aspect of physiology and metabolism. Internal timing of this biological clock, which is called a circadian rhythm, is achieved through a simple feedback loop in which four proteins turn each other, on and off. One set of two proteins works in morning to produce second set of molecules that accumulates during the day. In evening, this second set of proteins inactivates daylight-active proteins. While it may be possible that every cell in *Drosophila* has its own photoreceptor system, and can be directly stimulated by light and darkness, humans have a different regulatory system in which light perception is regulated by eyes. Rosbash asks, is circadian rhythm coordinated and synchronized within each human cell? Takahashi is intrigued by possibility that biological clocks may give us insight into basis of sleep disorders and affective disorders, such as bipolar disease and depression. These disorders all involve disturbances in the sleep-wake cycle, even though that may not be the root cause, Takahashi said, even process of aging involves a breakdown in circadian rhythms. He said: "Increasing the efficiency of clock will not reverse time, but it may improve a person's overall functioning."

Biological rhythms (biorhythm) are tied to physical environment, where organisms evolve. Humans have observed natural rhythms, including daily activity patterns of plants and animals for long. A biorhythm occurs on scale of a day and driven by an internal clock is called **circadian rhythm**. Organisms anticipate changes, like rising and setting of sun in the environment. An internal clock is used to anticipate important environmental events. Circadian clocks govern human activities from sleep starved students to active octogenarians. Body intrinsically follows a daily regimen of wakefulness and sleep. Internal clocks keep us mentally alert during the day and prompt us to rest during the night.

Rhythms or cycles are very common in nature, including four seasons and 24-hour rotation of earth. Our bodies have rhythms. Some rhythms of body and mind are tied to nature. Body responds to nature's cues to create ideal rhythms. Human circadian rhythm responds to morning light of a new day. This light cue body to produce cortisol, serotonin, other hormones and neurotransmitters. At sunset, body responds to dusk and ultimately night's darkness. As sun goes down, body secretes melatonin which decrease blood pressure as body prepares for eventual sleep. Circadian rhythms are cyclic and persistent patterns of behaviour, and are exhibited by most life on earth, from bacteria to largest mammal. These rhythms roughly follow 24-hour periods, reflecting the amount of time earth takes to complete a rotation. An animal in dark manifest periods of increased and decreased activity, which correspond with a 24-hour cycle. Repeated input from external stimuli reset circadian rhythms. Fluctuations in temperature do not impact circadian rhythms. Circadian rhythms are outwardly very similar in all species, but genes that make up clock mechanisms are quite different. Clocks probably evolved several times to perform very similar functions, so they are the example of "convergent evolution".

7.2 CIRCAANNUAL RHYTHM

The term circaannual (Circa, means about; annual means year) refers the rhythm usually less than 365 days. Annual rhythms may be of following types.

7.2.1 Breeding Cycle

Onset of breeding in many birds and mammals exhibit circaannual rhythm. In springs, a male chaffinch alone in an aviary begins to sing, continue singing at regular rate of three songs per minute. Sand pipers breed in artic regions nest and incubate in spring, when snow is still on ground. Eggs hatch when snow melts and insects are abundant to provide food for the young.

7.2.2 Hibernation

Some animals avoid unfavourable conditions by migration; others survive in unfavourable periods by entering a prolonged dormancy phase called hibernation at low temperature. Winter hibernation is found in *Rana tigrina*.

7.2.3 Aestivation

Dormancy period in unfavourable climate that occurs among insects is called **diapause**. Insects that avoid freezing by super cooling can survive very cold conditions in diapause, in which the body fluids freeze at temperature below 0°C. *Baracon cephi*, a Canadian wasp, may lower its freezing point to 46°C by increasing the glycerol concentration in blood.

7.2.4 Migration

Migration between the different habitats on a periodic or seasonal basis is exhibited by many species of butterflies, locusts, salmons, birds, bats and antelopes. Biological clock initiates migration and also controls its pattern. Migratory route is characteristic of particularly breeding population of white storks, *Ciconia ciconia* that breed in Western Europe, fly to their wintering grounds in Africa through a westerly route via Spain and Gibraltar. However, those that breed in Eastern Europe take easterly route. Young storks of Eastern Europe kept in captivity when released in Western Europe, fly in easterly direction, a characteristics of storks of Eastern Europe. Similar experiments with other species indicate that the direction of migration is controlled by biological clock.

Circaannual cycle of the golden-mantled ground squirrel has shown in Figure 7.3. Animals held in constant darkness and at a constant temperature nevertheless entered hibernation (black bars) at certain times year after year.

7.3 LUNAR CYCLE

Lunar orbit around earth takes 27.29 days. Because of earth's rotation and moon's movement, relative to earth, the moon rises 50.5 minutes later each day, i.e. lunar day is 24.84 hours long. Moon, sun, and earth are in same positions relative to each other every 29.5 days, the lunar or synodic month. Movements of sun and moon interact in a way, that at full moon, moon rises at dusk and sets at dawn, and at new moon, moon and sun rise and set together. Such relationships produce a lunar monthly cycle in duration of moonshine at night. Light intensity of moon varies with lunar phase at full moon 1.83×10^{-1} micro-watts/sq cm, at quarters

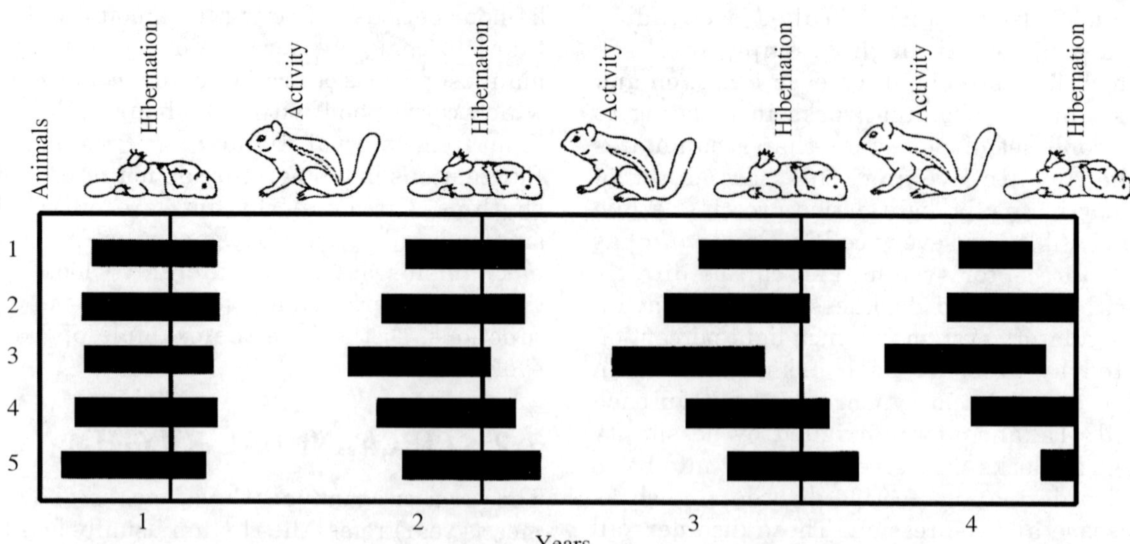

FIGURE 7.3 Circannual cycle of the golden-mantled ground squirrel in constant darkness and at a constant temperature nevertheless entered hibernation (black bars) at certain times year after year.

2.12×10^{-3}, and at new moon 1.8×10^{-4}. Heavy cloud cover reduces these values about 10 times. Intensity of moonlight is reported to be about 1/500,000 of sunlight.

Tides are caused by gravitational pull of the moon and the sun on earth. Because of the sun's great distance from earth and despite its much greater mass, its gravitational effect is less than half of moon. Lunar cycle of tide-producing forces rotates around earth each 24.84 hours producing two pairs of tidal maxima and minima at any place at each revolution. Many factors affect occurrence of tides. In some places, there are no tides, and in others up to four tides per day. As moon's orbit is at an angle to earth's equatorial plane, oscillating tidal asymmetry varies 28° north and south of the equator. Since lunar tide has a 24.84-hour cycle, and solar tide a 24-hour one, effects of moon and sun alternately amplify and oppose each other every 14.8 days, causing spring and neap tides (Smith 1968). Several secondary effects of lunar cycle occur in aquatic environment. Water pressure is a function of depth, and so fluctuates with tides, varying with daily and lunar-monthly tidal cycles. Similarly, air exposure, or water immersion varies on similar cycles. Forces like magnetism are known to exhibit lunar cyclic changes, and these forces are perceived in some fashion by organisms, but evidence is extremely scanty.

7.3.1 Occurrence of Lunar Rhythms

Environmental variables affected by moon may have lunar-daily (24.84 hours), semilunar (14.8 days), or lunar-monthly (29.5 days) periodicity. Most affected environmental variables by moon are light and tide. Tide appears to have greatest impact. Thus, lunar rhythms occur mostly in marine species. Lunar periodicity has relatively little consequence to terrestrial communities (Allee et al., 1949). Except marine animals, very few organisms show lunar rhythms of activity. Species exhibit lunar rhythms mostly in their reproductive behaviour. In many littoral animals, increase in activity occur with rising tide. Tide pool fishes leave rock pools to forage more widely. Sea anemones expand, when immersed by rising tide.

The ciliate *Conchophthirius lamellidens*, parasitic in a freshwater mussel, conjugates most freely after full moon. A species of *Pocillipora* breeds throughout the year, but with a lunar rhythm. In winter, breeding is related to full moon, and in summer to new moon. There is no correlation with tides since in winter, the lowest neap tides are related to the full moon, and in summer to the new moon. Response of animal appears to be on tidal amplitude, water pressure, or air exposure. Thus rhythm is an irregular lunar monthly one. A sea anemone *Actinia equina*, has a tidal rhythm in its cycle of expansion and contraction which was reported as a simple case of direct response to tidal stimuli. But rhythm persisted in laboratory for several days. Inter-tidal flatworm *Convoluta roscoffensis* raises to surface when tide is low, and retreats when tide rises. Movements do not occur at night. All marine polychaetes respond to lunar influences. Their periodicities and causation are variable. *Odontosyllis enopla* (Bermuda fireworm) breeds about 55 minutes after the sunset for about half an hour, only on nights when there is no moon in the sky during early part of night, i.e. from two or three days after the full moon, until the new moon. On such nights, worms swim to sea surface from their tubes on sea bed, apparently induced to do so by decrease in light at dusk. Lunar rhythm seems to be a response to decline in light intensity from sunlight to night sky without moon. Gonads maturation is reported with a lunar-monthly cycle, which is in phase with lunar-monthly cycle of a night sky without moonlight at dusk. Other species of *Odontosyllis* have similar cyclic breeding behaviour. In all these species, breeding is a response to lunar cycle of illumination.

The best known annelid with lunar spawning rhythms, *Eunice viridis*, spawns in immense concentrations on Pacific island coral reefs. Reproductive part (epitoke) of worm, detaches from benthic, tubicolous, vegetative part (atoke), and swims to surface. Swarming occurs, seven to nine days after the full moon, at low tide, when coral reefs are awashed and usually at dawn, regardless of the moon and cloud conditions. As moonlight is not essential for swarming, response does not seem to be directly related to changes in light, as occurs in *Odontosyllis*. Perhaps, breeding is related to lunar cycle in such a way that gonads become ripe at third quarter of moon in response to lengthening dark period after the full moon, as the moon rises later, and later in night will approach the time of new moon. *Eunice viridis* differs from *Odontosyllis*, in that its gonad maturation has an annual cycle, independent of lunar cycle except that gonads finally ripen apparently in response to changes in lunar illumination. Atlantic palolo, *Eunice fucata* are known to have similar breeding patterns to *E. viridis*. *Spirorbis borealis*, another benthic tubicolous polychaete, releases its larvae at the first and third quarters of the moon. This is due to occurrence of breeding at new and full moons. Thus, breeding which occurs at spring tides, manifests itself by neap tide release of larvae. Lunar periodicity in settlement of another tubicolous polychaete *Hydroides norvegica*, has interesting economic implications. Scientists found that it is better not to clean hulls of ships near spring tides because this worm exhibits spring tide (semilunar) periodicity in larval settlement. Ships cleaned at other parts of tidal cycle were found to remain unfouled longer, than those cleaned during spring tides.

A chiton *Chaetopleura apiculata*, exhibits lunar periodicity in breeding activity. *Mytilus edulis* matures during new moon period, and spawns following neap tide. Oyster *Ostrea edulis,* spawns at about spring tides, and so do *Chlamys opercularis* and *Pecten maximus*. *Littorina neritoides* lives at about high-water mark of high spring tides, and releases its larvae when immersed by sea. It thus exhibits semilunar breeding rhythm. Lunar rhythms of various types have been claimed, for crab *Uca pugnax*. Chilling of *Carcinus maenas* would lead to replacement of a circadian (solar-daily) rhythm by a tidal rhythm, which 'though not phased with external tides indicates that ability to show tidal rhythmicity is deep seated in British *Carcinus*. Chilling of crabs, reared from eggs in laboratory, cause 'spontaneous' establishment of a similar tidal rhythm, confirming that 'tidal rhythmicity is deep seated', and suggesting that 'it may be inherited'. *Clunio marinus*, a marine tidal chironomid's larva live among intertidal algae. Adults emerge from their pupae, when water level is low. Female is short lived, and wingless. Emergence at low tide facilitates males to reach females to copulate and allows females to deposit their eggs among shore algae, without inter-ference from water. Chironomids seem to respond directly to 'tidal' fluctuations, and animals in laboratory maintain their rhythmic emergence in synchrony with their parental populations.

Californian grunion *Leuresthes tenuis*, a small, silvery, atherinid fish leaves sea to deposit its eggs in the sand of several Californian beaches. Spawning lasts from late February to early September only on 3rd or 4th nights, after full or new moon. It commences from 1 to 3 hours after peak of high spring tide, and lasts for about an hour and a half. Female swims to beach with an ingoing wave, usually accompanied by several males, digs her tail into wet sand, releases eggs about 2 inches beneath the sand surface. Accumulation of sand on beach, as tide falls, leaves the eggs buried by 8 to 10 inches of wet sand. Eggs are washed out of sand again at next spring tide cycle, and hatch rapidly under stimulus of wave agitation.

This cycle is highly adapted to the tidal cycle. Since spawning follows the highest of spring tide series, the spawning bed is not again disturbed by tides, until next spring tide cycle, about 12 days later. At this time, development is complete, and larvae are ready to hatch. In the meantime, eggs have a safe refuge buried in moist sand. On particular tide, at which spawning occurs, eggs are deposited as tide is falling, so that they are unlikely to be washed out by later waves on that tide. In fact, the beach is being built up over eggs as tide recedes. Environmental variables to which a grunion time its spawning movements and reject light intensity is not clear, since spawning is semilunar and not related to the setting of the sun. Tidal rhythm may be the causal factor, but posed a question 'How do the animals know which is the maximum tide of spring series, and when particular tide has passed its peak?' Semilunar cycle in ovarian maturation has been reported, but the basis for cycle and precise spawning response remain unknown. *Hubbsiella sardina* spawns in similar fashion on beaches of Gulf of California. It differs from *L. tenuis*, by spawning during day and night. *Hypomesus pretiosus* has a spawning peak at high tide, which appears to be a simple response to water volume on beach and the eggs are deposited in beach sand.

Spawning of *Galaxias maculatus* is less widely publicised than grunions. Larva of *G. maculatus* live in sea, juveniles migrate into fresh water, juvenile and adult life is spent in lowland streams and rivers, usually beyond tidal influence, provides well authenticated examples of lunar rhythms in non-marine species. Gonads mature mostly in summer, and breeding occurs mostly in early autumn. Mature fish migrate downstream to spawning grounds at full moon. Fishes were found to spawn at both full and new moon spring tides. Ripe adults move up on to grassy flats of river estuaries in immense shoals, penetrating to areas covered by water, only at spring tides and eggs deposited among grasses are eventually washed down among bases of grass clumps, where humidity remains high, and the temperature is stable and low. Eggs usually hatch at next spring tides that cover grasses. If low temperatures retard development, hatchings are delayed until later tides, as recorded elsewhere.

Silver eel, *Anguilla anguilla* is influenced by the moon to a high degree. Females live mostly in inland waters, and males in coastal waters. Deelder suggested that lunar light cycle may establish an endogenous rhythm so that migration occurs at appropriate time, regardless of night sky conditions. Eels migrate at the full moon only, when the water is turbid, but lunar influence on silver eel migration is disrupted by cloud. Tidal activity rhythm in *Blennius pholis*, an intertidal rock-pool species is described with peak activity about time of high tide (lunar-daily rhythm). This rhythm persists in laboratory and disappear over a period of several weeks. A very short-lived rhythm in another species, *Acanthocottus bubalis* is also recorded.

A 'definite rhythm' in quantity of herring caught off East Anglia is an example of monthly rhythm which reached its peak at about full moon. It has suggested that this periodicity, and those in other clupeid fisheries may reflect behaviour of either fish or fishermen, and periodicity are found to be less evident in recent years which is attributed to improvement in fishing methods. Lunar rhythms in other commercial fisheries have been reported. Commercial trawl fishermen in North-eastern New Zealand believe that at certain phases of moon, fishes are more abundant or at least are more easily caught. So, they fish intensively at these phases.

7.3.2 Causal Factors in Lunar Rhythms

Numerous polychaetes exhibit lunar rhythms, which are very variable in character. Apart from polychaetes, lunar rhythms occur haphazardly in various phyla. Tidal periodicity of *Clunio marinus* has clearly not resulted from exposure to millions of years of tidal fluctuations, since one population has an endogenous rhythm correlating with a wind-induced tidal rhythm. This seems to be a product of relatively short-term adaptation to the prevailing tidal conditions. Tidal rhythms in activity may be genetically fixed, and are not experimentally induced. It has been suggested that timing of 'biological clocks' does not depend on obvious environmental stimuli, like light and temperature, as there is no obvious response to these stimuli in some species, and as animal cycle sometimes continues. If environmental variables are artificially modified to subtle geophysical forces to explain claimed rhythm though not demonstrated, these forces have been reported to correlate with other more obvious environmental variables. Claims to found responses to a magnetic field in *Dugesia*, and also in *Ilyanassa* (Gastropoda), *Drosophila*, and *Paramecium* are on record. Careful analysis of Brown's work on crab *Uca pugnax* shows that some of the lunar responses, he has reported are mutually exclusive; e.g. in sea, *Uca* was reported to have tidal rhythm out of phase with local lunar rhythm. In laboratory at Woods Hole, it was found to transfer to a strictly lunar rhythm, but when transferred 3000 miles west to California, it maintained a lunar rhythm in synchrony, with Woods Hole lunar rhythm, not responding to Californian lunar rhythm.

The moonlight and the tide have most probable primary importance, and is able to uphold this view in a number of instances. Others do exist, where there is no apparent relationship between rhythms in animal activity and obvious environmental variables. Korringa believed that periodicity is called forth by sequence of neap and spring tides in localities where tidal amplitude is large, but where tides are small, other factors can be the cause of animal's cycle, as in animals sensitive to light, alternation of dark and moonlit nights correlate sometimes with sexual maturation. A species of crab releases its gametes on a semilunar cycle at spring tides in Europe, where there are large and regular semilunar tides. In North Carolina, one of the monthly spring tides is much greater than other and gametes are released only once each month. In Jamaica, tides are irregular, with no lunar periodicity in gamete release. Korringa's view does not explain spontaneous establishment of lunar rhythms in laboratory-reared crabs, or persistence of endogenous rhythms in animals held under stable laboratory conditions.

7.3.3 Selective Advantage of Lunar Rhythms

Adaptations to cyclic variations in conditions of environment are of obvious value to animals, enabling prediction of imminent events in their habitat of either catastrophic or beneficial character, and allowing compensatory behavioural changes to be made. This is particularly true of endogenous rhythms, with their great predictive value in producing behaviour harmonic with environmental rhythms. Existence of endogenous rhythms in many diverse animals show that there are ecological and evolutionary advantages. Some advantages can be observed readily, others can be deduced, but many are unknown. *Convoluta* receives an obvious and simple benefit by emerging from sand at low tide, and retreating again, as tide rises. Exposure allows receipt of sunlight for photosynthesis, and burrowing into sand prevents worms from being washed away by wave action. Foraging by *Blennius* at high tides is more profitable because foraging area accessible to fishes becomes larger. *Leuresthes tenuis* and *Galaxias maculatus* share advantage of being able to deposit their eggs in supratidal environment, where they are protected from egg predators.

Most animals that react to lunar cycle do so in a part of their reproductive cycle. Especially in species that broadcast their reproductive products into sea in haphazard fashion. Periodicity is of great value, in concentrating spawning activity, in both space and time. When spawning occurs at very low tides, especially in shallow water, there is a further advantage in having gametes concentrated in a reduced volume of water, so that chances of fertilization are further increased. Animals respond to these stimuli, with a background of 'species experience' that such stimuli are associated with near optimal conditions for imminent activity; or activity may increase survival. Lunar periodicity in breeding tends to produce concentrated breeding activity in stable breeding conditions year after year.

7.4 TIDALLY RHYTHMIC BEHAVIOUR OF MARINE ANIMALS

The best general hypothesis for control of spontaneous tidal and daily patterns of behaviour in coastal animals postulates an endogenous physiological pacemaker system, which generates approximate periodicity, together with environmental adjustment of clock(s), to the local time. Free-running endogenous rhythms of circatidal, circadian, circasemilunar and circalunar periodicity have been demonstrated in a number of species, in constant laboratory conditions, in some cases clarifying hitherto poorly understood aspects of behavioural repertoire of animals in sea. Entrainment of circatidal rhythmicity has been demonstrated using cycles of simulated tidal variables such as temperature, hydrostatic pressure, salinity and wave action. The crab, *Carcinus* shows increased locomotor activity after changes of salinity (halokinesis). Responses to 34% salinity entrain endogenous clock, but responses to salinities above or below 34% are purely exogenous, and do not persist in constant conditions after entrainment. Phase responsiveness of circatidal rhythms to pulses of tidal variables have been demonstrated in several species. Phase response curves show marked differences from those of circadian rhythms.

Endogenous basis of tidal and diel behaviour in marine molluscs and crustaceans involves matching spontaneous rhythms of neuroelectrical activity. Also, in decapod crustaceans, a peptidic Neurodepressing Hormone (NDH) modulates neuroelectrical and behavioural rhythmicity. NDH is produced rhythmically in eyestalk neurosecretory complex, perhaps partly under control of other clock components elsewhere in CNS. Physiological basis of circasemilunar, lunar (and annual) rhythms of behaviour has not been studied, but studies of synchronization of these rhythms have been undertaken. In some localities, it has been shown experimentally that light intensities equivalent to moonlight are sufficient to entrain such rhythms. In other localities, where moonlight is a less reliable cue, relative timing of tidal and daily variables has been shown to be important. So far, there is no evidence that synchronization is achieved by absolute differences between tidal variables at neap and spring tides. Numerous examples in plants and animals now, confirm that many biological rhythms are endogenous, free-running, at periods approximating to environmental cycles, when deprived of exposure to such cycles in the laboratory.

Circatidal rhythms of behaviour in marine animals, usually run free for only a few days in the laboratory, and in many cases, "activity" peaks relatively quickly become diffused. Differences may be due to predictability of 24-hours light/dark changes in the environment, compared with relative variability of tidal time-cues. Tide times and amplitudes vary considerably over short geographical distances, in response to local weather conditions. Matching this contrast between circadian and circatidal rhythms is the relative states of knowledge, concerning nature of biological clockwork, controlling such rhythms. Hormonal and neural physiological mechanisms, driven by clock-genes are well understood to control circadian rhythms. In contrast, whilst hormonal and neural mechanisms have been shown to be involved in some circatidal rhythmic processes, little is known about the genetic basis of such rhythms. Indeed, there is even debate, as to whether, or not true circatidal rhythmicity exists at the physiological level in coastal animals. Conflicting hypotheses that postulate control by truly circatidal clockwork, or by paired circalunidian clocks, coupled in antiphase remain to be resolved. Nevertheless there is considerable evidence that coastal marine animals do exhibit endogenously controlled rhythms of physiology and behaviour that match rhythms of ocean tides, and even lunar phase. These aspects of chronobiology have considerable implication for a range of applied marine sciences, from fisheries and aquaculture to conservation and management of coastal environments.

7.5 BIORHYTHMS AND COASTAL ZONE MANAGEMENT

For resident coastal animals, which reproduce by releasing planktonic larvae in inshore waters, the environmental management practices should take account of lifestyle changes of animals during development. When juveniles and adults of same species occupy same habitat, same management practices might reasonably be applied, irrespective of stage of development of animals concerned. Species that might be impacted by human activities are talitrid amphipod crustaceans, the so-called sand hoppers. They are most abundant supralittoral and inter-tidal detritivores on temperate beaches worldwide. Adults burrow in sand in strandline zone, during their inactive phase, usually by the day. Here, they are particularly vulnerable to mechanical raking of tourist beaches, when they could be exposed to visually feeding birds. During their normal active phase, at some stage of diel cycle, hoppers emerge naturally from sand, usually during darkness, to forage up shore, or down shore depending upon local tidal conditions. On tidal shores, they forage down shore, but on non-tidal shores, such as in the Mediterranean, they forage up shore at night, presumably avoiding visual predators, as they do so. During the daytime low tides, adults typically burrowed near the high water mark, whilst juveniles most commonly burrowed amongst algal fragments on damp sand on the lower shore. Study found that two life history stages kept actively apart, adults out-competing juveniles by cannibalism, when two stages were kept together in the laboratory.

Selection pressure for separation of two stages is considerable, since patterns of nocturnal locomotor activity in adults, and crepuscular activity by juveniles were found to be expressed as typical circadian rhythm by hoppers in laboratory. Inherited homing directions of juveniles and adults were complementary to each other, during the day and at the night. Thus, any attempt to understand the dynamics of a population of coastal animals, as a basis for the establishment of beach management practices, should take account of life-history changes, and marked differences in behaviour of adults and juveniles at particular times of the day.

Annual breeding cycles must be taken into account, when assessing status of animal populations, but between biological rhythms of tidal, or daily periodicity on one hand, and annual periodicity on the other, and account should also be taken of biological events of lunar (29.5 days) and semilunar periodicity, which are related directly to phases of the moon, or indirectly to the moon in response to neaps/springs cycle of tides. Several species of tropical and subtropical fiddler crabs of genus *Uca* exhibit such rhythmicity. During breeding season, females move in large numbers from estuaries, marshes and mangroves, to release their larvae at the sea's edge during spring high tides, which is around times of full or new moon. Such phenomena can be observed casually and are unlikely to be missed in assessing coastal communities, when formulating coastal management procedures. Critical sampling procedures would be employed to detect changes in spawning times that might take place as times of highest spring tides switch, from time of the full moon to that of the new moon, as they do every fourteen months or so on, the socalled syzygy inequality cycle of tides. Fortnightly variation in some coastal animal populations that occur, between neaps and springs, would remain undetected by inadequate sampling regimes. This has been shown for sand-beach isopod *Eurydice pulchra* in numbers of adults and juveniles in surf plankton that varied differently, over a lunar month. The juveniles emerged from sand to feed in rising tide waters, during each tide, but adult males and females emerged to feed only around times of high spring tides. Without such information, tow net catches taken only at times of neap tides would collect mainly juvenile isopods and grossly underestimate the size of adult population. Management practices that might ignore one phase of an animal's development in this manner, may clearly have an adverse effect upon survivorship of species as a whole. It is clearly necessary not to rely on spot sampling to obtain biological data, and over-generalizations, concerning distribution and behaviour are to be guarded against. Without scale-specific information on animal movements, marine conservation and management strategies are unlikely to achieve their objectives.

A key requirement for stock assessment by fisheries biologists is an understanding of predictability of spawning times. Such information is crucial to provide advice for sustainable management of commercially important fish species. The question using long time-series of spawning dates from fisheries records for southern North Sea plaice (*Pleuronectes platessa*), Norwegian herring (*Clupea harengus),* Canadian Fraser River sockeye salmon (*Onchorhynchus nerka*), and Arctic cod (*Gadus morhua*). Each of these fish was shown to have a characteristic spawning time of a few days only, each year, and for three of the species, the spawning times were constant from year to year. Arctic cod did show a slight delay in spawning dates over time, but only by about eight days in seventy years. Spawning events typically coincide, approximately, with springtime production of phytoplankton, upon which the newly spawned fish larvae feed, but spring phytoplankton bloom shows greater variability than timing of fish spawning. This is concluded that since fish can only link their spawning times to plankton production cycles in an indirect manner, it is to their advantage, if they spawn at a relatively fixed season. It is argued that a fish population has the best chance of profiting from variability in timing of the production cycle. Physiological clock mechanism controlling spawning can be envisaged as having evolved against background annual periodicity of environmental events in the ocean. Seasonal changes in day-length and temperature are less well defined in the sea, than on land, and may therefore, be minor synchronizers of fishes' annual biological clocks. However, it is not unreasonable to postulate that circaannual spawning rhythms of higher latitude species of fish have evolved against seasonal changes in day length and temperature, together with averaged seasonal timing of plankton production over evolutionary time, even though timing of production cycles is more variable than spawning. However, at the present time, annual spring and autumn outbursts of plankton production in temperate seas can perhaps most conveniently be regarded as exogenous events, dependent most heavily upon responses to sea temperature and weather conditions, particularly wind strength and direction that influence stirring of the sea and upwelling of nutrients to surface.

In further support of this interpretation of control of fish annual spawning rhythms, a comparison is made between precise spawning times of plaice, cod, herring and sockeye salmon on one hand, and irregular spawning times of Californian sardine and Pacific tuna on the other. The last two species breed at low latitudes, their seasonality is much less pronounced than in higher latitude breeding grounds of first four species of fish. In chronobiological terms, these observations suggest that at low latitudes there has been little evolutionary advantage in acquisition of endogenously controlled circaannual breeding times, such as have been selected for higher latitude species, subjected to marked environmental seasonality.

Of further concern in fish stock assessment is a need to understand the patterns of behaviour of fish and other commercial species over different timescales. Annual migrations between breeding and feeding grounds have long been recognized to have processes of retention and recruitment of commercial species, in tide-swept estuaries. Early studies demonstrated how populations of oyster *Crassostraea virginica* in James river estuary (Virginia, U.S.A.) are self-replenished, despite intensive commercial harvesting and risk of seawards transport of free-swimming larvae. Larvae were shown to predominate near estuary bed, where risk of seawards flow would be reduced during ebb tides, but to swim actively upwards into water column during inwardly flowing flood tides. This behaviour was shown to retain larvae near adult oyster beds within estuary. Similar observations have also been made on species in inshore coastal seas, where tidal streams are strong, with clear evidence emerging that in several species endogenous clock control tidal rhythms of dispersal and recruitment behaviour. In offshore waters, too, there is evidence that vertical migration rhythms in the case of daily periodicity assist dispersal of larvae by utilization of different ocean currents at different depths. For example, phyllosoma larvae of rock lobster (*Panulirus cygnus*), commercially fished in Western Australian waters, and are commonly found up to 200 km or more offshore, yet many clearly return to re-stock adult population along coastline. Off-shore dispersal appears to result from upward swimming at night into surface waters that are influenced by westwards-blowing winds. When winds reverse and blow eastwards by the day, larvae sink 30–60 m below surface, thus, reported to avoid wind-affected surface water layer. Larvae, eventually, return to Western Australian coastline by exploiting deep inshore flowing ocean currents. In many examples cited above, advances in chronobiology theory have been invoked to explain behaviour of recruiting larvae or adults, as they have also helped to explain longstanding problem, concerning control of vertical migration rhythms of oceanic plankton and fish in general. Indeed, there is now recognition of underlying multi-oscillatory physiological processes in organisms concerned, which are therefore, not solely driven by external rhythmic stimuli, as originally thought.

A classical example of the phenomena of gated rhythm is seen in *Eunice viridis* which is fished for food around the Samoan Islands. During spawning, they release epitokes in water which forms the basis of a local food fishery for a few days.

Furthermore, behavioural differences over tidal and diel time-scales are critical, in terms of catchability of bottom-living species and, consequently on commercial stock management. For example it has for some time been recognized that variable catches of epibenthic crustaceans in commercial trawls are in some cases attributable to rhythmic patterns of behaviour, by which animals intermittently swim upwards off the bottom, or hide in burrows. Notably, variations throughout the day in commercial catches of Norway lobsters (*Nephrops norvegicus*) in European waters were confirmed and shown to be partly related to endogenous circadian rhythms of burrowing behaviour. When caught from continental shelf depths of 10–184 m off west coast of Britain *Nephrops*, maintained in artificial burrows in constant conditions in laboratory, exhibited an endogenous nocturnal rhythm of locomotor activity. Curiously, peaks of activity were expressed at times, when the prawns normally remained mainly within their burrows, as if in nature, they were concerned with burrow excavation and maintenance at those times. They ventured out only occasionally at night, and the occurrence of this endogenous rhythm appears to explain, why so few of the prawns are caught in trawls by the fishermen at night. Nature of spontaneous burrow-orientated activity of *Nephrops* at night seems to permit them to avoid capture by sheltering in their burrows when trawls are hauled over seabed. The largest catches of *Nephrops* are usually taken in the day, particularly, after dawn and before sun set. Daytime catches are not dependent upon an endogenous rhythm of emergence; evidently at these times prawns forage intermittently outside their burrows for food according to their hunger-state. In many localities, lower catches around noon are probably result of increased satiety, at that time, which reduces number and distances of feeding excursions away from vicinity of burrows. However, in shallow waters reduced catches around midday may occur because light suppresses emergence, particularly in view of poor light-adapting responses of retinal-shielding pigments of prawn. More recent studies of *Nephrops norvegicus* have been carried out in Western Mediterranean, where commercial trawling takes place at depths greater than 50 m on continental shelf, and where largest catches are taken at dusk and dawn. Catchability and behaviour of prawns at 100 m depth on continental shelf and at greater depth of 400 m on continental slope was compared. Trawling at 100 m yielded greatest catches around dusk and dawn, though with some prawns captured at night, and at 400 m greatest catches were taken by the day. Moreover, in laboratory experiments, prawns collected at 400 m depth, exhibited endogenous circadian rhythms of locomotor activity with peaks occurring during subjective night. As in Atlantic coast populations, daytime catches of prawns at 400 m depth are not dependent, upon the circadian rhythm of emergence from their burrows, which again did not coincide with times of greatest catches of prawns in trawls.

Moreover, an understanding of daily and tidal movements is also necessary in framing fish management strategies for north sea plaice *Pleuronectes platessa*. When migrating, between spawning and feeding grounds, this flatfish swims up into water column, when tide flows in migratory direction, but remains on seabed, when tidal stream reverses. Such studies, together with others concerned with vertically migrating epibenthic species, emphasize implications of chronobiology in understanding and managing benthic trawl fisheries. (Naylor 2005)

7.6 CHRONOBIOLOGY AND AQUACULTURE

Chronobiology began with interest in specialized phenomena, but has grown to demonstrate that biological processes of daily and/or tidal rhythmicity are fundamental to physiological makeup of eukaryotic organisms. Pittendrigh (1961) proposed that terrestrial organisms may be considered to comprise suites of endogenous circadian oscillators, and that optimum performance of an organism's physiology depends upon correct environmental synchronization of its various internal oscillatory processes. The hypothesis was tested and confirmed in insects (Saunders 1972). Insect species, cultured in non-24 hour environmental cycles, exhibited detrimental effects on growth and development, assumed to derive from internal desynchronization of their biological clocks. Tests of hypothesis, using marine crustaceans cultured in laboratory, were carried out by Dalley (1980), who compared development and survival of prawn *Palaemon elegans* in 24 hour and non- 24 hour cycles of light and dark. Results indicated that growth was retarded and survival to juvenile stage was significantly reduced in 8 h:8 h and random L : D conditions compared with the 12 h:12 h controls. In a similar study, using brown shrimp *Crangon crangon* survival, though not growth and morphological development was adversely in non-circadian cycles. In Palaemon, sex ratio was also shown to vary according to light/dark regimes, in which prawns were cultured. In 12 h:12 h L:D, the ratio was around normal of 1:1 male:female, but prawns cultured in 8 h:8 h L:D and random light/dark cycles were pre-dominantly male. It was previously known that amount of light received by a crustacean during development influenced sex ratio, as it has showed and studied further in marine amphipod *Gammarus duebeni*.

A daily light/dark cycle in which light predominates, induces maleness, whilst femaleness is most common in dark-dominated cycles. However, experiments with *Palaemon*, with equal exposure to light and dark, clearly indicate that non-24 hour cycles also reduce survivorship, particularly of females. As in insects, this reduced survivorship is assumed to be related to mismatch between imposed environmental cycles and prawn's endogenous biological clockwork. Given this interpretation of the relationship between environmental and endogenous rhythmicity, with its potential impact on survival and growth, and widespread occurrence of endogenous biological rhythms in animals and plants, implications are clear for artificial culture of crustaceans, molluscs and fishes for commercial purposes. In aquaculture studies, particularly of fish, research effort has concentrated extensively on assessing the effects on growth of various light and dark intervals, within standard 24-hour photoperiod (Stefansson et al., 1991). This is, generally, thought to be a more important aspect of light controlling fish development than, say, light intensity, or spectral composition. With fresh-water trout, experimental modification of seasonally-varying cycle of the day length has been used for production of year round supplies of eggs for commercial fish farms. Also, similar photoperiod manipulation is the tool most used by farms, to control growth, reproduction, and smoltifiation of Atlantic salmon, though physiological mechanisms underlying such commercial manipulations are still poorly understood. Increasing awareness of likely occurrence of endogenous rhythms in cultured fish has, however, prompted studies, for example of flexibility in rhythms of feeding (Boujard et al., 1995). Thus, clear evidence of circadian rhythms demonstrate feeding in rainbow trout. Fish fed at midnight showed lower growth performance and nutrient retention than fish fed at dawn. Endogenous circadian rhythms of feeding in rainbow trout and European catfish were found to be less responsive to phase-setting, in new light/dark cycles than those of sea bass and carp. Such findings have clear implications for aquaculture industry with regard to development of intermittent feeding regimes, to avoid waste from overfeeding. Understanding limits of entrainment of endogenous rhythms would also be important, in attempts to enhance fish growth in "daily" light/dark cycles of less than 24 hours. Evidence available from invertebrates suggest that exposure to "days" much shorter than 24 hour, may result in disruption of circadian physiology of aquaculture species. However, it would be of interest to investigate survival and growth of fish in light/dark cycles slightly shorter than 24 hour periodicity, within lower limit of circadian variability of species concerned. A priori one could postulate that exposure to shorter "days" might speed up growth rates provided underlying circadian physiology is not perturbed.

A further important application of chronobiology in aquaculture is emergence of evidence of endogenous circaannual rhythmicity in salmonoid fishes (Dustan and Bromage 1991). When rainbow trout's were kept for up to 51 months in a constant schedule of 6 hour light and 18 hour darkness (L:D 6:18) at constant temperature and constant feeding rate fish exhibited free-running circaannual rhythms of gonadal maturation and ovulation, which were self-sustaining for three circaannual cycles. Role of photoperiodic manipulation, against back-ground of circaannual and circadian periodicity needs to be fully evaluated, as a basis for increased aquaculture yields

7.7 INHERENT PERIODICITY

Richter developed a hypothesis, which he called "shock-phase hypothesis" to account for these cycles. According to this hypothesis, each component (typically a cell) in any organ or brain structure functions with an inherent periodicity. This periodicity may be different for each cell type, and individual cell may normally be synchronous in their activity, giving rise to a characteristic functional rhythm in each organ. In course of evolution, this primitive cellular synchrony gives way, especially in humans to functional asynchrony, among constituent cells in an organ. Benefit of such lack of synchrony is presumed to be that performance of organs and structures as a whole becomes more consistent over the time, which is useful for humans, who have presumably been emancipated from need to express strong rhythms. Exposure to an acute stressor or trauma may serve to synchronize the out-of-phase cells, and result in synchronous output with a periodicity characteristic of particular cell type involved. This periodicity might then be maintained, leading to persistent rhythmicity in symptoms, such as swelling in joints, or acute attacks of mania. Shock imposed synchrony might gradually dissipate, or might be disrupted by medical or other interventions that restore a more healthful desynchrony among cells.

7.8 RELATION BETWEEN EXOGENOUS RHYTHM AND BIOLOGICAL CLOCK

The relation between exogenous rhythm and the biological clock may be considered under following heads.

7.8.1 Functional Mechanisms of Exogenous Rhythm in Biological Clock

Exogenous cycles influence and regulate biological clock primarily through conduction and coordination system by acting on nervous system. Animals receive information about environment through sense organs. Stimulation received by sense organs is transformed into electrochemical messages by neurosecretory cells, which act upon hypothalamus. Hypothalamus in turn secretes neurohormone, which influences secretion of various hormones form pituitary that ultimately regulates behaviours.

7.8.2 Regulatory Aspects of Exogenous Rhythms on Biological Clock

Nocturnal and diurnal rhythmic behaviour due to influences of day night cycle; hibernation, aestivation, migration due to influences of environmental factors; various modes of reproduction in adjustment with seasonal rhythm is in existence in various groups of animals.

7.8.3 Influence of Exogenous Cycle on Physiology and Behaviour

Effect of exogenous day night cycle

The microfilariae (larval forms of *Wuchereria bancrofti*) generally exhibit nocturnal periodicity. They live in deep-seated blood vessels during daytime, but at night reach small superficial vessels of the skin. Thus, they take the chance to be sucked by nocturnal *Culex*, which serve as the vector. An amazing biological adjustment occurs in some localities where the opposite situation occurs. Mosquito that flies in day time transfers the microfilariae.

Effect of seasonal cycle

Generally the breeding period of a population is restricted to a specific season when environmental conditions favour the growth of young ones. Same population when inhabits climatologically different areas exhibit variation in reproductive cycles. Different species when exist in same environment reached differently in having breeding periods during different seasons. Scientists injected blood serum of a hibernating ground squirrel into a nonhibernating squirrel and found that hibernation is induced. The basic physiological mechanism responsible for controlling hibernation is driven by an endogenous biological clock may be inferred from these experiments. Light influences metabolic process, growth and development of organism. Salmon larva metamorphoses rapidly in longer day length while *Mytilus* larva metamorphoses rapidly in longer dark period.

Effect of lunar cycle

The moon moves along a path similar to sun and rises fifty minutes later each day. Lunar cycle of 29.5 days influence various aspects of animals behaviour.

Tidal rhythm

Gravitational pull of the sun and the moon cause bulges in water of the oceans. Such bulges are called **tides**, which consists of a regular rhythmic rise and fall of water twice a day over a range of several metres. Shore- and fiddler-crabs exhibit tidal rhythm in natural conditions.

Comment

Circadian, circaannual, seasonal, lunar and tidal rhythms are developed under the influences of day, night, annual, seasonal, lunar and tidal cycles of environment initially. Such endogenous rhythms gradually deviate from exogenous rhythms and become independent.

Review Questions

Short Answer Questions

1. Define chronobiology.
2. What is zeitgeber?
3. Define circadian rhythm.
4. What is circadian pacemaker?
5. Name four genes associated with circadian rhythm.
6. Define hibernation.
7. Define aestivation.
8. Comment on the concept of "inherent periodicity".
9. Define "biological clock".

10. What are "ultradian rhythms"?
11. Why do animals undergo hibernation?
12. Why do animals undergo aestivation?
13. Define inherent periodicity.
14. What is tidal rhythm?
15. What is master clock?
16. What do you understand by the term "biorhythm"?
17. Give two examples of biorhythm.
18. What do you understand by exogenous factors?
19. What do you understand by endogenous factors?
20. Name the hormones associated with circadian rhythm.
21. Comment on suprachiasmatic nuclei.
22. State the function of hypothalamus.
23. What do you understand by circadian rhythm disorder?
24. What is the function of pineal gland in circadian rhythm?
25. What is cryptochrome?
26. What do you understand by "per protein"?
27. What do you understand by "double time protein"?
28. Name two core genes associated with the circadian system of mice.
29. What do you understand by lunar periodicity?
30. What is the full form of NDH?
31. What do you understand by semilunar periodicity?
32. What is "shock-phase hypothesis"?

Long Answer Questions

1. Elaborate various aspects of circadian rhythm.
2. State the molecular aspect of circadian rhythm.
3. Describe the importance of lunar rhythm in ethology.
4. Describe the importance of understanding of biorhythm in coastal zone management.
5. Describe the importance of chronobiology in aquaculture.
6. Describe the relationship between exogenous rhythm and biological clock.

Chapter 8

Migration and Other Rhythmic Behaviours

Migration is a largescale, seasonal and bidirectional movement of animals. Highly migratory animals are insects, fishes, whales, shorebirds and songbirds. Birds are especially preadapted, for migratory life, owing to their efficient mean of locomotion. Animals usually migrate to track resources that are unequally distributed in space and time. Migration is an adaptation to deal with a high degree of environmental seasonality. When animals continue to travel from their "home range", till they become responsive to resources is called **migration**. Birds is strategic to cope with seasonal variation in changes in temperature and abundance of food. Migration of European Sylvia warbler's is an excellent model for studying migration. Their migrant populations fly to various regions while non-migrant populations are sedentary. The Garden warblers travel from the Europe to Northern Africa. Blackcaps breed in Europe, parts of North Africa and western Asia. Populations from Central and North Europe migrate to Southern Europe and parts of North Africa Cross breeding experiments between Austrian Blackcaps (which fly southeasterly) and German Blackcaps (which fly southwesterly) shows that offspring migrate in intermediary direction. In Blackcaps from southern Germany, hybrid of migratory and nonmigratory populations develops migratory activity; extent of activity is intermediate to that of parents. Genetic mechanism appears to be influenced by multilocus system. Vector navigation is an inherent quality of juveniles to determine the direction and duration of flight. End of journey of inexperienced migrant appears to be genetically regulated.

8.1 BIRD MIGRATION

In birds, migration usually involves movements between a breeding site and another location, where they spend rest of the year. Arduous journeys birds undertake in flying high over the mountain, desert, and sea with rewards, upon reaching destinations are subject of interest. Every fall, about 5 billion land birds of over 200 species leave North America for Central and South America. Similar numbers depart Europe and Asia, for Africa. Birds have a sense of time. They maintain an internal clock that operates independently of environment. Clock is regulated by pineal gland, and suprachiasmatic nucleus in brain. Birds are highly mobile and among the swiftest living creatures. Flight gives them power to move in any direction for as long as they have the energy to keep going.

A streamlined body shape and a lightweight skeleton composed of hollow bones minimize air resistance and reduce energy requirement help birds to become and remain airborne. Well-developed pectoral muscles attached to furculum power the flapping motion of wings. Long feathers of wings act as airfoils, to help generate lift for flight. Birds have a large, four-chambered heart, which proportionately weighs 6 times more than a human heart. This combined with a rapid heartbeat (resting heart rate of a small songbird is about 500 beats per minute; that of a humming-bird is about 1000 beats per minute) satisfies rigorous metabolic demands of flight. Unlike mammalian lungs, lungs of birds remain inflated at all times, with air sacs acting as bellows, to provide lungs with a constant supply of fresh air. Flight provides access to distant food resources, and for avoiding physiological stress associated with cold weather. Variations in patterns of migration are many. Some species move only a few kilometres up and down mountain slopes. Others travel hundreds of kilometres.

8.1.1 Evolution of Migration

Migration, as a part of animal's life, is characterized by movements and has been shaped by natural selection. No single theory is widely accepted for evolution of migration. Availability of food is the main driving force. Food abundance increases reproductive output, whereas a lack of food leads to death. A bird that finds food would have a longer life span and produce offspring more readily, than one that finds less food. By moving, from place to place, a bird can find more food, and thus, migration evolved. Many warmer climates have an over abundance of food. Most species of birds can tolerate colder temperature given that food is available.

Migration has evolved over many thousands of years. It began to evolve, when individuals that moved from one area to another, ultimately produced more youngs than those that remained in one area. It continues to evolve because of the changing environment. If environmental conditions favour migration, number of migratory birds increase. If conditions permit birds to stay in one place, sedentary type predominates. A good example of such adaptive behaviour is dark-eyed junco. Tendency to migrate disappeared, if it is no longer an advantage. In the early 1940s, some house finches from a non-migratory population in California were released on Long Island, New York. Once established in the east coast's far more seasonally variable climate, the birds began to develop a migratory pattern. Eastern house finch has since become partially migratory, and has spread throughout the Northeast. Some individuals are resident the year around, while others regularly migrate back and forth to the Gulf States.

Neotropical migrants are a group of birds whose migration patterns may have evolved over long periods of time. Their winter homes seem rich in food supplies and nesting locations. The tropical ancestors of these birds dispersed from their tropical breeding sites northward toward the United States and Canadian temperate zones. Seasonal abundance of insect food and greater day length allowed them to raise more youngs (four to six on average) than their stay-at-home tropical relatives (two to three on average). As their breeding zones moved north, birds continued to return to their ancestral home, as cold weather and the related decrease in food supplies made life more difficult. Supporting this theory is the fact that most North American vireos, flycatchers, tanagers, warblers, orioles, and swallows have evolved from Neotropical forms. In certain species, evolution of migratory behaviour over such long periods of time has resulted in a genetic disposition that favours migration.

8.1.2 Migration Patterns

Migration routes and patterns differ widely. Stereotypical migratory bird moves roughly north-south. Migratory birds must fit into territory, where there is less competition for resources. Summer months in respective hemispheres provide potential range of habitat and food, which migrants exploit to their breeding advantage. For this reason, most migrants travel roughly north-south. In North America, most migration routes are oriented north-south for 2 reasons. First, climatic conditions vary more consistently N-S than they do E-W. Migrants fly south for warmer winter weather and North for abrupt spring bloom of resources. Second, major topographical features like mountain ranges, coastlines, and deserts of North America are similarly oriented along N-S lines. In the old world, birds initially migrate E-W, in accordance with major topographical barriers, like Alps, Med Sea, African Deserts, then fly N-S, to reach latitudinal destinations.

Most of land on earth is concentrated in Northern Hemisphere, and for this, majority of migratory bird species travel north to breed, and return south for winter. Many seabirds breed in southern ocean. Some birds breed in far south of South America, Australasia and Africa, and migrate to northern wintering grounds. Some have distinct wintering grounds, while others roam far out at sea outside breeding season. Migration routes typically follow coasts, where water is shallow and sea life abundant, or ocean currents, which carry food from deep ocean upwelling. Geographic features and food availability tend to funnel migratory birds along general routes, called **flyways**. Mountain ranges, large bodies of water, river systems, ocean coastlines, and deserts create guides and obstacles that shape flyways and, to some extent, determine destinations. Scientists have identified three patterns of migration:

Complete migration

This occurs, when individuals leave breeding range during nonbreeding season. Many North American birds such as warblers, orioles, hummingbirds, and shorebirds exhibit this migration pattern. Complete migrants may travel incredible distances, sometimes more than 15,000 miles per year.

Partial migration

This is a common migration, characterized by seasonal movements, away from a breeding range by some, but not all members of a species. For example, song sparrows migrate south for winter, but some individuals remain in breeding area. Thus, migration is not complete across the species, instead it is only partial. Partial migrants, like complete migrants take advantage of seasonally abundant food.

Irruptive migration

Migrations, that are not seasonally, or geographically predictable and may occur for one year, but not again for many years. Distances and number of individuals involved are also less predictable, than with complete or partial migrants. Great Gray Owl, an irruptive migrant migrates south only occasionally. Numbers of these owls that migrate vary. Irruptive migrants are also known as food specialists. Many of these migrants may only eat certain seeds from certain trees. When seeds are available, irruptive migrants stay. When seeds are difficult to find, they move.

Migration of birds can be classified into

Permanent residents: Or just "residents," which are non-migrating birds, who remain in their home area all year round.

Summer residents: These are migratory birds, which arrive north during the summer, and return south to wintering grounds in fall.

Winter residents: These are migratory birds that have "come south" for winter.

Transients: These are migratory species, who can be seen only during migratory period.

8.1.3 Reasons for Migration

Reasons are complex and not fully understood. Migration is affected by food supply, wind and ocean currents. These make some routes and locations easier to reach. Birds breeding in temperate zone do not migrate and take advantage of spring bloom. They reproduce like crazy, from start of bloom, in early spring, to its end in mid-late summer. Their reproductive rates are very high. But their winter survival rates are very low. Thus, their populations are held below, what can be supported in summer. Birds, breeding in tropics do not migrate and have opposite pattern. They live year-round in relatively benign conditions. It is easy for them to find enough food to survive, but when time comes to reproduce, they must compete for food, with other species, and thus, they face a hard time for reproduction. Their survival rates are very high, clutch sizes are low, nest predation rates are high, and thus, their reproductive rates are very low. Migrants travel between these two regions, tropics and temperate zone, trying to get the best of both worlds. They try to be in temperate zone for big spring bloom and associated high rates of reproduction, and in tropics for relatively benign tropical winter months and associated high rates of survival.

8.1.4 Ages and Sexes as Factors in Migration Patterns

Usually, males winter farther north, than females. But in raptors, reverse is true. Usually, immature birds winter farther north in some species (shorebirds), but farther south in other (warblers). Nolan and Ketterson hypothesize three factors in operation. Greater migrational mortality favours young to fly shorter distances (wintering farther north). Males reproductive success is dependent on time of return to breeding grounds, so they winter farther north and as get back earlier survival is easier farther south. So adult females that are unaffected by the previous two factors winter the farthest south. Males usually depart north for breeding grounds, before females as they need to get back sooner to establish territories. Young in songbirds depart before adults on fall migration (showing that their route is innate/exploratory and not learned). Young in other species go with their parents or alter.

8.1.5 Fueling Migration

Besides laying down fat reserves, migrating birds also need to eat a lot to fuel their regular feather moults. Their feathers must be in tip-top condition for their long trips. Different species moult at different times; for most shorebirds it is just after breeding and before migration to wintering grounds. Fat stores more energy, than either carbohydrate or protein. Birds can double their weight in just a few days. Birds need to put it on before migration, and replenish it en route.

8.1.6 Timing of Migration

Timing of migration is a mix of internal stimulus, which results in a feeding binge to put on fat to survive the journey and then tendency to aggregate into flocks. Once pre-migration flock is gathered, feeding continues, while birds wait for suitable weather conditions. Thus, while birds' internal clock probably releases hormonal triggers at a fairly accurate date each year, availability of food and prevailing weather conditions decide, when migration starts. A 12-year study of common terns at Cape Cod showed that an average 75% of birds, and as much as 83%, returned to same area to nest in successive years. Factors that control the onset of migration can be divided roughly into external (exogenous) or internal (endogenous) factors. Seasonal timing of migration can be controlled by both at different times. Migration onset and seasonal timing are controlled genetically in some species. This would be an example of an endogenous factor. In other species, onset of migration is controlled by weather, food, or social factors. These would be examples of exogenous factors. Changes in climate (particularly ice age), and shifts in the positions of islands and continents as a result of tectonic drift serve as endogenous factors.

Photoperiod and circannual cycles are clearly important because even in captivity under constant temperature and food, migratory birds exhibit migratory restlessness. They get fidgety during night and bump into their cages to south or north depending on the season. External factors also play a role, especially, at finetuning migration. Many migrants wait at staging areas, for favourable weather conditions to initiate their flights—strong tail winds, clear nights, etc.

Soaring birds fly during the day to take advantage of rising columns of hot airthermal soaring. Aerial insectivores (swifts and swallows) also migrate by the day, so they can feed along the way. Most other passerines fly at night—when temperatures are cooler, winds are slower, and air is more humid.

8.1.7 Speeds

Most birds migrate at speeds very close to "speed of maximum range". Speed of maximum range is always slightly higher than speed of minimum power. These speeds vary from 20–40 mph; higher for species with high wing loadings (waterfowl and some shorebirds), lower for species with lower wing loading (passerines). Birds often fly faster, when on a migratory flight than they do during ordinary flight. Thus, distances of 200 to 400 miles a day are covered by the long distance migrants. Some birds, however, migrate more slowly. Robins coming up the Gulf coast average 13 miles a day. Most flights occur at between 600 and 5000 ft above the sea level, with an average height of 1525 ft above sea level. However, mountains may mean that greater heights are needed and heights over 10,000 ft above sea level are not uncommon.

8.1.8 Altitude and Distance Covered in Migration

Some bird species stay in same general location all year round, but move to a higher elevation in warmer months. Like north-south migration, this movement exploits territory and food supplies that are off limits in winter. Often, males and females winter at different latitudes. Bar-headed geese fly across the Himalayas at 29,000 feet. Species seen above 20,000 feet include whooper swan, bar-tailed godwit, and mallard duck. Most songbirds migrate at 500 to 2000 meters, but some fly as high as 6800 metres. Swans fly at 8000 metres and bar-headed geese at 9000 metres altitude. Raptors and passerines fly relatively low (800–3000 m). Shorebirds and waterfowl tend to fly much higher. Generally, birds climb as they get lighter.

Arctic Tern makes an annual round-trip of about 30,000 kilometres from Arctic breeding grounds to Antarctic seas. Tiny ruby-throated hummingbird makes 1000 km, 24-hour spring flight across the Gulf of Mexico from Yucatán Peninsula to southern coast of United States. Arctic tern flies a phenomenal round trip that can be as long as 20,000 miles per year, from Arctic to Antarctic and back is record for the longest migration on the planet. Other sea birds also make astounding journeys. Long-tailed jaeger flies 5000 to 9000 miles in each direction. Arctic terns can migrate as far as 20,000 miles per year. Sand hill and whooping cranes are both capable of migrating as far as 2500 miles per year, and barn swallows more than 6000 miles. Most songbirds migrate at 500 to 2000 metres, but some fly as high as 6800 metres. Various authors have tried to derive formulas for "gas mileage" based on depletion of weight, timing of arrival, and distance between points in migration. They vary considerably, but generally conclude that small migrants can cover about 2500 km in 100 hours, if 40% of their weight is in fat that will get them across any of world's major geographical barriers to migration.

8.1.9 Navigation

Studies have shown that migratory birds find their way over long distances. This implies that these birds must have a sort of mental compass and map. Migratory pathways used by birds are influenced by topography, geography and wind. These routes are not always same from year to year. Bodies of water, deserts, and mountain ranges are physiographic features of earth that can act as barriers. Wind is known as one of strongest factors in determining a migration pattern. Many birds rely on direction of wind to get them there. Many migratory birds take advantage of winds to make better progress. Some birds, particularly those flying across north sea, navigate at least partially using wind direction.

Birds find their way through a combination of features like rivers, coastlines, and mountain ranges. Monitoring earth's magnetic field, apparently with their visual system and with tiny grains of a mineral called magnetite in their heads, observing the stars, using the sun are noteworthy. Landmarks are useful as a primary navigation reference only if the bird has been there before. For cranes, swans, and geese that migrate in family groups, young of the year could learn the geographic map for their migratory journey from their parents. But most birds do not migrate in family flocks, and on their initial flight of south to the wintering range or back to north in the spring must use other cues. Yet birds are aware of landscape over which they are crossing and appear to use landmarks for orientation purposes. Radar images of migrating birds subject to a strong crosswind were seen to drift off course, except for flocks migrating parallel to a major river. These birds used the river as a reference to shift their orientation and correct for drift in order to maintain the proper ground track. Migrating hawks seeking updrafts along the north shore of Lake Superior or the ridges of the Appalachians pay attention to terrain below them in order to take advantage of the energetic savings afforded by these topographic structures.

Since the birds would maintain a constant direction even though the sun traversed from east to west during the day, the compensation for this movement demonstrated that the birds were keeping time. They knew what orientation to the sun was appropriate at 9 a.m. They knew what different angle was appropriate at noon, and again at 4 p.m. Melatonin secretion from light-sensitive pineal gland on the top of bird's brain is involved in this response. Starlings, homing pigeons, penguins, waterfowl, and many species of perching birds have been shown to use solar orientation. Even nocturnal migrants take directional information from the sun. European Robins and Savannah Sparrows that were prevented from seeing the setting sun did not orient under the stars as well as birds that were allowed to see the sun set. Birds can detect polarized light from sunlight's penetration through atmosphere, and it has been hypothesized that pattern of polarized light in evening sky is the primary cue that provides a reference for their orientation.

Artificial night sky provided by planetariums demonstrated that nocturnal migrants respond to star patterns. Franz Sauer demonstrated that if the planetarium sky is shifted, birds make a corresponding shift in their orientation azimuth. Orientation is not dependent upon a single star, but to the general sky pattern. The proper direction for orientation at a given time is probably innate.

Radar studies have shown that birds do migrate above cloud decks where landmarks are not visible, under overcast skies where celestial cues are not visible, and even within cloud layers where neither set of cues is available. Although iron-containing magnetite crystals are associated with nervous system in homing pigeons, in Northern Bobwhite, and several species of perching birds, it is unknown whether they are associated with sensory receptor for geomagnetic cue. An alternate hypothesis for the sensory receptor suggests that response of visual pigments in the eye to electromagnetic energy is the basis for geomagnetic orientation. It has been shown; however, that previous exposure to celestial orientation cues enhances the ability of a bird to respond more appropriately when only the geomagnetic cues are available.

Radar observations indicate that birds decrease their air speed when their ground speed is augmented by a strong tail wind. We also know that birds can sense wind direction as gusts ruffling the feathers stimulate sensory receptors located in the skin around the base of the feathers. Since there are characteristic patterns of wind circulation around high and low pressure centres at the altitude, most birds migrate, using these prevailing wind directions as an orientation cue. However, there presently is no experimental support for this hypothesis.

The sense of smell in birds was considered for a long time to be poorly developed, but more recent evidence suggests that some species can discriminate odours quite well. If the olfactory nerves of homing pigeons are cut, the birds do not return to their home, as well as birds whose olfactory nerves were left intact use to come. A similar experiment has demonstrated that European starlings with severed olfactory nerves returned less often than unaffected control birds even at distances as great as 240 km from their home roosts. And even more interesting, when these starlings returned to the nesting area the following spring, the starlings with non-functioning olfactory nerves returned at a significantly lower frequency than the other starlings. Considering the array of demonstrated and suggested cues that birds might use in their orientation, it is clear that they rely upon a suite of cues rather than a single cue. For a migrating bird, this redundancy is critical, since not all sources of orientation information are equally available at a given time, nor are all sources of information equally useful in a given situation.

Factors in the environment function to provide direct, proximal stimulation for the physiological preparation for migration and allow birds to navigate during migratory passage. Navigation requires knowing three things: current location, destination, and direction to travel to get from the current location to destination. Birds have successfully navigated for eons using environmental information. Birds must learn both the location of wintering area as well as the location of breeding area in order to navigate properly. Perdeck concluded that the proper direction of migratory flight was innate, that is, inherited in their DNA, since naive juveniles could fly that direction, and that birds were also genetically programmed to fly a set distance. But this navigation system is modified by experience. It is generally accepted that migratory behaviour evolved independently again and again in different bird populations. A single explanation to fit all cases perhaps should not be expected. Migratory birds navigating over long distances can determine their latitude on the basis of geomagnetic and celestial information, but longitudinal position is much more difficult to determine.

Scientists found that both adult and juvenile, white-crowned sparrows abruptly shifted their orientation from migratory direction to a direction leading back to breeding area or normal migratory route, suggesting that birds began compensating for west-to-east displacement by using geomagnetic cues alone or in combination with solar cues. Experiments suggest that, in contrast to what would be predicted by a simple genetic-migration programme; both adult and juvenile white-crowned sparrows possess a navigation system based on a combination of celestial and geomagnetic information to correct for longitudinal displacements. Results of study suggest that birds may in fact use declination—angle formed between magnetic North Pole and geographic north—to obtain longitudinal information. Geographic north can be determined by star positions late in summer, as night returns to high Arctic.

8.1.10 Hazards of Migration

Death during migration takes a heavy toll. It is estimated that half of all migrants heading south do not return to breed in spring. Predation and bad weather are two natural causes of mortality during migration. Collisions with tall buildings, windows, and other structures, being shot or trapped by hunters, and getting struck by automobiles are a few of the numerous human-made dangers. Continued loss and degradation of stopover habitat, however, is potentially the greatest threat of all. As birds fly south, they may find fewer places to rest. Neotropical migrants have a hard time, when they migrate and find fragments of forest or no forest at all.

8.1.11 Importance of Stopover Sites

Loss and degradation of stopover habitat not only can result in more birds dying while on migration, but it can also have serious repercussions in terms of nesting success. Birds heading north are already constrained by relatively short amount of time available to get breeding grounds, establish a territory, pair with a mate, and get on with further demands of raising the youngs. Late arrival, or arrival in poor conditions on breeding grounds because of inadequate food and rest en route, is likely to jeopardize a bird's ability to reproduce.

8.2 MIGRATION OF FISH

Many types of fish migrate on a regular basis, on time scales ranging from daily to annual, and over distances ranging from a few metres to thousands of kilometres. Fish usually migrate because of diet or reproductive needs; although in some cases reason for migration remains unknown.

8.2.1 Types of Migration

Potamodromous (Greek: Potatoes, river, and dooms, running) fish exemplified by carps, trouts migrate within fresh water only (Figure 8.1). Oceanodromous (Oceanus, ocean) fish migrate within salt water

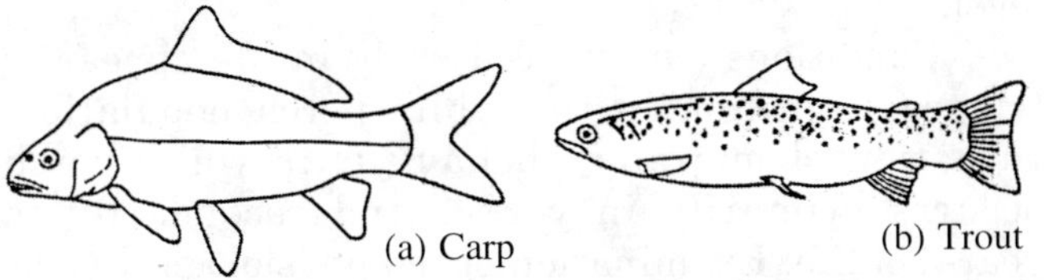

FIGURE 8.1 Potamodromous fishes.

only. Diadromous (Dia, between) fish exemplified by herring, mackerel and tuna travel between salt and fresh water (Figure 8.2). Anatropous (Ana, up) fish live in the ocean mostly, and breed in fresh water. Catadromous (Cata, down) fish live in fresh water, and breed in the ocean. Amphidromous (Amphi,

FIGURE 8.2 A tuna.

both) fish move between fresh and salt water during their life cycle, but not to breed. The best-known anadromous fish are five species of salmon. Salmon hatch in small freshwater streams, migrate to the sea to mature, live there for two to six years and on maturity, return to same streams, where they were hatched to spawn. Salmon travel hundreds of kilometres upriver. Other examples of anadromous fishes are sea trout, three-spined stickleback, and shad (Figure 8.3). Catadromous fishes are freshwater eels of genus *Anguilla*, whose larvae drift on the open ocean, sometimes for months or years, before travelling thousands of kilometres back to their original streams (Figure 8.4). An amphidromous species is bull shark, which lives in Lake Nicaragua of Central America and Zambezi river of Africa. Both these habitats are fresh water,

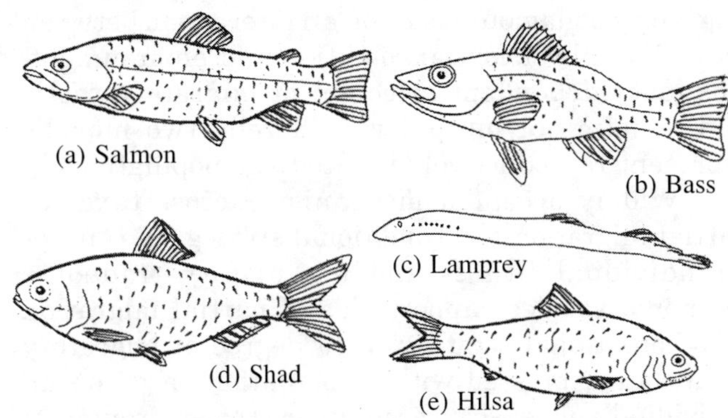

FIGURE 8.3 Anadromous fishes.

FIGURE 8.4 Amphidromous fishes.

yet bull sharks migrate to and from the ocean. Diel vertical migration is a common behaviour. Many marine species move to surface at night to feed, and then return to depths during daytime. Tuna migrate north and south annually, following temperature variations in the ocean. Gobies undertake such migration (Figure 8.4). Freshwater fish migrations are usually shorter, typically from lake to stream or vice versa, for spawning purposes.

8.2.2 Highly Migratory Species

Tuna undertakes migration of significant, but variable distances across oceans for feeding or reproduction, and also have wide geographic distributions. They are pelagic species, and do not live near sea floor, although they may spend part of their life cycle near shore waters. Salmon are capable of going hundreds of kilometers upriver. Atlantic salmon is found in the northern Atlantic ocean and in rivers that flow into the Atlantic. Coho salmon is an anadromous found in the North Pacific (Alaska, Kamchatka, Kuril Islands).

8.2.3 Partial Migration of Fishes

In such migration, one portion of a population is migratory and another portion is sedentary, remaining resident on the breeding ground over its lifetime. Partial migration within salmon populations include a conditional strategy, whereby an individual's genetic makeup allows for adoption of resident or migratory behaviour based on an interaction between individual physiological condition and environment, frequency-dependent selection of migratory tactic and genetic polymorphism, whereby two morphs represent reproductively isolated sub-populations.

Widely accepted mechanism across taxa for partial migration is a conditional strategy, a concept of individual fitness and life history tradeoffs. Salmonid studies support idea of partial migration as a conditional strategy, with degree of migratory behaviour expressed within population based on an individual's physiology, as influenced by environment, relative to a genetically-defined threshold. Within brown trout (*Salmo trutta*) and Atlantic salmon populations (*Salmo salar*) growth rate (or metabolism) early in life is identified as developmental threshold that triggers migratory behaviour. White perch (*Morone americana*) is a dominant and ubiquitous estuarine semi-anadromous species, completes its life cycle in fresh and brackish tidal waters. Conventionally, it was believed that all white perch move into brackish waters during late-juvenile to adult stage, and adults return to freshwater habitats in spring to spawn, with eggs and larvae developing in this environment. Recently, chemical tracers in otoliths of white perch identified divergent habitat use during the first year of life, with a portion of population remaining resident in natal freshwater region and a second portion of population dispersing into brackish water (salinities > 3) environments. White perch exhibit partial migration, with a portion of population remaining in natal habitat and another portion exhibiting denatant migration. Majority of individuals were migratory, moving into brackish waters during juvenile stage, remaining in this environment into adult stage, and returning to fresh water to spawn. Still, a detectable minority of individuals remained in their freshwater natal habitat throughout their lifetime.

Flexibility in life history in white perch is consistent with obligate partial migration and is similar to that found in several species of Salmonidae. Alternative life history tactics is thought to be governed by tradeoffs between costs of migration balanced against benefits of migration. Resident contingent fish exhibit slower growth, and are expected to have lower reproductive rates and fitness compared to migratory portion of population. Males tend to dominate the composition of resident fish, whereas females are more likely to migrate. Tendency for females to migrate is linked to growth advantage conferred to migrants within these populations and its consequences to reproductive success, such that larger females produce more eggs and thus have higher fitness.

Growth rate early in life is known as a controlling factor in expression of migratory behaviour in a number of fish species. In some cases, faster growing fish migrate and in other cases it is slow growing fishes that are migratory. Faster growing fish may disperse because they have the energy reserves necessary to migrate or in response to limited food availability relative to their high energetic needs. Alternatively, in some populations, slow-growing individuals initiate migration in response to low food availability or high population density that limits them from growing at an optimal or threshold level.

Conditions experienced early in life appear to trigger migration in the white perch population. Initiation of migratory behaviour in white perch occurrs primarily in year-1 and secondarily in year-2 of life. Examination of the physiological basis of migratory behaviour within this white perch population indicates that migratory fish grow slower early in life (larval period) compared to resident fish and subsequent to dispersal, migratory juveniles had higher growth rates. Evidence supports the hypothesis that the conditions experienced by white perch early in their life history have consequences to individual growth rates, and is the proximate factor determining migratory or resident behaviour of white perch.

Within fish populations that exhibit partial migration, individuals can shift between resident and migratory behaviour within their lifetime. Such behavioural change is thought to be related to changes in individual fitness and may be associated with changes in relative productivity of habitats. There was no evidence within white perch migrant population of becoming residents later in life. Benefit of a migratory lifestyle probably outweighs advantages associated with remaining resident. It is hypothesized that for these individuals, conditions in freshwater habitat became less advantageous to individual fitness later in life. Alternatively, resident fish may have become entrained into schools of migratory individuals during spawning season when resident and migratory individuals mix.

8.2.4 Migratory Movements

These are of two types; one along with the water currents called **denatant** and other against the water current called **contranatant**. Some species partly resort to both means. Movement can be caused by.

Drift: When fishes are carried passively by water currents. This type of movement is most commonly employed by larvae and rarely by adults.

Random locomotory movements: This may lead to a haphazard or random divergence of the species or it may lead to aggregation of the species specially if there are differences in environmental fields such as light, temperature, etc.

Orientated locomotry movement: When movement is caused by a particular stimulus leading to migration of species either towards or away form source. Speed is an important factor in migrating fish.

8.2.5 Speed of Fish

Speed during migration is influenced by a number of physiological and environmental factors. Maximum speed is ten times of a fish's body length per second. This speed can not be maintained by the fish for more than one minute. They slow down to regain stamina and may again attain the maximum speed. This speed can be sustained over long period of time. It is three times of the body length per second. Herring of 25 cm size can attain maximum sustainable speed of $25 \times 3 = 75$ cm/sec. In a cod of 80 cm this speed is $80 \times 3 = 240$ cm/sec. Migrants like salmon cod and eel breed in one area, but grow up and feed in another area. Distance between the feeding and spawning grounds may be over 700 miles.

8.2.6 Causes of Migration

Some scientists regard migration in fresh water species to arise basically as a result of insufficient food supply in rivers, whereas in marine species, it is due to more favourable conditions, and better protection for developing eggs in rivers than in the open sea.

According to Heape, fish mainly migrate for three primary reasons thereby giving rise to

1. Gametic migration (spawning/ breeding migration)
2. Alimental migration (feeding migration)
3. Climatic migration (wintering migration). Later Myers added another category
4. Osmoregulatory or protective migration (for water and mineral balance)

All the above mentioned types of migrations may occur in separate, well defined and widely separated areas or any two types may occur in the same area or a restricted area. White fish migrate between feeding and spawning ground, but wintering migration is not seen, whereas in gudgeons feeding and spawning occurs in the same area and only wintering migration to be noted. All four types of migrations form a link in a single cycle which may be repeated once or several times depending on biological conditions of each species.

Gametic or spawning migration

Spawning migration is undertaken by a fish to ensure better survival and proper development of their eggs and larvae. Just prior to migration most of the up river or contranatant (against water current) migrants either stop feeding completely or their food intake is drastically reduced. As a result the energy requirement comes from fat deposits of body. It is estimated that a male and female Chum salmon spend 25,810 cals and 28,390 cals. respectively during migration. Amongst Indian fish, *Hilsa ilisha*, which is found in the Bay of Bengal provides an explicit example of anadromous migration. During breeding season it ascends the Ganges river and is seen as far off as Allahabad, traversing almost half the width of the continent.

Alimental or feeding migration

Feeding migrations are brought about mainly due to shortage of suitable food supply in the wintering and/or spawning grounds. Food requirement by the body after exhaustive spawning or wintering conditions probably provides the stimulus for feeding migration, so that the fish have an access to food resources of different areas. Besides enjoying better food facilities, the individuals have a better chance of survival due to the faster growth rate, which ensure the juveniles to escape predation. Commercially important marine fish which show this kind of migration are the salmons, cods, herrings, tuna, important fresh water species like grass carp, Chinese roach, gudgeons, etc.

Besides regular horizontal feeding, some species of marine fish (mackerel, swordfish) and fresh water fish (whitefishes) show vertical migration which is correlated with movement of their food organisms. Mackerels are reported to rise towards the surface and sink along with movement of planktons on which they feed. Sword fish simultaneously move along with sardines which are their food species. Diurnal migration is observed in dace, gudgeous and razorfish in pursuit of rising amphipods and dipterous larvae. Among Indian species showing feeding migration are chanos which is common in coastal waters of the Arabian Sea. *Harpodon* (Bombay duck) is also abundant in the Arabian Sea off the coast of Bombay and migrates for feeding up to the Bay of Bengal after encircling the entire peninsular region. It migrates further up stream and has also been located in estuaries of river Ganges.

Climatic or wintering migration

Wintering migrations are initiated in fish due to inactive physical conditions and low BMR (Basic Metabolic Rate) either after feeding or before spawning. This depends both on the condition of the fish and environmental changes. These fishes therefore, need to move to areas with more favourable biotic conditions which provide better protection. This migration is undertaken by fishes which are prepared implying that their body can tide over the winter conditions. This is mainly achieved by hormonal and other physiological changes accompanying development of gonads. Since feeding stops, the nutritional demands of the body although low are met by accumulated food content deposited mainly as fat reserves. This ensures successful wintering. Wintering migration is commonly seen in migratory fishes, e.g. sturgeons, Atlantic salmon and in some semi migratory fish like roaches and perches. Among fresh water species, grass carp is known to move to the wintering grounds.

Osmoregulatory or protective migration

Spawning, feeding and wintering migrations can all be regarded as protective migrations as they ensure further life of the fish. Yet some mass movements may arise due to sudden unfavourable biological or environmental conditions like a stormy weather. These migrations are not cyclical.

8.2.7 Factors Influencing Migration

A fish may be influenced by a number of physical, chemical and biological factors during its course of migration. Some of the common physical factors are quality of water, depth, oceanic currents, temperature, photoperiod and intensity of light. Chemical factors chiefly include the hydrogen ion concentrations, dissolved oxygen, salinity and types of organic substances. Common biological factors influencing migration are sexual maturity and endocrine system, social response, response to predators and competitors and the biological clocks.

Several laboratory and field studies have been conducted to unravel the mystery by which a fish can select its course and subsequently maintain its direction during long migratory movements. Yet, very little convincing evidence is available. Some theories which have been experimented and debated at length are discussed in the following pages. These relate largely to studies on pacific salmon since it has greater commercial value in international markets. Many hypothesis have been proposed to explain this phenomenon, some of these are given below:

Olfactory hypothesis

Fish may recognize its surroundings and path of migration by detecting specific odour from a stream or water body. This odour may depend on soil, plants and springs at the bottom from which water body acquires its organic quality. This is also called **stream factor**. Interference with olfaction causes inability of the fish to retrace its original path and discriminate between streams, thus, confirming the view that olfaction contributes effectively to this behaviour of the fish.

A.D. Hasler has associated the olfactory hypothesis with 'imprinting' and regards it as sensory basis for homing. Fish (especially salmon) may learn to recognize the organic quality of water in the early days of its life cycle and this may become imprinted for the rest of its life. Imprinting can take place even at a later stage (smolt) up to even the 4th year. Accordingly "a fish smells its way home as would a fox hound". The 'paleo cortex', is therefore, a dominant part of fish brain as it receives impulses from the olfactory tissue. This region is much less significant in non-migratory fishes. Whether the factor responsible for a response to olfaction during migration is genetic or is environmental imprinting of olfactory cortex? Experiments with decoy odours support the conditional response concept. Eggs were hatched in a stream away form ancestral stream and 'marked' smolts then migrated to the sea. When the adult salmons returned to spawn them, however, ascended the adopted stream rather than ancestral stream. Some probable explanation for the role that olfaction plays during movements form river mouth to natal stream by salmon smolts which linger in estuarine waters for several weeks before ascending the river are:

(i) by identifying specific odour molecules of the river, (ii) by identifying a combination of odours of a river mouth, (iii) by detecting the organic nature of inland water, (iv) by identifying currents and sound vibrations.

Olfaction does not appear to have a bearing in open sea migrations, yet it has been suggested that air borne odour of the sea could also be a guiding factor in salmons.

Internal biological clocks

The oceanic phase of salmon migration is more intriguing since landmarks and olfactory, visual, auditory or tactile features do not seem to be of much use in the open sea Laboratory experiments on salmons show that the fish is able to adjust its course or orientation through internal biological clocks. These adjust to seasons (through latitudes) as well as to the time of the day (through longitudes) operate according to appearance of light.

Sun-compass mechanism

Distribution studies by scientists show that open water migrations in salmon may be directed by a sun compass mechanism. These may be directed either by the altitude of the sun or by changing diurnal azimuth of the sun or by a combination of the two. To swim in a single direction, the fish probably 'calculates' appropriate horizontal angles to the sun during different times of the day. This however, does not explain travel by height.

Ground water seepage hypothesis

According to this hypothesis, the assembly areas for spawning or otherwise are identified by marine fish with help of chemicals that enter the sea by ground water seepage. The hypothesis could, therefore, link some anadromous species with their fully marine counterparts, which spawn on or above continental shelf and show homing to areas with ground water seepage. Nursery and spawning areas can also be linked by their geological structure and ground waters. A report on Lofoten cod fishery states that "the idea very generally prevailing among fisherman is that cod fish seeks certain places on the bottom where there are said to be springs of fresh water which they drink in order to bring the roe to maturity".

Pheromones

Pheromones play an important role in identifying spawning area. They may deter further spawning in a particular area after a sufficient number of eggs have already been laid.

Celestial reference points

The use of celestial points in maintaining a steady course in mid waters was suggested to be useful especially in nocturnal travel. Nocturnal direction may also be assisted by gravity, magnetic field, and oceanic currents.

Tidal transport

This mode of migration saves energy and problems of direction are eliminated if the right tide is caught at the right time. Seaward migration of salmon smolts was earlier regarded as a passive movement of drifting with down-stream currents, but now evidence indicates that this type of movement too is oriented in response to environmental factors.

Gradient theory

This is based on the salmon's presumed ability to follow gradient of physio-chemical characteristics of water, mainly temperature and carbon dioxide. It has been shown that salmons swim in the direction of coolest waters and that high CO_2 tension repels migrating salmons. Certain objections have, however, been raised since neither of these factors accounts for stream selectivity of a salmon. It seems unlikely that the fish may experience sufficiently steep gradients to bring about migration, and changes in physio-chemical properties must occur often enough to prevent adaptation of the fish to that stimulus. The sensory capacity of salmon and other migratory fishes is, therefore, well established, but it is difficult to pin point whether a number of senses contributed to their success or a single factor contributes at a specific period during the life cycle. The importance of age and physiological condition of the fish as determinants of the manner and time of response has been highlighted by worker like Hoar.

8.2.8 Migration of Salmon

In fishes, migration of salmon is a well-known and classical example. Salmon are anadromous, i.e. they spend their adult lives at sea, but return to fresh water to spawn. There are nine species of salmon belonging to two genera *Salmo* (Atlantic salmon) and *Oncorhyncus* (Pacific salmon). The Atlantic salmon spawn mainly during November and December. They ascend the rivers as breeding period approaches, travelling several thousand kilometres in sea and then inland. On entering freshwater, they stop feeding and lose weight. As a result energy requirement for up river movement is provided by accumulated nutrient chiefly in the form of fat deposits in body. Their bright silvery colour changes to dull reddish brown shade. Body of male salmon becomes spotted with red, orange and large black spots. Such male breeding salmons are known as "red fish" and darker ripe females as "black fish". After selecting suitable spawning grounds, fish segregate into pairs. Shallow saucer-like depressions are prepared in river bed by females where spawning takes place. After spawning, return journey to sea begins, but few males survive to breed a second time. However, many female are able to reach sea and start feeding. They soon recover their normal conditions and silvery colour. Salmon does not usually sperm more than three times in its life span of eight or nine years. Sockey salmon (*O. Nerka*) hatch in streams of the United States and Canada. After the smolt stage of development, they swim down stream to Pacific Ocean. After two or three years in sea to exact stream where they were spawned, the return journey to coast is probably accomplished by means of celestial cues and internal biological clocks. Once at the coast they have to select the correct river and correct tributary stream within the river system. During smolting period, the salmon imprint upon its olfactory system the characteristic odour of their native streams. So during their return journey salmon are able to discriminate between the water coming forms their native stream and that from other tributaries (Figure 8.5).

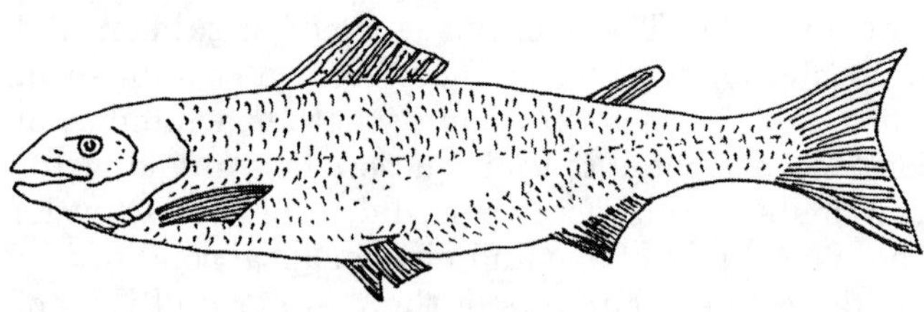

(a) Atlantic salmon (*Salmo salar*)

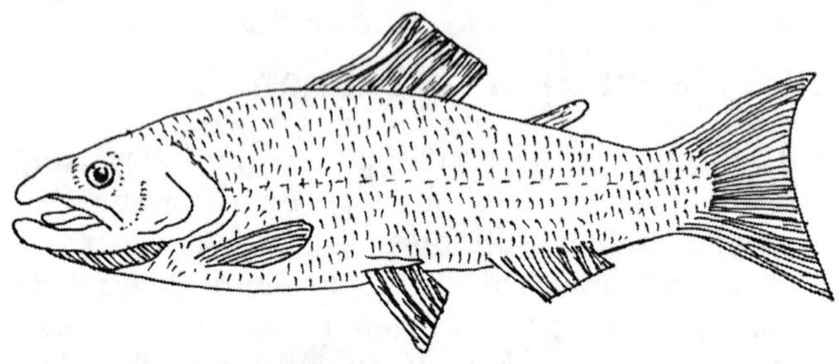

(b) Pacific silver salmon (*Oncorhynchus kisutch*)

FIGURE 8.5 Two common species of salmon.

8.2.9 Migration of Eel

European eel (*Anguilla anguilla*) and American eel (*Anguilla rostrata*) show catadromous migration. European eels are found towards eastern regions of the Atlantic ocean and in inland waters of countries near shores of Europe. Spawning occurs during spring seasons and eggs are laid at the depth of 500–700 metres with the temperature range between 10–12°C. On hatching, the larvae ascend to a depth of about 200 metres where the temperature is warmer (around 20°C). There larvae retain a yolk sac and are called protocephaline larvae.

Protocephaline larva transforms into toleptoceptialus larva after reabsorption of yolk content. Leptocephali are confined to a depth of about 100 metres Larva has a flattened leaf like glasses transparent body measuring about 5 cm. Larva swim by eel-like undulatory movements and is reported to show diurnal vertical migratory movements, coming towards the surface only at night. Leptocephalus larva migrate form Sargasso Sea to the nursery grounds of coastal waters by passively drifting with the warm water currents. They reach middle of the Atlantic ocean by their second summer and finally touch the coast of Europe by third summer. During autumn and winter of third year, leptocephalus larva metamorphoses to form “mini eels” or glass eels, called **elevens**. Elevens may remain in coastal waters for 1–2 years, blocking the mouth of rivers till they are strong enough to swim up the rivers. Male fish prefer brackish water and stay in the estuarine region, whereas females prefer freshwater and are seen to ascend upriver. Pigmentation gradually increases, transforming them into yellow eels. The eel spend 8–10 years on feeding and growing. They become sexually mature silver eels before starting the spawning migration to sea. During maturation of eel, some changes occur within them which trigger a movement covering a distance of 6000 kms. in search of more favourable conditions of salinity, depth and warmer temperature required by gametes (Figure 8.6).

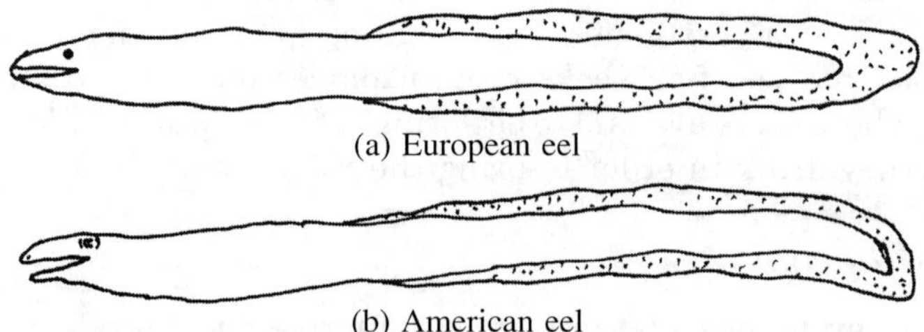

(a) European eel

(b) American eel

FIGURE 8.6 Two species of eel.

8.2.10 Migration of *Hilsa*

The Indian fish, *Hilsa ilisha* which is found in Bay of Bengal provides an example of anadromous migration. Recently this species has been reported to ascend up the Ganges and going up to Varanasi and Allahabad. During the monsoon months, *Hilsa* migrates to many large rivers of southern India. However, the details of its migratory habit are not known.

8.3 HIBERNATION AND TORPOR

Hibernation is the way that animals adapt to climate and land around them. Animals must be able to live through extreme cold, or die. Animals hibernate to escape that cold. They also do this, because it is really hard to find food during winter. Hibernation is a state of profound torpor, when metabolic rate is reduced to 1–5% of euthermic resting rate, at 37°C and homoeothermic is abandoned, so that core body temperature (Tb) falls to ambient. During hibernation, body functions are suppressed to low levels. In hibernating ground squirrels, with a core Tb of 5°C, heart rate is only 5–10 beats per minute, compared with euthermic rates of 350–450. Breathing drops from >40 breaths per minute to <1. By hibernating, ground squirrels save ~90% of the energy that would otherwise be needed to maintain euthermic, over the winter.

Animals that hibernate include bats, some species of ground squirrels and other rodents, mouse lemurs, the West European hedgehog and other insectivores, monotremes and marsupials. Even some rattlesnakes are known to hibernate in caves every winter. Birds typically do not hibernate, instead utilizing torpor. Hibernating ground squirrels may have core body temperatures as low as –2.9°C (27°F), maintaining subzero body temperature for more than 3 weeks at a time. Before entering hibernation, most species eat a large amount of food, and store energy in fat deposits, to survive the winter. Some mammals hibernate, while gestating young, which are born shortly, after the mother stops hibernating. Hibernating animals get their energy by gluconeogenesis. Until recently, no tropical mammal was known to hibernate. However, Fattailed Dwarf Lemur of Madagascar hibernates in tree holes for 7 months of the year. Malagasy winter temperatures sometimes rise to over 30°C (86°F). So hibernation is not, exclusively an adaptation to low ambient temperatures. Hibernation of the lemur is strongly dependent on thermal situations of its tree hole. If the hole is poorly insulated, lemur's body temperature fluctuates widely, passively following ambient temperature. If well insulated, body temperature stays fairly constant and animal undergoes regular spells of arousal.

8.3.1 Physiological Basis

Dausmann found that hypometabolism, in hibernating animals is not necessarily coupled to a low body temperature. Physiological details of deep or seasonal hibernation vary widely, between species. Entry to hibernation may be triggered by temperature, day length and shortage of food, especially in facultative hibernators (hamsters, chipmunks), or by endogenous circaunnual rhythms, as in some obligate hibernators (marmots, ground squirrels). In spite of the very low Tb, physiological control is maintained, due to a low level of metabolic activity.

Hibernation continues to be a mystery to scientists. Regulation of Tb in hibernators has traditionally been viewed as fundamental physiological process. Hibernation can result in deposition of fat in adipose tissue, which is an important source of energy and sites for fuel metabolism, changes in cell structure. Change in BMR observed in hibernators has been viewed as a passive response that is a consequence of hypothermia. During winter months, hibernating vertebrates maintain a very low metabolic rate. Supply of oxygen to tissues, such as heart, and liver under euthermic conditions are invariably linked to and dependent upon local blood flow and pulmonary function.

Studies performed on ground squirrels have allowed estimates of energy expenditure in hibernating and euthermic animals over similar periods. Use of hibernation to gain energetic advan-tage must be weighed, against a number of considerations, particularly animal size and behaviour, biogeographic distribution, and habitat. Hypothalamus, which lies below thalamus and above optic nerve chiasma and pituitary gland in brain, fulfils all of the functions. Syrian hamsters, which display pronounced circadian temperature fluctuations before hibernation, lose these circadian cycles on entry to hibernation, and start to regain them, shortly before arousal cycles. In hibernation, physical damage is not the only danger that cells face, recovering from low temperatures in absence of oxygen (due to a 90% drop in blood flow to brain) and energy supplies also poses problem. With true hibernation, the animal can be moved around or touched and not know it. Hibernation is different from regular sleep. With normal sleep, the animal moves a little, have an active brain, and can wake up very quickly. With true hibernation, the animal appears dead.

During fall, hibernating animals eat more food than usual. Their bodies will live off their body fat as they 'sleep' through winter. Animals get their winter nests, dens and burrows. Different animals hibernate in different kind of safe spots. When they go into hibernation, their bodies slow down, and enemies can get them easier. Animals try to find the safe place to spend winter away from these enemies. Food gives animals the energy they need to walk, run, hunt for food, and lots of other things. When an animal begins to hibernate, its body temperature drops very low, so that it almost matches the temperature outside.

Animal's heart beat and breathing slow down, too. Cold-blooded hibernators begin hibernation, when cold weather causes their body temperatures to drop. Cold-blooded animal's temperatures stay the same, as the air temperature around him. Cold-blooded hibernators will wake up, when air outside warms, or cools enough for them to be comfortable.

8.3.2 Genetic Basis

Differences in patterns of expression of a number of genes that manifest themselves during hibernation can be identified, against a background of no change. Processing of very large numbers of genes in this way became possible in late 1990s, with the advent of DNA chip technology. Such approach has led to identification of a few genes in a range of species, whose expression is either increased or decreased during hibernation. Some of the genes are associated with metabolic, respiratory, or control functions that might be linked to the maintenance of, or recovery from hibernation. Links revealed sometimes appear indirect or tenuous. It reveals changes in expression patterns in a handful of proteins, amongst a very large number, which do not undergo such changes, even those that might be predicted to do so.

Genetic analysis has revealed significant changes in expression of genes, whose function is not known, or not directly related to hibernation. Targeted approach seeks to investigate a possible role, for the molecules already suspected of participating in physiological regulation. An example is the enzyme arylalkylamine-N-acetyl-transferase (AA-NAT). This enzyme catalyses ratelimiting step in production of a hormone, melatonin, from pineal gland. Melatonin helps to reset circadian rhythms in a variety of physiological processes. Entry into hibernation requires neural and endocrine control systems that govern normal day/night (circadian) cycles of metabolism and Tb to be overridden.

Expression of AA-NAT, leading to elevated biosynthesis of melatonin, should increase just prior to onset and during hibernation. Circadian rhythms would then be interrupted, whilst the animal is torpid. Messenger RNA for AA-NAT protein biosynthesis increased in brains of hibernating animals as expected—an accepted indication that enzyme's activity had increased (Yu et al. 2001). Uptake of fatty acids by hibernating tissues has also been advocated, which is normally under tight control, should be deregulated in preparation for task of storing large reserves of lipid. Changes in activity of a key enzyme, acetyl CoA carboxylase are expected. This enzyme catalyses formation of malonyl CoA, a potent and major inhibitor of mitochondrial fatty acid uptake. In Richardson's ground squirrel, hypothesis was proved right. Acetyl CoA carboxylase in heart muscle was much reduced prior to and during hibernation, allowing more fatty acid uptake by mitochondria in cardiac muscle.

Remark

Many biologists believed that mammalian hibernation (endotherms) was a process in which thermoregulation was simply 'switched off', following receipt of a set of 'cues'. These cues included a declining Ta, a shortening day length, the extent of body fat and a lack of food, etc. With this model, the hibernator essentially becomes an ectotherm whose Tb follows the Ta quite closely and who is at great risk if the Ta in hibernaculum falls below freezing. Indeed, prior to the 1960s, many workers assumed that a poor thermoregulatory ability was a prerequisite for hibernation.

However, many animals over winter in hibernacula that are not very deep or well defended against cold. If they were indeed so dependent on Ta, there would be a very high mortality rate. In addition, the ability to hibernate occurs in species very closely related to nonhibernators. It is unlikely that thermoregulatory ability would vary so widely between closely related genera of rodents such as non-hibernating Siberian hamsters, hibernating Turkish hamster and European hamsters.

Hibernators can arouse at intervals throughout hibernating season without an apparent rise in Ta and that arousal could occur when animals were handled or disturbed in cold. These imply a remarkably high degree of control by the animal. It is now accepted that adaptive hypothermia occurs widely in both mammals and birds, but ability is scattered throughout different families. Even within single families, some species show torpor and some do not, suggesting that ability may have evolved independently many times. Species of birds and mammals that hibernate seem to have highly advanced euthermic ability and in most cases can control Tb closely down to 3–4°C, contrary to earlier views which assumed that hibernation was a manifestation of poor thermo-regulatory ability.

It may be mentioned that thermogenesis defines the warm–blooded life style and is the reason that many birds and mammals remain active in winter. There is only one example of poikilotherms that keep themselves warm in the winter by producing metabolic heat, and that is honey-bees (Southwick 1991).

8.4 FREEZE TOLERANCE

The ability to withstand the long-term freezing of body fluids has developed in animals like frogs, turtles, many insects, and a variety of intertidal marine molluscs and barnacles (Storey and Storey 1996). The driving force for freeze tolerance is probably an inability to mount an effective defence against inoculative freezing by environmental ice. Water permeable skin of frogs is no barrier to ice propagation and although frogs chill to 2°C may not super cool if they are sitting on a dry substrate, they begin to freeze in less than 30 seconds if they touch ice crystals. Since frogs need to hibernate in the humid leaf litter to keep from desiccating, they have virtually no chance of avoiding freezing if ice penetrates into their microenvironment. Studies of the physiology and biochemistry of natural freezing survival by frogs are revealing numerous secrets that are being applied in the development of improved cryopreservation technology for the freeze storage of mammalian cells, tissues and organs (Storey and Storey 2001).

8.5 TORPOR

Adaptive hypothermia occurs in at least six distantly related mammalian orders and in several orders of birds. There is a spectrum running from those species which can tolerate a drop in Tb by 2°C for a few hours, to seasonal deep hibernators, which maintain a Tb as low as 4°C for weeks on end. Torpor can be a flexible response that integrates the energy stores of an animal, with environmental conditions and biological requirements. Seasonal (deep) hibernators are likely to be small, and live in an environment, where there may be a large difference between Ta and Tb, and their food is likely to be absent or inaccessible for long periods. Most such animals are herbivorous or insectivorous. Size is a critical factor in the depth of torpor. For example, black and brown bears inhabit the same territory as several deep hibernators, but provided they have shelter, they can manage on stored energy reserves, mainly of fat for extended periods, by lowering their Tb by only 2–6°C. Availability of food is another factor. A number of small birds, for example, American goldfinch can survive in winter temperatures down to –60°C, remaining active and maintaining Tb with a huge (in excess of 500% over summer levels) increase in thermogenesis. Redpoll (*Carduelis flammea*) inhabiting northern United States show a nocturnal hypothermic torpor. If an animal is very small, quite short periods without food may present a problem, and nocturnal hypothermia and torpor can be important for energy conservation. Several species of tropical hummingbird undergo nocturnal torpor, even though difference between Ta and Tb is not huge.

Among the birds, torpor occurs in orders Apodiformes (hummingbirds and swifts), Caprimulgiformes (nightjars, nighthawks, goatsuckers and poorwills) and Coliiformes (mousebirds). In all hummingbirds, torpor, if it occurs, takes place on a daily (or more usually nightly) basis. They are able to re-warm themselves, independently of Ta, and show an increased thermogenesis, if Ta falls below 18°C during the time, when the bird is not searching for food. *Selasphorus rufus*, weighing only 3.0–5.5 g, has a summer range in North America that extends to Alaska, but it over winters in Mexico. While undertaking this huge migration, it undergoes overnight torpor, especially, when breaking its journey for a few days, feeding on nectar to rebuild its energy reserves.

There is no evidence for long periods of unbroken torpor in hummingbirds. *Phalaenoptilus nuttalli,* inhabiting southern USA, is perhaps the only bird studied so far that shows bouts of torpor at all comparable to those of seasonally hibernating mammals. Many birds, including doves and pigeons, enter shallow torpor (with Tb falling to about 32°C), when deprived of food. If food is available, pigeon responds to low Ta, by a large (up to 55%) increase in Basal Metabolic Rate (BMR). Indeed, a reduction in food supply seems to be a major factor in induction of torpor, in almost all the bird species studied. *Aeronautes saxatilis and Colius* sp. can tolerate a Tb of –20°C and spontaneously rewarm at low ambient temperatures. Among mammals, many groups contain species that undergo different degrees of adaptive hypothermia. All temperatezone bats, including the 15 UK species undergo daily torpor during certain seasons with some species remaining torpid for extended periods, as also does the European hedgehog. Ground squirrels and marmots enter prolonged torpor during winter hibernation when the food is scarce.

Review Questions

Short Answer Questions

1. Define migration.
2. Define complete migration.
3. Define partial migration.
4. Name two birds that exhibit complete migration.
5. Name two birds that exhibit partial migration.
6. Define irruptive migration.
7. What is potamodromous migration?
8. What is amphidromous migration?
9. Name two fishes that exhibit potamodromous migration.
10. Name two fishes that exhibit amphidromous migration.
11. What is gametic migration?
12. What is alimental migration?
13. What is climatic migration?
14. What do you understand by the term freeze tolerance?
15. Name an animal that exhibits torpor.

Long Answer Questions

1. Explain the reasons, purposes, and controlling factors of bird migration.
2. Give a detailed account of navigation.
3. Write an essay on fish migration.
4. Describe the migration of salmon.
5. Describe the factors influencing fish migration.
6. Discuss the phenomenon of hibernation.

Chapter 9

Socio-biology and Social Behaviour

Animals depend very much on experience to define their behaviour. Animals take advantage to live in complex social groups, where they all know each other and can learn from others' experiences. In such societies one can find the origins of what is called 'hierarchy', the 'pecking order', or again, the 'social status'. Since wisdom and experience are equated with age, it is very often that the older animals become leaders of a group. Among birds, crows and jackdaws live in social group, where pecking order is strict. Crows do not know instinctively their predators. Experience handed down from generation to generation dictates the birds to react at any give moment. For instance, if a younger, inexperienced bird makes an alarm call, the group may not react to it unless an older and experienced bird makes a move. It is the duty of the older bird to keep the flock disciplined and intact and in order. Older or high-ranking birds at times look after and protect the low-ranking members from those in the intermediate ranks. Konrad Lorenz opined that the low-ranking birds elicit less of an aggressive response in high-ranking birds than birds occupying intermediate ranks.

A complicated situation occurs when an unpaired female jackdaw, whose rank is lower because of her status, pairs with a male of higher rank. By virtue of pairing off with a male of higher rank, this female too acquires a higher status and can 'lord' it over females that had earlier enjoyed higher status than her. This behaviour occurs in primates like baboons and human beings, where a female of lower social standing can by marrying above her, acquire a higher status in keeping with her new role. In animals that live in social groups, there is a behavioural mechanism to prevent animals of higher status attacking and severely injuring animals occupying a lower status. Threat is enough to drive off intruders. In the social group the animals must live in harmony and yet be aware of their status, albeit subconsciously. Whenever a low-ranking or physically weak animal feels that it is about to be attacked by a dominating or stronger animal, it immediately adopts a posture of submission and remains inviolable while it is in that pose. Two dogs, when they meet on one or the other's territory, may well attack each other. The one that is on the other's territory senses its position to be weaker and less tenable and it adopts a submissive posture. For this it stands sideways, cringing before the dominating animal with its head turned away and its neck exposed. The aggressor could very easily grab the submissive dog by the neck and kill him, but some instinctive code of honour prevents this. Then, having ensured his dominance, he walks off proudly, leaving the other dog to slink away as honourably as he can. Such submissive behaviour is also seen in baboons and rhesus monkeys, when they're in their natural surroundings. They live in well-organized groups where the leader commands full submission from other members and in return he cooperates to protect the females and young ones from danger. It is also seen that the instinct to cooperate and protect the weak vanishes and fierce fights occur among members of the group. Dominant males often become tyrants and attack or kill the females and infants as well as submissive males. Thus, the whole society in captivity becomes violent and destructive.

Sociology aims to explain the social phenomena in terms of scientific principles and laws. Social phenomena involves social structure and social change. Social structure refers to the basic fact, of social order. Social change involves the evolutionary development of order. Thus, order and evolution are two sides of the same aim.

Socio-biology explores biological basis of social behaviour, including morality that helps promote survival and reproduction of group. It attempts to explain social behaviour, like mating patterns, territorial fight, pack hunting, and society of insects, considering evolutionary aspects. It deals with population that evolves and persists through time. Organism is simply DNA's way of making more DNA, and all behaviour is, therefore, selfish at its most basic level. We love our children, as love is an effective means of raising effective reproduce. Wilson stated these in his book "On human nature". Socio-biology seems to predict many characteristics of human behaviours, based on inheritance of behaviours through genes, and explains behaviour as a product of natural selection. The word socio-biology coined by John Paul Scott in 1946, became widely used, after it was popularized by Wilson. A rare genetic disorder, Williams's syndrome may lead scientists to genes for social behaviour. "We have known for many years that children with Williams's syndrome are more social than other children, in spite of their mental retardation and physical problems". Teresa Doyle of the Salk institute stated "Here we not only have shown hyper-social behaviour as a whole. Most symptom that follows a characteristics developmental course in Williams syndrome but we may be closer to identify the genes involved in regulates that behaviour". In human, moral systems exist, as they ultimately promote survival, and reproduction.

In 1970s, socio-biology attempted to explain human behaviour as adaptive. Dawkins introduced idea of meme, a unit of cultural transmission, analogous to gene, which describes cultural evolution independently of genes. This idea has recently been developed by Blackmore in her book, *The Meme Machine* (1999). Minds are adapted to environment of evolutionary adaptedness. Miller says that humans are unique in use of advanced technology, controlled energy, clothes, and sense-enhancements, like glasses, telescopes or microscopes, and advanced social organization, and advanced language. Many typical animal behaviours occur in humans. Mating and aggressive behaviour are inherited from animal ancestors by natural selection. No computer has general adaptable intelligence of humans, nor can they even reproduce themselves. This is caused by genetic mutation. Genetic and environmental influences underlie human behaviour, and this is consistent with Darwinian natural selection. This places human behaviour within a broad evolutionary framework.

Whitehead compared life to non-life. Whitehead says that life has rhythm. "Wherever there is some rhythm, there is some life. Rhythm is then the life, in the sense, in which it can be said to be included within nature." "The essence of rhythm," writes Whitehead is the fusion of sameness and novelty; so that the whole never loses the essential unity of pattern. A mere recurrence kills rhythm as surely as does a mere confusion of differences. Rhythm—the central dynamism and form of life cannot be analyzed as an object made of parts because it cannot be isolated from the milieu in which it is born and maintained.

9.1 LIFE

Langer extended Whitehead's idea of rhythm as defining feature of life to argue that rhythm can be analyzed into parts, she called "acts". According to Langer, an act is an inviolable, fuseable, and retrievable unit of activity. It consists of an initial tension and subsequent modes of meting out that tension. An act ends with resolution of tension. According to Langer, rhythm is a dialectical patterning of acts, whereby consummation of one act prepares the way for a succeeding act. Thus, in rhythm of respiration, an act of expanding, lungs induces its opponent act of contracting lungs that reinduces act of expanding lungs.

Life, according to Langer, is a patterning and play of acts. But what does it mean to say social life is a pattern, or play of acts. The words 'pattern' and 'play' name unspecified syntheses. The question is: How do acts comprise a pattern or a play. Langer calls "acts" are aspects of a single continuous rhythm. As Dewey (1936) showed, a rhythm only looks like it is made of separate acts because we cannot discern the common origin and continuities between them. In his example of simple reflex arc of physiology, perhaps the simplest vital rhythm of all he showed that what looks like separable component act of stimulus and response turns out, on closer analysis, to be a single unbroken and unbreakable rhythm.

Acts cannot be discriminated because rhythms are not actually assembled from them. Rhythm, again, is a pure form. It is not a whole made of parts.

9.1.1 Survival of Life

Life is the state of being alive. NASA stated that life is attributing of any system that is capable of replication, energy conversion, in order to offset entropy and subject to process of evolution. All life on earth replicate under auspices of evolution. Driving force for evolution of life is survival instinct, which is rooted in pain and pleasure principle. This principle states that every living organism always acts in what it considers being in its best self-interest, to avoid pain and to enhance pleasure. Organisms those perceive a threat, to their existence as a perception of pain can survive and replicate. To be selfish is ultimate imperative of all life.

Essence of evolution is survival of the fittest, through adjusting with changes in environment, in order to ensure survival and replication. Without this genetically imposed instinct for survival, an organism cannot perpetuate itself. In perpetual struggle for survival, all living organisms are constantly engaged in a battle for limited resources. Genetic survival mechanism makes it possible for any living organisms to exist. Without survival instinct, which forces organisms to perpetuate their genes, organisms would never have evolved, and not existed on the earth. There can be no life, without evolution, and no evolution of species without death of its individuals. DNA contains genetic instructions for development and function of organisms. Evolution explains the process, through which life forms develop, via. genetic mutation and selection.

Based upon insight from gene-centred view of evolution, Williams, Dawkins, Haig and others, concluded that the primary function of life is replication of DNA and survival of one's genes. Defining life in unequivocal terms, is still a challenge. One may say that life "feeds on entropy" which refers to process by which living entities decrease their internal entropy at the expense of some form of energy, taken in from the environment. It is agreed that life forms are self-organizing systems and environment maintains this organized state. Metabolism serves to provide energy, and reproduction allows life to continue over a span of multiple generations. Organisms are responsive to stimuli.

9.1.2 Fitness

Genetic information tends to change from generation to generation, to allow adaptation through evolution. These characteristics optimize chances of survival for individual organism, and its descendants. There are many ways to be fit, depending on exact situation. Fit systems tend to be intrinsically stable, adapted and adapting to their surroundings, capable of further growth and development, and being (re)produced in large quantities. Fitness is difficult to measure, relatively depending on situation, environment and moment in time. Purpose of living organization is to continuously increase future probabilities of encountering this same type of organization. Argumentation for this is found in variation and selection principles of evolution. "Self-actualization", Mallow's term for maximally developing all potentialities, for reaching highest psychological health and awareness is merely the implementation of fitness increase in mental domain.

Fitness of a system is defined as average number of instances of that system that is expected at next time step or "generation", divided by present number of instances. Fitness larger than one means that number of systems of that type is expected to increase. Fitness smaller than one means that type of system is expected to disappear. High fitness can be achieved in a very stable system, which is unlikely to disappear, and it is likely that many copies of that system will be produced by replication or by independent generation and more likely it is to be encountered on future occasions. The English word "fit" is eminently suited, for expressing underlying dynamic and its meanings ranges between two poles: (1) "fit" as "strong", "robust", "in good condition"; (2) "fit" as "adapted to", "suited for", "fitting". First sense, which may be called "absolute fitness", points to capability to survive internal selection, i.e. intrinsic stability and capacity for (re)production. Second sense, which may be called "relative fitness", refers to capability to survive external selection, i.e. to cope with specific environmental perturbations, or make use of external resources.

What is internal for a whole system, may be external for its subsystems or components. Concentration of oxygen in air is an external selective factor for animals; since in order to survive they need a respiratory system, fit to extract oxygen. Concentration of carbon dioxide is an external selective factor for plants. When we consider global ecosystem together, we see that concentrations of both oxygen and carbon dioxide are internally determined, since oxygen is produced out of carbon dioxide by plants, and carbon dioxide out of oxygen by animals. Survival of global system requires an internal "fit" of two halves of carbon dioxide—oxygen cycle: if more oxygen or carbon dioxide would be consumed than produced, whole system would break down.

Similarly, when we look at a crystal as whole system, we see it as a stable structure that is unlikely to disintegrate, i.e. it is absolutely fit and survives internal selection. However, when we look at molecules as the parts that make up crystal, we see that they must have the right connections or bonds, i.e. fit relations, to form a stable whole. Exact configuration of each molecule is externally selected by other molecules, to which it must fit. In this way, every absolute or intrinsic fitness characterizing a whole can be analyzed, as result of a network of interlocking relational fitnesses connecting the parts. In summary, a system will be selected if: (1) its parts "fit together", i.e. form an intrinsically stable whole, (2) whole "fits" its environment, i.e. it can resist external perturbations, and profit from external resources to (re)produce.

9.1.3 Characters of Social Life

Social life is most complex and effective survival strategy. A society is composed of group of individuals of same species. Social behaviour may be egoistic, cooperative, altruistic, or revengeful and meant for efficient food gathering, protection, communal raising of young, grooming, co-operative play, and aggregation during the time of migration, or hibernation. Such behaviour allows animals to live harmoniously in groups by establishing hierarchies of dominance, and to discourage fighting. Social behaviour found in mammals and birds is generally more complex than that of lower organisms.

Social life cannot be analyzed as a whole made of parts. It can be understood in terms of phases and moments. A "phase" is an objective principle of appearance. Social forms cohere. They are born, grow, metabolize, die, and relate recognizably to activities and things outside them. Some features of social life among animals come from instinct. Large social groups get power. Social life is commonly taken to mean "the shape and structure of something as distinguished from its material".

A social form though not a physical rather a dynamic structure or process, into which social activity is poured. This form is more like a waterfall, than a framework, which is visible with continuous activity, although unlike a waterfall. It is a "mode of existence, action, or manifestation of a particular thing or substance" (Webster's New Collegiate Dictionary 1979). A social form is an integral organism. Its visible appearance is given by exogenous forces and constraints, and also by inner forces that play among themselves. A social form is an appearance that lives with familiar qualities of life, like birth and growth. It has familiar dynamics of life, like division, tension, balance, coordination, rhythm, and harmony.

A social life has no space and time, and cannot be seen these two ways. There is nothing to see, as a whole or as parts. There is only the form or shape of activity. By keeping this difference in mind, we can comprehend social life. If we do not see social form apart from substance, we do not see its life. If we see only material "wholes", such as a couple, group, tribe, nation, or corporation, and material persons, who are its "parts", we miss the movement and activity that define life and define form. Social forms are logically and empirically distinct, from the material parts and wholes of people and groups.

9.1.4 Benefits of Social Behaviour

Such behaviour is the interaction among individuals, usually beneficial to one or more individuals, and is exhibited by a wide variety of animals, including invertebrates, fish, birds, and mammals. Many animals find it easy; to find food, when search as a group, especially, if food resources are clumped in certain places. Dolphin surrounds a school of fish and takes turn darting into centre, to eat fish that is tapped in middle. Many carnivores band together, when try to capture large prey. Lions hunt together; when hunting large prey, but often hunt singly, when hunt smaller prey. Large numbers of individuals can form a group, where their efforts are united for a common purpose.

Social behaviour of individual ants lacks flexibility, yet ant family is successful in biological terms and its success springs from rigid social organization. There are about a million times as many ants in the world as humans, and total mass of the ants is approximately equal to that of humans (an average human weighs about same as a million ants). Mammals, other than modern *Homo sapiens*, operate in much smaller communal organizations than ants. For hunting, packs of hyenas adjust their size and tactics, according to their prey. Cooperative groups of hyenas attack herds of zebras, as counterattacks can be anticipated. But, if prey is a single antelope, one hyena takes the chase. Male baboons and patas monkeys adopt role of sentinels, while their troops forage.

Animals live in social groups partly for protection. One baboon might not be able to fight off a leopard. A troop of baboons, often, is able to do so. Prairie dogs and large flocks of crows have some individuals acting as sentries. Schools of fishes and flocks of shorebirds travel in groups in highly coordinated way. Predators generally pick out a single individual in a group that they focus on and try to capture. A rapidly moving and turning school of fishes, flock of birds, or herd of antelopes make it difficult for a predator to remain focused on a single individual. If one individual is unable to keep up with group, predator focuses on it, and usually catches it. Social group make travel easier. Canada gees fly in a 'V' formation, to reduce wind they encounter. In this situation, lead bird perform most tiring job and several birds take turns, leading the 'V'.

Animals congregate in close proximity to one another in cold weather, to stay warm. Birds are known to huddle so closely, that they form a single large ball of birds. Animals share information, with other members of its social group, through communication informing about potential predators, organize attacks against the prey or about food sources. Communication is based on visual, chemical or auditory cues, its signals are individually species specific, or between species. Primate communication is most advanced, involving vocabulary to signal specific meaning. Many primate species use a complex vocabulary, where individual sounds communicate specific meanings. Vervet monkey, for example, can warn members of its social group, about the presence of specific predators, using a distinct vocalization, for different predators. Communication is more complex in humans, than in other primates. Human ability to learn a language is genetically programmed. Scents are primary means of communication in mice, and are important in mediating aggressive interactions and status differentiation among males. Scents also play an invisible role in priming reproductive physiology and development. Reproductive priming effects show how exposure to scents, age group and size influence sex hormone levels in reproductive cycle and development. Komodos dragons (*Varanus komodoenis*) are big lizards, naturally, found in Indonesia. Komodos have several reproductive tricks, e.g. strong sperm as 'backup' in case of poor reproduction conditions in future. Komodo have used a form of virgin birth, involving no fertilization and embryos produced were genetic clones of mother. Virgin birth is genetic clones of mother. Over the time, there will not enough variation, limiting natural selection and evolution. Virgin birth may be seen as an emergency measure to ensure species survival. In the long-run, normal reproduction with males is a key to ensure variation.

Different animals have different reproductive tactics. The placental mammals invest a lot of energy in developing baby inside mother's body. Neurobehavioural studies and molecular genetics have determined behavioural differences between two vole species linked to neural distribution of vasopressin 1A receptor in male brain. Vasopressin 1A receptor gene transfer including upstream regulatory sequence has enhanced male social affiliation in a non-monogamous species. Male affiliative bonding depends upon release of vasopressin and dopamine in ventral striatum enhancing reward value of odour cues that signal identity.

9.2 MODELS OF BEHAVIOUR

9.2.1 Optimality Models

The optimality models quantitatively predict consequences for a particular animal to behave in a certain way, when pitted against the environment. It assumes that animals should behave to maximize their fitness. These models consider separate, dependent, fitness benefits (B), and fitness costs (C) over a given range of values, in a decision variable (variable of interest). Solution to an optimality model attempts to find point where net benefits (B–C) are maximized. Individual solutions consider consequences for a focal animal regardless of what strategies other individuals are relying on. Consequences of behavioural decisions are described as equations, where success of decisions of one individual does not depend on what decisions other animals make. Decision variable is a variable which is aimed to optimize. Currency refers to the criterion used to examine outcome for different values of decision variable. Constraints are factors that limit relationship between decision variable and currency. Optimality models use calculus and logic to solve for a minimum or a maximum in a specific function. Such approach can be used to explore sensitivity to changes in values of decision variable, and to determine types of information that are relevant for the relationship. Precise predictions can be formalized for empirical testing.

9.2.2 Handicap Principle

When animals neither threatened, nor attacked, instead do something that seemed irrelevant to situation, the behaviour is 'displacement activity' (Tinbergen 1951), assumption being that stimulated animal has to do something. Males attack visiting females supposedly because males can not control their aggression. The behaviours correlated with the strength of social bond are behaviours that strengthen bond (Lorenz 1966). In 1950s, ethologists explained adaptive significance of the social behaviour by the benefit they conferred on other individuals (Lorenz 1966). In 1960s, a debate emerged about the importance of group selection in evolution.

Wynne-Edwards (1962) suggested that animals often reduced their reproduction in interests of the population, and it was asserted that individuals reproduce as much as they can. Maynard-Smith (1964) and Williams (1966) supported the point of view of individual selection. Most biologists eventually became convinced that group selection was rarely effective in the real world. This happened not because group selection models were illogical, but because under ordinary circumstances, such models can be exploited by social parasites. Handicap principle suggests that if an individual is of high quality and its quality is not known, it may benefit from investing a part of its advantage in advertising that quality by taking on a handicap in a way that inferior individuals would not be able to do, because for them investment would be too high. In 1990, Grafen formulated a mathematical model for the handicap principle.

Since 1990, handicap principle has been generally accepted as a mechanism that could explain evolution of reliability of signals. Many still believe that in many cases signals do not require handicaps, either because there is no incentive to cheat, when communicating parties are related to each other, or because signal evolved to be reliable without any investment. But there is an inherent conflict among all social partners: mates, parents and offsprings (Trivers 1974), and members of any social group. For that reason, an individual can never be sure at a particular moment, whether or not there is a conflict of interest between itself and any particular collaborator related or not. To be on the safe side, all signals demand reliability.

The handicap principle applies to evolution of all communication systems. Many signalling chemicals within the body are complex or noxious or have adverse effects on ordinary cells. Adaptive significance of complex or harmful chemicals as signals is to inhibit signalling by cell phenotypes that are not types of cells that should emit signal. Since all the cells have same genetic information, some of them perhaps a few millions out of billions of the cells could start signalling at wrong time, or transmit wrong information. For receiver cell, and for the whole organism it is important to be sure that information it receives is reliable. Hence, cells that develop into types that should signal also develop ability to cope with adverse effects caused by production or use of chemical signals. In other words, noxiousness is a handicap. Between organisms, handicaps in signals evolve to prevent cheaters from benefiting from using the signal. In multicellular body, signals evolve with handicaps to decrease possibility of their use by wrong phenotypes.

Several common misunderstandings about the handicap principle are as follows: First; some still assume that a handicap evolves to decrease fitness. This is not the case. Selective process by which individuals develop their handicap, increases their fitness, rather than decreases it. If 'cost' is measured by a loss in fitness then handicaps do not have a cost for honest signallers, since honest signallers increase their fitness by signalling. Only cheaters would decrease their fitness if they were to take on a handicap that does not match their qualities. Hence, efficacy of handicap lies in discouraging dishonest signalling. Second, investment in a handicap need not be very high, or very detrimental, as is often assumed. The investment is proportional to potential gain to cheaters from a giving signal. If potential gain is small, investment is small as well. Investment need not necessarily be in energy, risk, or material. It may be in information or in social prestige. It is not up to the signaller to decide how much to invest. It is the receiver of the signal who is forcing the signaller to invest in signal.

The handicap principle and complementary idea of testing the social bond provide ultimate explanations for, among other things, importance of displaying hesitation by the displacement activities and aggressiveness of a male towards its mate, as well as, reliability of signals. Handicap principle thus suggests that even when altruistic act benefits others, altruist gains directly from investing in its altruistic behaviours. There is, therefore, no need for any indirect-selection model to explain altruism.

Eusociality

Eusociality is common in sisters in bees, and wasps, which relies on their haplodiploid genetic systems, and in mammals (Sherman et al., 1991) in addition to crustaceans (Macdonald et al., 2006) based on their cooperative rearing of young and non-reproducing worker castes. One can find this evolved social behaviour in aphids and thrips, a beetle, snapping shrimp, and the naked mole rat. In eusocial species, non-reproductive workers care for the young of the reproductive queens. As such, workers are analogous to the somatic cells of an organism, which work for the transmission of their genes by proxy, via the germline cells. The members of a eusocial colony evolved to enhance the survival and reproduction of the larger unit. The colony consisting of one or more queens and workers has been called a super-organism, a new kind of organism built up of organisms of the old kind. It has a pivotal role in sociobiology.

The term eusocial refers to colonies with multi-generational members that cooperate in brood care. They are divided into reproductive and nonreproductive castes. Wheeler's classic paper called eusocial colonies super-organisms evolved by armed selection between colonies. Hamilton's (1964) inclusive fitness theory explains eusociality based on extra-high relatedness. The focus on genetic relatedness r is made as if social insect evolution explains without invoking group selection. West-Eberhard (1981) shows the degree to which between-colony selection was rejected as an explanation of eusociality in insects.

The evolution of eusociality falls within the paradigm of major transitions. Most traits associated with eusociality evolve by increasing the colony's contribution to the larger gene pool. Hamilton's rule calculates the conditions under which an altruistic act increases the proportion of altruistic genes in the total population. The fact that a trait evolves in the total population is not an argument against group selection. The Price equation showed that altruism is disadvantageous within kin groups. The importance of kinship is that it increases genetic variation among groups. Hence the importance of between-group selection compared to within-group selection. There are traits that evolve through within-colony selection. But they are forms of cheating that tend to impair the performance of the colony within individual organisms (Ratnieks et al., 2006). All social insect biologists should agree that regardless of the theoretical framework that they use genetic relatedness is the primary explanation for insect eusociality.

Hamilton's original theory states that the extra-high sociality of insect colonies can be due to extra-high relatedness among workers, at least in haplodiploid species, when groups are composed of single queens who have mated with a single male. Hamilton's rule gives the impression that the degree of altruism should be proportional to r. This perception was a principal reason for the erroneous early acceptance of collateral kin selection as a critical force in the origin of eusociality. Research led to a complicated story in which genealogical relatedness plays at best a supporting rather than a pivotal role. Traits that cause an insect colony to function as an adaptive unit seldom increase in frequency within the colony and evolve only by causing the colony to outcompete other colonies and conspecific solitaires, through the differential production of reproductive outcomes. If colonies start with small numbers of individuals, a single female mated with a single male, then ample genetic variation among groups and modest genetic variation within groups occur. This is only one of many factors that can influence the balance between levels of selection. Consider genetic variation for traits such as nest construction, nest defense, provisioning the colony with food, or raiding other colonies. All these activities provide public goods at private expense. All entail emergent properties based on cooperation among the colony members. Slackers are more than solid citizens within any single colony. But colonies with more solid citizens have the advantage at the group level. The size of group-level benefits and individual-level costs influences the balance between levels of selection. The size of group-level benefits and individual-level costs influence the level between the level of selection. The size of group, as well as the partitioning of genetic variation within and among groups, happen. Ecological constraints are more important than genetic relatedness in eusociality in mole rats (Burland et al., 2002). The same is true of the eusocial invertebrates (E.O. Wilson and Holldobler 2005). The ancestors of most eusocial insects-built nests and remained to feed and protect their brood. Such a "progressive provisioning" was the key preadaptation for the origin of eusociality in the Hymenoptera. It is the multigroup population structure provided by this ecological niche. The size of shared benefits that brought these species up to and over the threshold of eusociality is more than genetic relatedness. The reproductive division of labour must be a form of high-cost altruism that requires a high degree of genetic variation in groups to evolve. This is only true if heritable phenotypic variation exists for worker reproduction and if reproductive workers are not suppressed by the queen or other workers. Reproductive suppression is common in eusocial species. To understand its evolution, we need to study the policing and reproduction traits in conjunction with each other (Ratnieks et al., 2006). Suppressing the reproduction of others can be favoured by within-group selection, but it can take many forms that vary in their consequences for the reproductive output of the colony compared to other colonies. Between-group selection is required to evolve forms of reproductive suppression that function well at the colony level, but the amount of genetic variation among colonies does not need to be exceptional. That need is reduced. Further when the trait favoured by group selection is a form of phenotypic plasticity that enables single genotypes to be reproductive or nonreproductive—which, in fact, is universal in social insects (Holldobler and E.O. Wilson 1990). In eusocial insects, the evolution of distinct worker castes is a "point of no return," beyond which species never revert to a more eusocial, social, or solitary condition (E.O. Wilson and Holldobler 2005). The colony becomes a stable developmental unit, and its persistence depends on its ability to survive and reproduce, relative to other colonies and solitary organisms. The hypothetical mutant reproductive worker would be favoured by within-colony selection. Genes have additive effects on phenotypes. So that phenotypic variation among groups corresponds to genetic variation among groups. More complex genotype-phenotype relationships enable small genetic differences to result in phenotypic differences. These occur at the level of groups no less than individual organisms (D.S. Wilson 2004). A single mutant gene in a colony founded by unrelated individuals has powerful effects on phenotypic traits. It affects caste development, which might provide a strong advantage to the group.

Single eusocial insect colonies often have a population structure of their own, which can be spatial or based on kin recognition. Social insect biologists spend much of their time studying the mechanisms that enable a colony to function as an adaptive unit. The Wisdom of the Hive (Seeley 1995) alludes to Walter Cannon's (1932) writing. The Wisdom of the Body, which described the complex mechanisms of organisms. Social interactions enable an insect colony to make complex decisions which are comparable to neuronal interactions. This enables individual organisms to make decisions (Seeley and Buhrmann, 1999). These interactions did not evolve through within-colony selection. But colonies with the most functional interactions out-competing other colonies.

Most biologists devote their research careers to working out the intricacies of how their favourite organism functions. How it processes energy, transmits information from one place to another, regulates metabolites, gets rid of waste, builds itself, defends itself, and reproduces itself. If one wants to step back and consider what it means to be an organism, it is necessary to broaden one's scope and consider the full range of organisms. But an even broader strategy is to consider entities that have organismal properties but are outside the realm of standard organisms. Man-made artefacts have long been mined for fruitful analogies about organisms and functions.

Eusociality is a term coined to cover ants, bees, wasps, and termites that have properties of overlapping generations. Consider the famous honeybee waggle dance. This dance, performed by returning foragers, tells other workers the direction and distance of rich food sources. The colony benefits by exploiting the hard-won knowledge of those foragers who find food bonanzas. The dance is celebrated as a rare example of symbolic communication between individual organisms, but it can also be viewed as a part of a sign mechanism of the larger super-organism that regulates work according to supply and demand. If the supply of food is great, there will be more waggle dancers, stimulating more foraging to harvest it. But that is not the only change necessary. Foragers, with their knowledge of valuable food sources, do not waste time processing the food but hand it off to another set of bees inside the hive. If a forager has trouble finding a processor bee, she begins a different dance, the tremble dance, which both activates bees to become processors and inhibits waggle dancing. The result is a negative feedback system that allocates workers to foraging and processing tasks according to need. Links in the system include the needs of the brood and the degree to which storage capacity is filled. Such regulatory feedback systems operate in every aspect of social insect colony functioning, as they do in other organisms. Besides the clear similarities between organisms and super-organism colonies, there are some differences that show us that entities with organism-like functionality and integration can operate in unfamiliar ways. For example, the cells of organisms differentiate into specialized types, while social insect colonies have at most a few differentiated castes. Instead, much of the division of labour is carried out by means of a temporal specialization, often with the youngest adults tending the brood, the older ones carrying out other activities in the nest, and the oldest ones foraging outside the nest. As cells are more fixed in function, they are also more fixed in space. Social insects, in contrast, are not connected, and their colonies give us examples of organismal entities that are dispersed in space. A final important difference is the lack of centralized control in social insect colonies. Despite the controlling image conveyed using the term "queen", there is nothing like a colonial brain. No individual perceives the conditions of the entire colony and gives instructions. Instead, actions are usually organized by simple rules.

Everyone has small pieces of information. These details are integrated into the colony. A returning forager doesn't have information on the number of foragers and working processors. She perceives an indirect effect of numbers. Colonies serve like organisms in terms of their extent of integration and cooperation for the benefit of the whole. Members of social insect colonies sometimes appear like a single queen and her progeny. Selection may favour a gene causing self-sacrifice. The relatives of the queen might bear copies of the gene, but the aid must be large enough to expand in this area. Most ants, bees, and wasps forage outside their nests to bring food back to their helpless young. Foraging entails the risk of predation. The death of the adult is a way to compensate for the chance that the aided relative does not carry a copy of the gene. So, honeybee workers can gain by stinging large vertebrate intruders. In this way, they may save their whole colony of relatives who pass on the genes for stinging. The barbs on their stingers increase the effectiveness of stinging by anchoring in the victim. They also disembowel and kill the worker. This selection via effects on relatives is known as kin selection. There is nothing like a colonial brain. No individual perceives the state of the entire colony and sends out instructions. Instead, actions are usually organized by simple rules. Different individuals each have small pieces of information that are integrated by the colony. A returning forager doesn't know how many foragers and processors are at work. Instead, she experiences an indirect effect of those numbers—the time required to offload her nectar or pollen—and acts. Members of social insect colonies sometimes appear like a single queen and her progeny. Selection may favour a gene causing self-sacrifice. The relatives of the queen might bear copies of the gene, but the aid must be large enough to expand in this area (Queller and Strassman, 2003). Most ants, bees, and wasps forage outside their nests which entails the risk of predation.

9.2.3 Eusociality in Insect Societies

The highly complex societies formed by some insects are important for studying the causes and consequences of sociality. These groups are called **eusocial** (truly social) as they share 3 key criteria: cooperative care of young; overlapping generations (parents and offspring cohabitating); and reproductive division of labour, often culminating in caste development (Wilson 1971). Eusociality in ants first described by Wheeler (Wheeler 1928) is widespread in numerous other insects, notably in bees, wasps, and termites, and in other taxa like snapping shrimp and rodent. The coordination that makes an eusocial group successful depends on most group members forgoing personal reproduction. This allows workers to focus their efforts on specialized tasks, creating extraordinary efficiencies of scale. Helping behaviour in insects may be a product of both coercion and voluntary actions. Evidence of coercion comes from honeybees, where individuals cannot choose their reproductive role and develop into workers or queens based on nutrients they receive from others during larval stage. In honeybees worker voluntarily act to maximize their own fitness by rearing relatives although they are unable to mate and produce diploid offspring, but are capable of laying unfertilized eggs that become males. This rarely occurs when queen is alive as workers police other workers, destroying any eggs laid. Such behaviour maximizes fitness of policing worker by ensuring that her energy is devoted to raise daughters of queen (her sisters) and not her more distantly related nieces (Ratnieks and Visscher 1989).

Evolution of social behaviour at its most complex degree is found in eusocial animals. Eusocial species live in colonies. Small number of animals in colony reproduces. Non-reproductive colony members give resources, defense, and collective care of young. Eusocial animals are ant, termite, some wasp, some bee, few aphid and thrip species, naked mole rat and Damaraland mole rat, and many reef-dwelling shrimp. Individuals in colonies are related to each other. Relatedness can be greater than 0.5 due to unique genetics of some insects or inbreeding. Hamilton's rule and kin selection provide partial explanation for evolution of eusociality. Colonies often produce large number of offspring, such that even when relatedness is low the indirect fitness of non-reproductive workers may be greater than if they had capacity to reproduce independently. In eusocial animals, high productivity resulting from communal life and efficient division of labor in workers occurs in environment which is well defended against natural enemies. In nearly all eusocial species, colonies are protected through structural means (termite nests in wood), with venom (wasps, bees), or by both. Social and altruistic behaviours need a broad view of Darwinian fitness and an understanding that animal performs behaviours that are responsive to short-term and long-term consequences for their fitness.

Eusocial adults live in groups, and exhibit co-operative care for juveniles, reproductive division of labour and overlap of generations (Wilson 1971). Primitively morphological difference between reproductives and non-reproductives was absent in

TABLE 9.1 Example of Social Behaviours Studied from a Molecular Perspective

Behaviour	*Organism*	*Gene*	*Molecular function*
Foraging			
Rover versus sitter phenotype	*Drosophila melanogaster*	*Foraging*	Protein kinase G
Roamer versus dweller phenotype	*Caenorhabditis elegans*	*Egl*-4 (egg laying defective-4)	Protein kinase G
Division of labour: onset age of foraging	*Apis mellifera*	*Foraging*	Protein kinase G
Division of labour: onset age of foraging	*A. mellifera*	*Malvolio*	Manganese transporter
Division of labour: foraging related?	*A. mellifera*	*Period*	Transcription cofactor
Division of labour: foraging related?	*A. mellifera*	Ace(acetylcho-lineesterase)	Acetylcholine esterase
Division of labour: foraging related?	*A. mellifera*	*IP (3) K* (inositol1, 3, 5, triphosphate kinase)	Inositol signalling
Division of labour: exocrine gland function	*A. mellifera*	Royaljelly protein	Secreted nutritive protein
Foraging specialization: nectar versus pollen	*A. mellifera*	Protein kinase C	Protein kinase C
Social feeding	*D. melanogaster*	*npf*(neuro-peptideF)	neuropeptideY (NPY)homologue
Social feeding (aggregation)	*C. elegans*	npr-1(neuro-peptide receptor family1)	Receptor for NPY
Mate recognition and courtship			
Vocal learning, vocalization	*Taeniopygia guttata*; *Homo sapiens*	FOXP2 (winged helix/ forkhead protein)	Transcription factor
Vocal learning, song recognition	*T. guttata*	*zenk*(*Zif269*/*Egr1*) *NGFIA*/*Krox24*);	Transcription factor: other functions others
Pheromone mediated communication	*Mus m. domesticus*	*V1R, V2R* (vomeronasal receptor families 1 and 2)	G-protein receptors
Pheromone mediated communication	*D. melanogaster*	*Gr68a*	G-protein receptor
Pheromone mediated communication	*Bombyx mori*	*BmOR1*	G-protein receptor
Male courtship	*D. melanogaster*	*Fruitless*; others	transcription factor; other function
Male courtship; timing of mating	*D. melanogaster*	*Period*	Transcription cofactor
Female receptiveness (lordosis)	Rodent	Estrogen responsive genes	Various functions
Post-mating behaviour			
Refractoriness to mate, ovipositioning, decreased longevity	*D. melanogaster*	Genes for seminal protein	Various functions
Monogamy, parental care	Rodent	*V1aR, OTR*	Vasopressin and oxytocin receptors
Maternal care	*Rattus norvegicus*	*GR*	Glucocorticold receptor
Attachment to mother	*M.m. domesticus*	*Orpm*	Opioid receptor
Maternal care, pup retrieval	*M.m. domesticus*	*Dbh*	Biosynthesis of norepinephrine, epinephrine
Social hierarchies			
Territorial versus non-territorial males	*Haplochromis burtoni*	*GnRH1*	Gonadotropin releasing hormone
Dominant versus subordinate males	*Procambarus clarkii*	*5HTR1-2*	Serotonin receptors
Dominance interactions			
Aggression	*M.m. domesticus*	*Maoa*	Monoamine oxidase
Aggression	*Macaca mulatta*	*5HTT*	Serotonin transporter
Subordinate behaviour	*M.m. domesticus*	*DVl1*	Wnt-receptorsignaling pathway

eusocial organisms. Advanced eusocial organisms show different morphologies for reproductive and non-reproductive individuals and specialization within the non-reproductives. In subsociality, social behaviour between parents and offspring (Plateaux-Quénu 2008) is found. In parasociality, social behaviour among members of the same generation is found. Most eusocial animals are arthropods; a few are chordates. The order Hymenoptera (Figures 9.1–9.3) is the well-known animal group among eusocial species where this social system has arisen many times. Hymenoptera with a haplodiploid sex determination system may contribute to kin selection, favouring altruistic behaviour. Termites are highly evolved social insects which depend on mutualism with cellulose digesting protozoa and bacteria. Young individuals acquire these symbionts via anal trophallaxis. Within a colony, there is a queen and king, in addition to soldiers and workers. Termites (Figure 9.4) perform complex social behaviours including nest construction (Ladley and Bullock 2005) and territorial defense (Adams 1987).

FIGURE 9.1

FIGURE 9.2

FIGURE 9.3

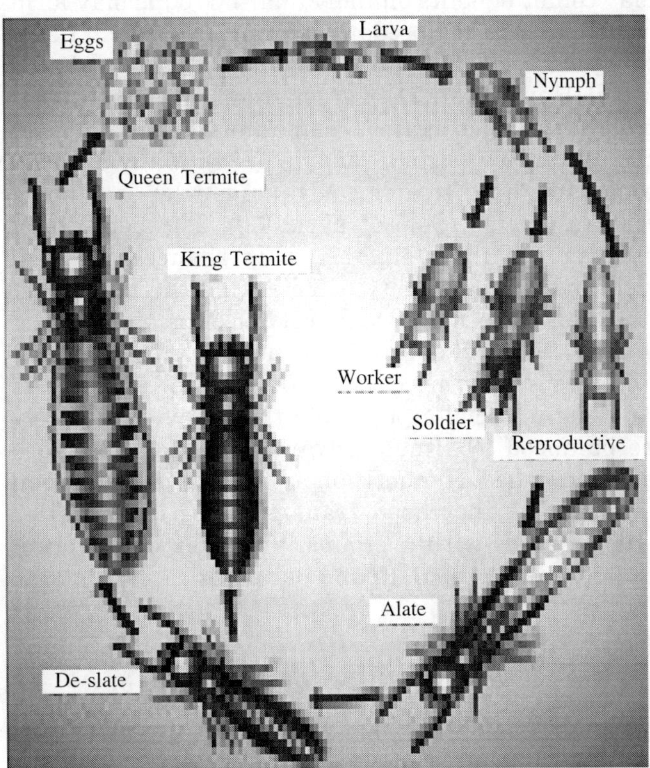

FIGURE 9.4

Recently discovered eusocial organisms are species of shrimp, aphids and thrips. Two separate origins of eusociality within the *Synalpheus shrimps* (Duffy et al., 2000) are predicted. These marine shrimps live in groups as internal parasites on tropical sponges. The host sponges with heterogeneous distribution may have contributed to the evolution of eusociality as dispersal to establish new colonies is riskier than remaining within the natal nest. Thrips are small insects in the order *Thysanoptera* (Crespi et al., 1992). Of about 5000 species, about 300 form nests in plants called galls. Of these, six species can be called eusocial. They have morphologically different soldiers which defend the galls from kleptoparasites. Social aphids also live in galls, feeding from the plant tissue (Aoki and Imai 2005). They are diploid, but can reproduce parthenogenetically and several species described have robust soldier morphs (Stern and Foster 1996). Among vertebrate eusocial species are the naked mole-rat and the Damaraland mole-rat (Burda et al., 2000). Both species are diploid, highly inbred and live in harsh deserts. Most individuals help to raise siblings or close relatives that are born to a single reproductive female (the queen) (O'Riain et al., 1996).

Eusocial organisms are subject to both the benefits and costs of group living. Living in a group may confer benefits on individuals. Groups may form as defence against predation, forming a "selfish herd" in which each individual has a less chance of being killed (Hamilton 1971). Larger groups in many species gain advantages against competitors, like colonies of the ant *Azteca trigona* (Adams 1994). Many eusocial organisms have strategic advantage when acquiring food in groups (Solé et al., 2000). The benefits of staying within the natal group reduce dispersal risk and the possibility of inheriting the natal nest as observed in naked mole rat societies (Jacobs and Jarvis 1996) and in primitive ants (Molet et al., 2005) and wasps (Monnin et al., 2009). There can be costs to living in social groups, and these must ultimately be balanced by fitness advantages. Disadvantages are increased competition for resources between individuals, increased transmittance of parasites and diseases within groups, and easy detection of the group by predators and parasites.

Evolution of eusociality

Eusociality started through a pathway with solitary organisms acquiring benefits to group behaviour, eventually leading to a "point of no return" (Wilson and Hölldobler 2005) wherein certain individuals no longer have the physical ability to reproduce and only gain evolutionary fitness indirectly. The selective forces at work during the inception of eusocial behaviour may be different from those that maintain advanced eusocial colonies (Hölldobler and Wilson 2009). Major contributing factors during the inception of eusocial behaviour are as follows.

Two routes by which a gene can copy itself in future generations are through producing offspring or indirectly, through the reproduction of close relatives. The sum of direct and indirect reproductive gains is called *inclusive fitness*. Thus, it is possible for selection of eusociality over solitary behaviour if indirect fitness levels exceed direct fitness levels. An altruistic act is one which benefits a recipient at a cost to the performer of the act. Hamilton's rule (Hamilton 1964) states that altruism is favoured if $r > C/B$, where B is the benefit to recipient of altruistic act in terms of lifetime reproductive success and C is the cost. The coefficient of relatedness, r, ranges from 0 to 1 and is the proportion of alleles shared between two individuals identical by descent. Eusociality depends on high levels of altruism within groups as individuals increase the fitness of others at a cost to themselves. Two well-known mechanisms by which r can be elevated are haplodiploid sex determination and inbreeding. Haplodiploidy may explain why eusociality has arisen multiple times within the *Hymenoptera*. In haplodiploid organisms, the relatedness between full sibling sisters (r = 0.75) is greater than that between a mother and her offspring (r = 0.5), thus weighting Hamilton's rule in favour of raising sisters rather than offspring. Inbreeding results in offspring which share a larger proportion of alleles, thus increasing r. This is common in organisms which do not disperse very long distances from the natal nest or are prone to mating with siblings (Ciszek 2000).

Delayed benefits: An intermediate step towards eusociality may have "hopeful reproductives"—that, workers which have the option to stay and help or to go out to start their own nest. The decision depends on hierarchical position within the group, territory or food resources and other conditions. Florida scrub jay offspring are known to stay at the natal nest (Breininger et al., 2010), helping to raise siblings until there is an opportunity to replace their parents. Primitively eusocial wasp colonies, such as *Polistes*, are commonly inherited by dominant workers on the death of a queen (Monnin and Ratnieks 1999).

Multi-level selection: Natural selection occurring at the level of the individual, family (related individuals, known as kin) or group is known as multi-level selection. For eusocial organisms, the traits of the colony interact with the environment to determine colony-level fitness and can be described with models of multi-level selection. Whether models of multi-level selection or inclusive fitness models are the best approach for exploring the origin and maintenance of eusociality remains as a point of debate (Nowak et al., 2010), as empirical evidence for inclusive fitness in groups is limited (Seeley 1997).

Ecological and life history contributions

Nesting behaviour has been described as a possible prerequisite for the development of eusociality, as it creates situations conducive to cooperative brood care (Anderson 1984). When nest construction is dangerous or territories or spaces are limited, "fortress defenders" can co-operate to defend this valuable resource (Queller and Strassmann 1998). In the case of termites, thrips, shrimp and aphids, the protected nest is also the location of food in a patchy environment. Parental care can also be an important life history component.

Research on eusocial behaviour

The origin and evolution of eusociality is broad and interdisciplinary as it deserves particular attention of genomics, caste determination/division of labour, and collective decision making. In 2006, the first eusocial insect genome sequence (*Apis mellifera*) was completed, beginning a new era of "sociogenomics" (Robinson et al., 2005). Sequencing is underway for several species of ants (Smith et al. 2009), with the goal to understand the genetic underpinning of behaviour, longevity, sociality and communication. Central to the concept of eusociality is division of labour. The interplay between genetic and environmental contributions to caste is being unravelled in several species (Schwander et al., 2010), including the utterly bizarre situation in harvester ants. Social insects provide excellent inspiration for the burgeoning field of biomimicry, wherein everything from task allocation (Krieger et al., 2000) to nest architecture (Turner and Soar 2008) has been shaped by natural selection and can serve as inspiration to solve human problems. Army ants swarm in search of prey, while other species follow them for their own survival. Army ants have a reputation for annihilating everything in their path as they march through the jungle. But the most complete study of its kind has found that army ants are creators of the whole world, not destroyers. The maintenance of eusocial behaviour might be too costly because of an increased likelihood of disease and parasite transmission.

An explanation for the evolution of eusociality is that helpers are physiologically constrained to find helping as their only realistic option. This sub-fertility hypothesis was tested in a facultatively eusocial hover wasp (*Liostenogaster flavolineata*) (Field and Foster 1999). The convincing results showed that females do not become helpers because of some unconditional physiological constraint. No unequivocal support for the sub-fertility hypothesis in facultatively eusocial Hymenoptera lacking morphological castes does exist. Quantitative tests of Hamilton's rule in facultatively eusocial insects lacking morphological castes have generally assumed that all females have the equal reproductive potential. The direct reproduction forfeited by females that become helpers has been estimated as the reproductive success of other females that choose to nest alone (Stark 1992). Another possibility is that helpers are low-quality females for which the costs of helping are low (Craig 1983). As per the sub-fertility hypothesis, some females are physiologically constrained to obtain higher inclusive fitness by helping than by nesting independently. Source of variation in phenotypic quality is likely to be the quantity and quality of food that females receive as larvae. Mothers might manipulate some offspring into becoming helpers by giving them less food (Alexander 1974). An important food quality in this context is the ability to mature and lay eggs. A female that could not lay eggs would be unable to nest alone, but might still be able to obtain indirect fitness by helping relatives. Hover wasp build small nests that are usually initiated by single females in moist, protected places. Newly emerged adult females may disperse or become helpers on their natal nests. There are no morphological castes. One female is usually behaviourally dominant on multi-female nests and probably lays most or all of the eggs. *L. flavolineata* has relatively large colonies (1–11 females) and a well-defined dominance system. Samuel (1987) found that a significant fraction of helpers were "permanently sterile", in the sense that they never developed mature eggs. These females spent the least time on their nests, and Samuel suggested that the basis of their sterility might be low reproductive potential determined before emergence (Turil-lazzi 1996). Newly emerged *L. flavolineata* females that leave their natal nests are not larger than those that stay and help. Since large size might be indication of high fecundity, this also suggests that low reproductive potential is not the basis of helping. Females that become helpers are not unconditionally constrained to do so. Under some circumstances, they are capable of laying eggs, and of surviving and maintaining their nests alone. The "permanently sterile" helpers observed in *L. flavolineata* (Samuel 1987) are capable of maturing eggs and may simply be females that die before they get the opportunity to oviposit. The sub-fertility hypothesis seems most likely to apply to helpers in species with large colonies that have morphological castes (Queller 1996). Such helpers may be genuinely incapable of maturing eggs and/or of nesting successfully alone.

9.3 SOCIOGENOMICS

Sociogenomics explain social life in molecular terms. It deals with all aspects of genome structure, genome activity and organismal function (Robinson 1997, Robinson 2002). Sociogenomics is predicated on two of the most important insights in biology to emerge. First, as argued by Wilson in socio-biology. Social life is a biological basic and, is therefore, influenced to some extent by genes and forces of evolution. The second realization is that the molecular functions of many genes are highly conserve across species, even for complex traits. Explaining social behaviour at the molecular level can help us to understand how complex and highly derived patterns of social behaviour have evolved from simple ancestral behaviour, the evolutionary relationship of apparently similar behaviours across distantly related taxa, and to integrate mechanistic and evolutionary analyses, which at present are largely fragmented (Robinson et al., 2005). Findings so far reveal two emerging themes. First genes involve in solitary behaviour are also used for social behaviour, indicating that molecular insights form simple behaviour can be used to generate candidate gene for more highly derived patterns of social behaviour. Second the genome is highly sensitive to social influence and the social regulation of the gene expression is a potent influence of the behaviour.

Advances in molecular biology and genomics now make it possible to identify genes and pathways that are involved in social evolution. *Csd* gene encodes a protein that regulates complementary sex determination in the honeybee which gives rise to haplodiploidy. Haplodiploid induced asymmetries in relatedness between offspring and sisters in social insect leads to the evolution of kin selection.

General protein 9(Gp^9) regulates queen number in colonies of the fire ant *Solenopsis invicta. The b* allele possibly in conjunction with alleles at genetically linked loci medicates a green bread gene effect. Workers (non-reproductive) with a *b* allele (B*b*; *b* is a recessive lethal) favour queens that also carry this allele, where BB workers favour BB queen. Because Gp^9 encodes a putative odorant-blinding protein, its alleles might have a differential effect on production and perception of pheromones. Gp^9 sequence analysis indicated that monogyny preceded polygyny in the *Solenopsis genus.* This study has been used in phylogenetic approaches to infer evolutionary processes at both molecular and behavioural level.

Major Histocompatibility Complex (MHC) is a family of 50-genes, known in mammals for their inter individual variability. MHCs are involved in systems of behaviour and cell-cell recognition. Mesoderm-specific transcript (Mest) is regulated by genomic imprinting and is expressed in mice only forms paternal allele. Heterozygote knock out mice that inherit a mutant allele from the paternal germ line are smaller than wild type, which is consistent with an evolutionary theory of imprinting that is based on competing interest of the two parents predicated bisexual selection theory. Vasopressin receptor la (*VlaR*) gene encodes a receptor for vasopressin, a neuro-peptide that is involved in reproductive behaviour in vertebrates. Monogamy is extremely rare in mammals and monogamous vole species have a different distribution of this gene in brain from polygamous species, owing to variation in *Vla* promoter. *VlaR* gene transfer from monogamous prairie vole (*Microtus ochrogaster*) increased the monogamous habits of otherwise non-monogamous meadow vole (*Microtus pennsylvanicus*), although this manipulation undoubtedly affects expression of other genes in brain. These results indicate that changes in regulation of even a single gene can drive rapid evolution of a social behaviour. Vitellogenin (Vg) is used as a yolk protein in invertebrate and some vertebrate linkages. Worker honeybees although mostly sterile produce vitellogenin. Expression of both mRNA and protein is highest in 'nurse' bee, which use it to produce brood food rather than eggs. These findings indicate that genetic regulatory pathway for reproduction in solitary insects has been used in the evolution of insect societies.

Social microbes with their short generation times and tractability in the laboratory are proving especially useful for identifying genes that are implicated in social evolution as indicated below.

Contact site (*csA*) gene encodes a hemophilic cell adhesion molecule and as predicted by Haig seems to function like a green beard gene. When food is scarce in the lime mould, *Dictyostelium discoideum* shows reproductive division of labour. Free living individual cells aggregate into a slug, with reproductive spore cells positioned on top of non-reproductive (altruistic) stalk cells. Wild type cells are more likely to form stalk cells (that are more altruistic) than *csA*. Knockout cells preferentially allow wild type rather than *csA* knockout cells to form spore cells.

Transcription factor that is encoded by this ***dif*** *insensitive mutant (dimA)* gene illustrate that pleitropy can promote altruism. *dimA* is required by *D. discoideum* to receive DIFI extra cellular signal that causes cells to enter pre-stalk (non-reproductive) stage. *dimA* knockout cells ignore DIFI signal, evading altruistic fate of stalk formation. They should presumably have a reproductive advantage over wild type cells. In presence of wild type cells, *dimA*—knockout cells are excluded form spore group, so negating any advantage they might gain by evading the stalk fate. Cheating by dimA loss of function is prevented because both altruism and reproduction require dimA function.

DNA binding protein that is encoded by *asgB* (Group A signal (*asgB*)) is involved in growth and development in *Myxococcus xanthus*, which form multicellular aggregation such as slime moulds in response to starvation. Strains that have a mutation in this gene; cheat' and produce a higher proportion of spore cells relative to their representation in an aggregating population. Maintenance of reproductive division of labour in *M. xanthus* indicates that cheating is kept in check by effects of others to be discovered gene as in *D. discoideum*. Alternatively the persistence of cheaters could reflect balance of population dynamic forces (Robinson et al., 2005).

9.4 FEEDING STRATEGIES

All animals are heterotrophic. Almost every living species is eaten by the other. Food varies in its spatial distribution, seasonal availability, and predictability, may be well hidden or easily detected. Competition may, or may not exist for a particular food. Consequently, animals have a variety of feeding strategies, to meet these challenges. Some animals are food generalists, euryphagous and eat a wide variety of foods. Coyotes, opossums, and humans are the examples. Others are food specialists, stenophagous, feeding on a narrow range of foods. Everglades kite feeds on just one species of snail. Many feather mites survive on one species of bird.

Behavioural ecologists are concerned with theories of optimal foraging. Animals must gain more energy from their food, than they expend in searching, capturing, and consuming it. They must acquire specific nutrients, like certain salts, which provide no energy, but crucial for survival. Thus, feeding is associated with issues, as food choice, prey switching, sensory mechanisms for recognizing, and locating food, optimal search strategies, overcoming defenses of prey, and compromising between finding food and not carelessly falling prey to some other. Following are basic methods that animals use to acquire food.

9.4.1 Herbivore-plant Relationship

As plant population increases, herbivore population also increases, (Figure 9.5) and then exceeds carrying capacity. At this point, plant and subsequently herbivore population start to decline. According to the theory of predatorprey interaction, relationship between herbivores and plants is cyclic. When plants are numerous, herbivores increase in numbers. Reduction in plant population, causes herbivore number to decline. This suggests that population of herbivore fluctuates around carrying capacity of food source. Keystone herbivores keep vegetation populations in check, and allow for a greater diversity of both herbivores and plants. When an invasive herbivore or plant enters the system, balance is lost and diversity can collapse to a monotaxon system.

FIGURE 9.5 The goat (grazer) populations.

9.4.2 Strategies and Theories

Grazers/browsers tend to be large herbivores that need a lot of food to maintain their metabolism, or herbivores that have a short amount of time to eat as much as possible, before reproducing, like many insects. Several theories attempt to explain the relationship between animals and their food, such as Kleiber's law, Holling's disk equation, marginal value theorem, and optimal foraging theory.

Marginal value theorem

Marginal Value Theorem (MVT) states that optimally foraging animals exploit resources distributed in patches and must decide when to leave a patch, to start searching for a fresh one. Animal is assumed to have evolved to optimize a cost/benefit ratio, when searching and manipulating for food is a cost, while food is a benefit. Decision taken by animals appear to be based on an expected transit time, among patches and an observed intake rate a within each patch. Examples include bees visiting flowers, birds eating berries.

The theorem predicts that individuals will stay longer, as the distance between patches increase, and when the environment as a whole is less profitable. The MVT has been criticized on the grounds that few foragers are optimal. Nevertheless; it predicts behaviour that is compatible with many types of real forager behaviours.

Kleiber's law

Kleiber's law explains relationship between size of animal and its feeding strategy. It states that larger animals need less food per unit weight than smaller animals. Metabolic rate ($q0$) is mass of animal (M) raised to 3/4th power:

$$q(0) = M\frac{3}{4}$$

Mass of animal increases at a faster rate than metabolic rate. Many feeding strategies are employed by herbivores. Some do not belong to one specific feeding strategy, rather employ several strategies. The different types of feeding strategies are shown in Table 9.2.

TABLE 9.2 Types of Feeding Strategies

Feeding strategy	*Diet*	*Example*
Frugivores	Fruit	Ring tailed lemur
Folivores	Leaves	Koalas
Nectarivores	Nectar	Honey possum
Granivores	Seeds	Hawaiian honeycreepers
Palynivores	Pollen	Bees
Mucivores	Plant fluids	Aphids
Xylophages	Wood	Termites

Optimal foraging theory

This theory predicts animal behaviour, while looking for food, or shelter, or water, and meant for assessing distribution of animal within a habitat. It may be used for browsing behaviour of an animal, while looking for food, and animal's specific location and movement within habitat, and its interaction with other animals of same species. Controversy arises that theory is circular and untestable, and that animals do not have ability to assess and maximize their potential gains.

Holling's disk equation

This equation models predicts that as the number of prey increases, amount of time a predator spends handling prey also increases and efficiency of predator decreases. In 1959, Holling proposed an equation to model rate of return, for an optimal diet.

$$\text{Rate } (R) = \frac{\text{Energy gained in foraging } (Ef)}{\text{Time searching } (Ts) + \text{Time handling } (Th)}$$

or $$R = \frac{Ef}{(Ts + Th)}$$

where

s = Cost of search per unit time
f = Rate of encounter with items
h = Handling time
e = Energy gained per encounter

In effect, this would indicate that an herbivore in a dense forest would spend more time getting handling (eating) vegetation because there was so much vegetation around, than an herbivore in a sparse forest, who could easily browse through the forest vegetation. Herbivore in sparse forest would be more efficient at eating than herbivore in dense forest.

9.4.3 Attacks and Counter-attacks

This is measured relative to another plant that lacks defensive trait. Plant defenses increase survival and/or reproduction (fitness) of plants, under pressure of predation from herbivores. Defense can be divided into two main categories, tolerance and resistance. Tolerance is ability of a plant to withstand damage, without a reduction in fitness. This can occur by diverting herbivory to non-essential plant parts, or by rapid regrowth and recovery from herbivory. Resistance refers to ability of a plant to reduce amount of damage it receives from an herbivore. This can occur via avoidance in space, or time, physical defenses, or chemical defenses. Defenses can either be constitutive; always present in plant, or induced; produced or translocated by plant following damage or stress.

Physical or mechanical defenses are barriers or structures designed to deter herbivores, or reduce intake rates, lowering overall herbivory. Thorns as found on roses, or acacia trees are the example, as are spines on a cactus. Smaller hairs known as **trichomes** may cover leaves or stems and are especially effective against invertebrate herbivores. Some plants have waxes, or resins that alter their texture, making them difficult to eat. Some plants sequester silica inside their tissues. Chemical defenses are secondary metabolites that deter herbivory. Chemical defenses can be divided into two main groups, carbon-based defenses and nitrogen-based defenses. Carbon-based defenses include terpenes and phenolics. Terpenes are derived from 5-carbon isoprene units, and comprise essential oils, carotenoids, resins, and latex. They can have a number of functions that disrupt herbivores, such as inhibiting ATP formation, molting hormones, or nervous system. Phenolics combine an aromatic carbon ring, with a hydroxyl group. A number of different phenolics, such as lignins are found in cell walls, and are indigestible, except for specialized microorganisms. The tannins have a bitter taste, and bind to proteins, making them indigestible. Furanocumerins produce free radicals, disrupting DNA, protein, and lipids, and can cause skin irritation.

Nitrogen-based defenses are synthesized from amino acids, and primarily come in the form of alkaloids and cyanogens. Alkaloids include caffeine, nicotine, and morphine. These compounds are often bitter, and inhibit DNA, or RNA synthesis, or block nervous system signal transmission. Cyanogens get their name from cyanide, stored within their tissues. This is released when plant is damaged, and inhibits cellular respiration and electron transport.

Plants have also evolved features that enhance probability of attracting natural enemies to herbivores. Some emit semiochemicals, odours that attract natural enemies. Others provide food and housing to maintain natural enemies' presence. Herbivores have three primary strategies for dealing with plant defenses: choice, herbivore modification, and plant modification. Feeding choice involves which plants an herbivore chooses to consume. Many herbivores feed on a variety of plants, to balance their nutrient uptake, and to avoid consuming too much of any one type of defensive chemical. This involves a tradeoff, however, between foraging on many plant species to avoid toxins, or specializing on one type of plant.

Herbivore modification occurs, when various adaptations to body, or digestive systems of herbivore allow them to overcome plant defenses. This might include detoxifying secondary metabolites, sequestering toxins unaltered or avoiding toxins through production of large amounts of saliva to reduce effectiveness of defenses. Herbivores may utilize symbionts to evade plant defenses. Some aphids use bacteria in their gut, to provide essential amino acids, lacking in their sap diet. Plant modification occurs, when herbivores manipulate their plant prey to increase feeding. Some caterpillars roll leaves to reduce effectiveness of plant defenses are activated by sunlight.

9.4.4 Adaptation Dance

The back and forth relationship of plant defense and herbivore offense can be seen as a sort of "adaptation dance", in which one partner makes a move, and other counters. This reciprocal change drives coevolution between many plants and herbivores, resulting in "coevolutionary arms race". Escape and radiation mechanisms for coevolution presents the idea that adaptations in herbivores and their host plants are the driving force behind speciation. While much of inter-action of herbivory and plant defense is negative, with one individual reducing fitness of other, some are actually beneficial. This beneficial herbivory takes form of mutualisms in which both partners benefit in some way from inter-action. Seed dispersal by herbivores and pollination are the two forms of mutualistic herbivory in which herbivore receives a food resource, and plant is aided in reproduction.

9.4.5 Impacts of Herbivores

Environmental degradation from white-tailed deer (*Odocoilus virginianus*) in the U.S. alone has potential to change vegetative communities through over-browsing and cost for forest restoration projects upwards of $ 750 million annually. Agricultural crop damage by same species totals approximately $ 100 million every year. Hunting of herbivorous game species, such as white-tailed deer, cotton-tailed rabbit, antelope, and elk in the U.S., contributes greatly to the billion-dollar annually hunting industry. Ecotourism is another major source of revenue, particularly in Africa, where many large mammalian herbivores, such as elephants, zebras, and giraffes help to bring millions of dollars to various nations annually.

9.4.6 Satisficing

It was suggested that the concept of satisficing may be a useful alternative hypothesis to examine. It was also intimated that satisficing may have a role to play in foraging theory. The Scottish word "sat'sfic'ng" (satisfying) was revived by Nobel laureate Herbert Simon (1956, 1957, 1983), to denote problem solving and decision making that sets an aspiration level, searches until an alternative is found that satisfies aspiration level criterion, and selects that alternative. These theories are also known as theories of **bounded rationality** because they are theories that incorporate constraints on information-processing capacities of actor. As applied to ecological contexts, there are two distinguishable, but related aspects to satisficing theory:

1. *Minimum requirements*

The decision-maker is satisfied after meeting some minimum requirement. Additional increases in rate of energy intake above a threshold produce no further increase in fitness. This aspect of satisficing is subject of debate and probably has little heuristic value to ecologists. Herbers suggested that this constitutes support, for satisficing, because animals stop hunting, when they are satisfied by some prey intake criterion. Herbers further suggested that laziness occurs, when prey is abundant or when predators can catch them easily. This aspect of satisficing, as evinced by Herbers, too readily fits facts, simply because any animal that does not feed continuously may be considered to have achieved satisficing criterion. Herbers' (1981) statement is readily countered by alternative explanations. Two plausible explanations are:

Scientists have shown that inactivity of animals can as readily be interpreted as an optimal strategy in animals that feed on large prey, and thus, experience considerable day-to-day variation in food intake, which restricts them to relatively rich habitats. In face of high predation pressure, animals that are less able to detect predators, while feeding may reduce time spent in feeding to increase vigilance. In doing so, they would be recorded as inactive, even though this may be optimal strategy (one that maximizes long-term fitness) under these conditions.

2. *Information or time constraints*

The lack of complete information on an issue, or an information-processing constraint on behalf of decision-maker, necessarily, limits ability of an animal to make an optimal decision. Ecological examples of this situation might be a predator searching for prey, and a parasitoid searching for a host. It is deemed unlikely that an animal can either remember or know all possible combinations of choices that it may make to lead to an optimal strategy. Decision-maker is, thus, constrained to satisfy a minimum requirement, or to do the best, it can under those constraints. Time constraints may cause similar problems for a decision-maker.

9.4.7 Grazing

The grazers including snails, grasshoppers, geese, rodents, kangaroos, and hoofed mammals (Figure 9.6) crop grasses and other ground plants on land, or scrape algae and other organisms from water surfaces. Grass and algae are palatable, and offer little or no resistance to being eaten. They are adapted to survive grazing and quickly replace lost biomass, but nutrient poor. Grazers, therefore, have to consume a large quantity of it, and spend a larger percentage of their time in eating and while eating becoming vulnerable to attack. Grazing mammals therefore tend to form herds.

FIGURE 9.6 An Indian bison.

9.4.8 Browsing

The terrestrial browsers including caterpillars (Figure 9.7), tortoises, grouse, giraffes, goats, antelopes, deer, pandas, koalas, and monkeys nip foliage, from trees and shrubs. In aquatic habitats, browsers like slugs, sea urchins, parrot-fish, ducks, and manatees feed on algae, aquatic plants, and corals. Browsers depend on less abundant food. They tend to form smaller groups, or to be solitary and secretive.

FIGURE 9.7 A caterpillar.

9.4.9 Burrowing

The herbivores, such as bark beetle fly and moth larvae called leaf miners, and wood-boring termite

Burrowing Behaviour —*By* FATIK B. MANDAL | *Posted*: NOVEMBER 13, 2012

Burrowing behaviour found in various higher animal groups like Amphibia, Reptilia and Mammalia are interesting from the evolutionary point of view. A brief account would help understand the importance of this behaviour.

Burrowing has evolved many times in the animal world. Moving through soil or sediment is widespread in the animal world. The moles and other burrowers among mammals show convergent adaptation to subterranean life in their communication behaviour. Reduction or loss of eyes in them is also noticeable. Limbless amphibians called caecilians and certain reptiles like burrowing skinks are also adapted to burrowing behaviour.

In frogs, it occurs in two routes. The 'hind feet first' route involve a less marked transition. The 'head first' route involves more specialization. The caecilians like *Dermophis* translate itself like a worm by developing a hydrostatic skeleton and by exhibiting about complete separation between the vertebral column and skin and muscle. The spine is thrown into curves like snake during locomotion, but when it separates from the skin and muscle, the latter moves freely and provides points of contact like that of the earthworms. Several tendons also form a crossed helix.

Tuatara and some turtles make burrows. Worm-like burrowing is found in several squamates comprising lizards, snakes and amphisbaenians. Agamids like *Agama etoshae*, gerrhosaurids and lacertids exhibit sand-diving and move through sandy substrate. They show modifications in having cylindrical body, fringes or spines on their toes, and adaptations for breathing. Sand-diving lizards that inhabit the Namib Desert include *Pachydactylus rengei, Angolosaurus skoogi, Meroles anchietae* and Riopa sundevallii.

The burrowing skinks are perhaps most specialized among fossorial lizards, displaying loss of limb as found in Scleotes of Southern Africa, and the entirely limbless condition in *Typhlacontias* of Southern Africa. Burrowing skinks have a flattened snout, smooth overlapping scales for movement. They generally tend to lack eyes, eyelids and opening of external ear. Limbless lizards include the semi-fossorial 'slow worms' including *Anguis fragilis.*

Among the snakes, the atractaspids have evolved adaptations for burrowing into sandy soils with smooth scales and fangs that swing sideward to inject prey within their tunnels. In the group amphisbaenas, the loss of limb and morphology of head and skeleton are suited for an underground existence. Examples include the limbless *Rhineura floridana, Anops kingii* and *Bipes bioporus*, which have tiny forelimbs. Convergent burrowing mechanism in many animals follows a particular burrowing cycle, which despite divergent anatomies relies on the same type of movements in higher vertebrate groups and believed to be the result of convergent evolution, the unifying concept in biology.

Printed from http://www.articlesbase.com/science-articles/burrowing-behaviour-6300119.html

burrow into their food, eating a tunnel as they go. In the sea, unusual clams and crustaceans called shipworms and gribbles, respectively, burrow through wooden piers and ships, causing enormous destruction. Earth-worms and many marine worms burrow in soil and sediment, eating indiscriminately, digesting organic matter, and defecating indigestible sand and other particles. The burrowing animals get benefit of being surrounded by food, and are less exposed to the predators.

9.4.10 Filter-feeding

The filter feeders can be active; creating their own currents, or passive; letting water currents do the work. Sponges use flagella of collar cells, to create a current. Food settles out at collar cell. Tunicates use water-currents generated by cilia, to bring food to pharynx, which is highly perforated, to allow water to pass, while food is retained on mucus coating. Similar scheme is used by clams. Lophophorates use mucus-covered tentacles, in combination with ciliary currents to trap food. Some species use nets of mucus. Tentacles are also used by some filter-feeding echinoderms, like crinoids. Arthropods use setae for filtration. They also weave silk nets to filter material from water. Many arthropods place filtering structures on legs or other appendages. Filter feeders include the largest organism that ever lived, the blue whale, which feeds by swallowing a huge mouthful of water, and then spitting it out through baleen.

Filter-feeding uses anatomical devices that act as strainers to remove small food items from water. Sessile filter-feeders, like barnacles, oysters, fan worms, brachiopods, and tunicates, pump sea water and strain planktons from it. Other filter-feeders are mobile. Herring swim with their mouths open, letting the water to flow through their gill rakers, and strain small particles of food from it. Flamingoes take in mouthful of water and mud, and then force water through fringed edges of their bills, which serve as strainers that retain food such as, brine shrimp, aquatic insects and plankton in the mouth. Filter-feeding is more common in the ocean than in the fresh water because plankton is less concentrated in the fresh water. Of all feeding modes, this might be among the easiest. Instead of requiring predatory or precision action, filter feeder just exposes its filters, which take many forms of food particles and sieve them through their filters to get desired food particles. A type of jellyfish use filter for feeding. Using a fine web of tentacles, they catch small food particles. Then, these tentacles slowly turn in a corkscrew motion to bring prey to jellyfish's mouth. As jellyfish's tentacles contain stinging cells, they paralyze the small prey on contact.

9.4.11 Suspension and Deposit Feeding

Organic matter including living and dead plankton and bits of dead animal, plant, and algal tissue settle to the bottom in aquatic habitats. Suspension feeders pick this material from water as it falls, and deposit feeders consume it, after it settles on the bottom. Deposit feeders would be gatherers and brushers. Deposit feeders are generally feeding on detritus; dead and decaying animal or plant material. Coprophagy is common among deposit feeders. Deposit feeding consists of consuming food particles in soil. Most deposit feeders are detrivores. Many terrestrial arthropods, including beetles and mites are deposit feeders. Earthworm might be the archetypal example of a deposit feeder, as it consumes large quantities of soil and plays an integral part in breaking down of the dead plant matter into humus. Insects and their larvae, which burrow through living or dead plants and animals or faeces are also considered deposit feeders.

Deposit feeding is a feeding strategy that only works in fertile areas, with a lot of pre-existing life. Top layer of soil is targeted, typically within 6 inches of surface, as this soil contains food particles that have not been completely broken down yet. These food particles are detritus. Detritus break down to chemically neutral state, humus. Humus has a black colour, due to its high carbon content. Terrestrial deposit feeding is practiced by fiddler crabs. They pick up little balls of dirt, reach them to their mouths, and pick out any edible material, including colonies of microbes. Then, the spheres are discarded as quickly as they were picked up. These little dirt spheres can be found, wherever fiddler crabs dwell. There are marine species that practice deposit feeding as well, which burrow through the ooze on ocean floor. These include polychaete worms, some bivalves. Detrivores are certain animals that are an extremely important part of the food chain.

9.4.12 Scavenging

Animals belong to a community of scavengers feed on organic refuse like manure (dung beetles, flies), leaf litter (snails, millipedes, earthworms), and dead animals (blowflies, vultures, hyenas, storks). The family name of vultures is Cathartidae, is from Greek 'katharos', meaning "to cleanse."

9.4.13 Fluid Feeding

The rarest of feeding modes consist of sucking fluids out of another plant or animal. It is used by herbivores, carnivores, and omnivores alike. Well-known fluid feeders are hummingbirds, aphids, spiders, ticks, leeches, vampire bats, and mosquitoes. Fluid feeding is defined as getting nutrients by consuming the fluids of another organism. Although the "fluid" in question may be botanical in origin, when it involves animals, the fluid is blood. But fluid feeding is obviously a successful mode, and has probably been around, since insects first crawled onto land, about 428 million years ago. Fluid feeders need some way to pierce protective wall (skin or plant walls) getting in the way of fluid, then some proboscis-like appendage for sucking out all fluids. Aphids are specialized in doing this with plants, which is why they can be found in such numbers on certain plants, especially those that produce an abundance of sweet fluid.

Hummingbirds, the smallest bird, with the highest metabolism, are fluid feeders of flowers. Their long beaks are used to drink nectar, out of flowers. The most despised fluid feeders are those that feed on human blood, spreading infections, and some of the worst diseases. Blood-sucking ticks are "the foulest and nastiest creatures." They spread various diseases, including Lyme disease, which can be fatal if left untreated. Mosquitoes annoy us with their buzzing, and spread malaria and other diseases which afflicts half a billion people every year, and kills between one and three million peoples every year.

9.4.14 Bulk Feeding

Bulk feeding consists of consuming chunks, or entire body of other animals or plants. Most mammals, birds, fishes, and reptiles are bulk feeders, usually consuming plants, but sometimes other animals as well. Carnivores specialize specifically in hunting, killing, and eating other animals. Some animals, like humans, are omnivorous; eating both plants and other animals.

9.4.15 Eating Nectar, Fruits, Pollen, and Seeds

Sweet nectar rewards bees, flies, moths, butterflies, and bats that spread pollen from one flower to another. Fruits attract birds, monkeys, fruit bats, bears, elephants, and humans to eat them, and spread the indigestible seeds.

9.4.16 Predation

Predators are animals that depend on killing other animals outright. Animals have evolved defenses like hard shells, toxins, ability to fight back, or simply running, or flying away, against predation. Predators have evolved a range of strategies for capturing the prey. They may hunt in packs (wolves), collaborate to ambush prey (lions), stalkers (solitary cats), use lures to attract unsuspecting prey (snapping turtles and anglerfish), employ camouflage so their prey does not notice them until it is too late (praying mantids), and use snares (spiders, jellyfish). Predation places common selective pressure on organisms that feed in this way. In order to feed on other organisms, they must be found, caught, killed, and ingested.

Finding prey (Figure 9.8) often involves sophisticated sensory structures. Most familiar sense used is sight. On land, predators are often told from herbivores by presence of forward-directed eyes, with good overlap of visual field, for stereoscopic vision, a necessity in gauging distance to prey. Herbivores usually have lateral eyes, with wide peripheral vision. In aquatic systems, however, this system breaks down with only top predators, such as alligators; having much in way of stereoscopic vision. Other predators, such as bass, are simply too vulnerable to the predation of themselves, particularly when young. Sound and related disturbances in water are frequently used by predators to find their prey. Familiar with hearing as a tool in locating prey, the fishes also have lateral line to help them sense disturbance in surrounding water. Marine mammals often use sound waves that they themselves generate to locate prey by bouncing sound waves off the prey. This sonar is a highly advanced sense, and porpoises at least are able to discriminate between objects very close in size with their sonar. Their sonar sense may have better resolution than vision, under all but clearest conditions. Sound waves that are generated have a short wavelength and are above normal hearing range for humans. Many marine mammals have a "lens" of fat or oil in their head that serves to focus sound waves, making this sense highly directional as well. Recent evidence indicates that stoneflies can "hear" swimming movements of their mayfly prey.

FIGURE 9.8 A tiger in search of prey.

Sense used in water is olfaction. Olfactory senses are strong among most aquatic organisms, with possible exception of aquatic insects, which perhaps have not developed these senses to their fullest potential yet. Great feats of homing attributed to salmon, eels, and sea turtles in returning to river (or beach) of their birth, to mate and reproduce, is due in part to an ability to chemically distinguish between different sites. Sharks are known to follow blood trails of their prey for miles. Crayfish in tests responds to vanishingly small dilutions of extracts from their food sources. Olfaction in water is at mercy of currents. In a stream, a predator cannot detect scent of a prey organism, just slightly downstream, while it may be able to detect other prey far upstream. Turbulence in water may either confine scent to a narrow, concentrated stream, or spread it into a wide, diffuse band. Former makes it easy to home in on prey once stream is detected, but chance of detecting odour is slight. In latter, it is easy to detect scent, but hard to home in on it to find prey.

Electric sense detects weak electrical energy, given off by all living things. Because water conducts electricity, this sense is particularly well developed among aquatic organisms. Use of electrical senses is found in both vertebrates and invertebrates. Duck-billed platypus uses electrical sensors in its bill to locate food in mud. Sharks use electrical sense to make final biting lunge for their prey. Knifefish; take electrical sense further, generating their own stronger electrical currents, and using these to "probe" their surroundings. Electrical current is generated by modified muscle cells, and for system to work properly, depends on body being kept in a straight line. Thus forming selective pressure for unusual form of swimming, exhibited by knifefish, which uses its anal fin for propulsion, eliminating need to bend spine.

Prey may be captured, using a number of structures. Many organisms simply engulf their prey with mouth. Seahorse and its relative pipefish also use suction to capture their prey. Other organisms use tentacles or other appendages to seize the prey. Among latter, preyingmantis like forelegs of mantid shrimps is particularly interesting. One species which uses knobs on ends of forelegs to crack open mollusc shells can hit with an impact equivalent to a .22 caliber bullet. Dragonflies and damselflies use an extensible labium to capture their prey.

Various structures are used, to tear or cut up prey, before it is swallowed. Octopuses and squids beak tear the prey, and inject a poison. Marine snakes inject a poison, similar to that of cobras. Arthropods generally have tearing or cutting structures, associated with legs. Examples include chelipeds of lobsters and crayfish, or gnathobases of horseshoe crabs. Starfish prey on bivalve molluscs and wrap themselves around clams, and prey it open just enough to evert their stomach into shell cavity, where it excretes a poison and digestive juices that digest unlucky mollusc, within its own shell. Coneshell gastropods are quick enough to lunge at small fishes, and overcome them with a toxin, and some snails use their radula to bore through hard outer cases of barnacles and bivalves.

Avoiding predation

There are three main tactics which the prey can try to increase search time, increase handling time, or actively repelling the predator. The first two rely on some economic-ecological principles, the first being that time is like money (economics), or energy (ecology). If predator requires too much time, to successfully capture and eat a prey item, it will leave that species alone in future. It does not do the individual any good, but it may help out that individual's offspring.

Characteristics of predators

Predatory birds possess outstanding eyesight and often hearing, as in case of owls. Many species of mammals have a very keen sense of smell that helps them locate prey. Many predators are very fast, and use their speed to help to capture their prey. Cheetahs, predators of the African savannas are world's fastest runners. Falcons, predators of other bird species are world's fastest fliers. Dolphins and barracudas are very fast swimmers.

Prey defenses

Most species possess several lines of defense against predators. First line of defense is to avoid being detected by the predator. One way to do this is to minimize noise production and any visual cues that predator might use to locate prey. Frogs and crickets, usually, stop singing as predator approaches. Camouflage colouration that blends into background, making it difficult for visual predators, to find prey. Many moths, common prey for birds, look like bark of trees, on which they rest during the day. Snowshoe hares, primary prey for lynx, have brown fur in summer, but white fur in winter, when their northern environment is covered with snow. As predators often use prey movements, to detect them, many prey remain as still as possible, when a predator approaches.

Many prey species are very fast runners, swimmers, or fliers. Many preys successfully deter a predatory attempt, by fighting back. Moose is able to use its hooves as lethal weapons, against much smaller wolves, and wolves generally give up once they realize moose is healthy and a formidable adversary. Some animals have morphological and behavioural adaptations that make it difficult for the predator, to get their prey into their mouth. Many fish and insects have spines that prevent a predatory fish, or bird from being able to eat them. Some prey, like puffer fish, make them-selves larger if threatened, again making it more difficult, often impossible for the predator to ingest the prey.

Many preys use social behaviour as a predatory defense. Many species of fishes and birds travel in groups, such as schools of fishes, and flocks of birds. These schools and flocks, often move very quickly, in a highly synchronized fashion. These groups provide protection for individuals in the group. Most predators have to single out, and focus on a single individual, in order to successfully capture a prey. Fast-moving and synchronized flocks and schools are believed to make it difficult, for predators to accomplish this. In some cases, a group of prey is able to successfully fight off a predatory attack, whereas an individual prey probably would not be able to do this. Although a baboon on its own would probably succumb to a predatory attack from a leopard, a group of males in a baboon troop can usually ward off such an attack.

Some prey are easy for predators to find, to capture, and to ingest. Yet they seldom fall prey to predators, because they employ a final line of defense-toxicity. They are poisonous. Poison dart frog of rain forests of Central and South America is an example. A beetle, *Brachinus* sp. stores quinones and hydrogen peroxide.

Some species of fishes swim and highly defend themselves against their predators. Small, brightly coloured frogs are easy to find, catch, and eat, but they are very poisonous, and most birds quickly learn, to avoid them. Indigenous people discovered that these frogs contain a potent toxin and learned to extract toxin from frogs. They then dipped tips of their arrows in this toxin before going out on a hunting expedition. Ironically, by using toxin that had evolved as an antipredatory defense, people became more effective predators.

Evolution of predator-prey relationships

Because cost of being caught and eaten by a predator is great, intensity of natural selection on prey species has been very high throughout evolution. Selection pressure on prey is probably higher than that on predator. If a fox fails in its attempt to catch a rabbit, it just misses lunch. If a rabbit fails in its attempt to escape from a fox, it loses its life. Because of intensity of selection on prey species, variety and effectiveness of antipredatory defenses is especially impressive. It is believed that predators and their prey have coevolved. This means that as predators developed adaptations that enabled them to capture prey more successfully, selection pressure on prey intensified, resulting in selection of more effective antipredator adaptations. These more effective antipredator adaptations are promoted selection of more effective predatory adaptations. This reciprocal ongoing evolutionary cycle among predators and prey is sometimes referred to as an evolutionary arms race (Biology Encyclopedia Forum 2008).

9.5 PACK BEHAVIOUR

Social behaviour may be organized in many ways. The famous one is "pecking order". In "Pecking order" structure, each member of group occupies certain position. The highest ranking bird pecks every other members of the flock, without retaliation. The second highest accepts pecking only, by the highest member and in turn can peck all lower ranking birds. This system is known as "dominance order". Two separate dominance orders within each pack are a male order and a female order. The highest ranking member of each order occupies "alpha" position, followed by "beta" individual and so forth until last position called "omega".

Very few individuals are considered "equal". Alpha position is occupied by a mated pair. Ultimate dominant individual, either male or female, directs activities of a pack. In wolves, sometimes alpha male refrains from breeding alpha female. He allows a lower ranking male with her instead. In wolf packs, alpha male and female are parents of subordinate pack mates. Very rarely are strange wolves allowed to enter an established wolf pack. As cannies are "first cousins" to wolves, this explains why entry of a strange dog to pack results in fight escalation. In well established packs other possible classification of social ranking are: (1) Mature subordinate animals, (2) Outcasts who rank so low that they avoid main pack members and (3) Juveniles, who do not become part of pack nucleus, until they are much older. Dominance orders cross sexual lines in immature animals and do not divide into male/female orders until sexual maturity. Older a pack is more stable its social structure becomes. When alpha male dies, grows old or weak, it results in competition for alpha position, and may disrupt social stability of a pack. Most conflicts within a pack are not severe.

Leader of pack initiates play pattern in which direction pack travel; when to rest and when to hunt. A well established leader rarely has his authority challenged. He/she directs pack activities, and also takes initiative in reacting to intrusions. Leader is neither despotic nor democratic, but a combinations of both. Dogs within each pack, generally, interact predictably and social structure of groups is maintained. Much of the behaviour is directed towards goal of either maintaining one's social status, or possibly raising it. A dog's social status is unusually, established in early life, but circumstance may change this position.

Any drastic disturbance, like loss or addition of pack members trigger status rearrangement. Every member is watchful and interested in all socially important happenings within the pack. In particular status, quarrels are never a private affair between two individuals. Whole society participates in final outcome. Role of omega dog is crucial to stability of day to day pack life. Usually this animal is outcast and is not allowed to join in pack activities. Omega position offers a way for wolves to disperse energy. If omega strays from allotted territory or attempts to join in feeding, the pack will persecute omega until an order is restored. Energy is released during confrontation, and this is immediately followed by a period of peace.

9.5.1 Communication in Pack

Certain posture and gesture express inner state of a dog. Other dogs notice these patterns and respond in way, depending on their own particular feelings. Patterns of expression evolve to help in holding the pack together, and to reduce aggression among its members. Harmony within a pack is displayed by submission. Rolling over, spreading the legs, and submitting the tender skin of stomach and genital area to pack inspection is example of submissive behaviour. Sometimes a simple lowering of head is enough to communicate submissiveness. Submissive behaviour serves as an expression of canine friendliness. Submission is defined as an effort of inferior to attain friendly or harmonic social integration.

Dominant animals often respond to this display with tolerance, friendliness, and superiority. Two types of submission are active and passive. Active submission feature friendliness, fostered by a friendly and tolerant response from dominant pair. Passive submission is a demonstration of inferiority and helplessness. Response of dominant animals is usually a show of self-assertion and less tolerance. Both active and passive submission patterns from food-begging to eliminative behaviour in pups are seen.

Without a doubt, vocal communication between dogs is extremely important. Basic types of vocal communication are whimper, growl bark and howl. Whimper indicates either sub-missiveness, or a friendly greeting. Growling is a sign of aggressiveness. Barking either declare alarm, or display stance for a possible challenge. Barking also has varying degree of sound. Bark/whine points out a desire to obtain something, or use as a warning. Howling is actually used to locate other pack members, or to call a "meeting". Howling of group is often preceded by whines and wags of tails. Howling is considered a community activity or a happy social event. Wolves are known to run almost any distance to join in. This is why many individuals believe that howling is used to assemble pack.

Baboons are "highly intelligent, curious and social animals". They live in groups, for reasons, such as for companionship, affection and survival, and are similar to humans. In addition to the sentimental side of baboons, it is not un-common to see baby baboons clinging to parent baboons everywhere they go. "Drawn to nourishment and close contact provided by their mother's milk, young baboons even have tantrums, when their mothers attempt to wean them." Apart from that, adult baboons are protective of baby baboons too. Although, they may lose out to their predators in terms of size, they continue to put up a brave fight. Male baboons often work together to scare off their predators, such as cheetahs, or leopards. In fact, baboons are omnivores. Not only they feed on fruit and plant parts, such as seeds and roots, they also feed on animals, such as birds and rodents.

9.6 AGGRESSION

Aggression is "a physical act or threat of action by one individual that reduced freedom, or genetic fitness of another actions caused, or threatened to cause injury to another individual" (Wilson 1975). This is not a unitary concept. Many phenomena were classified as "aggression". Moyer in late 1960s made a first approximation of 8 classes of aggression, distinguished by a complex of contextual elements, of which eliciting stimulus is most important. In almost all cases, "aggression" is an "expensive" behavioural pattern involving energy and risk of injury, and evoked as the only option for eliminating a contributor to organism's overall cumulative and potentially damaging stress load, social dominance, depostism and pecking order linear, or arborizing hierarchy. Each form of aggression may reflect ensembles of more or less open, or closed genetic programs. Moyer never doubted that his list of "forms of aggression" would evolve in a self correcting fashion as analogies and homologies were detected.

Conflict ubiquitous in animal sometimes results in direct violence, injuries, and death. Displays, posturing, and trials of strength resolve conflicts. Male fig wasps, *Darnels* sp. after hatching inside fig attempt to decapitate their brothers that hatch in same figure, attack them with mandibles (Hamilton 1967). Male elephant seals (Miring angstroms) kill rivals in fights over access to females (Hayley 1994). Male fallow deer, Dama dama employ violent head-on "jump clashes" at rut at the start of the breeding season (Jennings et al., 2005). Rival uses aggression to maximize chances of success in conflict over access to females. Animal compete for most highly-prized act, the mating. Individuals conflict over access to limiting resource like ownership of food and shelter or territories containing such resources. Natural selection produces weapons or large body size to enhance resource holding potential (RHP) or to defend territories. Non-injurious fighting are roaring contests in red deer, Cervus elephas (Clutton Brock and Albon 1979), claw waving in male fiddler crabs, Uca mjoebergii (Morrell et al., 2005) for territory and "shell rapping" in hermit crab (Pagurus bernhardus) for gastropod shell (Mowles et al., 2010). Dangerous fighting harm the species (Huxley 1966) but if aggressive behaviour results from natural selection, avoiding dangerous fighting is unlikely to occur "for good of species". Game theory helps to models of evolutionary processes that result in aggressive behaviour. Contests as "games" are played out between 2 or more alternative strategies to determine which alternative should be favored by natural selection. If one strategy adopted by most, the population is resistant to invasion by other alternative in game. It is the evolutionarily stable strategy (ESS). The first game theory model of fighting, the Hawk-Dove game (Maynard Smith and Parker 1976) pits a Hawk strategy against a Dove strategy (always use non-injurious display if rival is another Dove and always withdraw if rival is the Hawk). To find the ESS, the first stage is to specify the average payoffs (benefit in fitness, denoted as *E*) for individuals playing Hawk or Dove against opponents playing Hawk or Dove. This is best represented by a payoff matrix (Table 9.3).

TABLE 9.3 The Payoff Matrix of the Hawk-Dove Game

			Playing against
	Hawk	*Dove*	
Pay-off to	Hawk	½ $(V - C)$	V
	Dove	0	½ V

An individual that plays Hawk strategy against another Hawk has 50% chance of winning and 50% chance of losing. Average payoff of playing Hawk against another Hawk includes half the value of resource, from 50% of contests that they may win. Since they only withdraw when injured, we have to add half of cost of getting injury from 50% of contests that they will lose. An injury have negative fitness consequence in contrast to positive fitness consequence as gained from winning resource. Average payoff is thus half of positive value of resource (V) plus half of negative value of getting an injury (C), which can be written E = half $(V - C)$. When Hawk fights Dove, the Hawk always win without injury. So, average payoff to Hawks fighting Doves is V. Average payoff to Dove fighting a Hawk is 0 (zero), as Dove always loses but avoids injury by running away. For a Dove fighting another Dove, It can be assumed that there is 50% chance of winning, but there is no chance of receiving injury as Doves never try to injure the rival. Mean payoff for playing Dove against another Dove is half V. We can plug in values for C and V and calculate the payoffs in each cell of matrix, given the values of C and V. Let's assume that $V = 50$ and $C = -25$ (so in magnitude $C < V$). For Hawk against Hawk $E = +12.5$, for Dove against Hawk $E = 0$, for Hawk against Dove $E = 50$, and for Dove against Dove $E = 25$. Injurious fighting occurs when benefits of winning resource are high in comparison to the cost of injury. In elephant seals, receiving an injury or being killed is not good, but challenging the dominant male for chance to mate is the only way for reproductive success. In fitness terms, failing to reproduce is worse than getting injury. In contest like fight over food, shelter, or territory, the cost of getting a serious injury would be as bad as losing resource. Various models attempts to explain how natural selection produces protracted and repetitive contests including the Sequential Assessment Model and Energetic War of Attrition among others. One possibility is for each individual to gather information about the RHP of the opponent, compare this to their own RHP, and give up as they are weaker. This is called **mutual assessment** which forms the core of Sequential Assessment Model. Another option is to keep fighting until the individual threshold of costs, which build up due to non-injurious agonistic behaviours or receiving injuries, is reached. The first individual to cross its threshold make decision to give up through a process of "self assessment," as in Energetic War of Attrition (Briffa and Sneddon 2010). Aggression occurs in almost all animal taxa and takes wide variety of forms encompassing aerial displays in butterflies (Kemp 2002), battles between rival armies in ants (Batchelor and Briffa 2010) through to biting and gouging in elephant seals (Hayley 1994).

Predatory aggression

Provoked by natural prey, generally studied in one to two form and may also include cannibalism.

Anti-predatory aggression

Defensive attack by prey on predator (e.g. mobbing).

Dominance aggression

Stimulated by proximity of an unfamiliar male that dominate, in order to establish, or maintain a hierarchy, or by a familiar male those steps out of place within an established hierarchy.

Isolation-induced aggression

Sustained lack of normal stimuli allows organism to maintain relationship by means of "normal", less aggressive measures. Additional stress lowers threshold for any stimulus that might evoke aggression.

Fear-induced aggression

Such aggression is stimulated by confinement, or being backed into a corner. Implication is that alternative ways of coping with a fearsome stimulus are unavailable, thus, tipping cost/benefit factors towards expressing expensive aggression.

Irritable aggression

This is elicited by any 'aversive' stimulus. "Irritation" implies the organism is already stressed by other factors that lower the threshold for expressing aggression against additional stressors. The irritant(s) may function by adding to a cumulative stress load, and evoking aggression that may eliminate this (or another) source of stress, thus lowering the load.

Territorial aggression

This is elicited by an intruder in focal animals' territory. Territories are space occupied more-or-less exclusively, and defended by aggression or display.

Parental aggression

This type of aggression is elicited by proximity of agent threatening young.

Parent-offspring aggression

Disciplinary actions against young possibly related to weaning aggression (Wilson 1975) can result in dispersal from the nest.

Sexual aggression

Elicited by stimuli those are related to sexual responses, and are directed against prospective mates.

Instrumental aggression

Learned aggression reinforced by some consequence of a former manifestation.

9.7 SOCIAL DOMINANCE

A priority of access to an approach situation or of leaving an avoidance situation that one individual has over another is known as social dominance. In practice 'dominance is always relative. It applied only to an individual, relative to one, or more other individuals. An individual may be dominant in one context, and subordinate in another. Competition is a general method of establishing them, but dimensions along which animals compete can be variable.

9.8 ALLOPREENING vs. AUTOPREENING

Allopreening has been reported in at least 43 avian families (Harrison 1969). Its functions are poorly understood. It may function in plumage maintenance in individual or sexual recognition in monomorphic species, or reduces or redirects aggressive tendencies that might otherwise cause one bird to attack another (Rothstein 1977) (Figure 9.9). Most owls can probably recognize the sex of other individuals by sexual differences in pitch and repertoire of vocalizations (Forsman 1976). In most instances described by Sparks (1965), allopreening occurred in situations of obvious aggression. For instance, when a Red Avadavat (*Amandava amandava*) attempted to perch beside a potentially aggressive bird in a flock of avadavats, it assumed a solicitation posture and allowed the more aggressive individual to allopreen it, sometimes quite roughly. Submissive posture assumed by a subordinate bird informs a more aggressive individual that the subordinate is not aggressive. The agonistic tendencies of the aggressive individual are thereby thwarted and redirected into a ritualized form of aggressive behaviour, allopreening. This interpretation suggests that allopreening has evolved as a ritualized form of aggressive pecking or biting behaviour (Harrison 1964). By sub-limating aggressive tendencies, it facilitates coexistence, and thus, ultimately serves to maintain the pair bond

FIGURE 9.9 Allopreening in birds.

in birds, such as owls, that live much of the year in pairs. Cowbirds (*Molothrusater*) deviate from this general pattern in that it is usually the dominant individual that solicits allopreening, and solicitation is most commonly directed towards other species rather than conspecifics (Rothstein 1977).

Spotted owls partially or entirely close their eyes while allopreening as a "cutoff" mechanism by simply removing the threatening individual from view, thereby reducing the stimulus to flee. In the case of owls, however, a more probable explanation is that the eyes are closed to prevent accidental injury, while two birds are in close contact. Harrison (1969) accounted for the lack of aggression (associated with allopreening) in some species by postulating that, through constant repetition of allopreening behaviour, birds refine or "facilitate" their performance to the extent that, at a later stage, any agonistic tendencies are expressed as allopreening. In owls, the gentle "nibbling" associated with allopreening could be ritualized biting behaviour that has become so modified that all appearances of aggression have been eliminated.

Individual birds spent most of the time sitting/resting and auto preening, but activities of shorter duration, such as different kinds of locomotion, chewing on objects, and acrobatics were often seen, as well. Sitting/resting was observed as the predominant behaviour in macaws. The behaviour could last up to several hours or only a few seconds, interspersed with other behaviours. Particularly in the middle of the mid day rest, during the hottest hours, they are reported to rest for many hours with only few and short intervals of other behaviours. During this period the eyes often were found closed and they put the head backwards over the shoulder tucking the beak into the plumage of back and often fluffing the plumage. Sitting/resting and autopreening were often interspersed with locomotion of short duration. Locomotion was most frequently observed at the beginning or end of a rest period. The birds faced in the same direction or opposite each other while allopreening. Often one of the birds crouched on the branch and raised the feathers on its head and neck, similar to begging behaviour of juveniles (Christiansen and Pitter 1993). Macaws allopreen all parts of plumage. During simultaneous allopreening they either preened each other in exactly corresponding parts of the body or they preened different parts dependent on where it was convenient in relation to their position. The macaws were often seen sitting opposite each other on the branch preening each other's under the tail coverts and around the cloaca. In between they grabbed the tail feathers and pulled them through the beak. When the macaws alternated between allopreening and autopreening, the autopreening was often carried out as if the macaws imitated each other and preened exactly the same part of the body. The mated pairs used allopreening as a greeting ceremony as well, e.g. when the female returned after a long stay in the nest hole or after a long rest without any interaction between the two.

Harrison (1964) suggested that allopreening was often seen between neighbours in social groups of birds and reduced aggression between the members of a group. Allopreening never extended outside the mated pair in Canary-winged parakeet (Arrowood 1960), or in Budgerigars (Trillmich, 1976), and pairs in Red-fronted macaws. The spreading of allopreening probably reduced aggression to a low level between all the pairs of a flock. Harrison (1964) stated that most allopreening in birds is restricted to head and throat, while especially preening under the wing is exceptional. Hardy (1963) suggested that allopreening directed to the head, wings and tail areas in pairs of orange-fronted parakeets (*Aratinga canicularis*) is the strongest behavioural device for maintenance of the pair-bond throughout the year. This may also be the case with the red-fronted macaw.

Simultaneous allopreening was by far the most common type of allopreening among the macaws and according to the interpretations of Harrison (1964), this indicates that there is no rank difference between the members of a pair. Allopreening among the red-fronted macaws very often passed into playing/fighting. This is very common among birds (Harrison 1964) and allopreening serves as an appeasement (Hardy 1963). Presumably allopreening replaces aggressive fighting in mated pairs. Social play has also been found in White-fronted Amazons *Amazona albifrons* (Levinson 1981) and in green-winged macaws *Ara chloroptera* (Deckert and Deckert 1982).

9.9 HIERARCHIES

It was first described in 1920's in chickens to reflect relative dominance ranks when pecking order was reported. Hierarchies can be linear, or arborizing. Space is key resources that are monopolized by socially dominant individuals. The great white shark (*Carcharodon carcharias*) is well known to be the largest predatory fish on the planet. It can reach 20 feet (6 metres) in length and weight up to 5000 pounds (2270 kilograms). These highly efficient predators have developed a unique way to increase their chance of getting their prey. When seals cross deep waters to approach or leave their island sanctuary, waiting sharks swim at great depths to increase their momentum hurtle themselves with great strength and ferocity towards animals at the surface. This speed and momentum is powered by a high metabolism which only cold temperate seas can provide; where great white sharks thrive. If the seal survives, chase continues. Behavioural study, a study of this particular unique feeding behaviour of the great white shark known only to those found in South Africa could possibly be attributed to the aspect of "learning". According to Martin, by employing such a strategy, known as "breaching" the great white sharks increase their chance of a successful single breach up to 80%. Also the seas off the coast of Cape Town, South Africa are known to be harsh, and seals which live there have adapted to the extremities of the nature by becoming expert sea-worty swimmers, under such conditions. Hence, this makes them harder for great white sharks to prey upon. All in all, great white sharks for a long time have not been creatures of researchers study only till recently. "This interesting feeding behaviour, known only to Africa's Seal Island, hence, makes it all the more intriguing for us to find out more about the great white sharks, the nature born killers, which we known so little about yet.

9.10 SOCIAL PLAY BEHAVIOUR

In his book *The Descent of Man and Selection in Relation to Sex,* Darwin wrote: 'Happiness is never better exhibited than by young animals, such as puppies, kittens, lambs, when playing together, like our own children'. When individuals play, they use action patterns like predatory-antipredatory behaviour, and mating that are used in other contexts. Individuals use various behaviour patterns, play markers, to initiate or maintain a play mood, by punctuating play sequences with these actions. Bekoff (1995a) found that play signals in infant canids were used nonrandomly, especially, when biting accompanied by rapid side-to-side shaking of head, which is performed during serious aggressive and predatory encounters, and can easily be misinterpreted if its meaning is not modified by a play signal.

Cheaters are unlikely to be chosen as play partners. Others can simply refuse to play with them and choose others. Individuals also engage in role-reversing and self-handicapping maintain social play. Each can serve to reduce asymmetries between interacting animals and foster reciprocity that is needed for play to occur. Self-handicapping happens, when an individual performs a behaviour pattern that might compromise her. Watson and Croft (1996) found that red-neck wallabies adjusted their play to age of their partner. When a partner was younger, older animal adopted a defensive, flat-footed posture, and pawing rather than sparring. Role-reversing occurs, when a dominant animal performs an action during play that would not normally occur during real aggression. A dominant animal might not voluntarily roll-over on his back during fighting, but would do so while playing. In some instances, role-reversing and self-handicapping might occur together. A dominant individual might roll over while playing with a subordinate animal and inhibit intensity of a bite. From a functional perspective, self-handicapping and role-reversing, similar to using specific play invitation signals, or altering behavioural sequences, might serve to signal an individual's intention to continue to play.

9.10.1 Fine-tuning Play

Playtime, generally, is safe time. Transgressions and mistakes are forgiven, and apologies are accepted by other, especially, when one player is a youngster. There is a certain innocence, or ingenuousness in play. Individuals must cooperate with one another when they play. They must negotiate agreements to play. Levels of cooperation in play of juvenile primates may exceed those predicted by simple evolutionary arguments. Highly cooperative nature of play has evolved in many species. Detail of play in various species indicates that individuals trust others to maintain rules of the game. The individuals of different species seem to fine-tune on-going play sequences to maintain a play mood, and to prevent the play from escalating into real aggression. There are subtle and fleeting movements and rapid exchanges of eye contact that suggest that players are exchanging information on run from moment-to-moment, to make certain everything is all right that this is still play. Play in most species does not take up much time and energy. In some species, only minimal amounts of social play during early in development are necessary to produce socialized individuals. Play is very important in social, cognitive, and/or physical development, and for training youngsters for unexpected circumstances. Short- and long-term functions of play vary from species to species, and among different age groups and sexes within a species. No matter, what functions of play may be, there seems to be little doubt that play has some benefits. Absence of play can have devastating effects on social development.

During early development, individuals can play without being responsible for their own well-being. This time period is generally referred to, as 'socialization period'. It is important for individuals to engage in at least some play. There is a premium for playing fairly, if one is to be able to play at all. If individuals do not play fairly, they may not be able to find willing play partners. In coyotes, youngsters are hesitant to play with an individual, who does not play fairly, or with an individual whom they fear. In many species, individuals also show play partner preferences, and it is possible that these preferences are based on trust that individuals place in one another.

9.10.2 Social Play and Social Morality

During social play, individuals remain in a fun, and in a relatively safe environment. They learn ground rules acceptable to others. There is a premium on playing fairly and trusting others to do so as well. Codes of social conduct regulate, what is permissible, and what is not permissible. Existence of codes may play role in evolution of social morality. Individuals may generalize codes of conduct learned in playing, with specific individuals, to other group members, and to other situations. Social morality, in this case behaving fairly is an adaptation that is shared by many mammals, not only by primates. Behaving fairly evolved, as it helps young animals to acquire social skills to mature into adults. Group-living animals may provide insights into animal morality. In many social groups, individuals develop and maintain tight social bonds that help to regulate social behaviour. Individuals coordinate their behaviour. Some mate, some hunt, some defend resources, and some accept subordinate status to achieve common goals, and to maintain social stability. Researchers thought pack size was regulated by available food resources. Wolves typically feed on such prey, as elk and moose, each of which is larger than individual wolf. Hunting such large ungulates successfully takes more than one wolf. So, it made sense to postulate that wolf packs evolved because of the size of the prey. Defending food might also be associated with pack-living. Mech (1970) showed that pack size in wolves is regulated by social, not food-related factors, and number of wolves who could live together in a coordinated pack is governed by number of wolves with whom individuals could closely bond and is balanced against number of individuals, from whom an individual could tolerate competition.

9.10.3 Sharing Intentions and Mind-reading

Perhaps one's own experiences with play can promote learning about intentions of others. Recipient shares intentions (beliefs, desires) of sender based on recipient's own prior experiences of situations, in which she performed play bows. Research suggests a neurobiological basis for sharing intentions. 'Mirror neurons' found in macaques fire, when a monkey executes an action and also, when monkey observes the same action performed by another monkey. Neural imaging studies in humans suggest a neural basis for one form of 'social intelligence'. Or understanding others' mental states. Comparative data are needed to determine, if mirror neurons are found in other taxa, and if they might actually play a role in sharing of intentions between individuals engaged in an on-going social interaction, such as play.

Mammalian social play is a useful behavioural phenotype on which to concentrate in order to learn more about evolution of fairness, and social morality. There is strong selection for playing fairly, because most, if not all, individuals benefit from adopting this behavioural strategy. Numerous mechanisms have evolved, to facilitate initiation and maintenance of social play in numerous mammals, to keep others engaged, so that agreeing to play fairly and resulting benefits of doing so can be readily achieved (Bekoff 1995a, 1995b).

Play may be a unique category of behaviour in that asymmetries are tolerated more than in other social contexts. Play cannot occur, if individuals choose not to engage in activity and equality needed, for play to continue makes it different from other forms of seemingly cooperative behaviour (Bekoff, 2001; Bekoff and Allen 1998).

9.11 COOPERATION, CONFLICT, AND EVOLUTION OF COMPLEX ANIMAL SOCIETIES

Genes cooperate in genomes; cells in tissues; individuals in societies. In animal society collective action emerges from cooperation among individuals. Societies common in insects, mammals, and birds, and even in amoebas vary from eusocial insect colonies with single reproductive female supported by millions of non-breeding workers, to cooperatively breeding groups (vertebrates) with one or more breeders and few non-breeding helpers. Many species form temporary associations like flamingo (*Phoenicopterus minor*) colonies and zebra (*Equus quagga*) herds. African elephants (*Loxodonta africana*), snapping shrimps (*Synalpheus brooksi*) and superb starlings (*Lamprotornis superbus*) form permanent social groups.

Group-living provides benefits to individual member. Most animals have one pair of eyes. Animals living in groups benefit from many pairs of eyes for vigilance or forage. Living in groups confers costs to member. They become conspicuous to predators and competition for food increase. Individuals weigh the cost-benefit ratio of living solitarily versus with others. When benefits of living together outweigh the costs of living alone, animal forms group (Alexander 1974). Benefits of group-living include assistance to deal with pathogens (grooming), mating scope, better conservation of heat, and reduced energetic costs of movements. Costs of group-living increases attack rates by predators, increases parasite burdens, misdirect parental care, and increases reproductive competition. Many species form short-term, unstable groups (herds of wildebeest, colonies of gulls), some join stable long-term, social groups where interaction among members is altruistic. When ground squirrel emits an alarm call to warn nearby coyote, it draws the coyote's attention and increases its own chances of being eaten (Sherman 1977). When a meerkat forgoes reproduction and feeds young of another group member, it reduces the number of offspring it could produce during its lifetime (Clutton-Brock et al., 2001). Robert Trivers considered a hypothetical group of animals in which one individual is faced with scope to take a small risk to provide a large benefit to another (Trivers 1971). Although choosing not to help is the best choice for an individual's fitness in short-term, it could mean that individual will not receive reciprocal help from others when it is needed. This provides incentive for altruistic behaviour in situations where individuals interact repeatedly, which occurs when animals live in stable groups. Baboons live in large groups and form coalitions where such cooperative interactions are frequent.

9.11.1 Cooperative Breeding in Vertebrate Societies

Few vertebrates are eusocial. Many species live in complex, cooperatively breeding societies in which more than 2 individuals care for young. Alexander Skutch (Skutch 1935) made the first observations of cooperative breeding in birds. This is reported in various other taxa including mammals and fish. Cooperative groups often contain a variable number of non-breeding auxiliaries, or helpers, that aid in raising the offspring of others. Vertebrate helpers reproduce throughout their lives. In most cooperatively breeding species, helpers tend to be related to breeders and realize indirect fitness benefits of raising relatives. Kin structure of vertebrate societies is similar to that of insect societies, even though group sizes are smaller in vertebrates. Cooperative breeding is common in birds, like Florida scrub jays and white-fronted beeeaters, and mammals including many carnivores and rodents. Although kin selection likely underlies the evolution of cooperatively breeding societies, it fails to explain why some species are the social, and other closely related species are not. Environment might influence whether its members live socially. To breed, animals need various resources including territories, food, and mates. Without such resources, the likelihood of reproducing is low and so remaining with one's family to help raise relatives may be a better option for maximizing fitness than trying to breed unsuccessfully (Emlen 1982). In many birds, mammals, and fish, younger individuals often delay in finding their own breeding territories for few years until better chance arises, instead remaining at home to gain valuable experience and parenting skills besides indirect benefits of raising relatives. Cooperative breeding may be a conservative strategy to maximize fitness when breeding conditions cannot be accurately predicted (Rubenstein and Lovette, 2007) in hostile and unpredictable climates and can be a 'best of a bad job' strategy to maximize fitness, when scope for independent breeding are limited or breeding conditions are uncertain. Cooperation and sociality are widespread in animals. Altruistic behaviours like raising offspring of others instead of trying to reproduce can largely be explained by shared genetic heritage between interacting individuals. Most complex animal societies are families in which group members are related and share high proportion of their genes. Cooperative and often complex collective action that arises from such family groups is a product of interaction of individuals seeking to maximize their own evolutionary fitness. Genetic structure influences evolution of animal sociality. Studies in wasps suggest that difference in expression of relatively small number of genes may be linked to large social differences in closely related species (Toth et al., 2007). Social behaviour consists of a set of interactions among individuals of same species. Wide range of sociality occurs in animals. Some animals rarely interact with one another, when it comes to issues of parental care. Relatively asocial animals are mosquitoes and polar bears. Highly social animals live in large groups, and cooperate to do many tasks. Social groups are packs of wolves and schools of fish. The highly social animals include all ants and termites, some bees and wasps, and few other organisms.

Many social behaviours are adaptive. Being social increases an animal's fitness—its lifetime reproductive success. Adaptive ness of social behaviour can be seen in aggregation against predators as caterpillars feeding together on leaf, a herd of wildebeest, schools of fish, and flocks of birds. A landscape filled with solitary wildebeest offer easy pickings for lions. If wildebeests gather into a single group, risk of any single individual being eaten is reduced. In circumstance of an attack by a predator, chance of one individual being targeted are 100% for solitary individual, 1% in group of 100, and 0.1% in a group consisting 1000. The social costs are paid by the Wildebeests from their groups. Pastures may not provide sufficient food for each individual in group. Costs of social aggregation are smaller than benefits of defense against predation. Group living involves a balance of conflict and cooperation mediated by costs and benefits associated with living socially. When benefits of living socially exceed costs and risks of social life, social cooperation is favoured.

A fundamental problem in founding an evolutionary ethics is to explain how cooperation and altruism can emerge during evolution. "Weak" altruism can be defined as behaviour that benefits more to another individual, than to the individual carrying out the behaviour. "Strong" altruism denotes behaviour that benefits others, but at one's own cost (Campbell, 1983). Both are common and necessary in those highly cooperative systems, which Campbell calls "ultra social". Ultra sociality refers to a collective organization, with full division of labour, including individuals who gather no food, but are fed by others, or who are prepared to sacrifice themselves for defense of others. In animal world, ultra social systems are found only in social insects, in naked mole rats, and in human society. Highly developed systems of cooperation and mutual support can be found in all human societies. Yet, we still do not have a satisfactory explanation of how such social systems have emerged. Therefore, we also cannot determine how they would, or should evolve in future. Evolution seems to predispose individuals to selfishness. Natural selection promotes "fittest" individuals, i.e. one's that maximally replicate. Individuals need resources to survive and reproduce. Finite resources imply competition among individuals. Altruism tends to diminish fitness of altruists, since more resources will be used by the helped one, and less will be left for the helper. Without further organization, altruism tends to be eliminated by natural selection, yet all ethical systems emphasize essential value of helping others. Everybody will agree that cooperation is in general advantageous for group of cooperators as a whole, even though, it may curb some individual's freedom. The cooperation can increase the fitness of co-operators, when cooperators together can collect more resources, than sum of resources collected by each of them individually. This is only possible in a situation that is not a zero sum game.

For example, a pack of wolves can kill large animals, which no individual wolf would be able to kill. Yet, for each wolf separately, the best strategy seems to consist in letting other wolves spend resources, while hunting, and then come to eat from their captures. But if all wolves would act like that, all advantages of cooperation would disappear. Optimal strategy for collective system of all wolves is to cooperate, but optimal strategy for each individual wolf as a subsystem is not to cooperate. In general, global optimization is different from sub-(system) optimization. This is the problem of sub-optimization. Since evolution tries to optimize first of all at subsystem level, we need additional mechanisms to explain global optimization at the level of cooperating system. Perhaps, the most fashionable approach to this problem is sociobiology (Wilson 1975). Socio-biology can be defined as an attempt to explain social behaviour of animals and humans, on the basis of genetical evolution. For example, a lot of sexual behaviours can be understood, through mechanisms of genetic selection reinforcing certain roles, or patterns. Several genetic scenarios have been proposed to explain evolution of cooperation. However, none of these scenarios is really satisfactory.

For several decades, our understanding of cooperative behaviour has been shaped by two theoretical constructs, kin selection and reciprocal altruism. In recent years, however, the explanatory power of these theories has been challenged as researchers have suggested that the individual benefits and market forces shape the deployment of cooperative behaviour in nature. For example, (Clutton–Brock 2002) argued that helpers may derive direct benefit form helping to rear young in many cooperatively breeding vertebrates and (Noe 2001) proposed that market forces may shape exchange of services in a wide range of species (Shine 2004) examined the patterns of intervention of adult female baboons in agonistic disputes that occur within their social groups. Coalitions are potentially costly to actors, who expend energy and risk injury when they intervene in ongoing disputes, and are potentially beneficial to recipients who obtain valuable support against opponents. In primate groups support is typically nepotistic, but is not completely limited to kin. Both reciprocal altruism and individual benefits have been invoked to explain patterns of coalition formation in primate groups. Coalitionary aggression is thought to be closely linked to the evolution of social organization in primates. When ecological conditions favour collective defence of resource, selection is expected to favour investment in social relations with those who are likely to provide coalitionary support. The primary feature of social organization in cercopithecine primate groups including female philopatry, linear dominance hierarchies, acquisition of maternal rank and well differentiated female relationships are expected to be linked to the existence of alliances between females. Baboons, however, may constitute an exception to the proposed pattern. Although savannah baboon females are philopatric from matrilineal dominance hierarchies and establish strong social bonds, there are several sites in South Africa where coalitions do not occur.

9.11.2 Kin Selection and Living in Families

Altruistic behaviour benefits individuals even when it is not reciprocated. This was first realized by William Hamilton's during the study of evolution of the altruistic behaviours in living organisms (Hamilton 1963). He reasoned that extent to which an individual is willing to help another is determined by how much those individuals are related. Hamilton predicts that a person should be more willing to run into a burning building to rescue their own cousin rather than then save an unrelated stranger as relatives share many genes. Evolutionary fitness is determined by an individual's genes that enter next generation's gene pool, regardless of whether those genes come from that particular individual or from identical copies in its relative. According to Hamilton's rule, the donor receives direct cost C (in terms of lost direct fitness) for cooperating; whereas recipient receives an extra benefit B (in terms of increased direct fitness). The donor receives part of benefit B that is discounted by genetic relatedness r (degree of shared genes) between 2 individuals. This discounted part of benefit is equivalent to donor's indirect fitness gain. Hamilton realized that they also enhance their own indirect fitness by helping relatives. His argument known as Hamilton's Rule calculates percentage of genes shared between 2 individuals, and then uses this value to determine costs and benefits of cooperative act. Hamilton's insight, termed kin selection by John Maynard Smith was appealing to evolutionary biologists as it appeared to explain not only why many individuals in complex societies would forgo breeding to help their relatives reproduce, but how those societies might evolve (Maynard Smith 1964). Kin selection theory is relevant to social insect societies as many insects are haplodiploid where an individual's sex is determined not by presence or absence of a sex chromosome, but by number of copies of genome in an individual's cells. Diploid fertilized eggs containing 2 copies—1 from each parent—become females . The unfertilized haploid egg with single copy becomes male. Kin selection theory suggests that sex-determined differences in relatedness among individuals provide incentives for young females to stay at home and help raise their sisters. Initial excitement over this explanation as primary reason for sociality was tempered by observations of non-social haplodiploid species and discovery of eusocial diploid species like termites. In singly mated haplodiploids, females are more closely related(r = 0.75) to their sisters than to their brothers (r = 0.25). Haplodiploid females are more related to their nieces (r = 0.375) than diploid females are to their nieces (r = 0.25). In many haplodiploid species (honey bee) queens are polyandrous and mate with multiple males, creating broods in which most workers are not full siblings and instead are half-sisters (r between 0.25 and 0.5). When queens mate with many males, workers are closely related to queen's sons (their brothers, r = 0.25) than to their half-sister's sons (their nephews, r between 0.125 and 0.25). This creates an incentive for workers to police other workers by destroying their eggs.

Vampire bats share food not only because of anticipation of reciprocation. They likely share blood meals with their relatives and are inclined to share blood meals with kin than with unrelated individuals. Relatedness among individuals can be calculated. More closely related bats are likely to share resources. Individuals are far likely to perform altruistic acts for siblings than for nephews and less likely for third cousins. Mechanism behind the effect of relatedness on altruism is kin selection. Kin selection reflects how copies of an individual's genes are passes on through survival and reproduction of their relatives. Natural selection advocates that an individual act to maximize its fitness. Kin selection assumes that individual act altruistically to maximize fitness of its relatives. An individual's direct fitness is measured by copies of his/her own genes passed onto children, grandchildren, and so on, whereas indirect fitness is the measure of copies of his/her genes passed on through her cousins, nieces, nephews, and siblings. Selection favours an altruistic act when benefit of act (in terms of indirect fitness) exceeds cost of act (in terms of direct fitness). When individuals are so closely related, they have more relatedness (R) and altruism is more likely to occur. Relatedness is measured through a scale from 0–1 as it reflects proportion of genes shared by 2 individuals. Zero shows no relation in individuals. The coefficient r measured in other pairs includes full siblings: 0.5; parent-offspring: 0.5; grandparent-grandchild: 0.25; cousins: 0.125.In the 1960s, W.D. Hamilton formulated a rule now known as **Hamilton's rule**, in which relatedness moderate the probability of occurrence of altruistic acts. Altruism is favored when benefits (B) of altruistic act to recipient, multiplied by relatedness (r) to actor, exceed costs (C) to actor; this is expressed mathematically as $rB > C$. Besides vampire bat, paper wasp, ground squirrel, and wild turkey follow Hamilton's rule.

The first cogent explanation for the evolution of behaviour like alarm calling was provided by Hamilton (1964), who recognized that any process that would cause callers to interact selectively with other callers could facilitate the evolution of alarm calling behaviour. More specifically he suggested that selective interaction among kins which are descended from a common ancestor could promote the evolution of traits like alarm calling. He called the process kin selection. If callers live in a group composed of kins then they enhance the fitness of others that are likely to carry copies of the same genes. Even though the caller reduces her own fitness, her call increases the fitness of her relatives which may be callers themselves, in this situation. Callers are more likely to set benefit from calling than non-callers. So calling alters the relative fitness of caller and non-callers. At the same time because kin aggregate tighter, non-callers are less likely to find themselves in groups with callers than expected by chance alone. The same logic can be applied to any form of behaviour that is costly to the actor and beneficial to the recipient.

Many insect larvae, especially caterpillars are soft-bodied creatures. They rely on having a bad taste or poison to deter predators and advertise this condition with bright warning colours. Caterpillars with such colours are known as **aposematic** and are in stark contrast to other caterpillars that match the colour of leaves they feed on. Camouflaged caterpillars are called **cryptic**. Noxious *Datana* caterpillars have bright red and yellow stripes. Of course, unless it is born with an innate avoidance of a given type of prey; a predator has to kill and attempt to eat caterpillars to learn to avoid similar individuals in future. It is of no personal use to unlucky caterpillar to be killed.

However, animals with warning colours often aggregate in kin groups, so death of one individual is most likely to benefit its relatives and its genes will be preserved. Altruism is also seen in lions. Lionesses tend to remain within pride, whereas males leave. Lionesses within a pride are related, on average by $r = 0.15$. Females all come into heat at same time, one individual probably influencing others' estrous cycles by pheromones. The result is simultaneous birth of cubs. Females exhibit apparently altruistic behaviour of suckling of other females' cubs. As females are related, kin-selection hypothesis accounts for this behaviour. For a precise measure of natural selection, one should calculate an individual's contribution to the gene pool, called its **inclusive fitness**. Such fitness is the contribution of that individual, plus 0.5 times its number of brothers and sisters, 0.125 times its number of cousins, and so on. (Stiling 2004).

9.11.3 Altruism

Social behaviour based on genetic variation results in differential rates of reproduction among lineages of closely related individuals. Altruistic or social behaviours enhance fitness of other individuals in population at apparent expense of individuals performing them. Thus, altruism can be defined as sacrifice of fitness by one individual for benefit of another in related individuals (such as parent, and offspring) and can be the start of an evolution towards cooperation (even in unrelated species) (Odum and Barrett 2005). All offspring have copies of their parent's genes. So parents take care of their young are in the process for caring for copies of their own genes. Genes for altruism towards one's young will therefore become more numerous because off-spring have copies of those same genes. In meiosis, any given gene has a 50% chance of entire an egg or sperm. Thus, each parent contributes 50% of its genes to its offsprings. Probability that a parent and offspring will share a copy of a particular gene is a quantity r, called the **coefficient of relatedness**. By similar reasoning, brothers or sisters are related by an amount $r = 0.5$, grand children to grand parents by 0.25 and cousins to each others by 0.125 (Stiling 2004). The term inclusive fitness is used to designate the total copies of genes passed on through all relatives, nieces, nephews, and cousins, as well as sons and daughters. Selection for behaviour that lowers an individual's own chance of reproduction, but raises that of a relative, is known as **kin selection**. Hamilton proposed the fomula $rB - C > O$, to describe the spread of a gene for altruism by kin selection when donor sacrifices C offspring, for which the recipient gains B offspring, where r is coefficient of relatedness of donor to recipient. Therefore, one's own life is equivalent in genetic value to two sisters, two brothers, or eight cousins. Species, which live in colonies, exhibit extreme social relationships, and are said to be "eusocial". They include social insects, such as termites, ants, wasps, and some bees, and one species of mammals, the naked mole rat of Eastern Africa. The organization of social insects is based on the roles of certain groups, within the colony. There is usually one reproductive female, the queen, who may lay thousands of eggs in her lifetime. Most of the other insects in the colony are involved in the construction, maintenance, and defense of the colony. Certain groups of insects may have specific physical features, which relate to their roles. The queen bee is usually much longer, and has an enlarged abdomen for egg-laying. The worker bees are equipped with stings to defend the colony, and pollen baskets for collecting pollen, while the drones are stingless, and do not have pollen baskets as their only role is to mate with the queen before they die. Much of the insect colony behaviour is determined, either by instinct or by pheromones, released by the queen (Figure 9.10).

One fascinating aspect of some animal societies is the selfless way, in which one animal seems to render its services to others. In the bee hive, for instance, workers labour unceasingly in the hive for three weeks, after they emerge, and then forage outside for food, until they wear out two, or three weeks later. Yet, the workers leave no offsprings. How could natural selection favour such self-sacrifice? This question presents itself in almost every social species. The apparent altruism is sometimes, actually, part of a mutual aid system, in which favours are given, because they will almost certainly be repaid. One chimpanzee will groom another for removing parasites from areas where the receiver could not reach because later the roles will be exchanged. Such a system, however, requires that animals be able to recognize one another as individuals, and hence, be able to reject those, who would accept favours without paying them back.

HAMILTON'S FORMULA

Altruist's Cost < (beneficiary's gain) × (relatedness)

C = cost to altruist

B = beneficiary's gain

r = relatedness of altruist and beneficiary

$$C < B \times r$$

$$r > C/B$$

relatedness greater than cost/benefit ratio

Altruism benefits a relative carrying the same genes. Hamilton's formula (in graphic above) indicates that inclusive fitness increases if the altruist's cost is less than the benefit to the beneficiary × the coefficient of relatedness between them. Coefficients for several types of relatives are shown below.

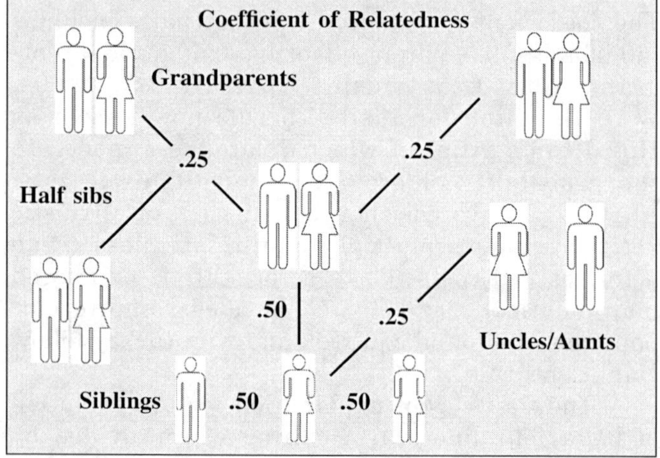

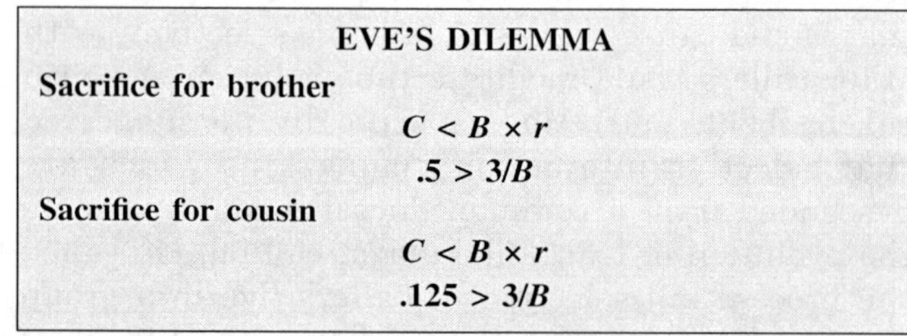

EVE'S DILEMMA

Sacrifice for brother

$$C < B \times r$$

$$.5 > 3/B$$

Sacrifice for cousin

$$C < B \times r$$

$$.125 > 3/B$$

This graphic illustrates several possible scenarios in which an altruist's sacrifice might lead to an increase in inclusive fitness.

FIGURE 9.10 Hamilton's formula.

An altruistic act is one that increases the welfare of another individual at the cost of an individual who performs the act as seen in ground squirrels, who warn other group members about predatory hawk. This brings the hawk's attention to individual giving warning call and benefits other individuals in squirrel's group. Altruistic acts are sharing nesting space and helping to raise offspring of an unrelated individual. Benefit of such behaviour is measured in its effect on animal's lifetime reproductive success. Natural selection operates against individuals who reduce their own fitness. Altruism, a powerful demonstration of natural selection at work decreases fitness of individual. Vampire bats returning from an unsuccessful foraging bout beg for food from successful individuals. It goes against interest of solicited bat to keep its own food, as it needs the nutrients to survive and reproduce, and giving up part of its meal is altruistic. Other members of this bat's own species are its competitors. In the early 1980s, Gerald Wilkinson demonstrated that vampire bats in Costa Rica shared blood with other bats sharing their roosts and bats did not share meals with all other bats equally. Some bats were more likely to interact with certain individuals than others. Bats are more likely to share blood with bats they are more likely to encounter in future. When there is more scope for reciprocation, bats are likely to share their meals. Bats do not share blood meals with other bats when there is little chance of return. Reciprocity enables existence of altruism as in long-term—benefits of altruism can exceeds costs of altruism. Relative cost of sharing food, when available, is less than potential future benefit of receiving food when hungry.

9.11.4 Various Examples of Altruism

Well-developed social behaviour is exhibited by ants, termites, bees, and wasps. Such species live in colonies, with thousands or millions of individuals. One benefit for these insects is that different individuals specialize in certain activities. Some are workers, who build colony and go out, looking for food that they bring back. Others are soldiers, who patrol the colony to protect it from possible attacks form other colonies. In many ant and bee colonies, all workers and soldier are females. Males are usually present in the colony, but do not contribute much. Queen's only job, for her entire life, is to lay eggs that workers care for. Substantial benefits exist in social group, along with some definite costs, to living closely with others of same species. First, one competes most with others that are most like oneself, and thus, a member of a social group always share or compete with others for resources. Second, due to numbers and close proximity of individuals in many social groups, disease spread through social groups rapidly.

Perhaps the most extreme sort of altruism is evolution of sterile caste in social insects where some females, known as workers, rarely reproduce, but instead help queen to raise offsprings, a phenomenon called **eusociality**. Differentiation of one species of social insect into different sized castes with huge soldier castes and tiny worker castes is noteworthy. Of the 263 living genera of ants known worldwide, 44 species possess caste systems. The peculiar system of castes primarily attributed to particular genetics of must social insect reproduction. The females develop from fertilized eggs and are diploid, which is a normal condition in most organisms. Males develop from unfertilized eggs and are haploid. Male gametes are formed without meiosis and every sperm is identical. Thus, each daughter receives an identical set of genes from her father. Half of a female's genes come from her diploid mother. Thus, total relatedness of sisters is: 0.5 (from the father) + 0.25 (from the mother) = 0.75. Such a genetic system is called **haplodiploidy**. Females are more related to their sisters than they would be to their own offsprings. Staying in nest or hive and producing new reproductive sisters are therefore advantageous. Queen is equally related to their sons and their daughters; $r = 0.5$ in each case. A queen should produce as many sons as daughters, to maximize her reproductive potential, a 50:50 sex ratio. If she did, then sterile worker females would spend as much time rearing brothers (to which they are related only by 0.25) as sisters. Average relatedness of a female to there siblings would then be 0.5, and the female would do equally well to breed on her own. From workers view point, it is far better to have more sisters and in this conflict with the queen they appear to have won because in any colony there are many more females than males.

Large social colonies exist in termites too, but theirs is not a haplodiploid genetic system, but a purely diploid one. Mole rats living under ground in South Africa have a division of labour based on castes, but animals are of course diploid (Jarvis et al., 1994). Only breeding female, queen suppresses reproduction in other females by producing a chemical in her urine that is passed around the colony, when its inhabitants groom after visits to a communal toilet. Other castes perform different work. "Frequent workers" do most of burrowing work. "Infrequent workers" consists of heavier individuals, some of them are workers; and even larger individuals the non-workers, rarely work at all. These are often male and may be a reproductive caste. A special dispersing morph, fat males are disinclined to mate with their own queen, but with a strong tendency to leave the home and mate with members of other colonies (O'Riain, Jarvis and Faulkes 1996). These males are equivalent to dispersing the queen of ant, bee and wasp colonies. Particular lifestyle of animals, not genetics that promotes eusociality is suggested (Alexander 1979). This is argued that in normal diploid organisms, females are related to their daughters by 0.5 and to their daughters by 0.5 and to their sisters by 0.5, so it matters little to them whether they rear sibs or daughters of their own. That mammals could exhibit a caste like society under the following conditions is also predicted:

1. When the individual of the species are confined in nests or burrows.
2. When food is abundant to support a higher number of individuals in one place.
3. When parental care is exhibited by the adults
4. When mother can manipulate other individuals.
5. When "heroism" is possible whereby individuals give up their lives and by so doing can save the queen.

These factors can account for eusociality in termites as well. In case of mole rat colonies, burrows become as hard as cement. A heroic effort by a mole rat blocking the burrow effectively stops a predator (commonly a raucousbeaked snake) because predators cannot rip open the surrounding substrate. Self-sacrifice by a "worker" becomes a genetic gain. Worker dies, but its genes are passed on by fellow nest mates many of whom share genes because they have the same mother, the queen. Queen manipulates the colony members; all workers and non-workers, whether male or female develop teats during their pregnancies—a testament to power of their pheromonal cues. Most abundant food in form of tubers of the plant *Pyrenacantha kaurabassana* weighing up to 50 kg provides food for a whole colony. Mole rats harvest tubers from below and often tubers remain only half eaten and can regenerate in time. Why eusociality has developed like burrowing rodent, in prairie dogs is open to speculation. That eusociality is promoted by haplodiploid mating systems and species, like termite and mole rate that defend fortress can be said. However not all species that could become eusocial have adopted such life system. Furthermore, even diploid termites and naked mole rates contain very closely related individuals because of in-breeding (Reeve et al., 1990) and this provides an additional reason that female might help queens within colonies (Stiling 2004).

Reciprocal altruism is widely practiced by social animals. This is the behaviour of one animal that helps another, with an expectation that help will be repaid so that benefits for all exceed the costs. Over a reasonable period of time there will be a net gain for everyone involved. The only problem is temptation to cheat by receiving help without returning it. In a stable group, breaking the code of behaviour will mean that sooner or later cheat will be denied help. Key to success of reciprocal altruism, as a tool for sustaining social cooperation is ability to take action against those, who fail to reciprocate. For spider monkeys and chimpanzees, strong emotional ties develop between females and their offspring, so that closely knit matrilines stabilize the framework, holding the community together. Females spend their time being pregnant, lactating, or caring for their offsprings. Males commit less time to social activity; they mate, groom, and roam. Male mobility provides a mechanism for interbreeding with different populations of the same species. Such mixing of the genes is a healthy feature from the viewpoint of evolution.

As in human societies, family relationships are important in monkeys and apes. When fights occur, animals tend to help their closest relatives. If an ape fails to join in a fight involving a relative, an obvious explanation is usually apparent. Apes seem to assess the chances of winning, and when odds are bad enough, they avoid fray. Choice of discretion over valor is, in this circumstance, likely to conserve genes for circumspection, and perhaps favour selection of wise rather than foolish siblings. Size of communities of monkeys and apes depends on both genetic and environmental factors of the species and territory. Distribution of food, water, and predators are all relevant. Where resources are plentiful and safe, groups often fractionate, but when threatened by deprivation, or danger, groups coalesce. This obviously mirrors human behaviour. Some species display considerable flexibility in social organization, breaking up into small, more efficient subunits for daily activities. The Hamadryad baboons congregate in large troops to sleep at night, and each morning they divide into small groups, that set out for the foraging grounds. Further dispersal then takes place into yet smaller units, consisting of one male with one or two females and their youngs. These compact teams collect food, until early afternoon, when they meet together at a watering hole, before returning to sleep with the combined troop.

Social interactions can be subtle and complex in communities of monkeys and apes. The most highly developed behaviour appears, when animals are dealing with each other individually, but their combined behaviour is also impressive. Communities hold on to their territory, and adopt firm policies on the extent to which their boundaries will be protected. There is naturally a cost to be paid for territoriality, and decisions seem to be determined by weighing the risks of warfare, against the benefits of the resource-supplies of food and water. Defense may involve aggressive vocalization, hostile displays, and direct physical fighting. Apes and monkeys also threaten by banging anything that will make a noise, and by displaying their erect penises. Eibl-Eibesfeldt describes how sentinel vervet monkeys develop an erection, if an unknown member of the same species approaches in a hostile way: "These 'guards', presenting their genitals, play part of living frontier posts. Apart from their genitals, they also make a display of a threatening face. In its origin, this behaviour can very likely be explained as a ritualized threat to mount".

A second kind of altruism is exemplified by the behaviour of male sage grouse, which congregate into displaying groups-leks. Females come to these assemblies to mate, but only a handful of males in the central spots actually sire the next generation. The dozens of other males advertise their virtues vigorously, but succeed only in attracting additional females, to the favoured few in the centre. Natural selection has not gone wrong here, however, males move further inward every year through this celibate and demanding apprenticeship, until they reach the centre of the lek. Altruism of honey bees has an entirely genetic explanation. Through a quirk of hymenopteran genetics, males have only one set of chromosome. Animals normally have two sets, passing on only one, when they mate. Hence, they share half their genes with any offspring, and the offsprings have hair their genes in common with one another. Because male hymenoptera have a single set of chromosome, however, all the daughters have those genes in common. Added to the genes, they happen to the share that came from their mother, the queen. Most workers are three quarters related to one another, more related than they would be to their own offspring. Genes that favours a selfless sterility assists in rearing the next generation of sisters, then should faster in the population than those programming the more conventional 'every female for herself' strategy.

This system, known as kin selection is widespread. All it requires is that an animal performs service of little cost to itself, but of great benefit to relation. Bees are the ultimate example of altruism, because of the extra genetic benefit that their system confers, but kin selection works almost as well in a variety of genetically conventional animals. The male lions that cooperate in taking over another male's pride, for example, are usually brothers, whereas the females in a pride, that hunt as a group and share food, are a complex collection of sisters, daughters, and aunts. Even human societies may not be immune to the programming of kin selection. The anthropologists consistently report that simple culture is organized along lines of kinship. Such observation combined with the recent discovery that human language learning is in part a kind of imprinting that consonants are innately recognized sign stimuli, for instance suggest that human behaviour may be connected more with animals behaviour, than was hitherto imagined. Both sun and sky bees automatically switch to a third navigational system, based on their mental map of the landmarks in their home range. Interaction within groups, exploiting a common resource, may be prone to cheating by selfish actions, that result in disadvantages for all members of the group, including the selfish individuals. Kin selection is one mechanism by which such dilemmas can be resolved. This is because selfish acts towards relatives include the cost of lowering indirect fitness benefits that could otherwise be achieved, through the propagation of shared genes. Kin selection theory has been proved to be of general importance for the origin of cooperative behaviours, but other driving forces, such as direct fitness benefits can also promote helping behaviour in many cooperatively breeding taxa. Investigating transitional systems is, therefore, particularly suitable, for understanding the influence of kin selection on the initial spread of cooperative behaviours. The role of kinship in cooperative feeding has been recorded using a cross-fostering design to control for genetic relatedness and group membership. Study animal was the periodic social spider *Stegodyphus lineatus*, a transitional species that belongs to a genus containing both permanent social and periodic social species. In *S. lineatus*, the young cooperate in prey capture and feed communally, providing clear experimental evidence for net benefits of cooperating with kin. Genetic relatedness within groups, and not association with familiar individuals, directly improved feeding efficiency and growth rates, demonstrating a positive effect of kin cooperation. Hence, in communally feeding spiders, nepotism favours group retention, and reduces the conflict between selfish interests and interests of the group. Social cooperative breeding behaviour is rare in spiders, and generally characterized by inbreeding, skewed sex ratios and high rates of colony turnover processes that, when combined may reduce genetic variation, and lower individual fitness quickly. On these grounds, social spider species have been suggested to be unstable in the evolutionary time, and hence, sociality is a rare phenomenon in spiders. Social spiders are unusual among cooperatively breeding animals in being highly inbred. In contrast, most other social organisms are outbred owing to inbreeding avoidance mechanisms. Social spiders appear to originate from solitary sub social ancestors, implying a transition from outbreeding to in-breeding mating systems. Such a transition may be constrained by inbreeding avoidance tactics, or fitness loss due to inbreeding depression.

Evolution of cooperation requires benefits of group living to exceed costs. Some components of fitness are expected to increase with increasing group size, whereas others may decrease because of competition among group members. Social spiders provide an excellent system to investigate costs and benefits of group living. They occur in groups of various sizes, and individuals are relatively short-lived, therefore, life history traits and Lifetime Reproductive Success (LRS) can be estimated as a function of group size. Sociality in spiders has originated repeatedly in phylogenetically distant families, and appears to be accompanied by a transition to a system of continuous intra colony mating and extreme inbreeding. Benefits of group living in such systems should, therefore, be substantial.

9.11.5 Altruism between Unrelated Individuals

Altruistic acts between unrelated individuals are often called **reciprocal altruism**. Unrelated individuals may occupy same territories as breeding individuals, and help parents to raise offsprings by foraging for additional food for the youngs as found in mongooses (Rood 1990). By comparing nesting or brooding success of breeding pairs with helpers, with those whose helpers were removed, has shown that helpers significantly increase parents' fitness. Motive in most situations, when habitat is usually saturated with breeders, and helper probably could not obtain a territory for themselves, seems to be a sort of reciprocal altruism. What they do is the helpers increase the size of territory of a breeding pair; later, they are able to carve off a fragment of that territory for themselves. Alternatively, they take over from one member of the breeding pair after that member dies.

9.12 COALITION

Primates female may sometimes benefit directly when they participate in coalitions (Chapais 2001). This may influence coaltionary activity in two ways. First, when dominance hierarchies regulate access to resource and influence reproductive success as high ranking females do. If the benefits of group membership are distributed unequally and low ranking individuals can leave the group, then high ranking individuals may need to provide incentives, such as coalitionary support for low ranking female to stay (Altman and Alberts 2003). Baboon and

FACTS OF HEPING BEHAVIOURS
Are They Selfish?

FATIK B. MANDAL

Helping behaviours like altruism and kin selection are very interesting. Altruism means sacrifice of fitness by one individual for benefit of other individuals. Altruists and recipients carry copies of the altruistic gene. An altruistic behaviour spread by kin selection reduces the organism's personal fitness, but increases the organism's inclusive fitness. Altruistic acts between unrelated individuals are called reciprocal altruism.

Helping behaviours, including altruism and kin selection, are very interesting from evolutionary point of view. Altruism is the sacrifice of fitness by one individual for benefit of other related individuals. Kinship ensures that altruists and recipients carry copies of the altruistic gene. An altruistic behaviour spread by kin selection reduces the organism's personal fitness, but increases the organism's inclusive fitness. Altruistic acts between unrelated individuals are called reciprocal altruism. When an organism provides benefits to another organism at a cost to itself, the behaviour is altruistic. Altruism is common in society with complex social organizations. This is a social behaviour that enhances fitness of other individuals in the population at the expense of the individual performing them. All offspring bear their parents' genes. Parents take care of their young to care for copies of their own genes. Each parent contributes 50% of its genes to offspring. Sharing a particular gene between parent and offspring can be expressed by r, the coefficient of relatedness. In the case of brothers or sisters, $r=0.5$; in grandchildren and grandparents, $r=0.25$, and in cousins to each other, $r=0.125$. The higher the value of r, the greater the probability that the recipient of altruistic behaviour will also possess the gene for altruism.

Inclusive fitness indicates the total copies of genes passed on through all relatives, nieces, nephews, cousins, sons, and daughters. For a precise measure of natural selection, calculating an individual's contribution to gene pool is required, which is called its inclusive fitness. Such fitness is contribution of the individual, plus 0.5 times its number of brothers and sisters, 0.125 times its number of cousins, and so on. Selection of behaviour that lowers an individual's own chance of reproduction, but raises the same in a relative is known as kin selection. Hamilton proposed the fomula $rB - C > 0$, to describe the spread of a gene for altruism by kin selection, when donor sacrifices C offspring, for which the recipient gains B offspring, and r is coefficient of relatedness between donor and recipient. Therefore, one's own life is equivalent in genetic value to 2 sisters, 2 brothers, or 8 cousins.

Lionesses within a pride are related on average by $r = 0.15$. Females all come into heat at the same time, probably due to pheromone's influence, which results in simultaneous birth of cubs. Females exhibit altruistic behaviour of suckling other females' cubs. As females are related, the kin selection hypothesis accounts for this behaviour. Kin selection predicts that animals are likely to behave altruistically towards their relatives, than towards unrelated members, and the degree of altruism will be greater in case of closer relationship. In birds, 'helper' birds are likely to help relatives to raise their young, than to help unrelated breeding pairs. Japanese macaques show altruistic actions like defending others from attack, which is directed towards close kin. Haplodiploidy means that females on average share more genes with their sisters than with their own offspring, as found in insects. So, a female may get more genes in the next generation, by helping the queen to reproduce, and hence increasing the number of sisters she will have, rather than by having offspring of her own. Kin selection provides an explanation of evolution of sterility in social insects, and this is a triumph of the 'gene's-eye view of evolution', which views evolution as the result of competition among genes for increased representation in gene pool. Individual organisms are mere vehicles of gene propagation.

Altruism seems anomalous from the individual organism's point of view, but from gene's point of view it makes sense. A gene wants to maximize the number of copies of itself. One way of doing that is to cause its bearer to behave altruistically towards other bearers of the gene, so long as the costs and benefits are satisfied. An altruistic behaviour spread by kin selection reduces the organism's personal fitness, but increases organism's inclusive fitness. Such fitness is defined as personal fitness, plus the sum of its weighted effects on fitness of every member of the population, the weights determined by the coefficient of relationship r. It is clear that natural selection acts to maximize the inclusive fitness of individuals in population. Instead of thinking in terms of selfish genes, thinking in terms of organisms may be applied.

Kin selection does not require that animals must have the ability to discriminate relatives from non-relatives. Many animals can in fact recognize their kin, often by smell, but kin selection can operate in absence of such ability. That kin selection is committed to 'genetic determinism', the idea that genes rigidly determine behaviour is wrong. So long as the behaviours in question have a genetical component, then the theories can apply. Kin selection also recognizes that all traits are affected by both genes and environment, and transmission of behaviours through non-genetical means, such as imitation and social learning, is possible. Kinship ensures that altruists and recipients carry copies of the altruistic gene. If altruism is to evolve, the recipients of altruistic actions have a greater than average probability of being altruists themselves. Kin-directed altruism is the way of satisfying this condition along with other possibilities too. Kinship remains as the most important source of associations between altruists and recipients.

Altruism between Unrelated Individuals

Altruistic acts between unrelated individuals are called reciprocal altruism. Reciprocal altruism explains cases of altruism among unrelated organisms. In such altruism, the cost of helping is offset by the likelihood of the expected benefit and it operates when individuals interact with each other more than once, and have recognizing ability of individuals with whom they have interacted. Example of reciprocal altruism is provided by blood-sharing in vampire bats. It is common for a vampire bat to fail to feed on a given night. Under such conditions, the starved bat is being fed by another bat through regurgitation, with expectation of such return from the beneficiary in future. In tropical coral reefs, various small fish species act as 'cleaners' and remove parasites from mouth and gill of large fish. In such interaction, the large fish gets cleaned, and the cleaner gets fed. However, when a large fish with cleaner in its mouth is attacked by a predator, it gives time to the cleaner to leave before fleeing the predator, rather than swallowing the cleaner. Since the large fish often returns to the same cleaner many times, it looks after the cleaner's welfare.

Unrelated individuals, occupying the same territories as breeding individuals, help parents to raise offspring by foraging for young, as found in mongooses. By comparing nesting or brooding success of breeding pairs with helpers, with those whose helpers were removed, it has been proved that the helper increases the parents' fitness. When habitat is usually saturated with breeders, and the helper probably could not obtain a territory for them, then such type of altruism occurs. What they do is, they help to increase the size of territory of a breeding pair; later, they use a fragment of that territory for themselves.

Altruism in Social Insects

Perhaps the extreme sort of altruism is the evolution of sterile caste in social insects, where some females, known as workers, rarely reproduce, but help queen to raise offspring, a phenomenon called eusociality. Peculiar caste system is primarily attributed to genetics of insect reproduction. Females developed from fertilized eggs are diploid, which is a normal condition in most organisms. Males developed from unfertilized eggs are haploid. Male gametes formed without meiosis are identical. Each daughter receives an identical set of genes from her father. Half of a female's genes come from her diploid mother. Thus, total relatedness of sisters is 0.5 (from the father) + 0.25 (from the mother) = 0.75. Such a genetic system is called haplodiploidy. Females are more related to their sisters than to their own offspring. Queens are equally related to their sons and their daughters; $r = 0.5$ in each case. A queen should produce as many sons as daughters to maximize her reproductive potential, a 50:50 sex ratio. If she did it, then sterile worker females would spend as much time in rearing brothers as sisters. Average relatedness of a female to there siblings would then be 0.5, and female would do equally well to breed on her own. For workers, it is better to have more sisters, and in this conflict with queen, they appear to have won, because in any colony, there are more females than males. Self-sacrifice by a worker becomes a genetic gain. The worker dies, but its genes are passed on by fellow nest match, many of whom share genes, because they have the same mother, the queen.

Altruism cannot evolve, if selection acts at the individual level as altruism has disadvantages to the individual. Altruism is perhaps advantageous to the group. In a group with some altruists, altruists sacrifice for group's interest, which has a survival advantage, than a group composed mainly of selfish individuals. Group selection allows altruism to evolve, in which the fitness of the group is increased. Groups composed mainly of selfish individuals leave behind only the altruists. The presence of a high proportion of alarm-calling monkeys has a survival advantage than a group with a lower proportion. Thus, the alarm-calling monkeys may evolve through group selection, although selection favours selfish monkeys.

Darwin argued that self-sacrifice might be beneficial at the group level. He suggested the evolution of altruism through group selection. According to Williams and Maynard, group selection is an inherently weak evolutionary force and is unlikely to promote altruism. Weakness of group selection as an explanation of altruism is 'subversion from within'. Altruists are exploited by selfish 'free-riders', who have a fitness advantage. In a group composed exclusively of altruists, a single selfish mutant can cause an end to this condition. The selfish mutant out-reproduces the altruists, hence selfishness eventually swamps altruism. Since the generation time of the individual is likely to be shorter than a group, a selfish mutant chance of spread is very high. From a Darwinian viewpoint, the existence of altruism in nature is at first sight puzzling. Natural selection leads us to expect that animals behave to increase their own chances of survival, and reproduction, not those of others. But by acting altruistically, an animal reduces its own fitness.

macaque group regularly split when they grow large (Dunbar 1987). So departures of sub-ordinates presents a real threat to dominants. Consequently, we would expect high ranking female to form coalitions more often, than lower ranking female. Furthermore, we would expect most support to be directed towards subordinates.

Second, animals may sometimes derive benefits form participating in coalitions. Female may use coalitions to preserve their own position in a dominance hierarchy by preventing instability in dominance relationship between their subordinates. Consequently, we would expect female to support dominants against subordinates, when they intervene in disputes between female rankings lower than themselves. We term this conservative support. Because high ranking females have a greater stake in preserving the status quo than low ranking females do, we would expect high ranking female to intervene more conservatively, than low ranking female.

Animals often evaluate these asymmetries by assessing communication signals that convey information about an opponent fighting ability. In anuran amphibians, difference in body size influences the outcomes of male—male contest. Large males usually defeat smaller males (Wagner 1989). The fundamental frequency of anuran acoustic signals is partly determined by the shape and mass of the laryngeal apparatus which is in turn related to body size (Duellman and Trueb 1986). Spectral call properties and body size are negatively correlated in many species.

Altruism poses a problem for evolutionary biologists because natural selection is not expected to favour behaviour that is beneficial to recipients, but costly to actors. The theory of kin selection first articulated by Hamilton (1964), provides a solutions to the problem. Hamilton's well known rule ($br > c$) provides a simple algorithm for the evolution of altruism via kin selection. Because kin selection recognition is a crucial requirement of kin selection, it is important to known whether and how primates can recognize their relatives.

9.13 MOBBING BEHAVIOUR

Mobbing behaviour frequently seen in birds, though it is also known to occur in other social animals, when a species turns the tables on their predator by cooperatively attacking or harassing it. For example, nesting gull colonies are widely seen to attack intruders including humans. Costs of such behaviours include the risk of engaging with predators, as well as energy expended in the process. Mockingbirds can effectively force a cat or dog to seek something less troublesome. One mockingbird might fly in front of the cat or dog, enticing it to lunge, while another pecks at the cat or dog from behind. While mobbing has evolved independently in many species, it only tends to be present in those, whose youngs are frequently preyed upon. Mobbing calls may be made prior to or during engagement in harassment.

The mobbing behaviour has functions beyond driving the predator away. It draws attention to predator, making stealth attacks impossible. It also plays a critical role in identifications of predators and inter-generational learning about predator identification. Reintroduction of species is often unsuccessful as the established population lacks this cultural knowledge of how to identify local predators. Mobbing can be an interspecies activity. Crows are frequently mobbed by smaller songbirds, as they prey on eggs and youngs from their nests, but these same crows will cooperate with smaller birds to drive away hawks or larger mammalian predators. On occasions, birds will mob animals that pose no threat.

Black-headed gulls aggressively engages intruding predators, such as carrion crows. Experiments on this species by Kruuk involved placing hen's eggs at intervals from a nesting colony, and recording the percentage of successful predation events as well as the probability of the crow being subjected to mobbing. Results showed decreasing mobbing with increased distance from the nest, which was correlated with increased predation success. Mobbing may function by reducing the predator's ability to locate nests, as predators cannot focus on locating eggs while they are under direct attack.

Effect of mobbing on the behaviour of an owl

When a large predator is seen, by small birds, they often mob it. When birds mob a predator, they gather near it, call loudly, fly close to it, and occasionally even strike it. Mobbing is sometimes carried out by individuals of one species, but frequently individuals of two, or more species are involved. Although, mobbing is risky, small birds benefit because they distract the predator, cause it to lose the element of surprise in an attack, or drive it from the area. If predator is a night hunting bird, like owls, birds may force owl to go, and look for a roost elsewhere. A study was carried out in Australia in 1989–1990 into the effects of small birds mobbing on owl *(Ninox stremia).* In one study, seven species of birds mobbed the owls. Two species, which did most of the mobbing, were noisy miner and spangled drongo. Diet of noisy miner is mostly nectar and some insects, that of spangled drongo mostly insects and some nectar. Noisy miners live in large family groups; spangled drongo's are more solitary birds.

9.14 SOCIAL LEARNING

Systematic differences have often been observed in the behaviours of populations in the absence of genetic or environmental differences between populations (Whiten et al., 1999). Chimpanzees use twigs or blades of grass in Gombe National Park to feed on termites and ants. These implements probe the passageways of insect mounds. When the residents attack an intruding probe, the chimpanzees extract the probe and eat any termites that are clinging to it. Variation exists among chimpanzee populations in the species of insects preyed upon and the materials used as probes. It also varies in how probes are prepared for fishing and the process of insect capture. Chimpanzees living at Gombe hold a long stick in one hand while using the other hand to wipe a ball of ants into their mouths. Chimpanzees at Tai Forest use a short stick to collect ants. They place it in their mouths, removing the ants with their lips and tongues. Chimpanzees in Senegal peel bark from twigs before using them as probes. The chimpanzees at Gombe do not peel the twigs before using them. Some specific behaviours of chimpanzees are traditions learned by one individual because of observing the behaviour of another individual (McGrew, 1992; Whiten et al.,1999). Some animals learn complex behaviours by imitating others of their species. Imitation, both as copying of a novel or otherwise improbable acts (Thorpe, 1956; Byrne and Russon, 1998), is a special type of SL. Learning is the process of doing an act after seeing it done (Thorndike, 1898). The former are the songs that adult male songbirds produce both to repel conspecific males and to attract conspecific females. Males from varied populations of some species create variants of a basic, species-typical song. Males create song dialects. The adults of the species sing that dialect (Marler, 1970). By imitation, birds learn song dialects. Although birds learn their songs by imitation, it has been concluded that learning depends on some non-imitative SL processes. Rats living in pine forests survive by stripping the scales from pine cones. They eat the scales that protect pine seeds (Aisner and Terkel, 1992). The rats first remove the scales from the base of a cone. Rats do not feed on pine cones, but they do gnaw through individual scales to access seeds. Interaction with cones started the development of behaviour in young black rats, making them efficient strippers of seeds from pine cones. Young rats snatch open cones from adults, and by interacting with the cones, they learn how to finish the job. The study of SL in Israeli rats has yielded two important findings. First, complex motor patterns transmit from one generation of animals to another. Second, the existence of a complex tradition in a population cannot be used to infer that a complex SL process, like imitation or teaching, is linked with its transmission

Classification of Social Learning

Such social biasing of learning by other individuals has been termed "local enhancement', defined as clear imitation resulting from directing the animal's attention to a particular object or to a particular part of the environment' (Thorpe, 1956). Local enhancement can be contrasted with imitation, defined as 'copying of a novel or otherwise improbable act' (Thorpe, 1956; Byrne and Russon, 1998) or 'learning to do an act through seeing it done' (Thorndike, 1898). In local enhancement, an animal learns only that it should interact with one part of the environment rather than with another. In true imitation, an animal learns about the behaviour in which it should engage. The distinction is fundamental to all academic discussions of SL in animals. Are two terms—local enhancement and imitation—enough to enable discussion of all animal SL? Humans and animals can learn from each other. One chimpanzee may observe another of its species using a rake to pull in food items that would otherwise be out of reach. Observing one chimpanzee, one can learn to use the rake to pull in food faster. Thus, it would if it had never seen a scope to observe another chimpanzee use a rake (Tomasello et al., 1987). Observer chimpanzees failed to imitate in the sense of copying the actions used by the demonstrator to get food. Observers developed their own techniques for using the rake. Observers seemed to learn that a rake was useful for acquiring food but did not learn much about the actual behaviour that a model uses when raking, and, as noted above, what is meant by imitation is "learning to do an act'. Tomasello et al., (1987) proposed that the observers were not so much copying the model's behaviour as attempting to create the results of the model's efforts, a process that they termed 'emulation."

Consider teaching, an activity that contributes to SL and is common in our own species. In local enhancement, imitation, or emulation, the model is passive. The observer extracts information from a model engaged in activities it performs without reference to the observer. In teaching, according to the most current definition (Caro and Hauser, 1992), a model modifies its behaviour in the presence of a pupil, often leading to some reduction in the efficiency of the model's performance. The model either encourages or punishes the pupil or provides the pupil with examples of behaviour or experiences so that the pupil acquires more information or skill than it otherwise would. The elements of this complex definition, distinguishing teaching from other activities that play a part in SL, involve costly modifications of the teacher's normal behaviour, resulting in accelerated learning by pupils. There are few plausible examples of teaching in the animal world, and even those few instances are debated. Feline mothers may meet the definition by delaying their killing and eating of prey and providing their young with incapacitated, live prey on which to practice predatory behaviour (Galef, 2003). There is contradictory evidence on whether experience with incapacitated prey facilitates the development of hunting behaviour in young cats. Killer whales teach their young to hunt seals, although this view is not generally accepted (Rendell and Whitehead, 2001). Chimpanzees may teach juveniles how to crack open nuts using a stone hammer and anvil (Boesch, 1991). It is worth quoting verbatim. Boesch's (1991) descriptions of one such instance provides a fine sign of both the strengths and the weaknesses of unusual field observations. According to Boesch, in this example, the mother corrected an error in her daughter's behaviour, and Nina understood this since she continued to maintain the grip demonstrated to her (Boesch, 1991). Some authors accept Boesch's interpretation of his observations, while others do not. Such complex interactions between a clear pupil and teacher have been seen only twice in many years of field study, making them difficult to interpret with certainty. The issue of whether animals teach is only one of many questions that are being debated by researchers who are interested in SL in animals.

Imitation in Animals

The discovery over decades that many different types of SL can bias behavioural development has made it difficult to determine whether any given case of SL is a product of true imitation or of some other less cognitively demanding process. Why should anyone care enough about the types of SL that animals use when using one another as sources of information about the environment to argue about it? Many believe that understanding similarities and differences in SL processes in humans and other animals will provide insight into similarities and differences in their mental processes. Indeed, the first laboratory investigations of SL in animals were undertaken to determine whether, like humans, non-human animals had access to mental representations that they could manipulate to solve problems (Thorndike, 1898). To imitate the behaviour of a model, an imitator must store a visual image of the model's behaviour and then match its motor output to that stored representation. the learning by birds of song dialect because of hearing the song of adults belonging to their social group. Such learning might be described as 'learning to sing a song from hearing it sung'. This is somewhat different from 'learning to do an act from seeing it done', which some hold to be the definition of imitation. Why should a distinction be made between seeing and hearing when defining imitation?

Learning to sing a song by listening may be simpler than learning to do an act by seeing it. To learn to sing like another, the singer needs to match the sounds it produces when singing with a stored representation of the sound in the form of a song. Thus, song imitation occurs within a single modality, audition. But to learn to do an act by seeing it done, a match must be made across modalities between a stored representation of a visual stimulus and kinesthetic feedback from a motor act. Such cross-modality matching is necessary. If an observer imitates an act performed by a model, the visual input to the observer while imitating is different from the visual input that the observer received while observing the model perform the act. When I see someone bow, I see it is different from what I see when I bow myself. When I whistle a song, what I hear is like what I heard when someone else whistled the same song (Heyes, 2001). So, when we discuss imitation in animals, should we be limited to instances where overt motor patterns are copied, or should we include bird-song learning? Most authors think that we should not include bird-song learning. For a behaviour to be considered a result of imitation, the imitated behaviour must be new to the imitator. If so, how does one know whether a motor pattern is new? How can you tell if an individual has learned to perform an act by seeing it done if it has been performed before? Regarding the second, an individual may never have used a rake to pull in food before, but it has grasped objects, used objects to move other objects about, sought food, etc. A novel act may be nothing more than a combination of acts that are already in an individual's repertoire. All a subject does when it imitates a familiar act is use the behaviour of another as a cue for which elements of its own behavioural repertoire it should try in the situation that it now faces. If one is interested in imitation as a tool to explore the cognitive capacities of animals, then cases in which observers use others' actions to cue their own behaviour are not very informative.

Importance of Social learning in Animals

SL is central to the development of adaptive behavioural repertoires in a variety of animals (Avital and Jablonka, 2000; Dugatkin, 2000). SL and natural selection are co-equals in producing adaptive behaviour. SL of one type or another is necessary for animals to maximize everything from the selection of a mate to the avoidance of potential predators. Although SL plays some role in the mate choices of guppies and quail and in the avoidance of predators by blackbirds and monkeys, claims of a major role for SL in the evolution of behaviour are recent.

The Roots of Culture

An intense area of debate concerns the degree of similarity between the traditions of animals and the cultures of humans. It is foolish to deny culture to chimpanzees (McGrew, 1992). Those in the opposing camp focus on apparent differences between chimpanzee and human SL. They argue that differences in the behavioural repertoires of various groups of chimpanzees tell us nothing about the processes causing the development of those differences. If one is interested in discovering true precursors of human culture, he or she should look for behaviours that are transmitted from one generation to the next by the same processes that support human cultures (Galef, 1992). Humans can teach one another. They do learn by imitation. Animals do not teach. They rarely learn by imitation. Most cases of human culture and animal tradition depend on different behavioural processes. The type of culture that animals do not teach or imitate could produce would be restricted. In human cultures, techniques or behaviours often become complex. Youngsters learn from elders. The young then spend a lifetime improving on what they have learned socially. then use those improvements as the starting point for the next generation. Such 'ratcheting' occurs only when learning involves social transmission and not when social transmission involves changes in attention to environmental stimuli. A local enhancer only increases attention to some aspects of the environment. A model for emulation indicates that a tool can be used to achieve a goal. In each generation, naive individuals whose learning was shaped by local enhancement or emulation learn for themselves how to behave in the part of the environment to which their attention is directed. And no ratcheting across generations can occur. The majority of those who study SL processes in animals in the laboratory believe that the differences between human culture and animal tradition are so great that they need different terms to describe them. They often want to restrict the use of the term 'culture' to humans and refer to 'animal traditions' when discussing population-specific behaviours of nonhuman animals.

9.15 RITUALIZATION

Many signals get their start as part of another behaviour. They take on a signalling function. Evolution favours modification of the incipient signal. Thus, it becomes stereotyped and facilitates communication. Julian Huxley (1923) called this ritualization. Let us illustrate a few classic examples. Three sources of raw material for signals are intention movements, displacement activities, and autonomic responses which are discussed below.

Intention Movements

Animals begin to behave in patterns with movements that prepare them to act. From these acts, one can estimate the intention of animals. They have been named intention movements (Heinroth, 1910). Before biting wolves pull back their lips and bare their teeth to improve their fitness. the bared teeth of an aggressive rival feel jaws clamp down. Avian displays originate with intentional movements for flight (Daanje, 1950). It is a hard task to judge how a bird could ever take off from this ritualized pose. During the ritualization process, the movement changed.

Displacement Activities

In these activities animals exhibit irrelevant actions with conflicting indecisive motivations in the face of an aggressor, conflicting motivations to fight or flee and may instead groom themselves. Displacement activities are incomplete actions. Courtship shows conflicting tendencies. Sex partners must come together for mating despite the aggressive tendencies that o keep them apart. The mock preening of the male courtship displays in many ducks originates in displacement preening (Lorenz, 1972). Whether a display is a result of displacement or whether it has a function that is unclear. Duck courtship includes other vigorous displays, such as rearing up and splashing down into the water, which ruffle a drake's feathers. Preening during courtship may not be a display at all but rather functional.

Autonomic Responses

Autonomic functions give birth to many displays (Morris, 1956). A change in the distribution of blood throughout the body occurs during stress or conflict. In turkey, jungle fowl the naked head may flush. fleshy appendages due to vasodilation may swell due. These are part of signaling in courtship. Visual signals also evolve from the respiratory system. Examples include the inflation displays of birds, in which the males fill pouches on their bodies with air to attract mates. A male frigate bird has an inconspicuous pouch on its throat when deflated. but when inflated, its enormous size and brilliant scarlet colour attract passing females. The erection of feathers and hair has a thermoregulatory function. Fluffy feathers or hair trap heat. Piloerection is a part of aggressive and appeasement displays (Smith, 1977). A zebra finch cannot escape when a dominant individual fluffs its feathers as an appeasement signal (Morris, 1954).

Materials for Displays

E.O. Wilson (1980) points out that "ritualization is a pervasive, opportunistic evolutionary process. This can be launched from any behaviour pattern, anatomical structure, or physiological change." For example, inconspicuous predatory behaviours have been ritualized. Food exchange has also been ritualized. The touching of bills together is likely to be derived from the parental feeding of the young. It has taken on a variety of meanings for different species of birds. It is common in courtship and appeasement displays to establish or maintain bonds. Portions of the courtship display of the fiddler crab Uca beebei may have evolved from the movement of the male while entering his burrow.

9.16 TERRITORIALITY

Territoriality, a type of intraspecific, or interspecific competition, which results from behavioural exclusion of others from a specific space, and the defended space is called **territory**. This is exhibited through songs and calls, intimidation behaviour, attack and chase, marking with scents, and is very costly for animals. The proximate reasons for it vary. For some animals, this is to acquire and protect food sources, nesting sites, mating areas, or simply attract a mate. Ultimate cause lies in increasing survival probability, and reproductive successes. In a territory, an animal secure a habitat to forage food, and increase overall fitness. In times of depleted resources territoriality increases. Territoriality often forces less fit animals to live in sub optimal habitats, thus, reducing their reproductive success. Reproductive and nutritional benefits from territoriality comes at a cost, which often interfere with other activities as parenting, feeding, courting, and mating. Thus, territoriality may not be a benefit for all animals. When resources are abundant and predictable, it is disadvantageous to defend territory. If resources are scarce and undependable, it is advantageous to exhibit territoriality. In selecting a territory, size and quality play a crucial role. Territory size, generally, tends to be no larger than organism's requirement for survival, because increase in territory size causes increase in energy expenditure. For some animals, territory size is not the most important aspect, but rather it is quality of the territory due to food availability, and superior nesting sights. Animals depend on these features to ensure their superior fitness.

Animals invest a lot of time and energy in defending territories with vigorous fight. When a rival challenges a territory holder, owner almost always wins the contest. Territory could be attributed to an evolutionary stable strategy, which assets that rules for the behaviour are controlled by an inherited proximate mechanism, such that differences between individuals in their strategies are likely to be different in their genes. Territory plays an important role as a mechanism of population regulation, ensuring success of fit animals, and eradication of less fit animals. It also plays a fundamental role as an indicator of carrying capacity and serves as an indicator of how much habitat is necessary to support viable populations.

Most birds are territorial, in the sense that they defend some area, even if just a nest site for at least during part of their annual cycle. Tigers also actively defend their territories.

9.16.1 Costs and Benefits of Territorial Defense

The time, energy and risk of injury are the costs, and improved access to resources (food, nest sites, or roost sites) represent benefits of territory defense. The birds defend territories, if benefits of defense outweigh costs. This idea of 'economic defendability' was first proposed by Jerram L. Brown (1964). If a territory is 'economically defendable', energy available in a territory should not be less than energy expenditure. Gill and Wolf (1975) tested this hypothesis on golden-winged sunbirds. Sunbirds defend feeding territories during non-breeding season and feed mainly on the nectar.

9.16.2 Classification of Territories

Type A, or mating, nesting, and feeding territory

An area within which all activities, such as, courtship, mating, nesting, and foraging occur. This is also called an 'all-purpose territory'. This type of territory is found in many song birds.

Type B, or mating and nesting territory

An area within which all breeding activities occur, but most foraging occurs elsewhere. For example in case of territory defended by male red-winged blackbirds. Male red-winged black-birds establish and defend 'Type B' territories, with clearly delineated boundaries during the breeding season. Many activities occur within territories, but males and females forage outside the territorial boundaries. Extent of off-territory foraging varies among nesting habitats and locations.

Type C, or nesting territory

A nest plus a small area around it constitute such a territory as found in the type of territory defended by the colonial water birds.

Type D, or pairing and mating territory

This type of territory is defended by males in lekking species.

Type E, or roosting territory

Roosting territories typically include foraging areas and roost sites. This may be equivalent (in terms of location), or nearly so to 'Type A' territory.

Type F, or winter territories

Many birds including non-flocking migrant passerine species establish resource-based territories that are defended for at least part of winter. Winter territories indicate that habitat availability and food resources are limiting factors during winter. Investigations on wintering migrants suggest that dominant individuals maintain stable territories throughout the winter, while some subordinate birds are forced into a non-territorial strategy in search of resources. Floaters (non-territorial) are found within wintering populations, which are indicators of habitat and/or food resource limitation and provide insight into a population's local distribution.

Radio-telemetry was used to measure hermit thrushes' movements and territoriality. Thrushes saturated suitable patches in the study area. Hermit thrushes established and maintained territories, using same agonistic behaviours, described for breeding birds. A few non-territorial birds (14%) moved among occupied territories, but most were faithful to a larger neighbourhood, apparently awaiting a territory vacancy. Behaviour of hermit thrushes support the emerging view of competition for spatially mediated resources on wintering grounds, such as food or covers, which contribute to limiting populations of migrant passerines.

9.16.3 Territoriality and Reproductive Success

Many birds defend territories only during breeding period. These territories include food or nesting sites which are necessary for successful reproduction. In polygynous species, differences in quality of territories may influence number of mates, that a male obtains. For example, male *Spiza americana* with better quality territories may attract more mates, than those with lower quality territories. In socially monogamous species, territory quality influences reproductive success.

9.16.4 Population Density and Territorial Behaviour

As population density increases, territories are compressed, but to a certain point. Limit to number of territories in an area results from a limit to number of breeding birds in that area. In a given area, some birds live without territories. In many species, males removed from their territories are usually replaced by conspecifics. This indicates presence of 'floaters', in at least some populations and suggests, that territorial behaviour may limit population density. Jerram L. Brown (1964) suggested that, in any given area, habitats available to breeding birds range from very high quality to unsuitable (for breeding). At low population densities, all birds occupy high quality habitats, and no birds are prevented from breeding. At higher population densities, some birds are excluded from high quality habitats, than those occupy lower quality habitats, but no birds are prevented from breeding. At still higher population densities, when high quality and lower quality habitats are 'saturated' with territory-holders, some birds occupy unsuitable habitats. Birds in unsuitable habitats do not breed and become 'floaters.'

Fretwell (1972) proposed that habitats vary in quality, ranging from good to poor. Suitability of habitats decreases with increasing population densities. When bird densities are high in good habitats, a bird may occupy a lower quality or poor habitat, but are not necessarily prevented from breeding. They simply settle where they can attain the highest fitness. For many species, Brown's model may best explain influence of territorial behaviour on population size. For example, a population of song sparrows on an island off the coast of British Columbia was studied for several years. Male song sparrows are territorial, but there is not space for every male to hold a territory. Non-territorial males are non-breeders, submissive, and move through area as 'floaters.' Proportion of floater males increases as number of territorial males increases. Floaters in other birds population suffer higher mortality because of inferior habitats they occupy. At higher densities, other factors may also limit population growth.

Number of successfully fledged young per female declines as number of breeding females increases that may be due to decreased brood size or to decreased success in raising youngs in the nest due to food limitation. Proportion of juveniles survival also declines as total adult population during fall season increases. In starlings and other perching birds, survivorship of the young both prior to fledging and immediately after is strongly related to size and weight of fledglings. Food limitation occurs as population density increases, causing females to provide smaller amounts of food to each nestling, so that they are fledged at a smaller average size.

Review Questions

Short Answer Questions

1. Define socio-biology.
2. Who coined the word socio-biology?
3. What do you understand by survival strategies?
4. What do you understand by the term fitness?
5. What do you mean by social life?
6. Why do animals live in social group?
7. Comment on the concept of "virgin birth".
8. Define sociogenomics.
9. State the "marginal value theorem".
10. State the "Kleiber's law".
11. State the "optimal foraging theory".
12. What is "adaptation dance"?
13. Name two insects which exhibit social behaviour.
14. What is pecking order?
15. What is dominance order?
16. What do you understand by the term "alpha position"?
17. Define aggression.
18. Define irritable aggression.
19. Define allopreening.
20. Define autopreening.
21. Define social dominance.
22. What is altruism?
23. What do you understand by reciprocal altruism?
24. What is weak altruism?
25. Which are cooperatve behaviours?
26. What is strong altruism?
27. What do you understand by the term "ultra sociality"?
28. Comment on the term "conservative support".
29. Define territoriality.
30. Comment on the mating territoriality.

Long Answer Questions

1. Discuss the benefits of social behaviours.
2. Describe the "handicap principle" in brief.
3. Discuss the molecular perspective of social behaviours.
4. Describe predator-prey relationship from ethological point of view.
5. Write a note on pack behaviour.
6. Discuss aggression from ethological point of view.
7. Give a comparative account of autopreening and allopreening.
8. Write an essay on altruism.
9. Write a note on mobbing behaviour.
10. Discuss territorial behaviour in brief.
11. Describe the history of socio-biology.
12. Describe the history of sociogenomics.
13. Describe the Hamilton's rule and concept of kin selection.

Chapter 10

Graviperception, Magnetoreception and Quantum Biology

Gravity has influenced the phylogenetic development of life forms for the last four billion years. The primitive ancestors of living beings acclimatized to gravity (Anken and Rahmann, 2002), and that trend has shaped the evolution of organisms to some extent. Primitive lifeforms used gravity, which was present in a fixed direction. The gravity helps them control their posture and orient themselves. Orientation and the ontogenetic development of animals are impaired in the absence of their interaction with changing gravity. The self-organizing dynamics of biological systems satisfy the need for general relativity and quantum mechanics (Conrad, 1997). The interaction between the negative influence of gravity, the cytoskeleton, and the protective diffusion results in the mean cell size of single cells (Anken and Rahmann, 2002; Moroz, 1984). Increased gravitational force reduces the cell's size. The cell size increases when the gravitational force is low in combination with weightlessness (Anken and Rahmann, 2002). Protonplasmic motion, the specific gravity of cell components in relation to the ground plasma, cytoplasmic viscosity, and a thin and elongated cell body provide protection to large cells against internal sedimentation. The impact of gravity becomes especially clear with the evolution of multicellular animals (Anken and Rahmann, 2002). Animals showed evolutionary radiation after the Cambrian explosion with the development of all invertebrate animals. In the Precambrian, anti-gravity systems were not elaborate. Thus, animals grew larger and were able to cope with the impact of gravity. They exhibited directed locomotion. All these happened at the stage of their movement from water to land (Anken and Rahmann, 2002). Directed locomotion helped animals predate other animals or escape from predators. Exercising directed locomotion, especially gravity, was one of the significant morphogenetic factors of evolution (Vinnikov, (1995).

The early ancestors of fish living about 400 mya coped with the terrestrial impact of Earth's gravity. The muscles and body structure of a bluefish withstand the force of gravity. The early fish are believed to have evolved some strength to overcome gravity-based inertia and drag. Fluid distribution in large land vertebrates is influenced by forces induced by gravity (Anken and Rahmann, 2002). Eventually, adaptive countermeasures involving changes in structure and function evolved. Studies in snakes suggest four countermeasures to cope with stress induced by gravity in lower vertebrates (Lillywhite, 1996). Enhanced arterial blood pressure due to gravitational stresses results from the impact of height and posture on the vertical blood columns, which are located above the heart. The anatomical organization evolved under habitat and behaviour-induced gravitational influence. Natural selection along with gravitational stresses shaped morphology for controlling the compliance of developing an anatomical antigravity suit. Natural selection in tall vertebrates produced components that are active in high-gravity-induced stress.

An aquatic snake is unaffected by the impact of gravity. As their heart is located at a distance from the head (Anken and Rahmann, 2002). The heart of a terrestrial snake is located near the head. In tree-dwelling snakes, the hearts remain almost behind the head. This allows blood supply to the brain even when climbing (Kirsch and Gunga, 2001). In the giraffe, the gap between the heart and the brain is about 2.8 m. Bipedal walking stands against the ability to cope with gravity. The short hindlimbs of young chimpanzees are unable to lift the body's center of gravity high enough. They are forced to expend more energy during bipedal walking (Kimura, 1996). Elongated hindlimbs optimize energy economy in our bipedalism. We consume 90% of the energy needed for locomotion to lift the legs (Kirsch and Gunga, 2001). The fossil Argentavis had high values for stress and strain in level flight. The strength of gravity was much less (Anken and Rahmann, 2002). Argentavis was not able to do so because of its flapping wings, but they soared as present-day condors. Both active, directed locomotion and the active maintenance of equilibrium require sensory functions during movement (Anken and Rahmann, 2002). Many animals maintain their bodies with the long axis horizontal. We are an exception. A fish may dive downward. A person may change his normal orientation by lying down. But in no case does any loss of equilibrium happen. Gravity is a suitable cue for orientation and postural control.

10.1 GRAVITY AS A CUE

Tactile stimuli predominate during an animal's movement over a solid surface. Proprioceptors play a role in spatial orientation in vertebrates and arthropods (Anken and Rahmann, 2002). The sensilla under gravity weigh down and stimulate mechanoreceptors in the arthropod's spatial orientation. Becoming detached from the ground, they orient themselves in space by keeping their backsides up. This is called the dorsal light response in fish. Visual cues maintain equilibrium. Fish and amphibians optomotor reflexes (Sebastian et al., 1996) bring about shifts in the image of the environment on the retina. Gravity receptors, the genuine organs of equilibrium, serve to determine the orientation of posture and movement in space. Gravity transforms into a biological signal in different ways. incorporated ion channels containing synthetic membrane bilayers respond to gravity (Schatz et al., 1996; Hanke1996). *Paramecia* perceive gravity through their heavy bodies (loxodes).

10.2 GRAVIPERCEPTION IN UNICELLULAR ANIMALS

In *Paramecium*, a pressure gradient between the membrane and the surrounding aqueous medium regulates gravikinesis (Anken and Rahmann, 2002). when the animal swims This gradient remains at equilibrium. Vertical tilts result in the opening or closing of mechanosensitive ion channels. This causes local depolarization, which in turn activates cilia for propulsion. Another ciliate, *Loxodes*, perceives gravity through intracellular organelles called Müller bodies. It consists of a membraneous pouch containing a "statolith" of $BaSO_4$ (Neugebauer and Machemer, 1997).

10.3 GRAVIPERCEPTION IN MULTICELLULAR ANIMALS

All multicellular animals exhibit active locomotion. They use so-called statoliths or otoliths. This remains in specialized organs responsible for transforming the acceleration into a body-owned signal. Vision may be poor or absent in a species, but other members have functional visual systems. In all animals' Mechanical stimuli are effective. Unicellular protozoans, about all invertebrates, and vertebrates use heavy bodies to orient towards the direction of gravity. Sense organs have shifted their positions during phylogenetic development (Anken and Rahmann, 2002). Gravity is functional in a biradially symmetric jellyfish. They have one statocyst at the apex of the body. Balancers support the calcareous statolith. A transparent dome encloses the whole structure. A pair of ciliated furrows emerge from each balancer, each of which connects with a comb row. Each balancer innervates the two comb rows. Tilting causes the statolith to press on one balancer. The resulting stimulus elicits a beating of the comb rows (Anken and Rahmann, 2002). Thus, locomotion and postural control result.

In the Mollusca and Arthropoda, there are developed sense organs. Pairing statocysts are the gravity-sensing organs of *Aplysia*. It consists of supporting and receptor cells that form a sac that contains calcium carbonate inclusions called statoconia. The receptor cells are hair-cell-like neurons. The cilia of cells are mechanosensory (Wiederhold, 1974). The statoconia remain in constant motion due to the constant beating of the cilia. Gravity pulls the statoconia down, obstructing the beating of the cilia at the bottom of the statocyst, which enhances membrane conductance to Na^+ and the creation of an action potential (Wiederhold, 1976).

Three semicircular canals are located next to each other for detecting the angular acceleration in both inner ears of vertebrates. The movement of the statolith bends the hair cell cilia, which open or close ion channels (Hudspeth and Gillespie,1994), changing the electrical current of the respective sensory cell. Signal transduction at the level of the brain thus causes a motor response (Anken and Rahmann, 2002). The brain integrates information from the vestibular organs plus proprioceptive and visual cues for spatial orientation, tactile control, and postural control. (Clément and Berthoz, 1994). In animals with a CNS, a close relationship exists between the developed sense organs and a specialized region of the brain, which results in behaviour. In the case of vertebrates, this is the cerebellum. This organ regulates postural control. It is especially large and efficient and needs to orient itself in all three directions of space. A visual cue is needed for correct postural control. Thus, normal behaviour is affected. Alternate gravity may have a strong influence on aspects of development since all systems should be adapted to normal earth gravity. Understanding of the role of gravity on the (ontogenetic) development of animals as well as their behaviour is also required.

10.4 BEHAVIOUR AND DIFFERENTIATION OF ANIMALS AT ALTERED GRAVITY

Microgravity syndrome (Sieving, 1997). is a prognosticated complex arising in reduced-gravity environments such as the surfaces of the Moon and Mars? It is accompanied by muscle atrophy, cardiovascular deconditioning, and bone demineralization.

10.4.1 In Invertebrates

Microgravity does not impair the morphogenetic development of various invertebrates. The shell of freshwater snails (Biomphalaria) develops in microgravity (Becker et al., 2000). This shows that the basic mineralization processes in molluscs are unaffected by exposure to 16 days of microgravity. In *Loligo* and *Octopus*, altered gravitational forces may not lead to morphogenetic aberration (Anken and Rahmann, 2002; Marthy, 1994).

10.4.2 In Vertebrates

Most studies on the effect of altered gravity on mammals focused on biomedical aspects. Such aspects include the physiology of the respiratory, intestinal, endocrine, immune, and muscular cardiovascular systems, in addition to the calcification of the skeleton (Kirsch, 1993); Moore et al., (Eds.); Gerzer and Hansson (Eds. 1997); Berthoz and Güell. (Eds.), 1997).

Amphibians, Reptiles and Birds

Earth gravity thus seems not to be required for the early ontogenetic axis formation in *Xenopus* (Aimar et al, 200). In the amphibian developmental stages, microgravity results in non-inflated tracheae and lung buds due to failure to inflate their lungs in a timely fashion (Anken and Rahman, 2002). Microgravity induced a malformed lordotic tail (Wassersug and Yamashita, 2000) with optometric responses (Souza et al., 1995) and retarded larval growth. The tadpoles raised in microgravity receive less vestibular information (Duprat et al., 1998). Microgravity does not influence egg maturation, fertilization, or embryonic development in Salamanders and frogs. Microgravity might retard larval growth. Few studies of changing gravity have been done with reptiles and birds. disorientation responses of various vertebrate animals exposed to microgravity produced by parabolic planet flights (Mori, 1995; Anken and Rahmann, 2002). Visual function in turtles compensates coordinated performance (Suda, 2000). In older chick embryos, all the tissues are formed. Microgravity seems to have no effect on normal eye development (Barrett et al., 2000).

Fish

Fish can be characterized by an absence of body weight-related proprioception. Alternate gravity cannot impair fertilization and development (Ijiri 1995; Anken and Rahmann, 2002). Visual and vestibular cues are used for postural equilibrium maintenance and orientation. In 1935, the DLR was described in which a fish tilts its back towards the light source. The DLR expresses a balance between the tilting force and the vestibular righting response (VRR) (Watanabe et al., 1991) The function of the DLR depends on the visual performance of an individual (Hilbig et al., 1996; Anken and Rahmann, 2002). During microgravity, fish exhibit abnormal swimming behaviours (Von Baumgarten et al., 1972). This behaviour has the same source as motion sickness in humans. The spinning movements induced by microgravity in some individual fish show a possible source of illusionary tilts. Asymmetric vestibular maculae at rest cause asymmetric shearing forces on sensory epithelia. A normal posture would then need neuro-vestibular compensation for the asymmetric discharge rates. In microgravity, there would be no weight differences in the otoliths from one side of the body to the other. The primary discharge rates should no longer be asymmetrical. As predicted by the asymmetry hypothesis, behaving individual fish larvae after hyper-g and in microgravity during parabolic plane flights indeed revealed a larger otolith asymmetry than behaving ones (Anken and Rahmann, 2002. The existence of a feedback mechanism adjusting the size, asymmetry, and Ca-content of fish inner ear otoliths towards altered gravity is on record.

Animals have the ability to both cope with and use the gravity vector for orientation. Mechanisms for the latter are well investigated. There is a high tolerance for altered gravity in arthropods. The presence of responses in fish and amphibians in most early developmental stages and a wide range of responses in birds and mammals (Nindl et al., 1996). The reasons for this varying range of gravity tolerance have not yet been disclosed.

10.5 MAGNETORECEPTION*

Life on earth estimated to have originated about 3.5 billion years ago. The earth's magnetic field (EMF) has been present since that time. Life has evolved under the influence of EMF which did not vanish altogether during periods of geomagnetic polarity reversals. The EMF continues as a reference frame for orientation. Animals use the EMF to move onward, amidst troubles due to fog, rain, and clouds. Biophysics help understand the magnetoreception (MR) by living beings and biological processes (Yan et al., 2021; Dyer et al., 2021). Spatial orientation, along with the EMF, are the basic needs for the evolution of MR. The EMF supplies directional and positional information to animals and hence is used for navigation. The direction of magnetic force at various locations of the earth differs due to the shape of the field lines. At the North and South Poles, the magnetic force is vertical while at the equator, it is horizontal. And everywhere in between, it stays at an intermediate angle to the surface. With a duration of a few thousand years, a global magnetic field reversal is long enough for individuals to slowly adjust to the changing conditions (Leonhardt and Fabian, 2007).

The earth's rotating iron core creates EMF. The magnet at the center of the earth shapes the EMF lines and supplies a signal to organisms. Appearing from the Southern Hemisphere, such a line curve encircles the globe. In the Northern Hemisphere, the line again enters the earth. The intensity of EMF varies, being the lowest at the equator and the highest at the poles. So, animals face a sharp inclination angle while moving northward from the equator. The EMF diverts the charged particles transported by its currents and protects the biosphere from the solar wind. The response of animals to EMF, called geomagnetotaxis, is a negative-positive orientation (Mandal, 2015). Organisms adapt to and alter the environment; exploit the environmental signals for purposes, like navigation and orientation. The dipolar character of EMF existed the last two billion years (Evans, 2006). For animals EMF serves as a source of directional information (Wiltschko and Wiltschko, 2005). The horizontal part of EMF work in darkness and gives the compass reference to animals to adjust the directional reference, and inform about polarity, inclination angle, and intensity. In insects, sense organs transform stimuli into nerve impulses that reach one of the central ganglia to bring change or to keep the existing behaviour. our present knowledge of MR, an EMF-induced behaviour, along with its adaptive significance as revealed from studies with fruit flies, bees, and ants is given here.

Mesmer affirmed the impact of universal gravitation on our bodies. He introduced the idea of animal gravitation in1766 in his doctoral thesis. Later, the idea changed to animal magnetism with Mesmer's experience of magnets (vide Crabtree, 1988). Thus, the history of animal magnetism goes back more than 250 years. The concept of animal magnetism has been debunked for a variety of reasons. In the last seven decades, information on MR has enriched as a transdisciplinary study. Insects have a magnetic sense. Frisch found the honeybee as a model animal for neuroethological studies. The revelation of the bumblebee's dance language in 1949 affected ethologists. Animals need to know their present location and the direction of travel to reach their terminus. Many animals exploit the properties of magnetic particles for navigation, orientation, and migration. The big question is whether animals have such a mental magnetic map to recognize regions by their characteristic magnetic fingerprints (Dennis et al., 2007).

---

*Mandal, F.B. and Chakroborty, B., 2022. Magnetoreception in Fruit Flies, Bees and Ants. *Acta Scientifica Malaysia* (ASM)6(1): 10–16.

Three viable hypothesis about the physical mechanism to answer the question are: (i) first one is based on magnetite in mechanosensitive structures, (ii) second one is based on a radical pair mechanism, and (iii) third one is based on induction (Kirschvink and Gould, 1981; Brown and Ilyinsky,1978; Schulten et al., 1978; Kalmijn, 1981). The first two hypothesis are widely accepted. The magnetite-based MR model holds the view that animals sense the EMF through magnetite that informs the animals about the magnetic map (Kirschvink et al., 2001). A Cry-based process also called chemical (radical-pair) MR. MR depends on light and magnet sensitive photochemical reactions. In chemical MR, the animal senses the axial alignment, inclination angle of the EMF lines and thus directional magnetic information for magnetic compass orientation. Studies involving transgenic animals have been conducted in insects (Gegear et al., 2010). The results point to an involvement of the cryptochrome protein in MR through the radical-pair mechanism, at least in magnetic fields ten times stronger than the natural ambient field.

Magnetite, the most common magnetic iron oxide functions in MR. Organisms perceive and transduce magnetic cues, and transfer them to the brain for interpretation and to create effective MR. The central complex called navigational heart of the insect brain mediates movement and steering. Inclination of field lines and the precise distinction between the geographic North and geomagnetic North impact route. The other affecting variables are the polarity of field lines and the strength of magnetic field (Lohmann et al., 2007). Total intensity denotes the size of the local magnetic field vector at any point on Earth. Ants and honeybees are well-studied for magnetite-based MR hypotheses (Kirschvink et al., 1997; de Oliveira et al., 2010). In the presence of blue light, the retinal Cry is hypothesized to be the only magnetosensors. Nocturnal insects like bees and ants seem to use a magnetite-based magnetic sense. But, both magnetite-based and Cry-based MR could occur at the same time in an insect species (Wiltschko and Wiltschko, 1995; Johnsen and Lohmann, 2008; Nordmann et al., 2017). Spatial variation in intensity and inclination of the EMF helps to detect the map. Light can reach the magnetoreceptors located at a peripheral site of the animals.

10.5.1 Role of Sensory Cues in Insects' Magnetoreception

To go ahead into the process of MR, it appears to me to provide a brief account of sensory cues. Indeed, MR cannot function in the absence of sensory cues. Animals need to know the "map" step and the movement of direction (the "compass" step) for navigation. Compasses are typically arranged in hierarchies. In the absence of celestial information, magnetic cues serve as backup. For animals to calibrate celestial cues and to recalibrate between magnetic and celestial compasses as situations require, magnetic information is important. Magnetic compasses are mostly magnetite crystal based. These crystals can give the direction of the goal and distance (Heinze and Reppert, 2011). Other compasses use paramagnetic interactions between visual pigments and the short wavelength. Odours are in use as cues in some cases (Gould, 1998). Besides, the polarized skylight creates a celestial pattern that serves as a minor compass cue (Reppert et al., 2004).

During first migratory phase, insects use stars, polarization pattern and the moon's disk as compass cues. The moon reflects sun rays like a disc rather than a sphere, as it seems to the naked eye. In fact, the moon is found far away and as a result all the light rays we receive as being parallel. The expected position of the sun at any point of time relative to the butterfly's migratory direction can be atoned for with the sun's movement (Mouritsen and Frost, 2002; Froy et al., 2003). Butterfly responds to wind turbulence and require partially for crosswind drift (Chapman et al., 2010; 2015; Reynolds et al., 2010). Magnetic and celestial visual cues include stars, the sun, the moon, and celestial polarized light which in turn mostly depends on the sun, and to some extent on the moon. Using these cues as a compass insect choose a desired course (migratory phase 1). Visual landmarks together with magnetic and olfactory cues detect the location after reaching the destination (migratory phases 2 and 3) in nocturnal migration. The monarch butterfly lacks map sense. The disk of the sun plays a major role as a cue (Stalleicken et al., 2005).

Monarch butterfly can sense the EMF (Guerra et al., 2014). Their use of EMF besides the sun's disc as a compass is yet to be found (Mouritsen and Frost, 2002; Stalleicken et al., 2005; Warrant et al., 2016). The moon is a less genuine cue for a long-distance nocturnal navigator than the sun is for a non-nocturnal navigator. Birds rely on EMF as a genuine compass cue and Bogongmoths have used the same (Wiltschko and Wiltschko, 1972; Cochran et al., 2004). They measure the visual optic flow of landscape characteristics and require this drift to control the flight (Srinivasan et al., 2006). The dim pattern of polarized light around the moon, the disc of the moon, and the constellations of stars serve as nocturnal compasses in long-distance navigation. Most such cues cannot function smoothly due to changing celestial positions throughout the night. Lunar cues vary in their duration and brightness at various times of the month. MagR (a protein) binds with iron to form a rod-shaped complex along with Cry and act as a compass needle. Associative learning helps insects detect magnetic fields (Phillips and Sayeed, 1993). Bees get cues about the direction of the food source from the waggle dance (Wajnberg et al., 2010).

Learning behaviour have reviewed the learning process in insects. Path integration, pheromone trails, and responses to stimuli are innate behaviours that ease learning for navigation (Perry et al., 2017). Insects do not undertake long journeys during a change in the surrounding environment. They travel close to the nest to get information on key features of the environment for future use. Visual cues and wind direction play a role in navigation. Path integration has a key role in visual learning. Ants learn faster using bimodal cues (visual and olfactory) than ants with a single cue (Steck et al., 2009; Buehlmann et al., 2020). For orientation and navigation, the insects use multimodal information. Coordinating the interaction between the multimodal strategies and information sources directs the route to meet the needs. This shows that small brain of an insect is dynamic in spatial cognition (Merlin et al., 2012; Buehlmann et al., 2020). Various insects take positions in water or air and comeback to a place of safety by using spatial memories. Hymenopteran life histories need spatial memories to equip their young (Collett and Collett, 2002; Buehlmann et al., 2020).

10.5.2 Role of Memory in Navigation and Orientation

There is a functional connection between the presence of magnetite in abdomen and MR in bees (Liang et al., 2016). Magnetic cues function in absence of celestial cues in learning of visual landmarks or in noting patterns (Collett and Baron, 1994). When light and chemical cues are not found, bumblebees follow the trained direction to the food source (Chittka et al., 1999, Wajnberg et al., 2010). A shift of 90° magnetic fields in total darkness changes the flight orientation of *A. milliner*. In *Schwarziana quadripunctata* the flight direction in daylight at the point of exit of underground hives is on record (Esquivel et al., 2007; Wajnberg et al., 2010). Experimental results in the presence of magnetic nanoparticles suggest that EMF serves as an orientation cue (Lucano et al., 2006). Only the reversed vertical field affects the inclination of the light trajectory. This shows that bees perceive the EMF whether the EMF is still in a pointed down or up direction. Bees can find the southern and northern hemispheres (Esquivel et al., 2008).

Gravity is supposed to supply the natural basis for this reference line. Bees with non-functional gravity receptors cannot perform waggle dances, supports this idea. In case of long distances, the local geomagnetic field (LGMF) is uniform and stable. The bees inform the position of food through orientation of a vertical comb in respect to LGMF and the LGMF affects (Lambinet et al., 2014). Factors like landmarks, pheromones, gravity, the sun compass, and polarized light, and vibrations help ants to orient. these factors should remain concealed for the sole use of a magnetic compass. The experiment by attaching the iron filings to body parts of ant *Myrmica* retinoids was conducted. The effect of EMF on the orientation was noted only when the iron was attached with antennae. The pedicel of the antenna was the most sensitive to EMF (Vowels, 1954; Wajnberg et al., 2010). The strongest proof for using a magnetic compass is the study of the re-oriented polarity of EMF. A group researcher placed a solenoid on a foraging trail of black-meadow ant (*Formica pratensis*) (Çamlitepe et al., 2005).

Studies with biomagnetism in several insects like leaf cutter ants (*Atta colombica*), foraging weaver ants (*Oecophyllas maragdina*), *Solenopsis substitute, Pachycondyla marginata*, and *S. invicta* are recorded. Such studies concluded the presence of a compass based on SPM particles and the involvement of nanoparticles in long-distance orientation (Riveros and Srygley, 2008). The strongest saturation of magnetic materials in antennae, use of nocturnal orientation cues, vibrational or otherwise, are recorded in ants (Riveros and Srygley, 2008; Wajnberg et al., 2004). Ants show sensitivity to the position of the Sun, polarization of celestial light, the geometric pattern formed by the tree branches on the celestial ceiling, the landscape of the near horizon and EMF in their navigation and orientation. Involvements of neurons that respond to magnetic field stimuli have not been confirmed in insects (Weinberg et al., 2010).

10.5.3 The Mechanism of Magnetic Sensing

The magnetite hypothesis assumes that neurons having magnetite crystals are connected to mechanosensitive structures. In the models of the crystals tend to align with the external magnetic field and thereby generate torque (Walker et al., 2002; Kirschvink, 1992b). The torque tries to rotate the magnetic inclusion, which is mechanically linked by cytoskeletal filament structures to mechanically gated ion channels in the plasma membrane of nerve cells. The torque causes ion-channels to open, allowing the exchange of ions through the membrane, and finally producing a signal (for example, a change in the spontaneous firing rate). The signal can be interpreted by the brain and then used for decisions in behaviour or navigation. Magnetite-based MR can be temporarily disabled with a strong magnetic pulse that is short enough to remagnetize magnetite crystals without rotating or moving them (strength 0.5 T, pulse length 0.5 ms;) (Walker and Bittermann, 1989; Kirschvink and Kobayashi- Kirschvink, 1999).

The radical-pair hypothesis posits certain biochemical reactions that are sensitive to weak magnetic fields, such as the EMF. A spin-correlated pair is made of two radical pairs, in which each radical has an unpaired electron with spin being either parallel or anti-parallel with respect to the other unpaired electron in the radical pair. Since each electron-spin has its own magnetic moment, the radical pair reaction can be influenced by magnetic fields. A radical pair can be generated by short- wavelength light in the candidate molecule, the cryptochrome (Cry) (Ritz et al., 2000). Cry has been found to be expressed in great concentrations in the retinal ganglion cells of night-migratory songbirds (Mouritsen et al., 2004).

10.5.4 Magnetoreception in the Fruit Fly

In adult and larval fruit flies, light wavelength influences the magnetic compass and orientation. When tested in a uniform arena of short- wavelength light (450 nm), the fly becomes trained towards a light gradient under UV light. They showed orientation towards the trained magnetic direction of the light gradient. Trained flies change their orientation by 90° in long wavelength light (> 450 nm). Such a change happened with respect to the learned magnetic direction (Phillips and Sayeed, 1993). It shows the presence of a light-dependent magnetic compass that responds to both short and long wavelengths of light. The spectral dependencies of magnetic compass orientation in flies are compatible with a Cry-based mechanism (Phillips et al., 2010).

10.5.5 Magnetoreception in Relation to Light

EMF activates Cry and activates voltage-gated calcium channels (VGCCs) in the fruit fly. The VGCC hypothesis states that EMF releases calcium ions. The calcium channel blockers protect against negative effects like oxidative stress (Pall, 2013). Calcium and VGCCs play a role in neurotransmission at excitatory synapses (Atlas, 2013). Only the pathway of light-dependent activation of CRY (by EMF) in the clock neurons leads to an increased action potential. This happens due to increased calcium release at synapses. Crys are the key photoreceptor molecules and produce magnetically sensitive radical-pair products (Mouritsen and Ritz, 2005). The Cry is used by the fruit fly to respond to and orient to the EMF. An EMF affects the Fe_3O_4 of magnetite and produces a transducible signal. The fly does not respond to EMF in the wavelength of light above 420 nm. Filtered light influences the action spectrum of Cry, having a size of 350 nm in plateaus and 430–450 nm in plateaus (Van Vickle-Chavez and van Gelder, 2007). When the change in Cry happens due to a missense mutation, the fly loses MR.

The free radicals are produced by a tryptophan triad (Trp triad) composed of tryptophan residues—W324, W377, and W400 of the Cry. Trp triad photo reduces the flavin cofactor of cryptochromes by electron transportation in vitro. Cry serves as a light-independent transcription repressor or photoreceptor. EMF works on such free radicals (Gao et al., 2015). The change of tryptophan protein does not affect the MR in the fly and the fly does not respond to the weaker EMF (Gegear et al., 2010). The photolyases produce long-lasting radical-pair intermediates to alter the effects of the magnetic field in the ancestral Crys, (Giovani et al., 2003). After getting excited by light the cryptochromes generate an intermittent radical pair. Orientation and the strength of the external magnetic field are known to exert an influence the reaction rate of radical pairs (Liang et al., 2016). The function of Cry in fruit flies depends in part on earth-strength magnetic fields (Gegear et al., 2008). It can transduce magnetic field information into a biochemical signal.

Drosophila has a magnetic receptor, CG8198, or MagR, and a multimeric magneto-sensing protein complex (Qin et al., 2016). MR- related Cry forms a magneto-sensing complex that responds to iron-based and CRY-based systems and aligns with the EMF. This protein complex forms the basis of MR (Qin et al., 2016). The quantity of iron that is associated with the MagR/Cry complex would not produce a permanent dipole moment (Meister, 2016). Reactions within the Cry protein form two molecules each with a lone electron having the potential to entangle with each other. They may exist in a single state (where the spinning direction of one corresponds to the spinning direction of the other) or in a triplet state (where two electrons rotate close to in parallel). The direction of the magnetic field controls the existence of singlet or triplet states.

If it can be assumed that the singlet and triplet states of the radical pair are linked with reactions, then the products of such reactions should give cues for the direction of EMF. If these substances influence neural signaling from the bird's retina, then this mechanism explains the basis for MR (Offord, 2019). Fruit flies offer many advantages for MR study. A functional fly Cry gene as well as broadband illumination of short-wavelength (420 nm) light function are needed for MR. They jointly can find a magnetic field that is about ten times the intensity of the EMF (Gegear et al., 2008). Besides playing a role in the Cry-mediated effects of blue light, EMF influences the circadian rhythm (Yoshii et al., 2009). Innate preference or associative learning helps flies detect magnetic fields (Phillips and Sayeed, 1993). Cry-expressing cells perhaps represent the magneto sensory neurons as shown in the cell-specific gene knockdown of Cry. A subset of proteins interacts with Cry, binds with iron, and is expressed in the head of the fly.

The first evidence for Cry-based MR stays the training of male fruit flies to respond to a magnetic field depending on the ambient wavelength of light (Phillips and Sayeed, 1993). In the fruit fly, Cry regulates visual perception, light-dependent arousal, circadian photoentrainment, as well as UV avoidance. Feeling of the EMF depends on a light-dependent magnetic sense. Electron transfer causes a conformational change in Cry that releases the C-terminal tail (CTT) of Cry and frees the binding sites for downstream partners. When Cry is not exposed to light, the CTT functions as a repressor and blocks the binding site for downstream intermediates. Cry on exposure to light interacts with the core clock protein Timeless (TIM)for proteasomal degradation. Thus, the release of CTT regulates interactions to bring about behavioural changes. In Cry- mediated MR, the ROS is produced at the flavin reoxidation step on exposure of CRY to light.

Exposure of Cry1 to BL activates the protein and creates ROS and H_2O_2. Overexpression of the redo regulating protein catalase depletes H_2O_2, blocking the increase in action potential firing in the clock neurons. Light-induced increases in neuronal excitability cause the closure of voltage-gated potassium channels in the presence of HYPERKINETIC(HK), a Kv potassium channel subunit. The subunit is redo-sensitive due to an intrinsic aldo-keto-reductase domain. Thus, a photo-induced change in protein structure may perhaps enhance neuronal activity and activate the downstream signaling processes. Besides, such a change influences the cellular redox state, along with the interference in the functioning of the Kv channel. The latter mechanism appears to be intriguing as it is an HK- dependent mechanism and an identical redox-modulates sleep in fruit flies (Bradlaugh et al., 2021).

10.5.6 Magnetoreception in Bees

Bees are highly sensitive to EMF, especially for orientation and navigation. For this reason, most such studies rely on bees. Adult honeybees have an MR sense and use EMF directional information (Ferrari, 2014). Bees learn to respond to changing local magnetic fields (Gould et al., 1980). This response requires exposure to abnormal magnetic fields (Lindauer and Martin, 1972). Magnetic material in front of the abdomen, thorax, and antenna mediates the orientation. The magnetic grain starts to develop in pupa and is also found in adults (Gould et al., 1978). The highest (2.4 to 0.15 g mg^{-1} tissue) iron level appears to be present in the fat body of adult workers when they start to forage. Paramagnetic substances create extra magnetism (Gould et al., 1980).

Electron-dense materials in SD and SPM magnetite are known to be found near the cuticle (Schiff, 1991). Magnetite in the abdomen exhibits sensitivity to EMF. Iron, calcium, and phosphorus granules having a diameter of 0.32 0.07-Um remain in the cytoplasm of cells of the abdominal segment (Kuterbach et al., 1982). The size and number of the granules vary according to the age of the bee. The bee extracts the iron from both the pollen and the honey. Kuterbach and Walcott (1986) suggest that iron granules play a role in orientation and iron homeostasis. Severe fluctuations in the earth's magnetosphere may occur due to major coronal eruptions under the sun's influence. Such fluctuation disturbs a forager's homing ability and even leads to the loss of this ability. Biomineralization of iron helps understand the basis for MR. Tracking EMF direction using magnetite reveals that mechanical orientation of crystals alters signal transduction, which in turn alters the ion channels of the cellular membrane (Winklhofer and Kirschvink, 2010).

The bees form superparamagnetic magnetite iron granules in the trophocytes of the abdomen (Hsu and Li, 1994). They deposit iron minerals intracellularly. Biomineralization in *Apis mellifera* is completed in two steps. In step 1, the iron deposition vesicles (IDVs) enlarge due to fusion with one another. In step 2, dense particles (7.5 nm in diameter) produce the iron granules with the help of a layer beneath the membrane of IDVs (Hsu and Li, 1993). EMFs may expand or contract the superparamagnetic particles in an orientation-specific manner. Variations in particle size trigger the increase, resulting in the release of intracellular Ca^{++}. A neural response starts upon receiving the magnetic signal from the associated cytoskeleton. In bees, the responses of the proboscis extension reflex (PER) exist (Liang et al., 2016). PER is used to train the bees to associate an odour with it. The trained bees can associate with the magnetic stimulus. Bees do not recognize magnetic stimuli if a cut in the ventral nerve cord (VNC) occurs, showing the function of the VNC in signal transmission. But they respond to an olfactory PER task.

Bees find the fluctuations in static intensity as weak as 26 nT against the background EMF. MR decreases rapidly with increasing EMF frequency (Kirschvink et al., 1997). A honeybee can detect small fluctuations in the EMF and weak earth-strength magnetic fields. The disorder study explains the loss of homing potential of the forager bee (Ferrari, 2014). Bees exploit the GMF for aligning the combs within the hives and for orientation in foraging. In the case of unaltered polarity, they respond to the magnetic anomaly but do not respond to it in reversed polarity (Lambinet et al., 2017). Thus, they must have a polarity-sensitive magnetoreceptor. The waggle dance, foraging, and flight of *Apis mellifera* show the existence of MR. Bees find cues for the direction of the food source from the foragers' waggle dance (Wajnberg et al., 2010; Liang et al., 2016; Lindale and Martin, 1972; Wajnberg et al., 2010).

In the abdomens of *A. mellifera*, the magnetite keeps a residue of magnetization at the normal temperature (Gould et al., 1980). The absence of such particles in the head and thorax shows only a diamagnetic contribution (Takagi, 1995). The antennae have the highest amount of magnetic material in *A. milliner* (Wajnberg et al., 2010). Squid magnetometry confirms the presence of SPM magnetite in trophocyte granules in the abdomens. Purified iron granules (IGs) show high intrinsic coercivity (Hsu et al., 2007; Weinberg et al., 2010). The low hydration level of crushed abdomens decreases the average magnetic volume of SPM particles (Wajnberg et al., 2001). HF intensity in the abdomen is higher than in other body parts. Thus, accessing a small amount of magnetic material is difficult in body parts. New techniques show the amount of magnetic material in the body parts of *S. quadripunctata*. The properties of magnetic material in various bees differ.

Phosphorus and calcium remain in a clear, non-crystalline arrangement in IGs (Hsu and Li, 1993). The fat body of *A. mellifera* and *Scaptotrigon apostica* queens has iron granules. These granules originate from holoferritin, calcium, and phosphorus (Keim et al., 2002). Changes in IG size of trophocytes supply further support for MR function (Hsu et al., 2007; Wajnberg et al., 2010). A group of researchers could not de-magnetize the magnetic material of honeybees because it was in the form of superparamagnetic crystals (Gould et al., 1978; 1980). Honeybees can distinguish between the presence or absence of magnetic anomalies. They do not do so when a magnetic wire is attached to the anterodorsal surface of their abdomen (Walker, 1989). This shows the likely location of the magnetoreceptor in the abdomen. Response to EMF of varying intensity and frequency shows that bees can distinguish between alternating fields when the frequency still is below 10Hz (Kirschvink et al., 1997). But they need the stronger fields to do so when the frequency is raised. This finding supports the magnetite-based MR hypothesis.

After returning from foraging, foragers inform other bees about the location of food materials by showing a dance relative to the vertical direction of hive combs. The angle between the direction of dance and vertical points shows the angle between food and the sun. Worker honeybees travel up to 12 km to collect food away from their hive and memorize visual landmarks to find their way home. They get directional information from the sun. The bees use the spectral pattern in the sky to guess the position of the sun in the presence of a cloud. They measure the distance information and integrate it for navigation. Honeybees detect EMFs most likely through granules (magnetoreceptors) in their abdomens. Although Cry occurs in the honeybee brain, its use in MR is seemingly unknown. The cytochrome stays a potential MR effector in honeybees. Although iron granules have the potential to be magnetoreceptors, the association of the iron granules with MR has not yet been confirmed and is still not convincing because they respond to changes in the magnetic azimuth in total darkness.

Cryptochromes become ineffective in the absence of UV-A/blue light (420 nm). As bees become older, iron granules (suspected magnetoreceptors) become clumped. The state of iron granules and the age of bee's influence MR. However, the relationship of iron granules with MR is yet to be confirmed. The role of iron-containing cells in the neural system is yet to be confirmed. The cry functions, on the other hand, are light-dependent. The honeybee can detect the magnetic field in their hive in a dark environment. Such a detection mechanism could not explain the Cry-based MR in insects, but it can be easily explained by the magnetite-based MR. In *S. quadripunctata* bees, the highest amount of magnetic material is believed to be present in the head and antennae. Bees can discriminate between oscillating magnetic fields at frequencies of at least 60 Hz. with very good directional magnetic compass sensitivity. These data have been extensively reviewed (Kirschvink, 1982; Kirschvink et al., 2001).

10.5.7 Magnetoreception in Ants

Reversal of the local magnetic field reverses the orientation of *Oecophyllas maragdina* (Jander and Jander, 1998; Wajnberg et al., 2010). *P. marginata* also uses a magnetic compass for orientation during migration. They use the GMF to find the axial migratory routes. In *Solenopsis interrupta* and *S. substitute*, a higher magnetic content is found in the head and antenna than in other body parts (Acosta-Avalos et al., 1999; Abraçado et al., 2005). The abdominals of workers of *S. invicta* mediate the MR. Magnetic particles and magnetic sensors were extracted using magnetic precipitation methods from *P. marginata* (Acosta-Avalos et al., 1999; de Oliveira et al., 2010; Wajnberg et al., 2010). The antenna has a high concentration of pure Fe/O (or Fe oxides) particles as sensorial materials. Johnston's organ perceives the EMF and gravity in *M. ruginodis* and *M. laevinodis* (Vowles, 1954; Wajnberg et al., 2010).

Iron and other particles perhaps cover the proprioceptor in the pedicel– scape joint. Magnetic particles close to mechanosensitive tissues along the antenna reveal the physiology of MR (Vowles, 1954; de Oliveira et al., 2010). Ants collect magnetic materials from the soil. Ants explore the surrounding areas for food. During the return to the nest, they mark the way back with pheromones that are straight in the direction of the exit of the tunnel. Ants such as *Pachycondyla marginata* use the EMF for orientation (Acosta-Avalos et al., 2001). Tests showed that the most promising parts for MR in ants are the antennae (Acosta-Avalos et al., 1999; Wajnberg et al., 2000; 2004). A group researcher was finally able to find iron-rich crystals in the Johnston's organ in the antennae of *P. marginata* and found the magnetite, maghemite, and the relatively weakly magnetic goethite and hematite (Oliveira et al., 2010).

In principle, the magnetic moments of the magnetic crystals in the EMF can produce a mechanical moment that can be transmitted into a neuronal signal via a mechanosensitive structure (Kirschvink and Gould,1981; Shcherbakov and Winklhofer, 1999; Davila et al., 2003; 2005; Ferreira et al., 2005). The Johnston's organ, a mechanosensitive structure, works as a graviceptor or an acceleration receptor (for hearing or flight control) (Sandeman, 1976). With the recent finding of magnetic minerals, the Johnston's organ could also serve as a magnetoreceptor (Oliveira et al., 2010). The characteristic magnetic properties of these iron-mineral deposits have not been decided yet, and it remains unclear if the structure meets the basic requirements for MR.

Biophysics is unwinding the components of biomagnetic impact that results in further development of MR in terrestrial organisms (Kobayashi and Kirschvink 1995). The role of Cry helps unravel the mechanism of MR. Insects contain Type 1 (found only in fruit flies), Type 2 or both types of Cry. Type 1 Cry and both Crys play a role in circadian clock regulation (Zhu et al., 2005). *Drosophila*-like Type 1 Cry are sensitive to ultraviolet- A/blue wavelengths of light and serve as circadian photoreceptors (Ozturk et al., 2008). The slow drift of the geomagnetic reference frame accumulates with time, which helps in the periodic calibration of the magnetic compass. In the absence of familiar landmarks, organisms find a target in true navigation by using a map and a compass (Gould and Gould, 2012; Kirschvink, 1982). Biomineralization produces a variety of biomaterials in animals (Bauerlein, 2005). Insects have very few neural components. So, information from insects can reveal the vital components for navigation.

The role of associative links between long-term memories has been found in bees and ants in their MR (Merlin et al., 2012). Magnetic orientation is widespread among animals. The role of the neurosensory system and magnetoreceptors is enigmatic in insects. EMF serves as a cue for navigation and orientation in insects, but environmental factors like the position of the sun influence navigation and orientation remarkably. Further study is needed to understand how insects calibrate with EMP from various parts of the world and navigate in the absence of suitable environmental cues. Sensory mechanisms behind magnetoreception along with its genetic basis may be elucidated for furtherance of this field of neuroethology. MR studies have been carried out in few sample insects. These experimental studies have revealed the existence of two possible but hypothesized mechanisms of magnetoreception. Presently we cannot prevent the existence of other mechanism of magnetoreception in insects. Thus, it is safe to conclude that further studies are needed to gain information on the possible theories that could integrate the existing knowledge of magnetoreception in insects in a consolidated scientific framework.

10.6 QUANTUM PHYSICS IN BEHAVIOURAL BIOLOGY

Quantum physics and biology have long been believed to be unrelated disciplines. The life sciences have provided more refined explanations of macroscopic phenomena. Thanks to an improved understanding of molecular biology. Modern biological understanding relies on organic information processing. Quantum physics relies on photons, electrons, and atoms. We have now realized the need for quantum physics to understand biological phenomena, Erwin Schrödinger ventured across the disciplines in his lecture series "What is life?" (Schrödinger, 1944). He expected a molecular basis for heterodity. It was later confirmed to be DNA (Watson and Crick, 1953). In a reductionist view, quantum physics influences biology. Electrodynamics and quantum physics shape all molecules and determine molecular recognition. Quantum physics helps us understand the discrete molecular orbitals, van der Waals forces, and stability of matter. Photosynthesis, the process of vision, the sense of smell, or magnetic orientation are now hot topics in this context. Quantum physics includes a wide variety of phenomena. Most of them appear to be unusual as they violate our expectations of how nature should behave. Quantum physics derives from the discreteness of nature. The Latin "quantum?" asks the question "how much?"; even now, a quantum is "small." If single photons or single spins can trigger a chain of macroscopic phenomena in organic systems, then it is judicious to be in quantum physics.

The concept of reality has been changing since the beginning of the 20th century. Thanks to quantum physics. knowledge of mathematics can access the attribution of real objects in classical physics (Holtfort and Horsch, 2023). . Measurement can probably reveal some features. probabilities in the form of wave functions are linked with various other simultaneous states (Schrödinger, 1926; Holtfort and Horsch, 2023). The wave functions are believed to be the potential realities, not actual ones (Weinberg, 1995). The decoherence change the quantum world into macroscopic reality (Zeh, 1970; Holtfort and Horsch, 2023; Matthews1977). The classical world does not regulate social life. As wave functions, quantum mechanics govern social life. Various aspects of social life are in turn explained by various theories, including decision-making theory, psychological decision theory, and cognitive biases. Nobel laureates Kahneman and Tversky's Work (Kahneman, 2011) shaped these concepts. The special properties of cognitive biases and decision-making should be viewed from the perspective of quantum Darwinism to get an insight into the phenomena.

10.6.1 Quantum Theory

According to Planck and Einstein, light can be viewed as waves and particles. Thus, matter could be behaving like waves and electrons (Susskind, 2014). In the quantum area, the particles are viewed as separate from each other (Malin, 2001). But what about the waves? The wave functions for the potential outcomes are observable at the time of measurement (Schrödinger 1926; Holtfort and Horsch, 2023). The possible states of waves exist at the same time in superposition (Albert, 1992; Holtfort and Horsch, 2023). The wave collapses into particles. Most likely, all the possible outcomes do not go to zero but rather to one.

10.6.2 Entanglement

Entanglement is a system where several particles are viewed in quantum physics (Bengtsson and Zyczkowski, 2006). Entanglement of particles and non-locality exist over very long distances (Bell 1966; Gisin et al., 1989, 1999). in a two particles system, one particle moves in opposite and distant directions at the end of the interaction and communicates with the other. This results in the phenomenon that when one particle changes, the other particle changes in the same way (Gisin et al., 2008; Holtfort and Horsch, 2023). Each particle thus aware what is occurring to the other without any signal transfer (Hardy, 1998; Holtfort and Horsch, 2023; Matthews 1977).

10.6.3 Decision-making and Cognitive Biases

Decision-making theory analyze the result of decisions (Petersen, 2011). Social science can mark the difference between the normative and descriptive decision theory (Holtfort and Horsch, 2023. The normative theory based on rational choice theory gives a framework to behave rationally (Eisenführ et al., 2010; Holtfort and Horsch, 2023; Matthews, 1977). There are two ways of rationality (Simon, 1959). The path to the decision must be rational, ensuring that the goals and preferences are clear. The decision must be internally consistent. The construct of homo economicus (Bee & Desmarais-Tremblay, 2023) found its way through the concept of rationality (Walras, 1874).

10.6.4 Social Science, Decision-making and Quantum Physics

Quantum physics has the potential to reinterpret social science, consciousness, and decision theory. This requires quantum social science (Haven and Khrennikov, 2013) in the form of quantum vitalism. This can be traced back to the study of Bohr in the 1930s (Bohr, 1933; Holtfort and Horsch, 2023; Matthews,1977). This was the starting point and is now one of several ways of thought in the quantum social science. these research fields are so mixed up that they cannot be separated. For further analysis the focus should be given on quantum Darwinism and quantum decision theory. Quantum Darwinism has shown an influence of the environment on the quantum system, which can be channelized to the quantum decision system. Quantum mind and matter focus on the relationship between mind and body (e.g., Kriegel, 2004) and address the panpsychism (Dyson, 1979; Holtfort and Horsch, 2023). The quantum brain theory deals with quantum consciousness and the brain deal with (Igamberdiev, 2012; Holtfort and Horsch, 2023; Matthews1977). This theory hypothesizes amplification of the quantum processes at the elementary level which remain in superposition at the level of the organism and then, through downward causation, constrain which reaches the brain (Gabora, 2002;Holtfort and Horsch, 2023; Matthews1977).

Technology has now enabled research below the cellular level. Birds can exploit non-local connections with the earth´s magnetic field for navigation. Plants use quantum effects in photosynthesis (Loyd, 2011; Holtfort and Horsch, 2023). The quantum models of evolutionary processes are closely connected to quantum biology (Gabora et al., 2013; Holtfort and Horsch, 2023; Matthews1977). McFadden after applying quantum physics to DNA claims that mutations are not random. Quantum Darwinism deals with the transition of conceivable quantum system with its potential for variations to the limited set of classical states by a selective process. The quantum system reacts in a way which is adaptive to its environment. Environment exerts selective pressure on the states (Zureck, 2009). each quantum system consists of variations of the classical states that are selected out and represent this information. Quantum decision theory can be of immense importance for both social science and economics. Quantum models can explain systematic anomalies in our behaviour during uncertainty (Pothos and Busemeyer, 2013; Holtfort and Horsch, 2023; Matthews, 1977). The social and economic sciences, including cognitive knowledge, could be at the start of a revolutionary movement, like physics at the start of the 20th century with quantum physics. Quantum decision theorists have not yet ventured into theoretical exchange on a social and economic level (2015) but remain in the physical-scientific world (Matthews 1977). The arguments for a penetration of economic literature can be summarized as follows:

- Quantum decision theory (QDT) is a holistic view including brain, emotions, and subconscious (Yukalov and Sornette, 2009; Holtfort and Horsch, 2023).
- QTD calls into question the subject-object separation. No well-ordered mind is given that makes exact predictions in uncertain environments (Wendt, 2015).
- QDT challenges the idea of rational utility maximization by assuming a superposition of the mind (Lambert-Mogiliansky et al., 2009).

10.6.5 Quantum Superposition

Quantum physics teaches us about reality, information, or space-time. The wave-particle duality of light is the dual nature of light. Each individual realization is random within the predetermined probabilities. Quantum theory terminates the wavelike propagation of the quantum state; it is often described as a "collapse of the wave function." Diffraction and interference deviate even more from our classical expectations when we observe them as electrons, neutrons, or atoms (Cronin et al., 2009). The wave nature of matter is relevant for life science in a different context. The short de Broglie wavelength of fast electrons allows one to get high-resolution images in electron microscopy. Neutron waves are useful in crystallizing protein structures. While electrons and neutrons are able to prove the delocalization of entire molecules, such as C60, using far-field diffraction (Arndt et al., 1999). Even bodies, such as porphyrin derivatives, revealed their quantum wave properties in a near-field interferometer (Hackermüller et al., 2003). These experiments show that the large-scale coherence of biomolecules is observable.

Spin is a quantum way to turn around. Spins may also be brought into the superposition of two or more exclusive states, comparable to a single compass needle pointing both north and south at the same time. Spin is also often responsible for magnetism in biological systems on the molecular level. The spin of protons is also exploited to derive the structural and functional properties of organisms (Lauterbur, 1973). Magnetic resonance imaging of human tissue relies on the simultaneous response of many trillion spins in a macroscopic biological volume. Quantum superposition of energy states that a molecule is excited by a femtosecond optical pulse. Femtosecond spectroscopy has thus become a regular tool for characterizing biomolecular systems (Felker et al., 1982). Coherent superpositions of electronic and vibrational energy states are observed in such experiments. When biological systems are exposed to incoherent daylight, excitonic coherences may form between coupled neighbouring molecules. In photosynthetic complexes, excitonic coupling is believed to occur across several pigment molecules (Dahlbom et al., 2001). Coherence in photosynthesis is a major field of current research.

10.6.6 Life Science in an Interface of Quantum Physics

The single quanta of light have been relevant for illustrating fundamental quantum principles. They are also ubiquitous in the life sciences. The most efficient detection techniques for fluorescent biomolecules are sensitive at the single-photon level. Individual particles of light have direct relevance in biological processes as they may affect the structure of individual molecules, which, in turn, can transduce signals in living organisms. The retinal molecule can switch its conformation after the absorption of very few photons. thus, turning the human eye into one of the most sensitive light-detecting devices that exists. Between two and seven photons are usually to be perceived by a dark-adapted human observer (Hecht et al., 1942). Studies showed that test subjects could count the number of photons, which was only limited by quantum shot noise (Rieke and Baylor, 1998).

Single-photon detectors are of great interest for quantum communication. The octopus rhodopsin, chosen by evolution because it is well-adapted to the darkness of the deep oceans, is a useful component in such applications (Sivozhelezov and Nicolini, 2007). Single-photon sources are gaining increasing importance in quantum communication. Single molecules are relevant emitters (Lounis and Orrit, 2005). When talking about the quantum properties of light, we usually refer to its wave-particle duality, graininess, and quantum statistical properties. For example, photon bunching, antibunching, or squeezing (Glauber, 2006). Fluorescence correlation experiments with proteins (Sanchez-Mosteiro et al., 2004) have revealed both the discrete quantum nature of molecular energy states and the nonthermal quantum statistics of light. An excited single molecule may usually not absorb a second photon of the same wavelength unless the excited state has decayed. The emitted photons are released with a time structure that differs from that of thermal light sources. Photons emitted by a single molecule come in an "anti bunched" rather than a bunched time series. The characteristic energy of a quantum system relates to its spatial dimensions. Fluorescent quantum dots are used as efficient labels for biomolecular imaging. They allow one to follow the dynamics of marked receptors in the neural membrane of living cells (Dahan et al., 2003). Similar results have recently been achieved with nanodiamonds. Their nitrogen-vacancy centers exhibit strong and stable fluorescence. They are biocompatible, and they are sensitive quantum probes for magnetic fields on the nanoscale (e.g., Balasubramanian et al., (2008).

10.6.7 Quantum Tunnelling in Biomolecules

Living organisms make and break billions of chemical bonds. Thermal activation and enzymatic catalysis regulate the reaction rates. Electron transfer has a long history (Marcus, 1956). The oxidation rate of cytochrome in the bacterium was the first evidence for electron tunnelling, which was inferred from the oxidation rate of cytochrome in the bacterium. Electron tunnelling is a common process in photosynthesis (Blankenship, 1989), cellular respiration (Gray and Winkler, 2003), and electron transport along DNA (Winkler et al., 2005).

The enzyme alcohol dehydrogenase transfers a proton from alcohol to NAD. Proton tunnelling regulates other enzymatic reactions (Glickman et al., 1994) at a distance shorter than 0.1 nm. Turin (1996) suggested that they were able to "smell" quantum tunnelling. Linda Buck and Richard Axel were awarded the Nobel Prize for their study of the olfactory system of mammals. The transmembrane proteins that encode odor receptors were found in the olfactory epithelium (Buck and Axel, 1991). And each odorant can be detected by different sensors. Many smells can be explained (Zarzo, 2007) by guessing a "lock and key" mechanism. An odor molecule binds to a specific receptor combination based on its shape, size, and chemical groups. Turin suggested that smell is linked to the vibrational spectrum of molecules. New experiments have rejected this theory (Keller and Vosshall, 2004). Thus, quantum tunnelling is present in many biological processes.

10.6.8 Coherent Excitation Transfer in Photosynthesis

Photosynthesis is realized in plants, algae, or bacteria, which convert light into chemical energy. Photosynthesis involves many complex processes such as redox reactions, hydrolysis, long-range excitation transfer, redox reactions, hydrolysis, proton transport, or phosphorylation. In photosynthesis, energy conversion starts with the absorption of an incident photon. This is carried out by a pigment molecule, such as a chlorophyll, porphyrin, or carotenoid molecule. These molecules ensure high photoabsorption. They enable efficient excitation transfer from the primary absorber to the center of the reaction. The reaction center is a pigment-protein complex containing a "special pair." By becoming excited, it donates an electron to an acceptor molecule. Fast secondary processes trigger the release of protons that are transferred across the membrane. They then synthesized ATP from ADP.

10.6.9 Spin and the Magnetic Orientation of Migratory Birds

Various animals can derive direction information from the geomagnetic field (Wiltschko and Wiltschko, 1995). Some mammals perceive the EMF as a polarity compass and detect "north" and "south." Birds and reptiles rely on an inclination compass that differentiates between poleward and equatorward directions. The orientation in the MF needs the presence of visible light beyond a certain photon energy. An oscillating magnetic field (0.1–10 MHz) can disturb the bird's senses. Vision-based magneto sensing might originate in the light-induced formation of a radical pair (Schulten et al., 1978). Radical pair mechanisms are needed to explain the induced dynamic polarization in nuclei (Closs, 1969). When light falls onto a receptor in the bird's eye, it excites it into a singlet. The molecule may then transfer an electron to a neighbouring acceptor molecule. The formation of a pair of radical molecules usually starts in the singlet state. Hyperfine couplings with the molecular nuclei result in an interconversion between the singlet and triplet states. Spin is rather well protected from environmental influences on a short-term scale. It may be assumed that the spin pair remains quantum correlated, i.e., entangled in this process. Even the interference of a weak external oscillatory magnetic field noise is not able to destroy entanglement. The electron spins both in the presence of the nuclei and in the EMF, which causes the ratio between singlet and triplet states vary. Many chemical reactions are spin-dependent. The transfer of electrons from the acceptor to the donor should influence the ratio of molecular products that are formed in the bird's eye. Transduction from the radical pair to the neuronal correlates was proposed (Weaver et al., 2000). It depends on the size and temperature dependence of the system to create a sensitivity. The protein cryptochrome found in the bird's retina mediates the radical pair mechanism (Wiltschko and Wiltschko, 2006). Both the electron transfer from a photo-excited FAD along a chain of tryptophan molecules and the reverse recombination reaction are perhaps sensitive to the EMF (Solov'yov and Schulten, 2009). Maeda et al., (2008) showed that the radical pair mechanism in the EMF is strong enough to alter the chemical end products in a complex that was formed from a carotenoid, a porphyrin, and a fullerene C60. that magneto sensing is related to quantum-correlated electrons, Cai et al., (2009) used a sequence of short radio-frequency pulses to get active quantum control over the radical pair spins immediately after their creation.

10.7 QUANTUM PHYSICS AND THE HUMAN MIND

About two decades ago, Roger Penrose raised the question of whether classical physics alone could suffice to explain the enormous problem-solving capabilities of the human brain (Penrose, 1989). And he speculated that a combination of currently irreconcilable pieces of physics—quantum theory and general relativity—might open a new window to our understanding of human consciousness. Together with the consciousness scientist Stuart Hameroff, proposed a model that assumes that the human mind may exploit at least two conformations of microtubule as values of a quantum bit. The quantum nature of the proteins solves complex computational problems in the brain, while the act of consciousness would be linked to a gravity-induced collapse of the quantum wave function (Hameroff and Penrose, 1996). However, no one has ever been able to prepare and characterize a coherent macroscopic quantum superposition of two conformations. Even if it exists, decoherence is believed to be orders of magnitude too fast to make it relevant on physiological time scales (Tegmark, 2000). A collapse of the wave function is currently only one of many models to explain the emergence of classicality in quantum physics. The dynamics of the proposed gravitational collapse are neither understood nor observed. It may also surprise you that microtubules were chosen as the decisive agents in quantum consciousness. They are by no means special to the human brain but rather ubiquitous cell support structures. In spite of its potential deficiencies, the model serves a purpose in that it stretches scientific fantasy to its very limits.

10.8 MAY QUANTUM PHYSICS SPEED UP BIOLOGICAL EVOLUTION?

The idea starts with the question of how a complex protein or strand of DNA could have been formed by random trials and mere chance from primordial amino acids or a series of nucleotides up to the high degree of complexity that is required to drive self-replication and evolution. Whether a faster macroscopic quantum sorting mechanism was involved in finding the first self-replicating molecule on Earth (McFadden, 2000). Its realization on our early Earth must have involved thousands of atoms and molecules in a warm and wet environment. And under the condition that all sorts of molecules were available, the formation of the sample molecular structures was accessible. And that the molecules were delocalized over large areas in the given environment. The latter need is in variance with the findings of molecule decoherence experiments (Hornberger et al., 2003). This confirms that any measurement—be it a collision with other molecules, phonons, or photons—can destroy the quantum delocalization if the interaction retrieves position information. If we hypothesize that large molecules could be delocalized in a primordial soup, the fastest speed-up in Grover's quantum search still has only a "square root" advantage. And the number of combinations is still stupendous.

Quantum physics and the life sciences are both attracting increasing interest, and research at the interface between both fields has been growing. As of today, experimental demonstrations of quantum coherence in biology are still limited to the level of a few molecules. This includes, for instance, all quantum chemistry, tunnelling processes, coherent excitation transport, and local spin effects. Quantum biology has stimulated scientific reasoning and fantasies and triggered hypotheses ranging from "exploratory" and "visionary" over "speculative" to "very likely to be wrong." The status of research does not always allow one to draw a precise borderline between these classifications. Experimental facts are missing, theoretical understanding is still an enormous challenge, and scientists are arguing both in favour and against a variety of these ideas. Fascinating combinations of physics and biology can be understood already. Many interconnects between quantum physics and the life sciences has been identified and the status of experimental skills is great. But the complexity of living systems and high-dimensional Hilbert spaces is even greater. When we talk about quantum information, the discussion always circles around exponential speed-up. But in living systems, any improvement of a few percent might already make the difference in the survival of the fittest. Even if coherence or entanglement in living systems were limited to very short time intervals and very small regions in space—and all physics experiments up to now confirm this view—simple quantum phenomena might result in a benefit and give life the edge to survive. We still must learn about the relevance and evolutionary advantage of quantum physics in photosynthesis, the sense of smell, or the magnetic orientation of birds.

Review Questions

Short Answer Questions

1. What do you understand by gravity?
2. What is acclimatization?
3. How gravity helped the primitive life forms?
4. What do you understand by ontogenic development?
5. Gravity influences cell size—comments.
6. What is evolutionary radiation?
7. What do you understand by Cambrian explosion?
8. What do you understand by high gravity induced stress?
9. An aquatic snake is unaffected by the impact of gravity—explain.
10. What do you understand by proprioception?
11. Define gravikinesis.
12. What is statolith?
13. Name the gravity sensing organ in *Aplysia*.
14. What is microgravity syndrome?
15. What is the full form of EMF?
16. What is geomagnetotaxis?
17. Who introduced the idea of animal gravitation?
18. Who found honeybee as a model animal for neuroethological study?
19. What do you understand by radical-pair mechanism?
20. What is the view of magnetite-based MR model?
21. What is magnetite?
22. Comment on magnetic compass.
23. What do you understand by visual learning?
24. EMF serves as an orientation cue—explain.
25. What is the full form of VGCC?

Long Answer Questions

1. Give an account of graviperception in animals.
2. Critically examine the phenomenon of magnetoreception in insects.
3. Write a note magnetoreception in relation to light.
4. Discuss the relationships between quantum physics and behavioural biology.

Chapter 11

Sexual Behaviour and Parental Care

Love is the first criteria of social life. Love is division in unity and unity in division. Without love, no social form can be born, grow, and flourish. Love is a condition of becoming and being the whole. Play has the second importance in social life. Play is fantasy in reality and reality in fantasy through which social forms are created. Play between and within the sexes is a dynamic key of social life. Males contests in hope of being chosen by the females to mate. Sex is the principal dynamic in organizing social life. Play in sex is dynamic by which social forms grow and develop. Durkheim described play, when he explained the division of labour in society in terms as used by Darwin the "struggle for existence.

Sex, in fertile heterosexuals, results in production of babies. Sex is packed with psychological and liturgical power. Sex involves physical realities and powerful symbolism. Sexually charged love is especially bonding. In lemurs, females may form social groups. Males of some lemurs try to kill offspring of females that mated with another male. Animals form social groups during certain times of the year. Many bird species flock together in foraging groups in winter. These birds sought one another out in winter, set up breeding territories in spring, and go to great lengths, to keep same birds out of their territory. For many species, social behaviour is flexible and can be adopted or abandoned, depending on environmental condition and time of the year.

One study used House mice (*Mus domesticus*), to select animals that spent more time running in a wheel. Selected male and female mice made up experimental groups. Their performance was compared with control group. Several pairs of mice, from each of two groups, were allowed to breed, and mice were tested after 13 generations. Scientists placed individual mice from both groups in cages having a wheel. Each wheel was linked to an automated system, that recorded times the wheel in each cage turned. Offspring of 'selected' mice ran faster than offspring of 'control' mice after 13 generations of selection.

Peacock has an extravagant train or tail. When exposed in breeding season, train is a large, colourful and impressive signal to a female of his readiness to mate. Some feathers end in an apparent eye-spot in the tail. Eye-spots spread over the tail, and their number varies from male to male. Wild peafowl live in small groups for most of the year. In breeding season, males are solitary, establish a territory in forest, and advertise their presence by calling. Call being a harsh, unearthly cry. Peahens visit males at their display sites, select male to mate, lay about 10 eggs, and incubate the same.

11.1 THEORIES ABOUT SEX

Perhaps no other natural phenomenon has aroused so much interest other than sex. The insights of Darwin and Mendel have illuminated so many mysteries, but have so far failed to shed more than a dim and wavering light on the mystery of sexuality (Graham 1982). A parent that reproduces sexually gives only one-half of its genes to its offspring, whereas an organism that reproduces by dividing passes on all of its genes. Sex also takes much longer and requires more energy than simple division. Thus, origin of the sexual process remains one of the most difficult problems in biology (Crow 1988). Various incompatibility factors pass along more "costs" through sex. Some of which are life threatening that are automatically inherent in this "expensive" means of reproduction. However, "sex occurs in all major groups of life". Sex would not have been evolved and retained, unless it is advantageous. Evolutionary biologists have suggested four different theories, known as: (1) The Lottery Principle; (2) The Tangled Bank Hypothesis; (3) The Red Queen Hypothesis; and (4) The DNA Repair Hypothesis, in addition to (5) The theory of sexual selection.

FIGURE 11.1 A Peacock's tail evolves due to sexual selection.

11.1.1 Sexual Selection Theory

Sexual selection theory has been revived over last two decades through combined efforts of researchers. Such revival involves theoretical population genetics, experimental behavioural biology, primatology, evolutionary anthropology and evolutionary psychology. Presently natural selection theory serves as conceptual and rhetorical foundation for evolutionary psychology (Tooby and Cosmides 1992). On the other hand, sexual selection theory seems to guide more actual day-to-day research. Important works on sexual selection include the works of Darwin (1871), Andersson (1994), Zahavi and Zahavi (1975).

Darwin distinguished between male competition for female mates (which typically gives rise to weapons) and female choice of male mates (which typically gives rise to gifts and ornaments) in sexual selection process. He recognized that female choice (Figure 11.1) and male competition are often two sides of the same coin, because mate choice by one sex usually implies competition by other sex, either through direct "interference competition" (e.g. physical fights over the opposite sex) or through indirect "exploitation competition" (e.g. scrambles to find and seduce the opposite sex before someone else does). Darwin had no real explanation of why males usually compete harder for mates than females do; why males court, and female choose-though he offered a staggering amount of evidence that this pattern holds from insects through human. After Darwin, sexual selection received such a frosty reception from Wallace and others that it was virtually forgotten (Cronin 1991). Modern synthesis of Mendelian genetics and Darwinism in 1930s viewed male competition as a sub-class of natural selection, while continuing to reject female choice. Sexual ornaments were assumed to intimidate other males or were "species recognition weakness" to help animals avoid cross species mating. Perhaps the best recent theoretical and empirical review of sexual selection was done by Andersson (1994). Accessible introductions to human sexual selection include works of a number of scientists. Champions of Zahavi's (1975) handicap principle have emphasized selection for genetic indicators—also called "good genes", "good sense" or "healthy offspring" selection. Champion of R.A. Fisher's (1930) runaway process has emphasized selection for aesthetic displays, also called "good taste" or "sexy son" selection. In evolutionary biology, these different mate choice criteria are often considered competing models of how sexual selection works, but there is now sufficient evidence for each (Andersson 1994) that they can be considered well established, often complementary selective forces. Of course, mate choice favour many other important qualities including parental ability and resources, fertility (Singh 1993), optimal genetic distance and similarity in appearance, behaviour and personality.

Traditional approaches to the study of behaviour here typically assumed that behavioural patterns, especially elements of reproductive behaviour, are invariant within species. Recent research on a diversity of behavioural traits in a wide array of taxa provides evidence that genetically based geographic variation in behaviour is common. Comparisons of populations that display geographic variation in behaviour can offer substantial insight into mechanisms of adaptive divergence and initial stages of speciation and in behavioural evolution.

11.1.2 The Lottery Principle

This Principle was first suggested by Williams (1975). Williams' idea was that sexual reproduction introduced genetic variety to enable genes to survive in changing environments. He used lottery analogy to get across the concept that breeding asexually would be like buying a large number of tickets for a national lottery, but giving them all the same number. Sexual reproduction would be like purchasing a small number of tickets, but giving each of them a different number. The essential idea behind the principle is that sex introduces variability. Organisms would have a better chance of producing offspring that will survive, if they reproduce sexually. Sex brought the variations that could allow organisms to survive change (Zimmer 2000). According to Ridley; a sexual form of life will reproduce at only half the rate of an equivalent clonal form. The halved reproductive rate of sexual forms are probably made up for a difference in quality. Thus, an average sexual offspring is probably twice as good as an equivalent cloned offspring (Ridley 1993). Sexual reproduction provides the best defense against the rapidly reproducing, infectious species that threaten the survival of organisms. Species diversity results from combining different gene pools that favour the survival of those that are sexually reproduced over those that are asexually. "Diversity in the species helps an organism to maintain its competitive edge in nature's struggle of "survival of the fittest." Lottery principle suggests that sex would be favoured by a variable environment, although a close inspection of the global distribution of sex reveals that where environments are stable, sexual reproduction is most common. In contrast, in areas, where the environment is unstable, asexual reproduction is rife.

11.1.3 The Tangled Bank Hypothesis

This Hypothesis suggests that sex evolved to prepare offspring for complicated world around them. Darwin referred to a wide assortment of creatures, all competing for light and food on a "tangled bank." In any environment, where there exists intense competition for space, food, and other resources, a premium is placed on diversification. According to Zimmer, a clone specialized for one niche can give birth only to offspring that can also handle the same niche. But sex shuffles the genetic deck and deals the offspring different hands. It is basically spreading out progeny, so that they are using different resources. However, today we find organisms that reproduce asexually, as well as, organisms that reproduce sexually that lead to the obvious question—Why do some organisms continue to reproduce asexually, while others have "evolved" the ability to reproduce sexually? (Herrub and Thompson 2004).

11.1.4 The Red Queen Hypothesis

Van Valen (1973) suggested through the hypothesis that the probability of organisms of becoming extinct bears no relationship to how long they already may have survived. Cartwright put it: "It is a sobering thought that the struggle for existence never gets any easier; however, well adapted an animal may become, it still has the same chance of extinction as a newly formed species (Herrub and Thompson 2004). This is a "genetic arms race", in which an animal constantly must run the genetic gauntlet of being able to become more fit. They constantly have to "run to try to improve". The red queen hypothesis seems to be the favourite of biologists. Parasites have short lives and large populations in comparison to hosts they attack. Parasites are probably going to adapt to most prevalent genes complexes of their host, which means that there is, in general, a selective advantage to rare alleles and recombinations. This principle states that since every improvement in one species will lead to a selective advantage for that species variation normally continuously lead to increases in fitness, in one species, or another. Since, in general different species are coevolving, improvement in one species implies that it will get a competitive advantage on other species and thus, be able to capture a larger share of resources available to all. This means that fitness increase in one evolutionary system will tend to lead to fitness decrease in another system. Only way, that a species involved in a competition, maintain its fitness relative to others is by in turn improving its design. Obvious example of this effect is "arms races", between predators and prey, where predators can compensate for a better defense by the prey by developing a better offense. In this case, it might be considered that the relative improvements to be also absolute improved in fitness. Thus, relative progress is necessary, just for maintenance.

Comment

According to red queen hypothesis, cross fertilization is advantageous, as it allows for production of genetically variable progeny, some of which escape infection by parasites that are "tracking" common host genotypes (Hamilton 1980). Theory has gained empirical support from a variety of approaches, but seems to require that parasites have very severe effects on host fitness, which could restrict its generality. Parasites may select for clonal diversity, instead of sex per se, which could lead to replacement of ancestral sexual population by a set of genetically diverse clones. Both of these difficulties can be overcome by either rank-order truncation selection, against most infected individuals, or by stochastic accumulation of mutations in clones that are driven though periodic bottlenecks by parasites.

11.1.5 The DNA Repair Hypothesis

Bernstein and others suggested that the lack of ageing of the germ line results mainly from repair of genetic material by meiotic recombination during formation of germ cells. Thus, basic hypothesis is that the primary function of sex is to repair the genetic material of germ line (Bernstein et al., 1989). DNA can be damaged in at least two ways. First, ionizing radiation or mutagenic chemicals can alter the genetic code. Secondly, a mutation can occur via errors during replication process itself. These theories valiantly attempt to explain why sex exists now, but they do not explain the origin of sex.

11.1.6 Other Theories

In addition to four above-mentioned theories, the following theories have also mentioned in the literature:

Runaway sexual selection (Fisher)

Positive feedback between an ornamental trait in a male and female preference for trait can lead to very elaborate traits. Females prefer males with traits that confer an initial survival, or reproductive advantage. Male's trait becomes exaggerated as female's preference for it is enhanced. At some point, cost to male outweighs benefit.

Good genes model of sexual selection

External trait indicates desirable underlying genetic makeup. In male guppies, brightly orange coloured males are preferred. Brightness increases with carotenoid content of the diet. So, brightness reflects foraging ability. This is also true for bird colour, and length of tail feather. Female fruit flies, either choose from a large number of males or force to mate with a randomly chosen partner.

Handicap hypothesis

Apparently, some secondary sexual characteristics can be harmful traits in males, but becomes attractive to females, as they indicate male's capacity to cope with them (Zahavi), or as they represent valuable trait of enhanced inmmunocompetence, often reflected in unexpected ways, such as singing ability.

Hamilton and Zuk proposed that genetic cycles of changing resistance in hosts to parasites, degree of resistance are reflected in secondary sex traits, which could be a clue in mate choice. In barn swallows, adult ornament size is inversely related to parasite burden. Cross fostering tests determine that parasitism level is strongly heritable. Development of tail influences female choice of mate, and reflects parasite load. In bowerbirds, male mating success is negatively correlated with louse burden. Zahavi realized that an honest signal must be costly to produce, and called this handicap principle. Cost of producing a signal will be too high for low quality males. Hamilton and Zuk forged a link between quality of individual and their sexual signals. They argued that disease resistance is crucial in evolution of sexually selected characteristics. In essence, only males with genes for say parasite resistance would be in prime condition, and thus, able to express best sexual signals. Flip side of theory is that sick males look drab, in comparison with healthier rivals.

Sexual Selection

Sexual selection, an aspect of evolutionary biology, explains why male deer have large antlers. It also explains how male peacocks have fancy tails. Sexual selection theory provides an evolutionary sense of courtship. However, it has generally been agreed that sexual selection is now politically loaded. Darwin first documented his view of sexual selection in "On the Origin of Species (1859)". Darwin later elaborated this in The Descent of Man and Selection in Relation to Sex (1871). He felt that to account for some non-survival adaptations, natural selection alone was insufficient. However, the account that is provided here on sexual selection is based on the article "Beauty and Beast" written by J. Roughgarden in the American Scientist. (https://www.americanscientist.org/article/beauty-and-the-beast).

Darwin emphasized psychological continuity between animals and humans. Darwin suggested that females "choose" their mates. Darwin viewed the cumulative effect of female choice as causing the evolution of things like the peacock's tail, known as an ornament. From his latter book, it appears that female birds have long been selected over the more attractive males.

Thus, sexual selection is the study of evolution's secondary sexual characters since Darwin's claim of animal minds. Female animals' choice of male is based on an innate sense. Sexual selection is often wrongly synonymized with female choice, but the two are different. Female choice is only one process of sexual selection. The competition between males for access to females is also another process. Thus, the processes of Darwin's sexual selection involve both female choice and male-male competition, which govern the evolution of male ornaments and male armaments, respectively. Although they are distinct processes, Darwin thought they were reinforcing. The history of sexual selection is intertwined with alternative views about the varied importance of female choice versus male-male competition and animal minds as the primary mechanisms. Wallace disagreed with Darwin about sexual selection. Wallace agreed with the overall idea of sexual selection by female choice. but he did not support the claim that animals had minds that allowed such choices. Now the term "choice" has been replaced by the mechanical stimulus and response. Experiments showed that male fruit flies with eye-colour and eye-shape mutations sire more offspring when male flies with the wild-type eye colour and shape surround them. Thus, female fruit flies, in some sense, choose the males they mate with.

Mendel's laws and mathematical population Genetics was brought into evolutionary studies, including those focusing on anatomy, palaeontology, taxonomy, and systematics, from the 1940s through the 1960s in what is called the Modern Synthesis. The architects of Modern synthesis, such as E. Mayr and others, primarily showed interest in relation to speciation but not in female choice. Mating behaviour, including female choice specifically, was of interest as an "isolating mechanism" preventing hybridization between the gene pools of geographically separated species. In sympatric speciation, female choice was a matter of interest to some extent as a possible cause of speciation without geographical separation, in which a single gene pool divides into two because females evolve into two groups for some reasons. Subsequently, each of which selects a different type of male to mate with, and the architects of Modern Synthesis paid inadequate attention to sexual selection. The theory of sexual selection suddenly sprang to life in the 1970s, resuscitated by the writings of W. Hamilton, J.M. Smith, G. Williams, R. Trivers, G. Parker, and R. Dawkins. Historians of biology mostly refer to the long period of neglect as the eclipse narrative, in the long history of sexual selection. The eclipse may be due to animal minds (1871).

11.1.7 Cost of Sexual Reproduction

Ubiquity of sexual reproduction is puzzling, as this is associated with several costs. Such costs are associated with mating, or conjugation. Many species take time and energy to secure a mate. Many plants invest substantial resources in floral display, and nectar rewards. Sexual reproduction is often slower than asexual. During mating, individuals are less able to gather resources and evade predators (Howard and Lively 1994). In species, with separate sexes or mating types, costly sexual conflicts can arise. Seminal fluid of *Drosophila* contains toxins that reduce fitness of mated females. Sexual individuals have a lower reproductive output per capita. If sexual couples and asexual individuals produce same average number of offsprings because one sexual partner does not contribute resources to offspring, reproductive output per individual for asexual species is twice that for sexual species. Hence, twofold cost of the sex. Producing offsprings by randomly mixing genes, with those of another individual is risky. Mutant allele that shifts resources towards asexual reproduction would rapidly out compete and displace its more sexual kin. Sex should be an evolutionary dead-end, a relict that is observed only rarely. Contrary to this expectation, selection for frequent mating is predicted to be stronger in males, than in females. Males gain fitness from each extra mating. Female fitness gains may cease, as mating frequency increases. Presence of female mating costs may reflect sexual conflict over mating. In such conflicts, males may evolve traits for increasing their relative fitness that decrease fitness of females with which they mate, or attempt to mate. In *Drosophila*, females that mate at high frequencies suffer fitness costs (reduced longevity and reproductive success), as a result of actions of male seminal fluid, Accessory Gland Proteins (Acps).

Acp-mediated mating cost is potentially large. Acps mediate a variety of effects, like stimulation of female egg production, reduction of female receptivity, ensuring effective sperm storage, and promotion of male success in sperm competition that benefit males. Female mating cost that arises from Acp transfer by males may be a side effect of Acp function, or a direct effect that is selected to reduce likelihood of female re-mating, and/or to increase current investment in reproduction. Initially, natural selection may have caused females to evolve sensitivity to substances, such as Sperm Protein (SP), and allowed them to adaptively modulate egg production and receptivity, after sperm transfer. It is not clear, whether SP alone is responsible for female mating costs, or whether harm is caused by interaction of SP with other ejaculate molecules. SP binds to sperm and can be detected on sperm heads, several days after its deposition in female reproductive tract. There is no reduction in mating cost in females, continuously exposed to spermless males, which suggests that SP must be free from association with sperm. SP stimulate egg production by releasing of juvenile hormone, and this release stimulates oocyte progression in the ovary. Increased JH levels are negatively associated with lifespan in other insects as reported in literature.

11.1.8 Symmetric and Asymmetric Sexual Reproduction

Sexual reproduction is divided into symmetric and asymmetric. In asymmetric reproduction, a fragment of genome is transferred. Reproductive individuals have genomes that have already proved themselves in current environment. Genome mixing is a risky endeavour. Conditions that favour evolution of high rates of sex and recombination are often restrictive. Proximate effects of asymmetric and symmetric sex are different. Asymmetric DNA transfer creates potential for sex to evolve simply, because genetic elements that cause themselves to be copied and transferred to other individuals, can spread in a population, as long as they infect new cells faster than they kill their host, or otherwise reduce host fitness. In this case, proximate effect of sex is transfer of the genetic elements, which enables them to spread in a manner analogous to spread of a disease. Phages and bacterial plasmids need not be beneficial to their hosts, although they are likely to spread, if they increased host fitness. For example, by conferring antibiotic resistance, or tolerance to new environmental conditions.

Asymmetric sex might represent an accidental byproduct of mechanisms that are encoded in genetic elements that enable these elements to spread. Although eukaryotic sex is by and large symmetric, cytoplasmic elements are often transferred in an asymmetric fashion. Transposable elements spread from one genome to another after syngamy. If transposable or cytoplasmic elements arise that make their host cells reproduce sexually with partners that would otherwise reproduce asexually, such sex drivers would spread through population, as long as, driving element was represented more often among sexual offsprings. Sexual processes in bacteria are almost always asymmetric, including transduction, transformation and conjugation. Although most bacteria show some level of asymmetric sexuality, extent of genetic mixing varies from vanishingly rare, as in obligatorily intracellular *Rickettsia prowazekii,* to common place, as in *Deinococcus radiodurans.*

In symmetric sexual reproduction, two genomes fuse and subsequently separate each time, producing cells containing mixtures of genes, due to chromosomal segregation and recombination. Symmetric sex results in alternation of haploid and diploid phases in a life cycle. Although asymmetric processes (such as horizontal transfer) are impli-cated in gene transfer into eukaryotic genomes, such events are extremely rare, relative to rate of symmetric sex, as observed in majority of eukaryotic species. Few eukaryotes are exclusively asexual. A list of known asexual vertebrate species comprising fish, amphibians and reptiles which represent a tiny fraction of about 42,300 known vertebrate species. Asexual eukaryotes tend to be isolated on tree of life; only rarely is a genus or larger taxonomic group, composed entirely of asexual species. Ancient asexuals are bdelloid rotifers that show no signs of mating, or genetic mixing. *Giardia* is potentially even older asexual. The failure to observe sexual reproduction does not mean that it never occurs. Many species, that were previously considered asexual, have subsequently been caught in sexual act. Wide-spread occurrence of sex, despite its seemingly overwhelming costs, is known as paradox of sex.

11.1.9 Sexual Reproduction and Recombination

Genetic exchange allow repair of double-stranded damage, in which an undamaged gene copy is used as a template. In prokaryotes, evidence for repair hypothesis is weak, and in eukaryotes this occurs form a donor individual to a recipient. In *Drosophila* males, proper segregation occurs without chiasmata. Minor changes in number and position of chiasmata could have a substantial impact on the probability on that recombination occurs between a given pair of loci. Recombination rates cannot be strongly constrained, as they evolve in response to selection. Such rates often increase, after periods of strong artificial selection for other characteristics.

Problem of ensuring proper segregation during meiosis can be entirely circumvented by reproducing asexually. Cross-over frequency per chromosome is as uniform across the taxa as chiasmata stabilize pairing of homologous chromosomes during meiosis. Homologous chromosomes that are joined by too few chiasmata can prematurely slip away from each other on metaphase plate. Chromosomes, that are joined by too many chiasmata can be too tightly intertwined to migrate to opposite poles, during anaphase I. Either case can result in abnormality When sex increases variation in fitness in a population, such variation need not be favourable. Variation produced by sex would reduce fitness, compared with an asexual population, composed entirely of fittest individuals. Modifier alleles that promote sex are selected against, even though sex increases variation in population. Variation alone is not sufficient.

Biotic and abiotic conditions of an organism are not constant. Gene pool in a population is changing, through mutation, migration and selection. Most evolutionary hypotheses for sex stem from idea that sex generates greater variability, because chromosomal segregation and recombination break down genetic associations. It is thought that modifier alleles, that increase sex frequency and recombination are favoured as they improve ability of a population, evolve by increasing genetic variation, on which natural selection acts. This idea is simple and appealing. Theoretical models, however, have revealed several inherent problems. For example, sex need not increase genetic variation in a population. Genetic variation can be selected against, and evolution need not favour increased levels of genetic exchange, even when genetic exchange increases genetic variability and variability is favourable. Sex has no effect on genetic variation in a population, in which observed frequency of each genetic combination is equal to its expected frequency. If genes in a population are already well mixed, shuffling genomes further by chromosomal segregation and recombination will have no effect. There is a long-term advantage to sex and recombination, whenever linkage disequilibrium is negative. With negative disequilibrium, extreme genotypes are under represented and can be regenerated by genetic mixing, thereby increasing genetic variation in fitness, which improves response to selection.

11.1.10 Maintenance of Sex

Two hypotheses—mutational accumulation hypothesis, and red queen hypothesis, are suggested for evolutionary maintenance of sex. These theories seem to dominate present literature (Bell 1982). According to former model, sexual populations gain advantage over asexual populations, due to efficiency of recombination, in reducing population mutation load. Muller (1964) first suggested that asexual populations might be undermined, by mutation accumulation, reasoning that stochastic processes would lead to an inexorable decline in clone's fitness. This version of mutation accumulation hypothesis is known as "Muller's ratchet." Recent studies have extended Muller's basic idea, representing deterministic model of mutation accumulation. For both versions of mutation accumulation hypothesis, any advantage to sex depends strongly on genomic mutation rate, and relationship between mutation number and individual fitness ("fitness function"). Estimates of mutation rate vary and little is known about general shape of fitness function in natural populations. Moderate effects of parasites combined with mutation rates of 0.5 to 1.0 per genome per generation led to evolutionary stability of sex (Howard and Lively 1984).

Indefinite maintenance of sex requires other benefits. In prokaryotes, sex might not be maintained indefinitely; it might arise sporadically, after appearance of altered genetic elements that became infectious, and promoted their own transfer. In eukaryotes, symmetric sexual reproduction arose roughly one billion years ago. Cellular processes that are involved in syngamy, meiosis, and gamete production have evolved to be extremely complex, and involve hundreds to thousands of genes. Such nuclear genes are identified in *Caenorhabditis elegans*. This indicates that, rather than arising sporadically, sexual reproduction has persisted for most of the evolutionary history of eukaryotes.

11.1.11 Conflicts in Sex

Conflicts between females and males over reproductive decisions are common. In *Drosophila*, and other organisms, there is often a conflict over how often to mate. Mating frequency that maximizes male reproductive success is higher, than that which maximizes female reproductive success. Frequent mating reduces female lifespan and reproductive success, a cost that is mediated by male ejaculate accessory gland proteins (Acps). A single Acp sex peptide (SP or Acp70A), decreases female receptivity and stimulates egg production in first mating of virgin females, is a major contributor to Acpmediated mating costs in females. Females continuously exposed to SP-deficient males have significantly higher fitness, and higher lifetime reproductive success than control females. Hence, receipt of SP decreases female fitness, making SP first identified gene that is likely to play a central role in sexual conflict.

Sexual selection, in general, and sexual conflict, in particular, should affect evolution of lifespan and ageing. Virgin females, from populations, evolving under sexual selection had reduced lifespan, as predicted by sexual conflict theory of ageing. This reduction was due to increased baseline mortality, rather than an increase in age-specific mortality rates with age. It is suggested that system-specific idiosyncrasies may often modulate general effects of male-female coevolution on evolution of aging. Sexual conflict over mating rate is suggested to play a pivotal role in male-female coevolution, and females are predicted to reject superfluous mating attempts. Increased mating rate increases time to oviposition, and reduces likelihood of successful reproduction in a polyandrous spider *Stegodyphus lineatus*. Female mating rate negatively affected offspring body mass. Manipulated females produced fewer offsprings, than control females. Observed patterns imply a net cost of polyandry to females and suggest that natural mating rates can be suboptimal for females under natural conditions.

Selection favour evolution of high reproductive rate early in life, even when this results in a subsequent increase in rate of mortality because selection is relatively weak late in life. Optimal reproductive schedule of a female may be suboptimal to any one of her mates, and males may, thus, be selected to modulate female reproductive rate. Owing to such sexual conflict, coevolution between males and females may contribute to evolution of senescence. By using replicated beetle populations selected for reproduction at an early or late age, it has shown that males evolve to affect senescence in females in a manner consistent with genetic interests of males. 'Late' males evolved to decelerate sene-scence, and increase lifespan of control females, relative to 'early' males. This finding demonstrate that adaptive evolution in one sex may involve its effects on senescence in other, showing that evolution of optimal life histories in one sex may be either facilitated or constrained by genes expressed in other. Males may differ in their risk to females in terms of predation (on females or eggs), reproductive parasitism or harassment. If high quality bourgeois (territorial) males attract large numbers of predators, reproductive parasites harassing males or pathogens, then females must trade off the genetic benefits of mating with high quality bourgeois males against the cost imposed by enemies. High quality males may be riskier because they are more conspicuous or because they harbour higher densities of sexually transmitted pathogens or ectoparasites (Kokko et al., 2002). It is suggested that high quality males term of scramble competition on this model (Hamilton et al., 2005).

11.1.12 Effects of Diet on Sex

Pre-copulatory sexual cannibalism by females affects reproductive success of both sexes in different ways in some species. Females get benefit from a meal, and male face risk of not reproducing at all. This conflict predicts evolution of traits to avoid cannibalism, and ensure male reproductive success. Diet affects both lifespan and reproduction, leading to prediction that contrasting reproductive strategies of sexes should result in sex-specific effects of nutrition on fitness and longevity, and favour different patterns of nutrient intake in both sexes. Males and females share most of their genome, and intralocus sexual conflict may prevent sex-specific diet optimization. Both male and female longevity were maximized on a high-carbohydrate low-protein diet in field crickets *Teleogryllus commodious*, but male and female lifetime reproductive performances were maximized, and in markedly different parts of nutrient intake landscape has shown by scientists.

11.1.13 Communication in Sex

Signals produced by one sex and received by the other, play a critical role in mediating reproductive interactions. Signals can range from simply alerting animals to the presence or location of opposite sex conspecifics to a more complex advertisement of the signaller's individual identity and potential quality as a mate. Scent is a major signalling modality in many animals, providing genetically encoded information on species sex individual identity and kinship of owner, as well as, information on animal's current reproductive social and health status, all of which may influence mate selection. Olfactory and neural pathways involved in sex-biased responses to conspecific scent signals, an analysis so far suggests that both the main and accessory olfactory systems play an integral part in detection and procession of signals that coordinate sex recognition and reproductive behaviour. Main Olfactory Epithelium (MOE) detects airborne scents (largely volatile chemical components) potentially at some distance from the source. Accessory olfactory system detects volatile and involatile molecules that are pumped to Vomeronasal Organ (VNO) during contact with the scent source. These two systems detect at least partially overlapping sets of social chemo signals, which mediate different sexual and social responses through each system. There is conflicting evidence, particularly concerning the ability of animals to recognize the sex of conspecifics through scents detected only through the MOE. Knockout male mice lacking a trp2 cation channel (required for normal odorant-activated transduction in the VNO) appear to be unable to recognize the sex of conspecifics, abnormally showing the same mating and courtship behaviour towards both males and females (Ramm et al., 2008). Nevertheless, these animals retain some ability to detect urine scents through VNO. Detecting the reproductive status of potential sexual partners is a crucial component of sexual behaviour. Males should have access particularly in species with a short mating period. Ability to distinguish between sexually receptive and non-receptive females allow males to concentrate breeding effort at times, when they are most likely to be successful. Counting a female that is not sexually receptive is costly, because it takes time and energy away from other important activities and exposes males to a predation risk. Thus, to distinguish between sexually receptive and non-receptive females may provide greater fitness to males. In reptiles, female reproductive status may be advertised via visual tactile or chemical cues. Advertisement via chemical cues is probably the least costly because chemical cues are often a byproduct of other processes associated with physiological changes during reproduction. However, little is known about the role of chemical cues in communicating important reproductive information in reptiles. This is surprising as most squamate reptiles have well developed chemosensory system.

In snakes, chemical communication is essential for the display of normal reproductive behaviour. Cutting the vomeronasal nerve prevents courtship mating and recognition of sex in adder snake (*Vipera berus*) and garter snake (*Thamnophis* spp.). Chemical communication in reproduction is less clear in lizards. It is believed that lizards rely heavily on visual cues for most social interactions based on studies of highly territorial sexually dismorphic taxa, such as *Sceloporus* and *Anolis*, which exhibit elaborate push up and/or develop display. Chemical cues may be important for sex recognition in a number of skinks and geckos, but not in more visually oriented iguanid lizards. Cooper found that males of the broad-headed skink (*Eumeces laticeps*) were able to distinguish between post reproductive females and post reproductive females were injected with estrogen to induce receptivity. Role of visual system in reproductive behaviour of lizards is more poorly known. Australian southern water skink *Eulamprusheat wolei,* staged mating encounters before, during and after the female sexual receptive period. Male retreat site experiments were used to determine the ability of the male to detect female sexual receptivity via chemical and visual cues (Head et al., 2005).

11.1.14 Courtship Displays

Males of nuptial gift-giving spider *Pisaura mirabilis,* display death feigning behaviour-thanatosis as part of courtship prior to mating with potentially cannibalistic females. Thanatosis, a widespread anti-predator strategy, is exceptional in context of sexual selection. When female approaches a gift-displaying male (Figure 11.2), she usually shows interest in gift, but would sometimes attack male, and at this potentially dangerous moment, male can drop dead. When entering thanatosis, male would collapse, and remain completely motionless, while retaining hold of gifts. So, it is held simultaneously by both mates. When female initiate consumption of gift, male cautiously 'come to life' and initiates copulation. Death feigning males are more successful in gaining copulations, but do not have prolonged copulations. Death feigning may evolve as an adaptive male mating strategy, in conjunction with nuptial gift giving under risk of being victimized by females.

Courtship displays of animals especially of males have attracted the interest of both ethnologists and evolutionary biologists ever since Darwin

FIGURE 11.2 Courtship displays in birds.

(1871). Most early studies focused on the description of sexual behaviour patterns and their inter-sexual dynamics. Later studies sought to determine the function of specific behaviour patterns. Now, attention is being paid to the potential for sexual selection to influence the elaboration and complexity of sexual displays. Males tend to be competitive and charming, whereas females tend to be choosy largely as a result of sex differences in potential rates of reproduction which usually are greater in males than the females.

Male sexual behaviour patterns involve persuasion of female to mate perhaps by offering a nuptial gift, or by producing a stimulating courtship display. From a proximate perspective, persuasion involves a male's attempt to increase a female's sexual motiviation, or responsiveness above a threshold required for mating. Form an ultimate perspective, females that require more stimulation to reach this threshold can be regarded as being especially choosy, with respect to their willingness to mate with a particular male. There exists variation among female in their sexual responsiveness, which determines the extent to which each individual needs to be persuaded in order to mate. Variation in female responsiveness may thus impact variance in male mating success, as it is a male's ability to persuade individual female that increases his success, not his mean ability to all females. Variation in female responsiveness also may be an important determinant of variance in female mating success.

Salamanders were studied for exploring empirically the connection between male persuasiveness and female responsiveness as determinants of courtship and mating success (Sullivan et al., 1995). Most salamanders practice internal fertilization, but sperms are transferred indirectly by means of a spermatophore that usually is deposited on the substrate. Physical coercion by males of females to mate such forcible insemination almost certainly is impossible in such a system (Clutton-Brock and Parker 1995). However, evidence indicates that the behavioural displays and pheromones produced by courting males are persuasive in salamanders, serving to increase female's sexual motivation. In sala-mander *Desmognathus ocoee*, males of high persuasiveness enjoy enhanced mating success (Vinnedge and Verrell 1998).

Courtship in salamanders is a dynamic interaction between the sexes. Studies have revealed variation among females in sexual responsiveness towards different males. Significance of this for courtship outcome relative to variation in male persuasiveness has not been addressed specifically. Gershman and Verrel (2002) have tested the hypothesis that differences in aspects of sexual performance are associated with differences in male persuasiveness in salamander.

11.1.15 Origin of Mate Choice Mechanisms

Probably, the most fundamental form of sexual selection is mate choice, for various "indicators", or viability and fertility. Health, nutritional status, size, strength, social status, disease resistance, or overall vigour functions as indicator. Temporal variation is spatial variation in selection, and mutation pressure maintains heritable variation. Temporally varying selection may result from coevolution between ecological competitors, between predators and prey, and importantly, between hosts and parasites. Spatially varying selection in various geographic areas combine with migration, maintains heritable variation in a population. Mutation pressure helps in maintaining heritable fitness variation, as most mutations are harmful and give rise to excess of low fitness individuals. Sexually selected traits have higher heritability and genetic variance, than naturally selected traits despite strong directional selection.

Mate choice is behavioural outcome of mate preferences, which are usually "mental adaptations implemented as complex neural circuits, and constructed through interaction between genes and environment. Mostly, such systems function without conscious awareness, deliberation, or complex aesthetic feelings. Mate choice mechanism operate by rejecting some potential mates and accepting others. Sexual harassments of females are common in nature. "Successful rape" seems fairly rare, reported in ducks, squids, dolphins, orangutans, and humans. Mate choice evolves, based on rationale that random mating is stupid mating, and attempts to select genetic quality of mate. Ugly, unhealthy mates yield ugly, unhealthy offspring. By forming a joint genetic venture with an attractive, high quality mate, one's genes are likely to be passed on. Modern women, who deny "role of genes in human behaviour", tend to choose their sperm donors carefully. Mate choice may evolve through direct selection, for mate choice efficiency (better preferences lead to more or better offsprings) and through mutation, genetic drift and genetic linkage with another trait that is undergoing natural selection or genetic drift. The last 3 processes, generally, produce harmful changes in mate choice mechanisms, and unpredictability of such processes is important in explaining diversity of sexual selected ornaments across closely related species. Evolution of adaptive female choice by indirect genetic benefits relies on presence of genetic variation for fitness. Female choice, by genetic benefits fall broadly into good genes (additive) models and compatibility (non-additive) models where selection strength is dictated by genetic architecture of fitness. Major determinant of offspring fitness is genetic interaction between parental genomes.

Individuals do not mate indiscriminately. They choose their mates carefully. Mate choice should only occur, when the characteristics of the detected animal matches the characteristics of the choosing animal's internal template (Widemo and Saether 1999). Traditional models of mate choice assume preference for quality as fixed (Kokko et al., 2003). However, a sexually selected genotype that does best in one environment might have a lower fitness in a different environment (David et al., 2000). State-dependent reproductive decisions in response to social or environmental changes would be beneficial for fitness optimization (Qvarnström 2001). Males of species with female polymorphism exhibit plastic frequency-dependent mate choice (Van Gossum et al., 2005). Further, female morph frequencies may change rapidly (Fincke 1994), suggesting that flexible male behaviour is likely to be beneficial. Competition among males for access to females is intense with a high proportion of males never mating (Cordero et al., 1997). Male-male mating behaviour is probably costly, both in terms of time wasted and probable body damage to one or both actors. One often-used explanation for same-sex mating behaviour is that it only occurs because opposite-sex partners are absent or scarce (Bagemihl 1999). Hence, such behaviour should disappear once opposite-sex partners are available. However, in the field where males and females were abundant, males mount males regularly in *I. elegans* under natural conditions. Alternatively, some males may have an innate preference to mate other males. Rather than having an innate preference to mate males, males may adopt a preference for males in a critical period during early life (e.g. subadult phase after metamorphosis for damselflies) through imprinting (Kendrick et al., 1998) or at some point prior to adulthood (Hebets 2003). Although innate preferences or imprinting can explain why male-male mating behaviour was observed in all choice experiments, the switch in sex choice made by the majority of males cannot be explained in this way. A more flexible state-dependent explanation is needed to explain switches between the sexes in male mate choice in *I. elegans* (Van Gossum et al., 2005).

11.1.16 Mating Behaviour

Copulation behaviour is one of the more ancient social behaviours exhibited among metazoans. Male mating behaviour has been considered to be the most complex behaviour in *C. elegans*; and can be broken down into simpler sub-behaviours that allow cellular and molecular dissection. Male initially responds to hermaphrodite contact by placing his tail flush on her body; he begins moving backwards along her body until he reaches her head or tail, where he then turns via a sharp ventral coil. He continues backing until his tail contacts the vulva; at that region of hermaphrodite, he stops moving, inserts his spicules, and ejaculates into the hermaphrodite uterus. Male response behaviour is initiated when sensory neurons located in the rays of his tail contact a potential mate. Male stops forward locomotion, presses the ventral side of his tail against his partner's body, and begins moving backward, scanning his partner's vulva. Bilateral pairs of sensory rays of the male tail, numbered 1 (anterior) to 9 (posterior) mediate response and turning behaviour (Figure 11.3) (Barr and Gracia 2006). Recent theoretical models suggest

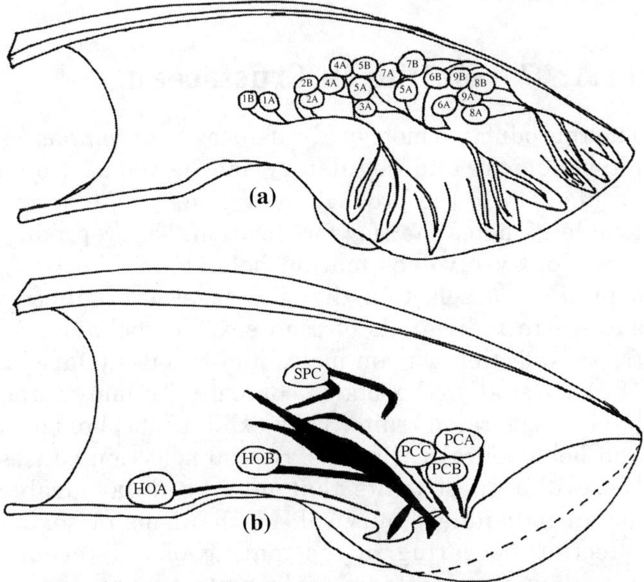

FIGURE 11.3 Male sensory neurons associated with mating behaviour. Diagram depicting the positions of ray neurons (a) and hook and Postcloacal Sensilla (PCS) and SPC neurons (b) In panel A, the male tail has 9 bilaterally arranged rays numbered 1–9 anterior to posterior (only one side is shown). Each ray contains the sensory dendritic process of an A type neuron and a B type neuron. Ray neurons are labelled according to what neuronal type they are, and with which rays they are associated. In panel B, sensory dendritic processes of the HOA and HOB neurons (asymmetrically located on the left side of the animal) and the PCS neurons are associated with the cloacal opening. The SPC proprioceptive neuron is physically associated with the spicule protractor muscles.

that males may respond to changes in paternity by adjusting their parental effort. Male's response depends on the availability of reliable paternity cues and relative costs and benefit of parental effort to male (i.e. its effect on the survival of young and alternative mating opportunities). Male breeding in pairs may be constrained because reduction in male parental efforts are unlikely to be compensated for by the female, and thus, survival of both related and unrelated young may decrease. In contrast, males breeding in cooperative groups may not have this constraint if other individuals in the group compensate for reduction in male parental effort. White browed scrubwrends *Sericornis frontalis*, breed in pairs and co-operative groups, typically with one female and two males (alpha and beta). The male parental effort relate positively to paternity for beta males, but not for alpha or pair males. Alpha males show paternity in all broods and always fed young. Beta males often had no paternity and sometimes did not feed young. Time spend near the fertile female was not and accurate predictor of the percentage of young sired in a brood. But it was a good predictor of having sired young in a brood (Whittingham and Dunn 1998).

11.1.17 Mating in Crustacean

The reproductive biology of crustaceans encompasses physiological and regulatory mechanisms. Some events lead to the production of gametes. Male and female gametes met for fertilization. The reproductive biology involves mating behaviour. Mating is a product of selection that increases reproductive success in individuals of each sex. For behavioural traits, selection acts on males and females (Darwin, 1871). Usually, the males compete for mates and have exaggerated traits. They exhibit morphological and behavioural variations. Sexual selection causes the evolution of traits that may be unfavourable for an individual's survival. Mechanisms of sexual selection operating in the mating of crustaceans are known. Aspects of male mating competition (intrasexual selection) and mate choice (intersexual selection) are interesting.

A difference in the action of sexual selection between the two sexes is called sexual dimorphism. The variance in reproductive success among individuals is greater in males than females. Where there is a sex difference, there is scope for selection (Skuster and Wade, 2003). Crustaceans pronounced sexual dimorphism involves body size, weaponry, and male attachment structures. It also includes exaggerated courtship displays and male aggression. Male fiddler crabs have large major claws. The claw-waving display attracts females (Blackwell et al., 2006). The shrimp live in a monogamous mating system exhibit considerable morphological dimorphism (Knowlton, 1980). Sexual selection operates in species of protandric hermaphrodite caridean shrimp (Bauer, 2006). Proximate mechanisms underlying sex differences in secondary characters. These are mating behaviours mediated by sex hormones. Androgens influence sexual and aggressive behaviour in vertebrates (Arnold, 2004). The androgenic gland handles the development of dimorphic traits, including behavioural traits.

Mating Competition

Competition occurs when two or more individuals are dependent on the same limited resource. There are two modes of competition (Nickolson, 1954). In a contest, one individual gains access to the resource and denies access to other. In scramble competition, everyone tries to maximize his share of the limited resource. But through direct interference from competitors. In fact, pure contests and scrambles are uncommon. They are two extremes in a continuum of competitive interactions (Parker, 2000).Due to anisogamy and sex differences in the investment of gametes and parental care, the male reproductive rate is lower than that of females (Parker and Simmons, 1996). So, females are the limiting sex. Males compete for access to receptive females. One sex is a critical resource for which individuals of the other sex compete. Ecological constraints and the availability of receptive females influence their distribution. This influences the number of potential mates for males and their ability to monopolize them. In some species females aggregate and female receptivity is asynchronous. There is a chance for a small part of the males to monopolize many females. They do so to maximise mating opportunities. In a polygynous mating system with intense male competition and high variation, male mating success occurs. Emlen and Oring (1977) classified mating systems by the behavioural means for individuals of limited sex to get mates. Christy (1987) noted eight kinds of mating associations in brachyuran crabs. They belong to three general categories. Males may (1) attract individual receptive females that they defend from other males; (2) defend resources that females need for breeding and mate with the females associated with these resources; or (3) compete to maximize the rate at which they encounter females. But neither defend females nor resources (Christy, 1987). These categories hold true for other crustacean groups. Each of them includes species with diverse competitive modes. In brachyuran crabs, the first two categories involve an element of contest and prevalence (Christy, 1987). In caridean shrimps, pure mate searching is predominant (Bauer et al., 2005).

Contest Competition

Contests for mating opportunities may be resource- or female-centred. They may manifest themselves in several ways. These are direct competitions for females, mate guarding, and resource defence. This also includes sperm competition and alternative male mating strategies.

Female-centred competition

Such competition is widespread among crustaceans. It ranges in intensity from intense aggressive fights over females to brief male-male aggressive interactions. The other is prolonged mate guarding to mate without attending to the female longer than needed for copulation. Male competition is related to the spatial and temporal distribution of receptive females and to the female reproductive life history. The morphology, ecology, resource defence, and female reproductive traits explain its mode of mating competition. Comparative accounts of mating behaviour in crustaceans are useful in this regard. On the temporal aspect of female availability, copulations occur when the female is soft shortly after moulting. The duration of female receptivity, during which females copulate hard-shelled eggs during the intermoult stage, is low. It does not extend for more than a few days. In several brachyuran crabs, copulation occurs after local decalcification of the hinge of the immobile opercula sealing the gonopore. This makes them flexible and mobile (Hartnoll, 1969). The temporal distribution of receptive females depends on the degree of synchrony of short receptivity episodes. The effect of the temporal distribution of females on male competition need a study on the breeding ecology of cancrid crabs. *Cancer magister* and *C. gracilis* are sympatric cancrid species. They resemble each other in many aspects of their reproductive ecology and biology. Both species inhabit open habitats where receptive females accumulate at breeding sites. In both species, many copulations are possible during a single receptive period. Sperm competition occurs. The combination of soft-shell mating and sperm competition in these crabs leads to the direct defence of mobile females with pre- and post-copulatory mate guarding. In *C. gracilis*, females produce many broods during a protracted breeding season. Female receptivity is *asynchronous*. In *C. magister*, females produce one annual brood. They are more synchronous in their receptivity period. C. *gracilis* males fight for the scarce receptive females and attempt takeovers of guarded females. In *C. magister*, males invest less in aggressive contests and tend towards scramble competition. The benefits gained from searching rather than staying and fighting for receptive females increase when another receptive female is found. Mating biology, ecology, and behaviour of grapsid crabs *C. lavauxi, H. crassa, and H. sexdentatus* are well studied. These show how different female-centred competitive modes evolved in related species that are similar in their mating traits. This is due to different combinations of influencing factors. Females in grapsid species mate during the intermoult period. Mating is limited to certain times when the gonopore opercula of the female become mobile. The temporal features of female availability in *C. lavauxi* and *H. sexdentatus* are similar. Despite their short and synchronous mating seasons, the OSR is male-biased in both species. The short receptive period of females decreases the number of receptivity episodes overlapping in time. Males of both species fight over receptive females. *C. lavauxi* males do not guard females. *H. sexdentatus* males exhibit post-copulatory mate guarding. This last for several days in a species with sperm competition. The difference in duration of sperm competition may be due to the different habitats occupied by these crabs. In intertidal habitats, *C. lavauxi* crabs occupy the high-intertidal zone. They face wave action during high spring tides and long periods of aerial exposure during low neap tides. Males do not guard females in harsh environments. *H. sexdentatus* crabs occupy the mid-intertidal zone. Their predators are protected from desiccation. This causes them to search for and defend receptive females for a prolonged period. The habitat difference influences the spatial distribution of receptive females. Thus, shaping the male competitive modes of these species. No data is available on female distribution within these habitats. *H. sexdentatus* females' ability to extend their receptivity period enables them to be more selective. This increases male competition and enhances mate-guarding. *C. lavauxi* and *H. crassa* crabs reveal that the two species differ in the temporal characteristics of female receptivity. In both species males compete for females. But do not exhibit mate guarding. *Helice crassa* crabs occur in open mud flats. They dig short-lived burrows. The lack of mate guarding in males is due to the high predation risk in open habitats coupled with surface mating. The latter restricts prolonged guarding on the surface. In species where mating is linked to moulting, post-copulatory mate guarding protects the moulted female from predation to prevent sperm competition. In the portunid crab *Callinectes sapidus*, males copulate with moulted females. They guard them during their vulnerable period. The duration of post-copulatory guarding increases under high predation risk (Jivoff, 1997). The effects of predation pressure on male mating show the importance of female reproductive traits in shaping male competition. Variation in mating competition also occurs between and even within populations of the same species. Variations in the duration of mate guarding occur due to density, local sex ratio, and OSR in various species. In some species, these factors may alter the mode of competition. In the crab *Uca beebei*, both sexes hunt for mates, and population density plays a role in determining which sexes hunt for mates. At high density, females hunt more. Whereas at low density, males hunt more. At a high population density, when females are the primary searching sex, males' competitive mode shifts from searching to attracting mates to the burrow using intense claw-waving signals. Predation decreases the amount of searching by gender. But it did not induce a switch from female to male searching (DeRivera et al., 2003). In species with a short breeding season and synchronised female receptivity, males adjust their mating behaviour to the rapid changes in the spatial distribution of receptive females and in OSR. The dynamics of explosive spring breeding in *Orconectes rusticus* are well documented (Berrill and Arsenault, 1982). The onset of copulation in crayfish occurs when the water temperature rises above 4 °C. During the first 10 days of mating activity, when OSR was about 1:1, males and females wandered over the substrate. The copulations were very frequent. Aggressive interruptions of copulations began 8–9 days after the onset of copulations. Now OSR starts to rise owing to the accumulation of non-receptive females. Females sequester themselves in shelters to incubate the extruded eggs. When receptive females became rare, feeding rather than competition for females became typical. Male behaviour and copulation stopped. Berrill and Arsenault (1984) removed all the females from a population 10 days after the onset of copulation. So, inter-male aggression stopped within 5 days. Whereas it continued in a control population in the presence of females as OSR increased with time. Male crayfish shifts from the scramble mode of competition. Both searching and aggressive competition diminish when the OSR become high. The above variations in the mode of competition can alter the spatial distribution of receptive females.

Resource-centred competition

Resource defence in crustaceans involves some refuge, such as burrows, crevices, or cavities. A well-studied example is the defence of breeding burrows in fiddler crabs (*Uca* spp.). Fiddler crabs are semi-terrestrial crabs. They inhabit dense mixed-sex colonies on intertidal sand and mud flats. They excavate individual burrows. In several species, receptive females search for burrows occupied by males. Males attract and guide females to their burrows by waving their enlarged major claws. In some cases, by also rapping this claw against the substrate. After the female enters the burrow, the male plugs the entrance. Copulation takes place underground, following oviposition. The male leaves the burrow in search of another female. Possession of a suitable breeding burrow is crucial to the male's reproductive success. Thus, males compete for breeding burrows and defend them from intruding males (Morrell et al., 2005). In the sand fiddler crab, *U. pugilator*, the most stable burrows are found in the supratidal zone, where burrow flooding and collapse do not occur. Males compete for burrows located near the beach. Large males exclude small ones from these burrows (Christy, 1983). More than defending one's own burrow, *U. mjoebergi* males defend the burrow of their neighbour against an intruder. The ally helped only a smaller neighbour when the intruding crab was larger than the neighbour when the ally was larger than the intruder. A sucky territorial coalition evolves when it is less costly for the ally to assist an established neighbour in negotiating territory boundaries with a new, stronger neighbour (Backwell and Jennions, 2004).

Resource defence occurs in species of a solitary nature, such as clawed lobsters and stomatopods. American lobsters, Homarus americanus, live in solitary shelters. They take refuge from predators and strong water currents. The males compete for territory and defend shelters. Most females mate after moulting. Mating may occur at any moult stage (Waddy and Aiken, 1991). Females approaching their moult search for mates and prefer to mate with dominant males. Males advertise their presence in the shelter and their dominance status using urine signals in combination with visual and behavioural cues (Buskmann and Atema, 2000). Shelter cohabitation occurs only in the context of mating and may last for up to a few weeks. It begins with intermittent cohabitation. Then it becomes continuous cohabitation before moulting and mating (Gosselin et al., 2003). The period of the female's exoskeleton hardening continues till the hard-shelled female leaves the shelter. During this time, the male provides protection to the female through shelter defence. A similar resource-defence mating system exists in stomatopod species that occupy cavities in rock or coral usually in those possessing the smasker type of raptorial appendages (Caldwell, 1991). Mating in *Gonodactylus* takes place in the cavity of either sex. The male assumes the role of cavity defence (Skuster and Caldwell, 1989) and in one who leaves the cavity after the female spawns, regardless of previous owners. The persistence of breeding pairs extends beyond the mating period in monogamous species associated with refuge. or symbiotically associated with invertebrate hosts (Correa and Thiel, 2003). In snapping shrimp, both partners take part in the defence of a specific refuge against same-sex intruders. Intrasexual competition is not restricted to males. Selection for resource defence cooperation (Mathews, 2002a) and selection of males for extended mate guarding influence many aspects. These include predation risk, sex ratio, and the cryptic nature of the female's moult cycle. All these are coupled with the post-mould receptivity period. All these have helped in the evolution of social monogamy in pair-living snapping shrimp.

Mate Guarding

Mate guarding can be defined as a temporary male-female association that exceeds the time required for insemination. Mate-guarding takes the form of a male carrying a female. Sometimes for a period of days, or a male performing non-contact guarding by attending to the female and in some cases caging her between. Pre-copulatory mate guarding is a male's competitive strategy to monopolize a female before copulation to ensure his priority access to the female. At the moment, he is ready to copulate and increase his mating success. This strategy evolves in species where female receptivity is time-limited and predictable (Parker, 1974). This may explain why it is common among crustaceans. From the male's point of view, guarding duration reflects a balance between a selective force favouring prolonged guarding to ensure the monopolization of the female. A contradictory selective force favouring minimization of guarding duration to increase the number of mating and decrease guarding costs. The decision to start guarding depends on the ability of the male to assess whether the female is in an appropriate reproductive state. Males may be attracted to receptive females by distant waterborne pheromones, both waterborne and contact stimuli, or visual cues. The male benefits from assessing the size of the female and her potential fecundity. Since fecundity contributes more to reproductive success than the number of mating opportunities. Guarding duration is influenced by factors related to the availability of receptive females and the intensity of competition among males. In snow crabs and isopods, males guard females longer as the OSR become more male-biased. Male hermit crabs guard females earlier and longer with more male-biased sex ratio. Male-female encounter rates were low, and gender differences between competitors were small (Wada et al., 1999). Mate-guarding by males may impose a fitness cost on females. When guarding costs for males and females differ, intersexual conflict over guarding duration may arise. The costs of pre-copulatory pairing due to predation by fish are higher for females than males in amphipod species. Female resistance to male guarding attempts observed in this species, as in various other amphipods and isopods, is viewed as a female behaviour that serves to shorten the duration of male guarding and reduce guarding costs for the female. When the female's ability to resist guarding attempts was reduced, guarding duration increased. While the reduction of the male's guarding ability had no effect on guarding duration. In species with sexual conflict, guarding duration is influenced by female interest. But female resistance by itself has imposed energy and fecundity costs on the female. Resistance has an upper limit at the point where its costs exceed its benefits. In theory, under conflicting interests, the promised guarding duration would be set somewhere between the optimal durations for the male and female. The main factors that determine it would be the relative power of the sexes and the sex ratio.

Post-copulatory mate guarding in males occur to avoid sperm competition. Sperm competition is an extension of male-male competition, in which the sperm of two or more males compete for the fertilization of ova. The occurrence of internal fertilization and their ability to store sperm cause sperm competition among brachyuran crustaceans (Diesel, 1991). Studies using molecular techniques confirm that many paternity and sperm competitions are common in crustacean species with external fertilization. The species are lobsters, crayfish, anomuran crabs, and ghost shrimp. Sexual selection in males via sperm competition favours adaptations that prevent the sperm of future males from reaching and fertilizing an inseminated female, as well as adaptations for displacing sperm from previous mating. These include morphological and physiological adaptations used for sperm delivery into the site of fertilization in the female's receptacle. This leads to hardening seminal plasma that forms sperm plugs, increased ejaculate size, and behavioural adaptations, i.e., post-copulatory mate guarding. Suck guarding is expected in species in which

(1) Female receptivity continues for some time after copulation.
(2) The interval between copulation and fertilization is not long.
(3) Mate selection efficiency is high, and
(4) The last male sperm has precedence (Simmons, 2001).

A function attributed to post-copulatory guarding by males is the protection of mates from predation. This is of great adaptive value, particularly for crustaceans, in which females moult before copulation. This could in part explain why post-copulatory guarding is common among brachyuran crabs in which copulation is linked to female moult (Portunids and Cancrids; Jivoff, 1997) and why it is absent in some species in which the females copulate in the intermoult stage, despite the occurrence of sperm competition. Wilber (1989) showed in stone crabs (Xanthidae) that males guarded post-moult females for a long duration under conditions of male-male competition than under conditions of predation risk. This concludes that in these crabs, post-copulatory guarding is driven by sperm competition. The finding that females survived predation only in trials in which guarding durations were the longest indicated that post-copulatory guarding prevents predation in stone crabs. The above lack of response in guarding males to predation risk has been specific to the type of predator used rather than a general lack of response to any predator (Wilber, 1989). Jivoff (1997) showed that post-copulatory guarding by male blue crabs prevent inseminations as well as predation on females. The effective male strategy arising from sperm competition in crustaceans, whether guarding, leaving a sperm plug without guarding, or both, is species-specific and depends on the relative costs of these strategies, such as the female's mating traits, encounter rates, and searching risks.

Alternative Mating Strategies and Tactics

In various crustaceans, males exhibit discontinuous variation in mating behaviour and morphology called alternative mating strategies. Such discontinuous variations are known in species in which sexual selection is strong and male contest competition is strong (Andersson, 1994). Males that are excluded from mating enjoy adopting an alternative strategy instead of engaging in a contest to get mates. There is still a debate over whether alternative strategies evolve and persist in natural populations and whether alternative strategies are determined or are conditional strategies that do not yield equal fitness (see Skuster and Wade, 2003). Alternative mating strategies in crustaceans can manifest as inflexible or flexible phenotypes. Skuster (1992) described three distinct male morphs in a marine isopod. *Paracerceis sculpa,* or which, breeds within spongocoels. These morphs differ in mating behaviour and morphology. The largest male possesses elongated uropods and defends harems with spongocoels. The smaller male resembles females in morphology and size and invades the spongocoels by mimicking the behaviour of females. The y-male is tiny and secretive, and he invades the harems by stealth. When isolated from females, the three male morphs did not differ in their ability to sire young. Relative fertilisation success among male morphs varied with the density of females as well as the frequency of other morphs. male morphs Witkin the spongocoels; males could defend females and get all fertilisations when there was only one female and one more– or y-male in the spongocoel, whereas, in the presence of two or more females, the fertilisation success of – and y-males increased and even exceeded the success of males (Skuster, 1989). The tendency of females to accumulate in spongocoels already containing gravid females sets the stage for the co-existence of discrete male morphs in this isopod species by creating a "mating niche" for the y-males. On the basis of these results and calculations from monthly samples of sponges in the field over 2 years and the life history and relative contribution of each morph to the population, Skuster and Wade (1991b) showed that the average mating success was equal among the three male morphs over time, and only a small fraction of the total opportunity for sexual selection was attributed to variance among morphs. In this, there is no selection favouring one male mating strategy, and conditions exist for maintaining genetic polymorphism in male mating behaviour morphs and morphology in isopods. That is, these morphs are distinct at a single genetic locus. In two caridean species, the freshwater prawn *Macrobrachium rosenbergii* and the marine rock shrimp *Hynchocinetes* typus, three distinct mature male morphotypes coexist, representing successive stages in the developmental pathway of adult males (Correa et al., 2000). In *M. rosenbergii*, the first adult stage, termed the Small Male (SM) morphotype, is smaller than the other morphotypes and possesses more delicate claws. Only part of the SM in the population grows to the next stage, termed the Orange Clawed (OC) Morphotype, characterised by orange claws and rapid growth, and transforms into the Blue Clawed (BC). Morph heritability type, which possesses long blue claws. The three morphotypes coexist with prawn populations in a density-dependent proportion (Karplus et al., 1986). The growth of SM individuals is suppressed in the presence of the BC morphotype, but when the BC is removed, some of them proceed along the same developmental pathway as described above (Karplus et al., 1992). The heritability of sexual variation in males was found to be negligible (Malecka et al., 1984). Although the developmental expression of phenotypic plasticity in *M. rosenbergii* reflects conditional strategies, it may arise from genetic architectures that are sensitive to social-environmental cues and allomorph type male's ability to adjust their mating phenotypes in response to their changing social environment (Skuster and Wade, 2003). In a similar manner, typus males first become mature as the female-like typus morphotype, after which they moult through various intermediate stages to the final robustus morphotype (Correa et al., 2000). In both species, the three morphotypes fertilise the female when isolated from her. When grouped together, a linear morphotype-related dominance hierarchy emerged, dominated by the largest morphotype (Correa et al., 2003). Two alternative mating tactics are practised by males of different species: the dominant morphotype courts and defends receptive females, and the subordinate morphotype engages in sneak mating and attacks his spermatophore on the female's sternum while the dominant male is trying to defend the female from takeovers by other males (Correa et al., 2003).

Alternative male strategies involving small sneaker males (minors) and large sneaker males (majors) were also described in the marine amphibian *Jassa marmorata* (Kurdsiel and Knowles, 2002). Juvenile amphipods that grow on high-quality food will mature later, at a larger size, and become majors, possessing claws (gastropods) with large thumbs, whereas those that grow on low-quality food will be minors. Heritability analyses did not reveal genetic differences between dimorphic males (Kurdsiel and Knowles, 2002). Alternative male tactics in *J. marmorata* are conditional tactics, whose expression depends on the interaction between individual genotypes and their environment. Alternative mating tactics are not associated with different morphologies and developmental trajectories. For example, in sand-bubbler crabs (*Scopimera globosa*) inhabiting sandy-muddy intertidal sones, males use two mating tactics: underground mating is employed by large (older) males, and surface mating is employed by small (younger) males (Koga, 1998). Most of the men who employed these tactics belonged to middle-class families. The reproductive success was higher for underground mating males, owing in part to the fact that before surface-mated females often mate again underground, and last male sperm precedence occurs. Most surface mating occurred in the water-saturated area where small crabs tended to forage on a rich diet, which they used to accumulate energy for growth and reproduction. Since the growth of bubbler crabs is indeterminate, surface-mating small crabs will grow and practise underground mating. Thus, the observed alternative mating tactics reflect a plastic behavioural polymorphism with age- and size-dependent ontogenetic shift between the alternatives. Alternative reproductive strategies can manifest in sexual changes, as was described in the protandrous applied shrimp, *Athanas kominatoensis* (Nakaskima, 1987). In this species, all individuals mature as males, and they are all capable of changing sex. Smaller and subordinate males change sex to become females, while larger ones remain males throughout their lives. Owing to direct intraspecific competition for mates, small and subordinate males would have a low mating success rate as males. Nakaskima (1987) suggested that subordinate males change their sex as the "best of a bad job", implying that their strategy is lower fitness. It is not clear whether the lifetime fitness payoffs of the two alternative strategies are indeed different and whether sex-changing and primary males are distinct and predisposed to their mating strategies.

Scramble Competition

Pure scramble competition, in which kick males search for and inseminate receptive females without denying access to competitors through aggressive interactions, is found in the lower crustaceans of the Anostraca order (Belk, 1991). Among higher crusts, which are the subject of this review, scramble competition in the form of pure searching was ascribed to carideans as the principal mode of mating competition (Correa and Thiel, 2003). A pure searching mating system in carideans involves the efficient location of mates and the transfer of spermatophores in brief interactions, immediately followed by the separation of the mate (Correa and Thiel, 2003). Theoretical studies have often considered sparring and guar as alternative options in mating competition. This mode emphasises high encounter rates between males and females as a key factor favouring pure searching in males (Wickler and Seibt, 1981). Male encounter rates are expected to be higher when females are clumped. Asynchronous female receptivity increases the number of receptive females that research can encounter. The tight spatial distribution of females combined with the asynchronous availability of receptive females will favour the male guarding of groups of females (harems). Thus, Schuster and Wade (2003) predicted that in most combinations of spatial and temporal distributions of receptive females, some form of guarding will persist, while male pure searching will favour a combination of moderate spatial and temporal distributions of females. Indeed, reported scramble-type competition in crustaceans often combines elements of contest competition. Orensans et al., (1995) suggested the mating system of the crab. *Cancer magister* tends towards scramble competition polygyny since female reproduction is synchronised and predictable in time, females are aggregated in space, and males scramble for possession of females. Once in possession, males guard their mates for several days. The orconectid crayfish *O. rusticus* has explosive spring mating, in which there is no mate guarding, but scramble competition gives way after several days to inter-male aggressive interactions over females (Berrill and Arsenault, 1982). In the fiddler crab, *Uca paradussumieri,* males search for burrows inked by pre-ovigerous females. It is helpful for males to be the first to locate a pre-ovigerous female in a burrow because they are usually able to defend it and mate with it. Hence, *U. paradussumieri* males combine both scramble competition and contest competition in their mating behaviour (Murai et al., 2002). Even in caridean shrimp, known for their prevalence of pure searching, some brief contest interactions among males have been reported (Correa and Thiel, 2003).

Mate Choice

A common observation in crustaceans is that large dominant males are not only more successful in contests than females but are also preferred by females. It is a convention that differential mating success among males can arise from both male-male competition and female mate choice, but it is often impossible to disentangle the effects of these two mechanisms of sexual selection. In the context of mate choice, male characteristics and female preferences co-evolve; sexual selection may be stronger than mate choice in male-male competition (Skuster and Wade, 2003). Male psychological traits that influence a male's competitive ability, as well as any traits indicating male quality, may become exaggerated in female culture. Females can use these traits as selection criteria. For mate choice to evolve, a female could get benefits by choosing certain mates over others, since mate choice is likely to have costs. Hypotheses for the evolution of mate choice emphasize genetic benefits for offspring viability and attractiveness as well as direct fitness benefits for female gains (Andersson, 1994). By choosing a large male, female American lobsters obtained direct benefits from the fertilization rate of ova and better protection during the vulnerable post-moult period (Gosselin et al., 2003). Female isopods (*Lirceus fontinalis*) that discriminated against energy-depleted males avoided reduced fertilisation success (Sparkes et al., 2002). Female mate choice in crustaceans takes various forms, and females use a variety of criteria, often in combination. Females usually prefer to mate with a dominant male, but on what trait of the male do they base their choice? There is no doubt that in aquatic crustaceans, either sex can use chemical cues to locate potential mates, but it is less clear that these cues are also used for choosing among potential mates. In the rock skrimp, the female choice of the large robust male morph over the small typus morph is based on chemical signals and not visual signals (Das and Tkiel, 2004); there is no evidence for the female choice of certain robustus individuals over others using chemical cues. In the blue crab Callinectes sapidus, females and males emit chemical signals to attract mates (Buskmann, 1999), but when mates are near and engage in mutual displays (Jivoff and Hines, 1998), these signals may function as visual cues for mate choice. In the American lobster, receptive females prefer to enter shelters occupied by more dominant males (Cowan and Atema, 1990); they first approach a shelter and then spend time attempting to enter while evaluating the male. Buskmann and Atema (2000) showed that when male urine release was blocked, the incidence of female approach and the time spent attempting to enter the shelter were reduced. Artificial release of male urine in the presence of a catheterized male in the shelter restored the female approach but not time spent attempting to enter, while artificial urine release in the absence of a male did not induce any response in the female. These results show that female lobsters use male urine-borne chemical signals to locate and choose a mate, but signals in combination with the urine-borne signal are also used in mate choice.

Although females are considered a limiting sex, there is growing evidence that the reproductive success of females can be constrained by sperm limitations. By preferring dominant males, females may sometimes reduce their fertilisation rate since sucking males are more likely to be sperm depleted than others because of their competitive advantage in obtaining many matings. Sato and Goskima (2006) recently demonstrated the occurrence of sperm limitation in populations of *Haplogaster dentata,* owing to many matings and a low sperm recovery rate in males. These authors also demonstrated that, on the basis of waterborne chemical cues, female stone crabs prefer larger males, but they also prefer to mate with males that are not sperm-depleted (Sato and Goskima, 2007). Males, in turn, were demonstrated to divide sperm among females and economise sperm to increase mating success (Sato et al., 2006). In species in which the reproductive success of a female depends on a resource, the characteristics of the resource that a male defends are likely to be used by the female to choose a mate. Active mate choice by females has been demonstrated in fiddler crab species, in which females hunt for males. Females sample several male burrows before selecting a male and entering the burrow to mate. While they search, males wave their enlarged major claws to attract females into their burrows. Females may base their mate selection upon male physical traits, male displays, and burrow characteristics. *Uca pugilator* females base their choice on burrow stability and not on male size, but by selecting a stable burrow They are likely to select a large male because sucking males are more abundant in locations where burrows are stable (Christy, 1983). In *U. annulipes*, females use at least two criteria for mate choice since there is no relationship between burrow quality and male sexual selection (Backwell and Passmore, 1996). Females entered burrows of up to 24 males and travelled up to 28 m before selecting a mate. They first sampled the larger males in the population, but the final mate acceptance appeared to involve an invariant Threshold criterion based on burrow structure. Backwell and Passmore (1996) observed that females decreased their acceptance criteria of willingness to sample and mate as the time to the end of the mating cycle decreased. The difference between *U. pugilator* and *U. annulipes* was accounted for by the lower burrow density coupled with higher predation risk in U. pugilator, which imposes higher searching costs and restricts prolonged sampling using many criteria. In populations of California fiddler crabs (*U. crenulata*) living in muddy-sand substrate at low density and low predation risk, male burrow characteristics vary. Female crabs conducted extensive searches and, like *U. annulipes* crabs, used many criteria to select males; they selected mates based on burrow characteristics that are important for successful incubation and release of larvae and on correlated male characteristics, such as burrows defended by males near the size and with small claws given to the female (DeRivera, 2005). It appears that in all the *Uca* species, the burrow is crucial for reproductive success; female mate choice depends on the physical features of the burrow.

In some species with male mate guarding, female resistance to guarding attempts by males was suggested to act as a means for mate choice and to reduce female costs of being guarded. It was suggested that by resisting mating attempts, females get information about males' vigour and condition. Female resistance in isopods and amphipods selects for large male sizes (Jormalainen and Merilaita, 1995) and for males with high energy reserves (Sparkes et al., 2002). Instead of probing a male's vigour, resource-holding power, or condition, a female can choose a mate on the basis of the variability in the expression of secondary sexual characteristics among potential mates. Elaborate male signals reflect the genetic quality of a male because they are condition-dependent signals reflecting the level of the male's resistance to parasites (Hamilton and Zuk, 1982) or because they are costly to produce and maintain, so low-quality males cannot express as exaggerated signals as high-quality males (Zakavi, 1987). The conspicuous courtship displays of male fiddler crabs are an example of the kind of signal on which females may base their behaviour. When a female approaches a male or a cluster of males, the males wave their enlarged claws up and down to attract the female into their burrows for mating. The major claw at its strongest may reach almost half the total weight of the crab, so waving is costly. Variation in wave rate is partly due to variation in male condition combined with the energy costs of waving (Jennions and Backwell, 1998). Studies revealed that, in addition to claw size and position (Oliveira and Custodio, 1998), female choice through the male's waving display in fiddler crabs is based on features such as display leadership in waving crabs (Backwell et al., 1999) and fine-scaled spatiotemporal features of the display (Murai and

Backwell, 2006). An example of a morphological secondary male character trait that may serve as a choice criterion for a female is the male ornament in the Australian red claw crayfish, *Cherax quadricarinatus*. Adult males of this species have a soft, uncalcified red patch on the outer surface of the cheliped. The location of this vulnerable soft membrane on this weapon used for fighting renders its dimorphic structure a handicap for this male. The red patch is expected to have evolved through signal selection according to the handicap principle (Zakavi, 1987) because (1) it is costly and only high-quality males can afford to bear a large and conspicuous red patch, and (2) it is a reliable signal reflecting the male's quality since the patch colour, derived from carotenoids obtained from the crayfish diet, may vary from bright red in males in good condition to pale orange in malnourished males. Female red claw crayfish mate when they are hard-shelled, and active cooperation of the females during copulation is required (Barki and Karplus, 1999). Thus, it is most likely that female mate choice occurs and that a reliable male signal is used in this context (Karplus et al., 2003a), but this possibility has yet to be validated.

A somewhat different form of mate choice is female copying, in which females imitate the mate choice of other females rather than focusing their mate choice criteria on the male. The sucking-a-choice strategy is exercised by females of the marine isopod *Paracerceis sculpta*, which kick breed in sponges. Female isopods do not base their choice on the features of a male or a sponge; they prefer to mate in spongocoels already containing gravid females. Females and males are attracted from a distance to sucking spongocoels, guided by chemical signals; males fight for access to the location, and those that are larger have intact uropods and are successful in takeovers (Skuster, 1990). Thus, a female may enjoy copying the behaviour of other females by mating with a better male. Female copying evolved in this species owing to the high predation risk, which makes extensive movement between breeding spongocoels to discriminate between male or sponge characteristics costly. As an alternative, females use the scent of other females as a sign of the quality of the breeding site. Female mate choice is not limited to selecting a male for copulation; it can also occur after copulation at the sperm level, in a process termed cryptic female choice. In the rock shrimp *Hynchocinetes typus,* subordinate males of the typus morph practise sneak mating and are often the first to seize a receptive female and attack spermatophores before the female is taken over by a dominant male of the robust morph (Tkiel and Correa, 2004). Given that subordinate males predominate in a dense male population, the costs of resisting aggression from these males may be high for a moulted female. Instead of this, the female is able to delay spawning when inseminated by subordinates, which reduces the fertilisation success of their sperm. Further, the female removes a large part of the sperm of the subordinate males (Thiel and Hinojosa, 2003).

Although mate choice is usually exercised by females, there is ample evidence for the occurrence of male mate choice in crustaceans. Studies on male choice have shown that choosy males show that they increase their reproductive success by means of their preference for more fecund females and/or for females that are closer to becoming fertile. Isopod males *(Idotea baltica)* given a choice between a large and a small female preferred the one that matured earlier for partial moult, not the large one, indicating that maturity was more important to the female size (Jormalainen et al., 1994). Such a male choice was based on the isopod *Lirceus fontinalis* on variation in the levels of moult hormone in the females. The males did not discriminate between reproductive and somatic moults (Sparks et al., 2000). Female blue crabs (*Callinectes sapidus*) have a prominent red marking on their cheliped dactyli, whose size and robustness are correlated with their body size, fecundity, and reproductive state. Male blue crabs were selected to choose females on the basis of the relative intensity of female marking coloration (Williams, 2003). Male preference for females that were closer to being fertilisable and/or for larger females has also been shown in other crustaceans, e.g., in gammarid amphipods (Dick and Elwood, 1990, 1996), in the stomatopod *Pseudosquilla ciliata*, a species with sex role reversal (Hatsiolos and Caldwell, 1983), and in decapods (Rahman et al., 2004 in a monogamous snapping skrimp, *Alpheus heterothallic*).

Sexual Dimorphism

So far, it is evident that the form and strength of sexual selection in mating competition and mate choice in crustaceans are linked with sexual dimorphism. The development of the theory of sexual selection by Darwin was inspired by his observations on a prominent outcome of sexual selection, sex differences in secondary characteristics. Referring to Crustacea, Darwin stated in *The Descent of Man and Selection in Elation to Sex*: "In this great class, we first meet with undoubted secondary sexual characters, often developed in a remarkable manner.

Unfortunately, the habits of crustaceans are very well known, and we cannot explain the uses of many structures' peculiar to one sex" (Darwin, 1871). There are many categories of sexual dysfunctions arising from the different mechanistic mechanisms of sexual selection (Andersson, 1994), e.g., in psychological and behavioural characteristics related to competition and attractiveness. Size dimorphism is common among crustaceans, and its degree and direction may vary among species with different mating systems. Male contest competition favours large male sexes and structures that enhance the male's ability to fight and guard females, as well as increased male aggressiveness. Studies have shown the advantage of a large body in contests. Body and cheliped size are usually correlated, so a larger individual also has larger chelipeds. The relative importance of body size and cheliped size in determining the fighting ability of males could be evaluated in species in which the male population comprises distinct male morphs of comparable body size that differ in cheliped size relative to body size (Guiaşu and Dunkam, 1998). A remarkable example is provided by the orange-clawed and blue-clawed male morphotypes of the freshwater prawn *M. rosenbergii*. Owing to their larger chelipeds, small blue-clawed males dominated large orange-clawed males in mixed male groups (Barki et al., 1992). Blue-clawed males having only a 10% advantage in cheliped length almost won pair-wise contests despite being 44% smaller in body mass than their orange-clawed opponents, and those having equal cheliped size were as likely as their opponents to win contests despite being almost half their body mass (Barki et al., 1997). The overriding importance of cheliped size has been shown in various decapod species in laboratory experiments, for example, inshore crab *Carcinus maenas* (Sneddon et al., 1997), hermit crab *Pagurus bernhardus* (Neil, 1985), crayfish *Orconectes rusticus* (Rutherford et al., 1995), and American lobster *Homarus americanus* (Vyee et al., 1997), as well as in their natural habitat (Morrell et al., 2005). Sexual dimorphism in these weapons may arise from female choice.

In species where kick scramble competition occurs, large body sizes may be less important for males. Rather, pure searching is likely to favour mobile males with small body sizes and sensory capability to detect and get receptive females (Andersson, 1994). Indeed, size dimorphism with smaller males and females is common among pure-searching caridean skrimp, owing as well to a fecundity advantage for large females (Correa and Tkiel, 2003). Lack of significant sexual dimorphism has been shown in species in which sexual selection is weak, for example, monogamous applied shrimp with persistent pairs (Correa and Thiel, 2003). Slight sexual dimorphism may also be evident in some species in which both contest (female guarding) and scramble, which are important components of male mating competition (e.g., *Cancer magister,* Orensans et al., 1995), occur because of contradictory sexual selection for large and small male size preferences. In some cases, scramble competition may favour large males, as was suggested in *Asellus aquaticus*, an isopod with pre-copulatory mate guarding (Bertin and Cesily, 2003). In this species, sexual dimorphism is evident in both body and antenna length, and both characteristics are correlated. Males with longer antennae were better able to detect and pair with receptive females and had an advantage in scramble competition. Thus's large body size in male isopods may also be favoured by scramble competition, to its obvious advantage in female guarding.

Sexual dimorphism with more developed structures in the male may also involve structures used for holding a female; for example, the posterior gastropods of the amphipod *Gammarus pulex* were necessary for the male to achieve successful copulation (Hume et al., 2005). Sexual dimorphism in sensory organs has been found in olfactory sensillae on the antennules, where they are sensitive to waterborne sex pheromones (Hallberg et al., 1997), and in sensillae on the second antennae, where they are sensitive to sex pheromones via contact (Bauer and Caskey, 2006). Male-specific olfactory sensillae are found in peracarid groups such as mysids, crustaceans, and some amphipods, but not in decapods. In Peracarida, both putative pheromone sensors are in dimorphic males and general olfactory sensors in the sexes, whereas in Decapoda, there is only one type of olfactory sensillum. Sexual activity in the latter group is observed in species and a number of sensillae (Hallberg et al., 1997), which may be correlated with antenna length. Related species and even populations of the same species may differ in degree or direction of dimorphism as a result of differing ecological and population factors.

Wellborn and Bartkolf (2005) explored two related amphipods with the *Hyeaella azteca* species complex. The larger of the species inhabits a fishless habitat with low predation risk, and mortality decreases. In contrast, in the smaller species inhabiting a fish-containing habitat, mortality increased with size. In the large species, pairing success and female preference increased with male size, and males were larger than females, whereas in the small species, pairing success and preference by females were similar for intermediate and large males, and males were smaller than females. Knowlton (1980) showed plasticity in sexual dimorphism in the snapping shrimp, *Alpheus armatus*, which lives in pairs symbiotically on a sea anemone. In one location where predation risk was lower, males had a greater tendency to leave anemones in search of a new mate, and large males had a larger major chela and more conspicuous uropod spines than males from a low-predation risk area. The degree of sexual dimorphism may be constrained by natural selection. Bertin and Cesilly (2003) analysed selection gradients for body size and antenna length in five isopod populations and found that sexual selection favours large males in all populations and long antennae only in a few of the populations. In a laboratory experiment (Bertin and Cesilly, 2005), the main determinant of pairing success was male body size at a high density and antenna length at a low density. They suggested that the variability detected in the pattern of sexual selection in *A. aquaticus* populations is related to population density, which influences the relative importance of contest (i.e., large size) and scramble competition.

Hormoval Regulation of Sexually Dimorphic Behaviours

The link between hormones and social behaviour in crustaceans has been investigated about aggression, which occurs in the context of mating competition, as reviewed in the section on Mating competition at the beginning of this chapter. Studies on the American lobster demonstrated that aggressive behaviour and dominance change over the moult cycle, increasing as the animal enters pre-moult stage D0 and dropping before moulting (Tamm and Cobb, 1978). This pattern of change in aggressive behaviour correlates well with changes in hemolymph titres of ecdysteroids secreted by the Y-organ (Snyder and Ckang, 1991). Injection of the moulting hormone (20-hydroxyecdysone) into lobsters exerted differential effects on the abdominal plastic flexor muscles (related to the escape response) and the dactyl closer muscle of the claw. This hormone is correlated with the variation in aggressive motivation over the moult cycle. The above effects of the moulting hormone may account for the aggressive resistance of early pre-moult females to guarding attempts by males, which diminishes in late pre-moult females in species in which the females moult before mating. Yet, sucking effects on aggressive behaviour over the moult cycle are not sex-specific, as they occur in juveniles and males. In insects, juvenile hormone, a sesquiterpene hormone produced in the corpus allatum, is known to regulate a broad array of processes, among them sexual behaviour, aggression, and dominance (Scott, 2006). The crustacean analogue to the insect juvenile hormone is methyl-farnesoate (MF), a form of JHIII missing an epoxide group, secreted by the mandibular organ. A relationship between MF and reproductive behaviour was postulated in the spider crab *Livinia marginata*; the most active males that gained all copulations in a competitive situation belong to a male type with an abraded carapace and large claws. These males had higher haemolymph concentrations and in vitro synthesis rates than MF than other male types (Sagi et al., 1994). Unlike the Y-organ, the mandibular organ, and the rest of the endocrine complex, the androgenic gland (AG) is the only sex-specific endocrine gland in crustaceans. The involvement and modes of action of androgens and their metabolites in the regulation of mating and aggressive behaviour are well established in vertebrates, but not in Crustacea. This gap stems from the fact that androgenic hormones (AGH) have not yet been identified in decapods. A relationship between androgenic factors derived from the AG and male mating behaviour can be postulated on the basis of studies in protandric shrimps showing the degradation of the well-established AG during sex changes from functional males to functional females (Kim et al., 2006), and on studies in gonochoristic decapods showing: (1) natural seasonal variations in the activity of the AG (i.e., degradation and proliferation) in cold temperate crayfish accompanying the transformation between the active (form I) males with large claws and inactive (form II) males with small claws (Carpenter and deRoos, 1970); (2) high activity of the AG in the active dominant blue-clawed morphotype compared to the other male morphotypes in the prawn *M. rosenbergii* (Okumura and Hara, 2004); and(3) functional sex reversal yielding neo-females capable of spawning in early AG-ablated juvenile males (Sagi and Kkalaila, 2001). Gleeson et al. (1987) found that male crabs (*Callinectes sapidus*) perform spontaneous courtship displays following eyestalk ligation, which would otherwise occur only in response to the release of female sex pheromones. Based on the co-occurrence of these displays with the hypertrophy of the AG, it was suggested that androgenic factors from the AG mediate the control of courtship behaviour. All the studies did not test the role of AG in the regulation of behaviour.

A skort report published more than 30 years ago (Caiger and Alexander, 1973) presented an interesting result suggesting a possible effect of the AG on dominance status in male crabs (*Cyclograpsus punctatus*). The androgenic glands of the dominant males in groups of four crabs were removed and implanted in the most subordinate individual within each replicate group. Dominant individuals dropped in dominance rank and usually became the lowest ranked in their groups, whereas AG-implemented subordinate individuals increased in rank and, in some cases, became the most dominant. This somewhat anecdotal but intriguing finding, suggesting an activation role for TKE AG in aggressive behaviour, deserves further examination. Studies on the red claw crayfish, *Cherax quadricarinatus*, provided direct evidence for the important role that AG plays in the mediation of male dimorphic behaviours and its known effects on primary and secondary morpho-anatomical and physiological characteristics. Implanting AGs into female crayfish at an early juvenile stage induced male-like agonistic and mating behaviour after they attained sexual maturity. The aggression of AG-implanted females was like that of intact females in contests with males (Karplus et al., 2003b). Aggression in contests between AG-implanted and intact females was lower than in contests with AG-implanted or intact pairs (Barki et al., 2003), like the reduced aggression exhibited in heterosexual contests. Elements of mating behaviour were exhibited in some of the encounters between AG-implanted and intact females, including male courtship displays by the AG-implanted female and, most importantly, a typical sequence of copulation that included antennae and chelate contact, overturning of the AG-implanted female on its dorsal side, and freezing in a male-beneath-female position with the ventral surfaces brought face to face (Barki et al., 2003). This case of false copulation occurred even though the AG-implanted female was not capable of inseminating the receptive female because of her lack of a masculine reproductive system. To further investigate the effect of the AG on agonistic and mating behaviour, Barki et al., (2006) employed the intersex model. Intersex individuals of the red claw crayfish are female (Parnes et al., 2003) but morphologically and functionally male. Having both male and female genital openings, testis, and sperm manipulations seemed to affect the feeding ability and dominance status of crayfish, as had been demonstrated in crabs by Caiger and Alexander (1973). The results suggested that the presence of the AG in the crayfish and the active secretion of the AGH play no role in the activation of male-like mating behaviour, but they may have consequences for the mating success of male crayfish through their effect on mating competition.

Thus the different aspects of sexual selection and the diversity of mating behaviour and mating systems in the Crustacea is interesting. Studies have investigated various factors influencing mating behaviour, strategies, and systems in crustaceans. Studies investigating the interplay of many ecological, reproductive, and life history factors underlying the observed diversity in mating behaviour and mating systems are still needed. Crustaceans offer an excellent model for investigating the complex network of relationships caused by the interplay of many factors, and for studying current issues related to sexual selection such as sexual conflict, mate choice (male and female), male sperm limitation, and sperm allocation. While the fascinating mating behaviour of crustaceans has attracted a wealth of behavioural studies on the ultimate processes shaping mating behaviour, little is known about the proximate hormonal mechanisms underlying these behaviours. Current advances in the identification of AGH in decapods (Manor et al., 2007) open a new avenue for exploring questions about the neuroendocrine mechanisms mediating the organization and activation of sexual behaviour in crustaceans.

11.1.18 Female Remating Behaviour

The female remating behaviour is predicted to evolve, according to net effect of remating on female fitness. In many taxa, females commonly resist male remating attempts, because of costs of mating. In seed beetle *Acanthoscelides obtectus*, early females are more likely to remate with control males, as they aged, while late females were more resistant to remating later in life. Female remating rate decreases with age, when direct selection on late-life fitness is operating, and increases when such selection is relaxed. Female resistance to remating can evolve rapidly, and such evolution is in accordance with genetic interests of females.

11.1.19 Sperm Competition

Gonads and genitals are clear expressions of sexual selection. They are most directly related to fertilization, but serve no survival function. The primary sexual characters, such as penises are necessary for breeding, and hence, are favoured by natural selection is misleading (Anderson 1994). Females of many species mate with more than one male. In such cases males evolve larger testicles, larger ejaculates, and faster swimming sperm (Smith 1984). Male North Atlantic right whale has 2000 pound testicles to pump out gallons of semen. In primates, testicle size increases with intensity of sperm competition across the species (Harvey and Harcourt 1984). Female chimpanzees are highly promiscuous. So male chimpanzees have evolved large 4-ounce testicles. Eberhard (1985) has argued that male genitals often function as "internal courtship devices" to stimulate females into accepting sperm from copulating male. Human female orgasm may function partially to suck sperm into the uterus (Baker and Bellis 1995). A common conclusion from the literature is that, most spermatozoa are not capable of fertilization. Sperm competition theory provides circumstantial evidence that the selection process involves mechanisms by which the quality of fertilizing spermatozoon is controlled, thereby ensuring that females and their offsprings receive high quality genetic material. Of ubiquitous occurrence of sperm selection mechanisms throughout the nature, some depend upon the self-selective propensities of spermatozoa, while others involve antagonistic selection imposed by the female reproductive system. Hypotheses have linked sperm selection to the inheritance of superior fitness traits. Sperm competition occur when spermatozoa from more than one male have the opportunity to fertilize eggs from a single female during the same fertile period (Parker, 1998). This occurs, when several males mate with a polyandrous female. In such situations, the male producing the best-quality spermatozoa gets an advantage over his rivals. Studies have demonstrated that large testes size, hence a greater sperm production capacity, is a feature of species which exhibit multimale mating systems (Harcourt et al., 1981). Social dominance is also a determinant in such mating systems, as age, body weight and behavioural differences influence the relative number of spermatozoa contributed by each of the males.

Paternity is still skewed, even if confounding factors such as sperm numbers and insemination timing are eliminated (Dziuk 1996). Elegant Heterospermic Insemination (HI) experiments with bull, rabbit and pig spermatozoa, where equal numbers of spermatozoa from two, or more males are mixed and inseminated in equal proportions, have shown that spermatozoa from individual males can be ranked in order of fertilization efficacy (Berger et al., 1996) suggesting that some aspects of sperm quality per se determine fertilization success. Thus, the meaning of the term 'sperm quality' implies (i) the existence of a positive correlation between sperm phenotypes and fitness of offspring that derives from that particular spermatozoon, or (ii) that some spermatozoa simply possess a 'fertilization advantage' over others.

Hypotheses attempted to explain the evolutionary benefits that females may derive from such high levels of polyandry and sperm competition (Keller and Reeve 1995). Females can somehow assess the genetic quality of spermatozoa from different males and choose those (the 'good' sperm or 'good' genes) that will confer genetic benefits to their offspring (Yasui 1997). Females could 'choose' spermatozoa with immunologically compatible characteristics, perhaps based on the major histocompatibility antigen they express. Pizzari and Birkhead (2002), considered that sperm function is influenced by 'fertilization efficiency genes', which provide fertility advantages when needed. This differs subtly from cryptic female choice models, since there is no overt necessity that fertilization efficiency correlate with. The chosen spermatozoon is the end product of a stringent selection process and, by definition it must possess all of the attributes that make it fertile. Urodele amphibians allow several spermatozoa to fuse with a single egg. These all then undergo nuclear decondensation to form pronuclei, generate asters and synthesize DNA (Wakimoto 1979). One of the sperm pronuclei fuses with female pronucleus, where upon accessory sperm pronuclei degenerate. Interestingly, before syngamy is finally achieved, female pronucleus executes a complex series of excursions within egg cytoplasm, which involves 'visiting' the various male pronuclei and possibly indulging in some kind of selection process.

11.1.20 Sex Differences

Trivers (1972) pointed out that since females invest more matter and energy into producing each egg than males invest in producing each sperm, eggs form more of a limiting resource for males than sperm do for females. Thus, male should compete more intensively to fertilize eggs than female do to acquire sperms, while females should be choosier than males. Males compete for quantity of females and females compete for quality of males. In short, males court and females choose. A man's reproductive success generally increases with his number of sexual partners in absence of contraception, whereas a woman reaches her reproductive limit rather quickly as her number of sexual partner's increases. This is because males can opt out of parental investment in a way that woman can not—nature can not enforce child support laws any better than modern Governments. Of course, women under ancestral conditions probably used abortion and infanticide to avoid maternal investment during difficult times, but they could not induce another woman to bear a child for them. Maternal investment was obligatory in hominids; paternal investment was not.

11.1.21 Sexual Selection in Primates

Sexual selection in multi-male, multi-female primate groups is intense because social context of mating is complex and dynamic. Sexes compete, both sexes are choosy; both sexes have dominance in relations and both sexes form alliances. Sexual relationships develop over weeks and years rather than minutes. Under these relentlessly social conditions, reproductive success came to depend on mental capacity for "chimpanzee politics" (de Waal 1982), "Machia-vellian intelligence" (Byrne and Whiten 1988) "special friendships" (Smuts 1985) and creative courtship rather than simple physical ornaments and short-term courtship behaviours as in most other animals. Primates and especially hominids are extremely "k-selected taxa" having slower development, larger bodies, fewer offsprings, higher survival rates and longer lifespans than more "r-selected taxa" such as insects, fish or rodents (Harvey, Martian and Clutton–Brock 1996). The more k-selected the species, the more important sexual selection usually becomes, compared to natural selection (Miller and Todd 1995). K-selection usually reduces relative energetic demands of reproduction on female and almost eliminates need for male help, because slow gestation spreads maternal investment over a longer period and small litters of large, well-developed offsprings are easier to care for.

11.2 MATING SYSTEM

A ratio of one male to one female is the expected sex ratio in most populations. In polygynous mating system, males mate with more than one female. In polyandrous system, females mate with more than one male. In monogamous systems, each individual (male or female) has only one mate (Figure 11.4).

11.2.1 Monogamy

When each individual mates exclusively with one partner over at least a single breeding cycle, and sometimes for longer, it is called monogamy. This is common among birds, about 90% of which are monogamous, as in birds eggs and chicks require a considerable amount of parental care. Most eggs are incubated by one parent for hatching. Chicks require continual feeding and without both partners most would starve. Monogamy is rare among other vertebrates.

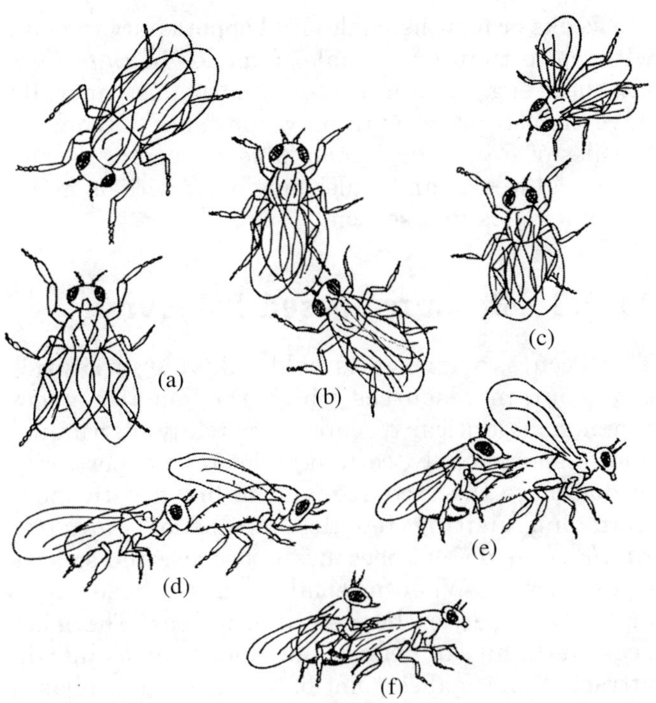

FIGURE 11.4 Mating in *Drosophila*. Sequence of courtship behaviour in *Drosophila melanogaster* males towards females. (a) male orientates towards the female (b) taps her (c) signs a species-specific courtship song (d) licks the female genitatia (e) curts his abdomen in an attempt to copulate (f) copulates the female.

11.2.2 Polygamy

In polygamy, individuals mate with more than one partner. Most mammals have polygynous mating system in which males mate with several females, but each female mate with only one male. In most of the fishes, female deposits eggs, and then the male fertilize them. Females are, thus, able to desert. When fertilization takes place internally in fishes, females provide parental care. Polygyny is influenced by spatial or temporal distribution of breeding females. When all females are sexually receptive at the same time, there is little opportunity for a male to garner all females for himself. Monogamous relationships are common in such situations. When female reproductive receptivity is spread out, over the weeks or months, males find opportunity to mate with more than one female. Females of *Bufo bufo* lay their eggs within a week, and males generally have time to mate with only one female, whereas females of *Rana catesbeiana* have a breeding season of several weeks, and males may mate with as many as six females in a season.

11.2.3 Resource-based Polygyny

This occurs, when males hold valuable territories that contain resources which the females need. When some critical resource is patchily distributed and in short supply, certain males take opportunity to dominate the resource, and to breed with more than one visiting females. Males of *Indicator xanthonotus* defend bees' nests, because the species feeds on wax. When a female comes to feed, male mates with her, exchanging food for sex. The more bees' nests he can defend, the more females he will attract. From male's point of view, resources-based polygyny is advantageous. From female's points of view, there are drawbacks. Although, by choosing dominant males, a female may gain access to good resources, she may have to share resources with other females. In *Marmota flaviventris,* males attract more females, when they defend the best burrow sites. For a female, sharing a burrow means less success per female. Although it is best for a female to be with a monogamous male, it is best for a male to mate with as many females as possible. A compromise is often evident, in which about two females are usually observed per territory.

Female defense polygyny

The female live in permanent groups, which the males defend directly (roosting groups in bats).

Resource defense polygyny

The females do not live in permanent groups, but are spatially concentrated at food, nesting sites, or other resources, which some males can control (impalas, vicunas).

Lek polygyny

The males compete for high dominance ranking, within a group, usually at a traditional display arena, the winner is often selected by many females (hammer-headed bats, white bearded manakin).

Scramble competition

The Polygynous females may be scattered, or clustered spatially; but if clumped, they cannot be defended because of high male density. Males race to contact as many receptive females as possible, without engaging in territorial defense (horseshoe crabs).

Polygynous mating in which harems are defended

Sometimes males simply defend a group of females as a harem, without bothering to command conventional resource-based territory. This is common, when females naturally occur in groups, or herds.

Polygynous mating involving displays

In birds and mammals in particular, males display in specific communal courting areas, called **leks.** Females come to these areas to find a mate. Females choose their prospective mates, after males have performed elaborate dancing displays. Most females choose the same male. Few successful males perform the majority of mating. In white bearded manakin, one male accounted for 75% of 438 copulations at a 150-M^2 lek containing as many as 10 males.

11.2.4 Polyandry

In systems, in which one individual mate with more than one individual of the opposite sex, the polygamous sex is male. Polyandry, the opposite condition, in which female is polygamous is rarer. For spotted sandpiper *Actitis macularia,* productivity of breeding grounds is so high, that female becomes rather like an egg factory, laying a number of eggs in 40 days. Her reproductive success is limited not by food, but by number of males. She can find to incubate eggs, and females compete for males and defending territories where males sit.

Polyandry is also exhibited in species, where egg predation is high as males need to guard nests and only females can replace the youngs. Polyandry occurs in *Jacana spinosa,* that uses its elongated toes to walk on floating lily leaves. Female jacana is large, to enhance her egg-laying capacity and have large territories, with several nests, each attended by a male. Males incubate eggs, while females move about their territory, mate with different males, lay eggs and defend territories against other females. Females also break up fights between squabbling males.

Prostitution polyandry

The females may mate with more than one male, to gain access to resources that males control and offer only to their mates (humming-birds). Polyandrous individual females may mate with more than one male per breeding season.

Polyandry without sex role reversal

Sperm replenishment polyandry

The females may mate more then once, to secure additional sperm, with which to fertilize a new clutch of eggs (fruit files).

Polyandry with sex role reversal

Resource defense polyandry

The female control resources attractive to more than one male. Males may provide greater parental care per offspring than females (fishes).

Male defense polyandry

Females provide neither resources nor greater parental investment per offspring, than male they compete directly to monopolize male mates in female aggregations (red-necked phalaropes). Coordination of breeding lizards can be exemplified by the green anole (*Anolis carolinensis*).

Costly polyandry

Polyandry is a widespread, costly behaviour that may arise, when females accrue benefits from multiple mating outweigh costs, or males manipulate females, against females' best interests. In a polyandrous spider *Stegodyphus lineatus*, females mate with up to five males, but behave aggressively towards additional males after first mating. Female aggressiveness may act to select for better quality males. Females may try to avoid superfluous matings. Maklakov and Lubin allocated females into single-mating (SM), and double-mating (DM) treatments. Double-mated females either accepted or rejected the second male. DM females laid more eggs, but did not produce more offsprings than SM and rejected females. Offspring of DM females were smaller at dispersal, than offspring of SM and RE females. Nest failure was significantly more common in DM females. Paternal variables did not influence female reproductive success, whereas maternal body condition explained much of variation. Polyandry is costly for females, despite production of larger clutches and suggests that multiple mating results from male manipulation of female remating behaviour.

11.2.5 Promiscuity

Both males and females may have more than one mate during breeding seasons. Dominant male with territories may have highest fitness, if they mate with as many females as possible, while female provided parental care to the resulting offspring.

Extra-pair matings in "monogamous" species

In many apparently monogamous species, it is not uncommon for mating to occur between individuals, which are not nominal mates. These are termed extra-pair mating. Extra-pair mating is done surreptitiously, when the nominal mate is not in attendance.

Although numerous hypotheses have been put forward, it is still difficult to explain why female should readily solicit copulation with multiple males, despite associated costs (Zeh and Zeh 2001). Recently interest has been given to a hypothesis concerning the role of parental genetic incompatibility in the evolution of polyandry. According to the genetic incompatibility avoidance hypothesis, an important function of polyandry is to enable females to exploit post copulatory paternity biasing mechanisms to minimize the risk and cost of fertilization by genetically incompatible sperm (Zeh and Zeh 1997). Genetic incompatibility can result from a variety of potential sources including dominances, over dominance, intra and intergenomic conflict, foetomaternal interactions and immune system function (Tregenza and Wedell 1998). In many cases, parental genetic incompatibility has fatal consequences for the developing offspring such that females incur cost of wasted investment in ova and embryos.

Promiscuous mating has been shown to decrease reproductive failure in experimental studies of polyandrous female pseudoscropions, crickets and cuis, although other studies found no evidence for genetic incompatibility avoidance benefit of polyandry. A comparative approach could also provide new insight concerning the likely benefits of polyandry from a life history perspective, that is in addition to potential energetic costs associated with wasted investment in failed ova and embryos, when females typically produce more than one offspring per reproductive attempt, it is possible that genetic incompatibility may also have important consequences for variation. This may be important for species such as mammals in which females invest heavily in their offspring, since increased variation in offspring mortality is likely to make it more difficult to control variation in offspring numbers, and hence, to achieve and optimal litter size. By biasing paternity of their offspring, polyandrous female mammals may thus benefit by reduction variation in reproductive failure caused by genetic incompatibility in addition to reducing reproductive failure per se (Stockley 2003).

11.2.6 Advantage to Males of Extra-pair Matings in Monogamous Species

If males are successful in fertilizing eggs through extra-pair mating, there would be a large benefit to their fitness, as they will obtain reproductive output without investing in parental care. When populations consists of males that do not attempt to cuckold, and if a mutant attempts to cuckold, then mutant male has higher fitness, than other males, as he will fertilize all eggs of his mate plus some number of additional eggs that he fertilized through cuckoldry. So, that cuckolding trait will increase in populations. Conversely let us consider a population that consists entirely of cuckolding males. On average, reproductive output of 3 cuckolding males will be equivalent to number of eggs produced by one female. For males, this output will consist of some fraction of eggs of their nominal mate that fertilized. If a mutant male appears in this population that is not cuckolding males, he will not be able to fertilize all eggs of his mate, and will not make up this shortfall by cuckolding other males. Although, this mutant male may suffer less from cuckoldry than other males, because they may be more vigilant. It is not possible to be vigilant all time due to constraints imposed by requirement to perform other activities, such as gathering energy, or defending his territory. Therefore, cuckoldry is a stable reproductive behaviour.

11.2.7 Extra-pair Mating in Females

Females can gain no direct reproductive benefit form extra pair mating, as this have no effect on number of eggs that female produces. Some cost may be associated with extra pair mating, if there is a risk of being rejected by her nominal mates, if the female is caught in the act. Thus, it would be more or less disadvangeous for the female to participate in extra-pair mating. An advantage for females in extra-pair mating comes, if long-term fitness of females is considered. If a population consists of non-cuckolding females, and a mutant cuckolding male appears, some genetic variations are expected of female's mating with the cuckolding male. Females that mate with the cuckolding male have female offsprings with average fitness. But their male offsprings have higher than average fitness. Net result is that females that mate with cuckolding males will have higher fitness (more grand-children) than females that do not mate with cuckolding males. Behaviour for females to be willing to mate with cuckolding males, and will thus increase in population. If population consists entirely of females that participate in extra-pair mating, and mutant female appears that does not mate, mutant females have lower fitness, because her sons will tend to be less likely to attempt extra-pair matings. End result is a population of "monogamous" individuals, in which males seek for extra-pair mating and females willingly participate. In this monogamous/ promiscuous mating system, there is clearly no net benefit, in terms of reproductive output of the population, but much time and energy is spend in extra-pair matings.

11.2.8 Female-Female Aggression and Maintenance of Monogamy

Female-female aggression occurs and potentially limits opportunities for the female chats during settlement. Such aggression is recorded in a number of avian species. Hypothesis exists that females who exhibit more aggression, are more likely to end up in socially monogamous mating situations. European starlings have a variable social mating system, with a relatively high degree of polygyny, compared to chats. Female-female aggression is likely to be in direct conflict, with optimal social mating system for pair male. Having additional social mates should increase male reproductive success, and males should attempt to interfere with aggressive interactions among females. Male disruption of female-female aggression is observed in insects, fish and birds. In addition to maintenance of socially monogamous pair bonds, avoidance of intraspecific brood parasitism may also be a selective force, favouring female-female aggression. This is particularly relevant for cavity nesting species, and waterfowl that tend of have higher incidences of intraspecific brood parasitism.

11.2.9 Mammalian Mating Systems

Females are constrained to provide parental care. Mating system is polygamous in most species. Details of mating systems tend to be similar. Most mammals are solitary, and female occupy home range. Size of a female's home range is determined by energy requirement to rear offspring. The critical resource, for males is not food, but females. Male territories, therefore, tend to be larger, than required, to meet male energy requirements, and included home ranges of a number of females. Males invest their time and energy in maintaining a large territory throughout the year. Presumably, this reduces competition for food, between intruder males and territorial male's offspring. Young males are unable to hold territories, as they are not large enough. Old males are unable to hold territories, because senescence decreases their ability. Reproductive life of males may be limited to a relatively short period during which number of offsprings they produce is similar to the number of offsprings reared by the females, but over a longer time period.

11.2.10 Social Mammals

Sociality is generally more evident in females, than males, due to obligate association of mother with their offspring, combined with male biased dispersal, leading to related groups of females in local areas, in certain species. Social groups may be more successful in foraging on clumped resources, or in defending their youngs through increased vigilance. When female groups are small mating, they may be functionally similar to that of solitary female mammal. Males hold individual territories with discrete groups of social females that he has exclusive access to. When female groups are very large, and group forages over large areas, it may be possible for a single male to exclude all other males form the area used by females. In this case, a number of males may jointly defend the area used by females. If these males share in mating with females, resulting mating system is likely to be promiscuous, rather than polygynous.

11.2.11 Leks

Leks are aggregations of males. In leks, males defend small territories (Figure 11.5) for display, and to attract females. Females choose mates from displaying males. Small numbers of males tend to obtain most of the mating. Female's choice depends on a male's position on lekking ground, or vigour of his display that provides females with information on survival quality of male or male's parasite resistance. Lekking is known for 42 out of 12,000 species of mammals, and birds and must be a response to an unusual set of circumstance. Leks are male aggregations usually at traditional sites that are visited by females primarily for fertilization. Lekking is important for understanding both sexual selection and benefits of group living. Leks offer particularly good systems for addressing questions of females mate choice, as costs of choice among lekking males should be low. Studies have shown female choice for certain male traits on leks (Andersson 1994). However, variation in these traits should diminish rapidly due to directional selection imposed by female choice. Nevertheless female choice persists and so does variation in male secondary sexual traits chosen by the females. Maintenance of variation in face of strong directional selection on a trait is a problem in evolutionary biology. Combination of strong female choice with persistence of variation in male secondary sexual characters on leks has led to this phenomenon being labelled the "lek paradox" (Taylor and Williams 1987). Here variation in male traits could be maintained despite strong directional selection and absence of direct benefits to female choice through nonlinear selection for exaggerated traits. Cyclic evolution of sexually selected traits occur due to intrinsically

FIGURE 11.5 The mutual display in penguin.

unstable Fisher's runaway process and conditions' dependence of sexually selected traits.

Lek polygyny

Lek polygyny is a mating system common in polygynous species of insects and birds in which the male provides no parental care to its offspring. This system is driven by females' pursuit of their mate, rather than males. Group of males performs displays to attract receptive females. Such group displays increase the ratio of visiting females per males. Visiting females compare the males' physiques and courtship displays to pick up the most attractive male as their mate (Alcock 2001). Few most attractive males will do majority of mating (about 99%), while the subordinate males do no mating at all (Sherman 1999).

Lek Polygyny mating system promotes a heavily skewed mating success rate among lekking males. Most individuals in a lek never receive a mating opportunity, but lek polygamy flourishes among various species of birds and insects (Sherman, 1999). This suggests that fitness of subordinate males must somehow be indirectly benefited by communal displays. Several hypotheses were proposed to explain the reasons behind lekking behaviour. Well established hypotheses include the Hotspot Hypothesis and Female Preference Hypothesis (Alcock 2001). Petrie et al., (1999) proposed an alternative hypothesis predicting that lekking behaviour is driven by kin selection. If all the males in a lek are genetically related, then males receive fitness benefits. The kin selection hypothesis was demonstrated by determining the genetic structure of various lekking groups of (Whipsnade Park) peacock population (Petrie et al., 1999). Peacocks aggregates at neutral display sites, use their calls to attract receptive peahens and on arrival of a peahen, peacocks cease calling and perform plumage displays. Peacocks with most appealling courtship displays have high mating success while subordinate peacocks have no mates. After a male's fourth year of lekking peacock establish permanent lekking site. Peacocks return to this same lekking site every mating season (Petrie et al., 1999). Using multilocus fingerprinting Petrie et al., (1999) compared genetic similarity of within and between lekking groups of Whipsnade Park's peacocks. Results supported the proposed kin selection hypothesis. Degree of band-sharing within the leks was higher than band-sharing between leks. Band sharing within leks is indicative of that of half-siblings. Subsequent hypotheses of why such lekking groups consisted of related individuals was proposed. It is suggested that related peacocks display together as their dispersal concentrates around their natal sites [Petrie et al., (1999)]. This hypothesis was rejected when peacocks of mixed relatedness were reared away from their natal sites but on reintroduction into whipsnade park population, joined leks with related peacocks. Peacocks' tendency to congregate with relatives is due to shared genetic preference of habitat choice. This was however rejected. The unique structure of peacock leks has driven by kin selection based on self-referent phenotypic matching (*Armpit Effect*, Petrie et al., 1999). It has been predicted that peacocks match heritable similarities of their own phenotype with other males (Petrie et al., 1994) as peacocks do not participate to rear their offspring. Thus, a peafowl's recognition of father must be genetically innate. Tendency to form leks with relatives occurs in the absence of social cues of identity. The subordinate peacocks forfeit chances of mating by cooperatively forming leks with genetically related individuals and subsequently increase the chance of transmission of their genes. Indirect benefits of fitness seen in peacock lekking displays outweigh costs of communal displays (Petrie et al., 1999). Four distinct genera of Peafowl are Pavo, Afropavo, Rheiinartia and Argusianus. Male peafowl do not participate in nest defense or rearing paternal care of young but it is observed in all 4 genera. Behaviours of Pavo cristatus in zoo settings may not reflect its evolutionary history. Hotspot hypothesis states that males form leks as females often visit certain "hotspots". The hypothesis predicts that males form leks because subordinate males congregate around highly attractive males to increase their chance of being noticed by receptive females. Female Preference Hypothesis predicts that males form leks as female like to visit large clusters of males consisting of various potential mates from which she can safely compare the quality of her mating choices.

Another possibility is that reproductive skew on leks is not as high as often thought as females also mate with low quality males. Variation in male traits persist because there is not strong directional selection on these. This may be due to a number of reasons. Simple errors in mate choice enable the maintenance of variation in mate quality. Females may not be as choosy as typically thought, if there is high predation risk on the lek. Female may also express specific preferences which could increase variation in mate choice. Further female may choose for compatible rather than good genes, so that each female has a different optimal mate. Although lekking males are typically not though to provide direct benefits to choosing females, the latter may benefit form reduced costs associated with choosing particular males.

11.2.12 Fecundity and Fertility, and Copulatory Plugs

In one kind of hemipteran insect, male pierces female body wall and injects sperm. They swim around, until they find eggs, Some things really bizarre happens, where males pierce other male, and sperm swim into victim's testes. Herma-phroditic leeches having barbs as penises go around stabbing to other leeches. All else remaining constant increase in fertility increases an organism's fitness. An organism can increase its energy capture rates by growing and increasing its future fertility. In juvenile phase, fertility is typically zero. Upon reaching a size, some allocation to reproduction increases fitness. Some energy during reproductive phase is diverted away from reproduction and allocated to maintenance. Natural selection on age of first reproduction and on adult reproductive rate will end to maximize total allocation of energy, to reproduction over the life course. Natural selection on offspring number and investment per offspring tend to maximize long-term production of the descendents which is as estimated by number of offspring that survive to reproduce themselves during an organism's life time, or if fertility affects production and survival of grand children. The female provides investment to the offspring greater than 95% in mammalian species. Male provides similar amount or more total investment, among most altricial birds, male brooding fish, and some insects, such as Katydids (Clutton–Brock and Parker 1992). One that does more investing sex is in short supply, resulting in operational sex ratios, and competition for mates occurs among members of sex that does less investing. The more investing sex is selected to be choosy about, when and with whom to mate. Less investing sex is selected to posses characteristics that increases its mating opportunities. Male resources are wasted on costly display, or handicaps, or on fighting rather than in offspring production.

11.3 HOMOSEXUAL BEHAVIOUR IN ANIMALS

Homosexual behaviour is generally, regarded as pathology, assuming purpose of sexual inter-course solely for producing the offsprings. About 1500 species of animals display homosexual behaviour including insects, like dragonfly and housefly, primates like bonobo, fish like salmon, birds like black swan and chinstrap penguins. Homosexual behaviour in chinstrap penguin, a characteristic of "classic pair-bonding behaviour" suggests that such behaviour is used as a tool for developing social bonds and connection that in turn helps in "resource acquisition and alliance building". In bonobo, 75% of sexual activity is not for procreationary purpose. Sex is mainstay of social life for bonobos, and sexual intercourse takes place between males, between females, and between males and females, making species pansexual. That two chinstrap penguins became a pair of homosexual couple for nearly six years in Central Park Zoo, Manhatten, is reported. They exhibit "ectastic behaviour", where they entwine their necks, vocalize to each other, and have sex. When offered females companionship, they reject them, and remain faithful to each other. Homosexual behaviour has been observed in a variety of animals including non-human primates, birds, reptiles, amphibians and insects (Yamane 1999) in addition to female western gulls, bottleneck dolphins, and macaques. In some species, homosexual behaviour is viewed as adaptive. Homosexual behaviour in female Japanese macaques may help to establish alliances between females. These alliances may benefit females in later aggressive introduction, with other members of the troops (Vasey 1996). The female weevils (*Diaprepes abbreviatus*) behaved sexual towards other females in order to attract larger, perhaps higher quality males (Harari and Brockmann 1999). Homosexual behaviour may be an artifact of artificial captive conditions in many insects and such behaviour has been explained as an error (Wang et al., 1996).

Sexual interactions between individuals of the same sex occur in birds with over 130 avian species worldwide. (Bagemihl 1999). Courtship behaviour often results in same sex mounting and even copulatory behaviour and for the same sex pair-bonding and long-term same sex association as reported as in black swan *Cygnus attratus*. Same sex sexual behaviour is observed for species that are primarily socially mono-gamous (Hunt et al., 1984) through to species that exhibit extreme polygamy (*Rupicola rupicola*). Males of precocial species may sexually imprint and later mount other males as in *Anser anser*. For altricial species, both males and females may display same sex preference in adulthood as a result of experience during early development. Sex ratio of conspecific and later social experience are reported to affect sexual imprinting preferences in altricial species.

CANNIBALISTIC BEHAVIOUR

FATIK BARAN MANDAL

Cannibalism, a possible disease transmission route and mortality factor is found in the animal kingdom including humans. Cannibalism, an intraspecific predation results in altered population dynamics. In the past, it was a common practice in human societies worldwide. Human killed and consumed following an attack on neighbouring villages, or consumed as part of funeral ceremony. The relation between cannibalism and spread of disease has drawn attention in case of *Kuru,* which spreads through cannibalism among the people in Papua New Guinea.

Historically, members of families or village often shared captured individuals in ritualized meals where the group size was often very large. Some human societies practiced cannibalism across groups and necrophagy within the group as in case of the fore people of Papua New Guinea, in which both intraspecific necrophagy and cross-group cannibalism were common. In *Kuru,* necrophagy could maintain and spread the disease within a village, while cross-group cannibalism could promote the disease spread on larger metapopulation scale. Cannibalism in humans dates back at least to the Neanderthals. Cannibalism might be part of the natural ecology of human societies for substantial nutritional gain.

Cannibalism is the major transmission mode of the protozoan, *Sarcocystis* in lizard .In many species, another form of cannibalism is the consumption of dead conspecifics, a common transmission route for many diseases. In the animal world, cannibalism is generally one-on-one interaction in which larger and stronger individual kills and consumes smaller and weaker conspecific. Cannibalistically transmitted parasites or pathogens usually have one or hosts, but in some species of *Gallotia* inhabiting the Canary Islands, parasitic transmission is directly cannibalistic. In the lizards, cannibalism is often partial; occur by consumption of autotomized tails. One individual host can infect several cannibals during its lifetime following tail regeneration. In group cannibalism, several smaller individuals can kill and together consume one larger conspecific as seen in social Hymenoptera and Isoptera and in some species where the mother serves as food source for its offspring. Sometimes, several adults kill and consume an infant together as in social mammals like lions, when a group of male acquires another male's harem. It also occurs in chimpanzees where male groups commonly attack conspecifics.

Diseases transmission by cannibalism is rare as the epidemiological conditions necessary for its spread, especially group cannibalism, are rarely found in natural populations. Cannibalistic transmission may play an important role in maintenance of blood-borne infections like simian immunodeficiency virus different types of hepatitis in Chimpanzee populations. Given the importance of social primates as reservoirs for human disease, this possibility deserves detail investigation.

ArticlesBase.com, 2012

11.4 SEX BIASED DISPERSAL

Birds and mammals both tend to have juvenile dispersal that is biased in favour of one sex. In mammals, juvenile males tend to disperse further from their natal area, than do juvenile females. In birds, juvenile females tend to disperse further from their natal area, than do juvenile males. Differences may arise, due to problems of inbreeding combined with differences in mating systems of birds, and mammals. Inbreeding generally reduces the viability of offspring, probably due to higher probability of deleterious recessive alleles occurring in homozygous state in inbreeding. This loss of fitness associated with inbreeding would produce selection for behaviours that avoided inbreeding. An obvious adaptation, that reduces inbreeding is dispersion as this reduces the likelihood of encountering and mating with the siblings, or close relatives. So, dispersions in general are adaptation for reducing fitness losses due to inbreeding.

11.4.1 Bird

Mating tends to be monogamous, and specific fecundity of males and females is similar. Territorial males may maintain their territories for extended periods of time. Juvenile females that do not disperse from their natal area may end up mating with up their fathers over extended periods of time and suffer a fitness loss. Juvenile females that disperse do not have this problem, and would probably not have any more difficulty in obtaining a mate, if they disperse. Same argument could be applied to males, although males have a better chance, to eventually obtaining a territory, if they remain close to their natal area, because local "knowledge" may be useful in obtaining and defending a territory.

11.4.2 Mammals

Male's reproductive life is relatively short. Females are less likely to mate with their fathers, or if they do only for a relatively short portion of their reproductive life span. Females seem to have less to gain (fitness wise) from dispersing than do female birds. Polygynous males, generally, have to wait for an extended period of time beyond sexual maturity, before they are able to obtain and hold a territory. Territories held by male tend to be occupied continuously, rather than seasonally. Juvenile male in their natal area are more likely to be driven out by resident males, as resident males are territorial throughout the year.

11.5 MATERNAL CONDITION ADVANTAGE

Under specific ecological or social conditions, fitness benefits of the offspring may vary differentially. When this is so, facultative manipulation of offspring sex gives potential to parents to fine-tune the number and quality of offsprings to prevailing circumstances, thereby maximizing parental fitness. "Maternal condition advantage" hypothesis predicts that females should adjust sex ratio of their offsprings in relation to the effect of their own conditions. Supporting data in this connection are scarce and mainly correlational, particularly in birds and mammals. In a study of *Anser caerulescens*, differences in performance between male and female offspring were found to be related to quality of mother, but no significant bias in sex ratio at hatching. In birds, sex ratio variations fit the predictions of sex allocation theories, initially thought to be attributable to constraints, imposed by avian sex determination process itself. In the bird seychelles warbler (*Acrocephalus seychel-lensis*), that lays a single egg, and adjusts primary sex ratio in response to changes in benefits, to be gained by producing daughters, who function as helpers at nest, or sons, which disperse. Maternal condition hypothesis are considered likely to have most widely applicable effects on sex ratio. This condition during egg laying is clearly a very important parameter, which may underlie the sex ratio variations reported in several correlative studies. Variation in this parameter is likely to explain patchy observations of sex ratio skews in non-experimental studies.

11.6 LIFE HISTORY EVOLUTION

Life history is associated with lifetime pattern of growth and reproduction. Salmon produce their all offsprings in a single reproductive event and is called **semelaparous**. Mammals are iteroparous, as they reproduce repeatedly in a relatively larger life span. Effects of mating systems, habitat selection, and dispersal can be brought under r-k continuum. MacArthur and Wilson (1967) originally proposed r-k classification; r-selected species are poor competitors with high per capita population growth rate r. Such species have high dispersability. They colonize new habitats before being out competed by k-selected species, which are good competitors, exist in mature habitat where they out compete most species. Other life history strategies can be classified in terms of growth, longevity, fecundity triangle. Insects are r-selected species, as they produce many youngs in a relatively short life time. Mammals, like elephants produced few youngs in their relatively large life time, are k-selected species, which are at risk of extinction. Four life history traits listed elsewhere are brood size (number of seeds, eggs, youngs, or other progeny), and size of the young (at birth, hatching or germination), are distribution of reproductive efforts, and interaction of reproductive effort with adult mortality especially ratio of juvenile adult mortality. Predictive theories summarized by Pianka (1970) and later noted by Odum and Barrett (2005) are as follows:

Species should reproduce only once in a lifetime, where adult mortality exceeds juvenile mortality. Organism should reproduce several times, where juvenile mortality is higher. Brood size ideally maximizes number of youngs surviving to maturity, averaged over the lifetime of parents. A ground nesting bird requires a clutch size of 20 eggs to ensure replacement. Bird nesting in a protected place requires a relatively smaller clutch size. In expanding populations, r-selected organisms breed at an easy stage, whereas in stable population, maturation is delayed. For example, in fast growing countries child birth begins at an early age, whereas in stable countries people postpone childbearing for later. To avoid risk of predation, scarcity of resources, or both, size at birth is large. On the other hand size of the young is small with increasing availability of resources, and decreasing predation of competition pressure.

For growing or expanding populations, in general, age of maturity is minimized. Reproduction is concentrated early in life. Brood size is increased, and a large portion of energy flow is partitioned to reproduction and a combination of traits is recognizable as the r-selection tactic. For stable populations, one expects the reverse combination of traits, or k-selection. Species level specialization result in a number of life history characteristics, which can be delayed on a fast-slow continuum. Mammals exhibit short gestation times, early reproduction, small body size, large litters, and high mortality rates. Organisms are capable of slowing down, or speeding up their life histories depending on environmental conditions, such as temperature, rainfall, food availability, density of conspecifics, and mortality hazards. Allocation to reproduction, as measured by fecundity and fertility, vary over short-term in relationship to food supply, and energetic output among plants, birds, and humans. The birds under variable conditions adjust clutch sizes that tend to maximize number of surviving youngs produced in life course. Calories restriction rates at young ages tend to slow down growth rates and leads to short adult stature, even when food becomes abundant later in juvenile period. Rates of senescence vary across different populations of grasshoppers, with those at higher altitudes and earlier winter senescing faster than those at lower altitudes as a result or differential selection as genotypes (Tator et al., 1997). Males and females parental investment vary in relation to local ecology. Katydid males provide female with a "nuptial gift", to support offspring production.

11.7 HORMONAL MECHANISMS IN PAIR-BONDING

Understanding the neurochemical pathways that regulate social attachments may help to deal with defects in people's ability to form relationships. Scientific tale of love begins with voles. Prairie vole is an associable creature. One of only 3% of mammal species form monogamous relationship. Mating, between prairie voles, is a 24-hour effort. After this, they bond for life, and prefer to spend time with each other, groom each other, and nest together. They avoid meeting other potential mates. The male becomes an aggressive guard of the female. When their pups are born, they become affectionate and attentive parents. Another vole, called montane vole, has no interest in partnership beyond one-night-stand sex. These vast differences in the behaviour are due to genes. Two vole species are more than 99% alike genetically. At times of prairie vole's sex, oxytocin and vasopressin are released. When release is blocked, prairievoles' sex becomes a fleeting affair, like that enjoyed by their rakish montane cousins. If prairie voles are given an injection of hormones, but prevented from having sex, they will still form a preference for their chosen partner. When this magic juice was given to montane vole; it made no difference. It turns out that faithful prairie vole has receptors for oxytocin and vasopressin in brain regions, whereas montane vole has not. Brain has a reward system designed to make voles do what they ought to. Without it, they might forget to eat, drink and have sex with disastrous results. Animals continue to do these things, as they feel good due to release of dopamine. When a female prairie vole mates, there is a 50% increase in dopamine level in reward centre of the brain.

When a male rat has sex, he learns that sex is enjoyable, and seeks out more of it. In contrast to prairie vole, rats do not learn to associate sex with a particular female. Rats are not monogamous. If gene for oxytocin is knocked out of a mouse before birth, that mouse will become a social amnesiac, and have no memory of other mice, it meets. Same is true, if vasopressin gene is knocked out. Salient feature, in this case, is odour. Rats, mice and voles recognize each other by smell. Exposure to opposite sex generates new nerve cells in brains of prairie voles in particular in areas important to olfactory memory. It is suggested that prairie voles become addicted to each other, through a process of sexual imprinting mediated by odour. Reward mechanism involved in this addiction has probably evolved in a similar way in other monogamous animals, to regulate pair-bonding in them, as well. Animals, which form strong social bonds do so because of location of their receptors for vasopressin and oxytocin. Evolution acts on distribution of these receptors, to generate social or non-social versions of a vole. Social groups, and society itself rely ultimately on these receptors. Phelps found great diversity in distribution of vasopressin receptors between individual prairie voles. He suggests that this variation contributes to individual differences in social behaviour.

Fisher suggests that lust, romantic love and long-term attachment are separate phenomena, with their own emotional and motivational systems, and accompanying chemicals. They have evolved to enable respectively, mating, pair-bonding, and parenting. Lust involves a craving for sex. Aftermath lustful sex is similar to the state induced by taking opiates. A heady mix of chemical changes occur including increase in levels of serotonin, oxytocin, vasopressin and endogenous opioids. "This may serve to relax body, induce pleasure and satiety, and perhaps induce bonding to very features that one has just experienced all with this". But final stage of love, long-term attachment allows parents to co-operate in raising children. This state is characterized by feelings of calm, security, social comfort and emotional union, as they are independent. These 3 systems work simultaneously to achieve results.

11.8 MATING SYSTEM EVOLUTION

The behavioural interactions between males and females are crucial, for elucidating selection pressures that shape mating system evolution. Although difficult to observe in free-living populations, behaviour of males and females, in both intra-sexual and intersexual encounters has a significant influence on mating system evolution. Theory has addressed mating system evolution, primarily in regard to polygyny, as a function of characters like sex ratio, parental investment patterns, or distribution of resources. The ability of one sex to monopolize access to either resources, or mates is a key component of polygyny models, based on resource distribution. In polygyny threshold models, selection favouring polygyny is generated by environmental potential. Female-female aggression proposes that limits choice of females, and therefore, provides a basis for explaining evolution of monogamy. Females are not always free to settle anywhere. Important limitation to female settlement is aggression by resident females. Female-female aggression appears to be important in shaping mating system, as it limits ability of both sexes, to exploit environmental potential, for polygyny and leads to monogamy as predominant system. Work on mating system evolution is concentrated in polygamous mating system. Less attention has been paid in explaining monogamy. Molecular genetic analyses have uncovered sexual infidelity, as a common feature of monogamy, particularly among passerine birds, creating a distinction between social and genetic monogamy. Genetic studies only document occurrence of Extra Pair Fertilizations (EPFs) by recording number of extra-pair offspring. Relatively less work has done on underlying behavioural interactions, either between males, and females.

11.9 SEX IN YELLOW-BREASTED CHATS

Yellow-breasted chats (hereafter referred to as chats) are Neotropical migrant passerines. They breed in scrub, and edge habitat. Chats are largely monomorphic, socially monogamous, and exhibit biparental care. Given high degree of male parental care during nestling stage, male help may be an important contribution towards offspring survival. Females may suffer a cost of sharing male help and benefit, from excluding secondary females from settling on their territories. Less is known about mechanisms underlying extra-pair mating systems. Male and female chats regularly leave their territories, and intrude on neighbouring territories. Males attempt to guard their territories from other intruding males. So, even though both males and females are known to engage in extra-territorial forays, consequences of these trips are not clear. In simulated extra-territorial intrusions, chats respond differently based on sex of focal individual, and species and sex of intruder.

11.10 SEXUAL BEHAVIOUR OF PENGUIN

Emperor penguins breed in winter. Females compete for males, and are notoriously unfaithful. Emperor penguins *Aptenodytes forsteri*, are largest of penguin family. They live in harsh environments of Antarctic, and are only warm-blooded animal to winter on open ice. These birds present unique behaviours. Males grow yellow tufts in areas surrounding their ears. Tufts are believed to be an attraction factor for females. Males compete for nesting space, and attract females through quaint rituals. They do not create nest, and put on mating displays that last for weeks, during which future parents learn each other's call. After copulation, and about 15 days later, a single soft-ball sized egg is laid. Female leaves to sea to feed well. Meanwhile, father cares for the egg. To protect it from frosty floor, for next 6–8 weeks father balances the egg on its two feet, under protective blubbery flaps of its underbelly, to keep it warm. During whole incubation period, father does not eat, and may lose a third to half of its body weight. Mother returns right before hatching, well-fed and with food for her offspring.

Two parents apparently find themselves through distinguishable calls. Mother then feeds chicks for the first time. As female broods young, the father is free to replenish his energy, only to return to jointly rear the chick along with the mother.

Newly hatched chicks are barely protected from sub-zero temperatures by their downy coats. Displaying similarly, between adult penguins and chicks, is formation of clustered groups called **crèches**. Within these groups, birds maintain temperatures of about 96 degrees F, in contrast to, minus 30 degrees wind. Anatomical preparation for copulation is not stimulated by mating displays. Penguins reach mating areas have seven times normal amount of luteinizing hormone, testosterone, and estrogen, suggesting that environmental cues activate gonodal preparation, specifically, corresponding to maximum size of gonads. Between copulation and egg-laying, levels of all hormones begin to fall, but LT remains slightly above average in incubating males, and in both sexes that are rearing their young. Quantity of food and climate is perfect for chicks to learn to fend for themselves. These two factors may have been enough evolutionary pressure, to give rise to these vital behavioural traits in emperor penguins.

11.11 PARENTAL CARE

Parental care is any behaviour that increases the fitness of offspring (Clutton-Brock 1991), and is likely originated and maintained for this function (Royle et al., 2012). Widespread biparental care has evolved few times in vertebrates. Few comparative studies exist in birds. In insects, parental care has evolved in more than 10 orders (Costa 2006). Female care for offspring is common in insects, male contribution to care are rare (Tallamy 1994). Biparental care associated with nests in insect has evolved several times independently and has importance in phylogenetic comparisons and comparative physiology of offspring care (Trombo, 1996). Two reasons why males provide less care than females is pointed (Queller 1997). First, multiple mating and sperm competition produce uncertainty about paternity of males and reducing the expected fitness gain for caring young. Second, a subset of males may be consistently like to mate than others if sexual selection favoured their traits. Whole mating system approach in studying paternity and paternal care and often lower probability of parentage for males make males less likely than females to provide care (Wright 1998). Interactions between male and female and relationships between parental effort and paternity is hypothesized (Houston and McNamara 2002). Wade and Shuster (2002) reanalyzed Maynard Smith (1977) model. According to Wade and Shuster (2002) if deserting males gain extra offspring, male parental care evolves whenever half the magnitude of indirect genetic effect of paternal care on offspring viability exceeds the direct effect of additional mating success gained by desertion. Sex roles regarding offspring care was reviewed. Adult sex ratio would generate differences in breeding systems (Kokko and Jen mons 2008). Loss of paternity will be generated by female-biased adult sex ratios (female multiple mating) and it drive female-only care. There are empirical reviews of paternal and/or biparental care of arthropods. An overview of pre-zygotic paternal investment, biparental care, exclusive paternal care by terrestrial arthropods with emphasis that paternal investment is correlated with certainty of paternity and male territoriality is given (Zeh and Smith 1985). The enhanced fecundity hypothesis in which paternal investment evolve as a trait increasing mating opportunities by benefit of care (Tallamy 1994).

11.11.1 Biparental Care in Insects

Biparental care in insects is reported in three orders: Blattodea, Coleoptera, and Hymenoptera. Most members in these orders make nests underground or in wood burrows and prepare food for young in nest before oviposition is finished. Nest-guarding by males against other males is reported in most species. Although, exclusive paternal care has evolved independently in some Heteroptera species (Tallamy 2001), biparental species is absent in Heteroptera. Female attendance in an ancestor appears to be required for evolution of biparental care. Paternal care can evolve the sexual selection of males with superior genes, and females can use nest construction or the act of guarding another female's egg as honest signals of paternal intent and quality (Alonzo 2010). Female choice for male allows male care to evolve despite low relatedness between male and offspring (Alonzo 2011). Such studies show evolutionary conditions in paternal care, but not in biparental care.

Food is considered to be a mover for biparental care because some food like rotten wood are difficult to eat for young, or are less efficient to defend from competitors without help by the parents. All species of biparental Blattodea, Coleoptera, and Sphecidae of Hymenoptera make nests in food of young or carry food to their nests before larval hatching (burying beetle). So, aspects of nests are common among biparental insects. For example, (1) nest has enough food for young before finishing oviposition, (2) females usually stay in nest, and (3) nest has a tough wall made up by soil or wood. About all biparental Coleoptera and Blattodea live in rotten wood or in underground nest. Most species collect food before oviposition, and species collecting food after ovipositon is rare (Rugg and Rose 1991). Paternity is assumed to be a prerequisite for maintaining biparental care (Kokko and Jennions 2003). However, few works have studied the relationship between biparental care and paternity in insects. The evolution of biparental care gets impeded by extra pair copulation (Suzuki, 2013).

Presence of sneaker males of Onthophagus taurus reduces the mass of provisioning and increases the rate of desertion by paternal males (Hunt and Simmons 2002). Thus, male care by 0. Taurus affects offspring size and confidence of paternity. If the nests of most biparental insects protect against intrusion by other males as in *Nicrophorus,* such nest increase the confidence of paternity and promote biparental care for males. If securing paternity promotes biparental care and nest-making, *Polistes* may be the exception (Suzuki, 2013).

Parental care results from a cost benefit relationship (Figures 11.6 and 11.7). This cost benefit tradeoffs appear to drive the parental effort of birds that are susceptible to predation pressure. Most (90%) of the parental care is contributed by females in mammals while remaining 10% is contributed biparentally. Birds on the other hand exhibit 90% biparental care, 8% female only care, and 2% male only care. In fishes, 80% does not provide care. Out of the remaining 20%, 50% are male only care, 30% female only, and 20% biparental (Gross, 2005) (Figure 11.8).

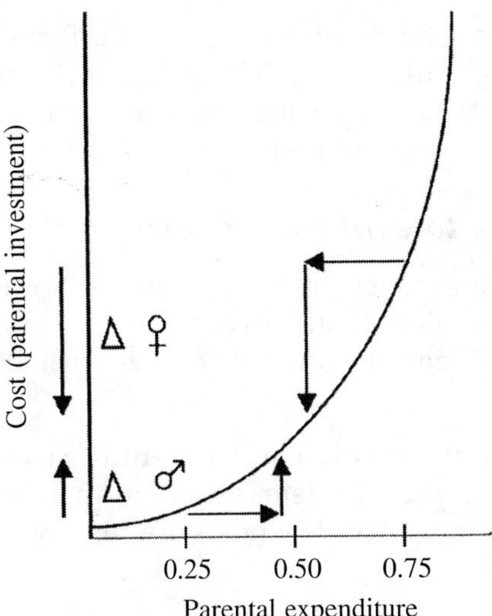

FIGURE 11.6 Relationship between parental effort and cost. Parental effort units are expressed as proportion of effort (time and energy) needed to rear one brood.

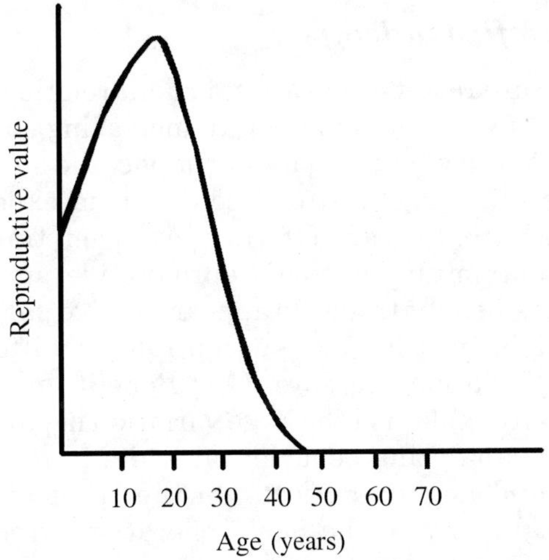

FIGURE 11.7 Relation between reproductive effort and age.

FIGURE 11.8 Male rhesus monkey aged 2 weeks restrained by his mother during an exploration.

Ecological factors

K and *r*, representing carrying capacity of environment and intrinsic rate of population growth, respectively.

K: Characterizes traits that are adaptive in a more stable environment. Population is near K probably intense competition, quality is favoured over quantity—fewer offspring, and more care for each.

r: Traits favoured rapid development in fluctuating environments, smaller bodies; off-spring gets less or no parental care.

Mammalian r strategies: Large litters are produced after a gestation period of less than three weeks. When food is ample; the young females can become pregnant, even before they are weaned. r-K is a continuum. In bird species r strategy is generally more pronounced.

Hypotheses to explain pattern

Paternity certainty hypothesis: Differences in certainty for males should favour care by males with external fertilization, and by female with internal fertilization.

Order of gamete release: Parental care lowers the total number of offspring one can have, and which sex shows care depends on which one is left holding the bag.

Association hypothesis: Members of one sex find themselves in closer proximity to embryos after fertilization, than do members of other sex—this predisposes them to become the care giving sex.

Cost-benefit tradeoffs

Parental care results from a cost benefit relationship. Energetic investments made into increasing current offspring survival will negatively impact the survival of future offspring. Natural selection is expected to adjust parental care efforts spent over time, to achieve maximum lifetime reproductive success. These costs-benefit tradeoffs appear to drive parental efforts of birds that are susceptible to different types of predation pressure. Cost-Benefit approach traditionally explained the huge variation in parental behaviour, analyzing behavioural traits in terms of positive and negative effects on the transmission of parental genes to the next generation. Presence of the various trade-offs between costs and benefit associated with parental care (Harshman and Zera 2007) gets support from empirical evidence although governing mechanisms is controversial. Costs imply a reduction in offspring number other than those that are presently receiving care (Trivers 1972). Benefits are increased fitness in offspring currently being cared for both derived directly from resources allocated to offspring, and indirectly from protection against predators. The more a parent spends on caring for an individual offspring, the less it spends in caring for other offspring. The fitness cost-benefit can be measured in terms of number of offspring which allows for comparisons between individuals. Selection pressures may act on both resource acquisitions and allocation in parental care.

Limiting resources

Resource Allocation required for parental duties may be constrained by negative effects they have on other fitness related traits (Roff 2002). Many allocated resources to parental care are considered limited resources that can be spent once (van Noordwijk and de Jong 1986). Parents should distribute these resources optimally to maximize their fitness, with 2 major tradeoffs: between current and future offspring and between quantity and quality of descendants (Stearns 1992). Energy and time are the resources used to exemplify the currency to be traded off. Energy and time can be combined into 'energy per unit of time' (Clutton–Brock 1991). Keeping more energy and time to parental care reduce the energy and time for self-maintenance and future offspring. Energy acquisition and allocation are complex traits affected by different factors. Such traits are difficult to measure. Animals (capital breeders) may provide care to the current offspring using previously accumulated energy stores or use energy gained contemporaneously (income breeders). However weighing up the contribution of both processes in the same individual is difficult (Stearns 1992). Macronutrients and non-energetic substances like essential amino acids, carotenoids, vitamins and minerals (micronutrients) may need to be traded-off between competing functions. Many micronutrients benefit offspring growth and development, and parental survival. In fish and bird species, carotenoids increase fecundity and parental care (Tyndale et al., 2008), but are required for parental immune or antioxidant defences (Pérez-Rodríguez 2009).

Non-linear relationships between fitness and resource allocation

Although simple monotonic relationships between resources allocated to parental care and fitness are reported, the common cases probably involve sigmoid-saturating relationship (Clutton-Brock 1991). In the diet of Argentine ant (Linepithe mahumile) queens the size of pupae (a fitness proxy) positively correlate with macro-nutrient availability (Aron et al., 2001). Benefits accrued from non-energetic micronutrients show concave trend. When allocation for offspring care increases fitness diminishes in parents. Hatching success is positively correlated with amount of carotenoid deposited by female in egg yolk in salmon (Oncorhynchus tshawytscha), although survival benefits decrease asymptotically (Tyndale et al., 2008). When thresholds exceeded, switch in physiological pathways allocation strategies change. Minimum food availability is needed in income breeders (Schradin et al., 2009) and a critical level of fat stores is necessary to start egg laying in capital breeders (Alisauskas and Ankney 1994).

Limitation of resource allocation trade-off

Current reproduction may divert resources away from maintenance, but reproduction directly alters physiological homeostasis, which causes somatic damages. The links between resource acquisition and metabolism may explain the trade-off between current and future reproduction. The resource required for offspring differ with resources needed for somatic maintenance of parents is the First problem of resource allocation models. Parents provide offspring with different food to that they use for their own maintenance (Cherel et al., 2005). Differences in currency occur in many trade-offs that animal confront during parental duties. This is known as common currency problem (Houston and McNamara 1999). Animals should weigh up benefits of these simultaneous goals, that is, energy collected vs. foraging time and mortality risk (McNamara and Houston 1986). Introducing state variables that characterize the current physiological state (Clark and Mangel 2000) can tackle this problem. State variable may be level of damage, which must not exceed a certain threshold, while variable to be maximized is fitness. Aside from limiting resources, several mechanisms underlying parental care are discovered when exploring the physiological complexities of organisms. This oxidative stress where imbalance between the amount reactive oxygen species produced through cell metabolism and state of antioxidant and repair machineries occurs ultimately leads to oxidative damage (Kirkwood and Austad 2000). Parental duties increase cell metabolism and reproduction and oxidative damage over time, thus accelerating sene scence (Metcalfe and Alonso–Alvarez 2010). Limiting substances like antioxidants or energy for repair mechanisms may be subjecting to the principle of allocation constraining parental care (Edward and Chapman 2011). ROS are very reactive (Kirkwood and Austad 2000) and in some parental activities above a certain threshold antioxidant and repair systems are inefficient and soma damage may be unavoidable. Oxidative damage explains the link between uncoupled life-history traits, between activities separated in time and not subject to direct trade-off. Costs and constraints of nutrient sensing signalling systems may be independent of resources, although current evidence is inconclusive. Environmental challenges imply trade-offs that are independent of limiting resources as those derived from risky, stressful conditions during care (Harshman and Zera 2007), and some are 'all-or-nothing' trade-offs. Mechanistic approaches have revealed that resource acquisition has intrinsic trade-offs in diet components. Study (Lee et al., 2008) on fruit flies shows that protein: carbohydrate ratio that maximizes egg production differs with ratio that maximizes life span. High ratios favour reproduction, but impair survival since organism suffers damage caused by sub-products of protein metabolism such as reactive oxygen species and nitrogenous breakdown substances (Lee et al., 2008). Since no diet maximizes both functions, a trade-off between reproduction and maintenance may be inevitable outcome of resource acquisition, rather than effect of energy allocation as is proposed by traditional models.

Cost-free resources and not involved resources

Distinction between costly and cost-free resources are critical to understand evolution of parental care since costly resources reduce parents' ability to produce other offspring (Trivers 1972). Production of particular form of cost-free care probably depends on its context-dependent′ benefits for offspring development. Female bird's deposit hormones in egg yolk, but cost for mothers are unknown and perhaps even non-existent (Gil 2008). Hormone deposition may have environmental or sex-specific effects on offspring fitness, which explain differences in hormone levels among eggs in single clutch (Gill 2008). Direct female control in testosterone deposition into eggs is under debate. Allocation of substances to offspring may influence offspring fitness, although sometimes this act should not be regarded as parental care as it is a by-product of parental environment. Mothers of many species passively transfer pollutants into eggs, which may be beneficial since they prepare the offspring phenotype for polluted environment (Ho and Burggren 2010). Here, selection has probably acted on offspring's developmental pathways rather than on parents' behaviour.

11.11.2 Benefits of Parental Care

Parents obtain benefits from their reproductive expenditure by increasing offspring survivorship during development (short-term benefits) or by improving offspring survival and fecundity in the long-term.

Short-term benefits of parental effects

In various species, parents improve offspring short-term survival by protecting descendants from harsh environments or by allocating limiting resources that favour their development (Clutton-Brock 1991). Many obtain long-term benefits, although short term effects are the most intuitive. Parents prepare and maintain suitable nesting sites or directly defend offspring from predator, brood parasite or conspecific. Orange-crowned warbler (Vermi voracelata) elevate the nest site when perceived risk of predation is high (Peluc et al., 2009). In fish, offspring are guarded and protected in one of parent's mouths (mouth brooding). Aggressive offspring protection is found in many taxa, while examples of birds being able to discriminate parasitic eggs by visual cues and reject them, thereby preventing offspring mortality. Parents improve offspring viability by regulating water, thermal energy, oxygen and both non-energetic and energetic nutrient. Social insects control temperature of nest by metabolic heat production, fanning and water evaporation. Clutch thermoregulation by parents is reported in reptiles and birds (Deeming 2004). In tree frogs,large water pools for egg deposition favours tadpole development (Brown et al., 2010). Parents of fish and some crab species enhance survival by oxygenating eggs via fanning (Green and McCormick 2005). Nutrients are given in many forms and mothers of spiders, frogs and fish produce non-developing eggs or egg-like structures to feed their offspring (trophic eggs). Nourishment of offspring by maternal body (matriphagy) is found in arachnids and some insects. The foetus of viviparous caecilians is known to scrape lipid-rich secretions and cellular materials from their hypertrophied maternal oviducts (Wake and Dickie 1998). Parents provide offspring with non-energetic compounds. The transfer of carotenoid and vitamin to eggs in many vertebrates protects embryos from oxidative stress induced by high anabolic activity. Males transfer substances via their sperm. Males of Australian field cricket (Teleogryllus oceanicus) produce sperm with certain proteins that can be absorbed by eggs and improve the embryo's chances of survival (Simmons 2011).

Long-term benefits of parental care

Parental care has a strong influence throughout an offspring's life span. Benefits may be delayed and become evident after care has ceased. Malnutrition permanently alters morphology, physiology and/or metabolism during adulthood and cause long-term effects on fitness. In zebra finches (Taeniopygia guttata) maternal micro-nutrient in egg (carotenoids) influence sexual ornamentation displayed by offspring during adulthood (McGraw et al., 2005). Lack of macro- and micronutrients as a nestling reduces reproductive capacities in adulthood (Blount et al., 2006). Parents of many passerine species give spiders to offspring despite their low energy content. Spiders contain high amounts of taurine, free sulphur amino acid that is required for brain development (Arnold et al., 2007). Bluetit (Cyanistes caeruleus) nestlings that were supplied with taurine later exhibited greater abilities in spatial learning than control birds (Arnold et al., 2007). Parents influence offspring fitness by affecting their brain development, thus helping in cognitive, perceptual, and learning capabilities in adult. In species with prolonged parental care, offspring may devote more time to learn how to forage and practicing social skills, and to being taught by their parents (Hoppitt et al., 2008). Early learning helps to develop anti-predator behaviour, defence against brood parasites, effective foraging, and mate choice in adult (Davies and Welbergen 2009) and, thus increases fitness (Mateo and Holmes 1997).

11.11.3 Parental Care and Phenotypic Adjustment in Offspring

Genotypes produce different phenotypes in response to distinct environmental conditions. In fluctuating environments with short-term predictability parents program offspring development to cope with particular situations (Uller 2008). Parents produce different offspring phenotypes by affecting developmental pathways or by providing morph-specific resources. Parental influence has long-lasting consequences due to phenotypic organization or epigenetic changes resulting from gene expression (Ho and Burggren 2010). Early programming results from both parental behaviour and plasticity in development pathways. Development pathways, in adverse environments, explain how early conditions affect offspring phenotypes without active parental effects (Monaghan 2008). Phenotypic adjustment of progeny by parents is based on 2 important assumptions: (i) that environmental cues experienced by parents predict the environmental conditions that their offspring will encounter and (ii) that phenotypic plasticity in offspring development is sensitive to signals produced by parents (Mousseau and Fox 1998). Exposure to signals during embryonic development likely to cause accommodation effects, since a disproportionately large part of phenotypic organization occurs during this relatively brief stage in offspring's life history (West-Eberhard 2003).

Pathogens

Mothers can transfer information about pathogens that offspring encounter ('Tran generational immune priming'; Grindstaff et al., 2003). Mammals transfer antibodies to offspring through placenta, colostrum and birds through egg yolk (Boulinier and Staszewski 2008). Parents of some invertebrates (mostly insects) transfer specific immune factors to their offspring (Freitak et al., 2009). In red flour beetle (Tribolium castaneum) offspring sired by males exposed to heat-killed bacteria were more resistant to a pathogen infection than offspring from non-exposed males (Roth et al., 2010). Seminal substances, genomic imprinting and/or micro RNAs in the sperm could explain these findings (Roth et al., 2010).

Predators

Many animals learn anti-predatory behaviour from conspecifics (Mateo and Holmes 1997), although it is controversial whether parents teach their offspring how to cope with predators. Parents also transfer such information via their eggs. In three-spined sticklebacks (Gasterosteus aculeatus), maternal exposition to dummy or natural predator before egg-laying has important influence on offspring anti-predator behaviour such that offspring of predator-exposed mothers exhibit closer shoaling behaviour (Giesing et al., 2011). Such effects may be mediated by maternal transfer of high levels of hormones with organizational effects (glucocorticoid; Giesing et al., 2011).

Other adverse environmental conditions

In many insects, females favour diapause in their offspring as response to short photoperiod, low temperature or scarcity of potential host, thus increasing the chance of survival (Mousseau and Fox 1998). In Bugula neritina, females inhabiting polluted environments produce larvae with high dispersal ability (Marshall 2008). In polluted environments parents of some avian species produce competitive and/or aggressive offspring by depositing testosterone in their eggs (Gil 2008). Parents prepare offspring for future harsh environmental conditions by acting on their epigenome. In Rattus norvegicus maternal care influences the stress tolerance of their pups by increasing gene expression in promoter region of glucocortocoid-receptor gene (Weaver et al., 2004). Epigeneomic changes persist into adulthood. Offspring unattended by mothers are likely to keep a low profile and respond quickly to stress, which may be advantageous when food is scarce and danger is high. However, this is less beneficial when food is plenty. Parents prepare offspring to the level of care they will receive. Hinde et al., (2010) found that foster canary (Serinus canaria) chicks grow better if they beg at a level similar to that of original chicks suggesting that mother's increase offspring fitness by matching offspring demands to parental capacity.

11.11.4 The Costs of Parental Care

Explanations for evolution of parental care are based on variations in cost of behaviour (Clutton-Brock 1991). Parents transfer the cost of parental care to current offspring; yet cost can be measured in terms of offspring sacrificed due to on-going care. If parents desert, cannibalize or reduce provisioning to current brood, the current offspring pay cost, and parents lose potential future benefits. Sometimes parental care imposes cost in terms of reduced numbers of broodmates. This cost rises as clutch size increases and is known as 'depreciable care' (Clutton-Brock 1991). An example is young/egg provisioning, which constrains clutch size in various species (Stearns 1992). Anti-predator behaviour benefits all offspring in brood (non-depreciable care, Clutton-Brock 1991) and costs depreciate future reproduction. Four approaches assesses the cost of parental care: phenotypic correlations between traits, phenotypic manipulations, genetic correlations and selection experiments (Reznick 1992). A huge body of literature is produced on the first 2 methods. Work on the latter 2 has been restricted for analyses of life-history traits like negative genetic correlation between growth and fecundity. It is difficult to classify the costs of parental care but may be divided into non-physiological and physiological costs. The former is related to resource acquisition from the environment with risk of exposure to predators, rivals, conspecific or inter-specific parasites, and from reduced time for future mating or reproduction. Physiological costs are linked to resource allocation but arise from trade-offs between parental care and homeostasis, whether they are based on limiting resources. Parental care entails reduced survival, few mating scopes and low capacity to invest for future offspring. All above-mentioned mechanistic costs are closely interrelated. Reductions in body energy stores or key micronutrients impair immune-capacity, favour stress and lead to more propensity for infection (Nordling et al., 1998), which reduces ability to escape from predators, thus increasing the risk of body injuries. This implies that selection may act directly or indirectly simultaneously with mechanistic costs (Moore and Hopkins 2009).

Non-physiological costs

Experimental evidence supports the positive correlation between infection risk and parental effort (Knowles et al., 2009), although causal relationship between infection intensity due to parental effort and future reproduction or mortality has to be conclusively demonstrated. Wild female collared flycatchers (Ficedula albicollis) rearing enlarged broods had higher levels of blood parasites (parasitaemia) than control birds. They were correlated to over winter survival (Nordling et al., 1998). The fitness of experimental females was not studied. In wild great tits (Parus major), females with enlarged broods had increased parasitaemia and poorer over winter survival rates, although parasitaemia and survival were not correlated (Stjernman et al., 2004). These correlations do not necessarily show causation. In a study of common eiders (Somateria mollisima) female survival was negatively associated with clutch size, but during an avian cholera epizootic outbreak, thereby suggesting that parental effort reduced resistance to infection and negatively affected fitness (Descamps et al., 2009).Parental care increases the risk of predation, and predation reduces fitness. Examples of increase of predation risk due to parental activities are common in invertebrates. Invertebrates carrying eggs suffer from high predation than non-carrying individuals (Li and Jackson 2003), perhaps due to their conspicuousness, low escape ability (Shaffer and Formanowicz 1996) and/or higher energetic value for predators. In pipefish (Nerophis ophidion), males carrying their brood in pouch suffer high predation rate in comparison to females seems to be related to their greater conspicuousness (Svensson 1988). Clutch or litter burden also impair escape capacity, which has been well demonstrated in vertebrates. In lizards and birds this effect seems to be mediated by the impairment of muscle condition. In birds, fat reserves required for egg production may impair take off and flight capacity, increasing predation risk (Witter and Cuthill 1993), although our knowledge the link between this loss of escape capacity and mortality has been demonstrated to date in reptiles (Cox and Calsbeek 2010). Parents suffer injuries while defending their reproductive investment from con-specifics or reproductive parasites. Burying beetles (Nicrophorus pustulatus) suffered more injuries when protecting their young without help from their mate (Trumbo 2007). Parental care may wear and tear integuments. Experimentally reared collared flycatchers showed enlarged broods, suffered greater wear on primary feathers. The intensity of feather damage was positively correlated to post-breeding mortality (Mërila and Hemborg 2000). Reproductive con-specific or inter-specific parasites impair parents' survival or future reproduction. In former case, examples again are found in birds. Experiments have so far found little evidence of any long-term cost of conspecific parasitism, a finding that is not particularly surprising since all these studies used pre-cocial species in which cost of rearing additional offspring tends to be lower (Lyon and Eadie 2008). In the latter case Hoover and Reetz (2006) reported reduced returning rates in prothonotary warblers (Protonotaria citrea) parasitized by brown-headed cowbirds (Molothrus ater). In certain species including insects and fishes that do not expend energy feeding their offspring—hosts may not necessarily suffer a cost when receiving eggs from con-specifics or inter-specifics.

Parental care consumes time that could be devoted to remating, conducting new reproductive events and/or self-maintenance. Trade-off between parental care and new mating opportunities has generated a prolific literature focused on the evolution of sexual conflict and biparental care. If time was dedicated to produce more offspring, it is shown in captive lace bugs (Gargaphia solani) that time invested in protecting eggs is traded against fecundity in subsequent clutches (Tallamy and Denno 1982). Water striders (Aquarius remigis) that bred once a year (univoltine life cycle) had time to recover lipid stores and survived the winter better than breeders that had 2 reproductive attempts per year (bivoltine cycle); the latter even had lower lifetime fecundity and longevity (Blanckenhorn 1994). Blue tits that produced a second clutch when the first was removed delayed their moult and produced the plumage with poor insulation capacity, and had lower over winter mortality and less reproductive success the next season (Nilsson and Svensson 1996). In the last 2 cases the evidence is merely correlational and could be confounded by energetic constraints.

Physiological costs of parental care

Energetic cost

Physiological costs have been primarily studied in terms of a loss of limiting resources like energy or nutrients. Using various techniques the allocation of resources can be estimated by measuring energy expenditure. The increase of energy expenditure during parental care is relevant in income breeders.

Most organisms stockpile energy in their bodies and changes in total body mass or growth rates may be used for estimating energy loss (Speakman 2001). To assess the state of body energy stores that accumulate macronutrients is the third option. In vertebrates an increase in energy expenditure associated with an increased intensity of particular parental care behaviour has ever been shown in mammals and birds. Studies were done on female rodents in captivity or in semi-captive conditions during gestation and lactation. Despite various studies, a link between energy expenditure in current care and parents' survival and/or future reproductive success is only supported by 2 avian experiments. Experiments reporting body mass loss or growth delay as cost of parental care are performed for fish, reptiles and birds, and have linked such costs to fitness. Some reptile and bird study also have shown changes in specific body energy stores, although only 2 have ever reported a link with fitness, probably due to technical limitations in assessment of body composition, which usually requires sacrifice (Speakman 2001).

Non-energetic micronutrients

In cases with calcium, carotenoid and methionine, a link with parental behaviour is established. For calcium, allocation to the egg-shell in oviparous species or milk and foetal bones in mammals is well studied. Calcium levels drop during gestation and lactation in mammals. Carotenoids are used in physiological functions, as well as in pigments of integuments. Egg yolk of fish, reptiles and birds contains large amounts of carotenoids that protect embryo from the effects of oxidative stress. Increased parental effort depletes maternal carotenoid levels are correlated in laying birds (Bortolotti et al., 2003). Methionine stimulates fecundity female fruit flies, but in a specific ratio with other essential amino acids (Grandison et al., 2009). When this proportion is not met, methionine becomes pro-oxidant, diminishing parental survival and reproductive success (Grandison et al., 2009).

Physiological stress

Parental care may lead to an exhaustion of energy stores, which in turn leads to physiological stress. Physiological stress may be triggered by other environmental stressors (Wingfield and Sapolsky 2003). Such a state provokes damage in parents. Assessing levels estimated this of heat shock proteins (HSPs), molecules that repair protein damage induced by various stressors (Sorensen et al., 2003). High HSP values are related to decrease fecundity in fruit flies (Sorensen et al., 2003). Only one study has ever related parental care and HSPs: in blue tits, parents whose brood was enlarged had increased blood HSP levels. In vertebrates, glucocorticoid levels in blood are the most analyzed proxy of physiological stress, high values revealing high stress levels. Experiments in birds and fish support the idea that an increase in glucocorticoid levels is a consequence of parental effort. Recent reviews have questioned the link between this effect and fitness and in fact we have found only one study that supports this assertion.

Oxidative stress

Cost of parental care due to oxidative stress is supported mostly in mammals, but the by some experiments on birds. In the latter case, zebra finches whose parental effort was increased by brood enlargement had less resistance to ROS at the end of reproduction than controls (Alonso-Alvarez et al., 2004). Oxidative stress generated during gestation in mammals compromises the life of the mother during birth a link between reproductive oxidative stress and fitness is supported by a limited number of experiments and correlations. When exposed to a pro-oxidant agent (paraquat), the female fruit flies that were experimentally stimulated to produce eggs died faster than non-breeder. In zebra finches a negative correlation between the numbers of breeding events and resistance to oxidative stress is reported. Here parental effort could have included mating effort. Male of 2 reef-fish species that protect their broods in their mouths suffer from hypoxia (Östlund-Nilsson and Nilsson 2004). Hypoxia could be an alternative cost of parental care but is probably associated with oxidative stress (Metcalfe and Alonso-Alvarez 2010).

Immunosuppression

Parental care may also lead to immuno-suppression as an indirect result of other physiological costs. Immunity is reduced through high energy expenditure, loss of body energy, micronutrient depletion, glucocorticoid- and oxidative stress (Perez–Rodriguez 2009) in vertebrates. Examples from other taxa are scarce (Fedorka et al., 2004). Impact of parental effort on immunocompetence is well supported on birds, in which incubation and brood rearing efforts were manipulated and capacity to establish innate or acquired immune responses were accordingly impaired. Immunosuppression is well known as process associated with implantation and gestation in mammals (Medina et al., 1993). Immunosuppression protects the embryo from maternal immune defences, although consequences for maternal fitness are still unclear (Speakman 2008). The link between this immunosuppression and fitness is only supported by a handful of studies.

The cost of regulatory systems

Endogenous (neuroendocrine) control systems involved in parental decisions may *per se* create constraints and costs (Lessells 2008). Selection may favour simple costless parental rules that are not always optimal, perform well on average (McNamara and Houston 2009). Signals involved in reproductive acts may have a negative effect on soma maintenance (Edward and Chapman 2011). Studies on fruit flies and nematode *Caenorhabitis elegans* suggest that negative effect of reproduction on longevity arises from a signalling pathway rather than from direct resource competition. Molecular signals activate the physiological mechanisms needed for reproduction that in turn generate damage (Barnes and Partridge 2003). It is still to be established whether neuroendocrine control system mediates or creates costs in parental care (Lessells 2008).

Discrimination

When in conspecific groups, parent-offspring discrimination is expected to develop as the mistake is costly, and the chance of exploitation is high. Parents and their offsprings should develop some form of signalling to identify each other. These signals can be in the form of chemical cues and the vocalizations. In some animals, the ability to discriminate is still accurate, after 3 weeks of separation, such as in the subantartic fur seal. Research has shown adaptation on both part of parental and offspring. Chicks of colonial birds would have much more variation in their vocalizations, compared to solitary relatives. It appears that these begging calls also elicit unwanted attention from predators. Additionally, in operant conditioning experiments, adult colonial birds are able to learn to discriminate between two chick calls much faster than adults of solitary spices.

Problems

In some species, discrimination is relaxed because the cost of mistakenly excluding the parents owning the offspring is much higher, due to imperfect identification versus the small chance of adopting a genetic stranger. This lack of discrimination sometimes allow exploitation by brood parasites. Brood parasites are animals that dump their eggs into the nest of other species to be raised by the host. It occur in birds, insects and fishes. Probably, it started as some members inadvertently sharing conspecific nests due to nest competition. Then, gradually drifted to related species, and finally to unrelated species. In most cases of brood parasitism, the parasitic chick always exceed in body size, in comparison to the host chicks. This is due to the fact that parents distribute care unevenly (more on this later), and most of the time favours the largest individuals.

Most parasitic chicks hatch earlier, and are also armed with behavioural adaptations, such as mimicking, begging calls, and ability to push out host chicks/eggs. So, how does host species deal with parasites? Some are unable to nest; others will destroy the eggs, or rebuild a new nest. There are costs associated with actions taken against parasitism. If the chance of parasitism is low, the cost of egg destruction behaviour might actually be higher due to error. Cost of starting a new nest is also high.

Parent–offspring conflict

Parents often distribute resources unevenly. Often more is given to the larger offsprings. Offsprings want to eliminate each, to gain extra resources from the parents. While this is beneficial to the offsprings, it is a reduction in reproductive success for the parent. Hence, the conflict. However, some parents will take preventative action to some degree. Other parents actually appear to behaviourally encourage this, via asynchronous egg laying, and producing differential hormonal loads for each of the egg. Such parental neglect is actually due to production of lager broods by some parents than they can support during a normal year, to achieve higher reproductive output on "good" years. In that case, parents just have to sit back, and wait for the stronger offspring, to eliminate the rest. Eliminated offsprings were most likely not going to achieve reproductive success anyway.

According to scientists, normal asynchronous brood of cattle egrets, actually, produce highest parental efficiency. Another problem, associated with parental care is, who cares. 90% of the parental care is contributed by females in mammals, while the remaining 10% is contributed biparentally. Parents are not identical to their children. The interests of the two may not always precisely coincide. Sometimes the parent may be best served by killing or with-holding care from some offspring, so that they may have many more offspring in the future. Originally, it was believed that there were more cases of maternal care, due to fact that female already invested more, in making egg, and to provide subsequent care would be logical. However, this did not make sense, when females of several species of fishes abandoned their expensive eggs, after mating had taken place.

11.11.5 Parental Care in Fishes

Most of the fish species utilize broadcast spawning, through which gametes are released into water without parental care. Goal of this reproductive method is to produce maximum amount of progeny, in hopes that as many offspring as possible will survive. Many fish species opt to employ parental care to supply their smaller number of offsprings, a greater chance of survival. Oral brooding and sex role reversal are two unique, successful methods of parental care. Oral brooding, although quite rare in nature, is found by many fish in cichlidae. After fertilization of eggs, cichlids, such as *Tilapia*, place egg clutch into the mouth of the female. Fecundity is significantly lower in oral brooders. However, eggs tend to be larger, and receive more nutrients. By placing eggs into the oral cavity, the female provides protection for eggs, and churning ability of female rotates eggs, and the egg is exposed to oxygenated water. Similar to oral brooding, sex role reversal in fish is rare, but is found in Sygnathildae. In sea horses, female inseminates male by inserting oviduct into male brooding pouch several times, to ensure fertilization. After fertilization, female departs, and male attaches itself to a nearby object, with its tail waiting for eggs to mature. Nature exhibits many alternatives that species have adapted to reproduce.

Parental care is any behaviour performed after breeding, by one or both parents that contribute to survival of their offspring. Parental care is interesting among fishes. Of some 250 families described, about 77 per cent fish show no parental care, another 17 per cent including fish species that care for the egg only, and less than 6 per cent contain species that are known to care for eggs and newly hatched young. Fish show all grades of parental care from random spawning and from deposition of large number of uncared eggs to protection of the youngs. Lack of parental care is correlated with production of great number of eggs and sperms. Two general types of variation in parental care behaviour exist among fishes. First either both parents, or one alone care for offspring. Thus, there are paternal, maternal and biparental species. Second, the eggs and newly hatched youngs are either maintained on substrate—that is on plant under stone in excavated pits and so on (these are called substrate-brooders or guarders) or carried about in the parent's mouth (these are called mouth brooders or incubators). Fishes have evolved many means of affording care to fertilized eggs, and young ones by one or both sexes.

Spreading eggs over aquatic plants

In fishes like pikes *Esox lucius*, carps *Cyprinus carpio*, *Carrassius autatus*, eggs are scattered usually over aquatic plants.

Depositing eggs in sticky covering

In many carps, eggs are laid with some sticky covering by means of which they are attached to each other and to stones, weeds, etc. In yellow perch *perca flavescens,* eggs are deposited in a rope of single mass.

Laying of eggs at suitable places

Suitable spawning ground is selected by anadromous fishes like *Salmo solar*. Acipenser *Oncorhyncus* dig excavation in gravel substrate, lay their eggs in pits, cover them with gravel and desert them. Sand Gobi *Pomatoschistos minutus,* lays its eggs in protected spot which are guarded by male who aerates them by his movement.

Nest building

A few fishes construct nest which may be complex or simple. Simple nests are merely hollowed out depression in bottom as in lung fishes. Males of many species such as darters, sunfishes and cichlids prepare a shallow basin like nest. All stones and rock crystals are carefully removed from the bottom. Eggs are laid in nest and male remains on guard till the young ones are hatched. *Protopterus* prepare a simple nest in the form of deep hole in swampy places along the river banks. After spawing, he guards the nest. *Lepidosiren* also prepares a nest in the form of a burrow and male develops highly vascularized filaments on its pelvic fins for aeration. Male bowfin *Amia calva,* of the great lakes of North America builds a crude circular nest among aquatic vegetation. Male stands on guard till the young ones are hatched. Young ones leave the nest only under protection of the father.

Before onset of courtship, male stickleback (*Gasterosteus aculeatus*, *Pygosteus pungitius*) builds a quite elaborate spherical or elongate nest by collecting plant fragments, rootlets and then binding them together with adhesive kidney secretions. Various activities of male such as probing, boring, sucking and glueing result in formation of a compact nest with an internal chamber to receive eggs. Male drive and induces the female into nest for laying eggs, then chases her away, enters the nest, fertilizes the eggs and guard them from intruders. The most elaborate cup shaped nest is made by *Apelts quadracus* attached to rooted plants close to bottom. After a clutch of eggs is laid, male builds an extension of nest up and over the eggs, with a concave upper surface to extension. A second clutch of eggs is laid on new nest floor and this procedure may be repeated several times, until the male has several clutches of eggs stacked vertically within a single multitiered nest. Floating nests are made by catfishes in which eggs are suspended in a mass of bubbles and mucus produced by fish. Male siamese fighting fish (*Betta splendens*) too builds a floating nest and sticks fertilized eggs to lower the surface of foamy nest. He stays on guard on this nest and fights till death to defend it. Male paradise fish *Macropodus,* also prepares a similar foamy nest.

Coiling round the eggs

Butter fish (*Pholis gunnellus*) rolls all eggs into a ball and curls around it.

Deposition of eggs by ovipositor in mussels

Female *Rhodeus amarus* deposits eggs in the siphon of a fresh water mussel by means of very long urogenital papilla. Male immediately sheds the sperms on opening (of mussel) over the eggs (Figures 11.9 to 11.17).

Egg brooding in mouth and intestine

Female *Tilapia mossambica* broods the fertilized eggs in her mouth and allows the young to take refuge in her buccal cavity in time of danger, for some days after hatching. In *Galeichthys felis*, the male carries egg in mouth for a period of nearly six weeks. Eggs of this oral incubating fish are large and relatively few in number. During this period, brooder fish do not take any food, thus, exhibiting great degree of self-sacrifice. *Tachysurus* keeps the fertilized eggs in its intestine till hatching occurs.

Brood pouches

Male seahorse and pipefish carry eggs in a brood pouch on abdomen. In *Hippocampus*, fertilized eggs are transferred by female into brood pouch

FIGURE 11.9 A shallow basin-like nest of sunfish.

FIGURE 11.10 The male *Amia calva* guarding its circular nest.

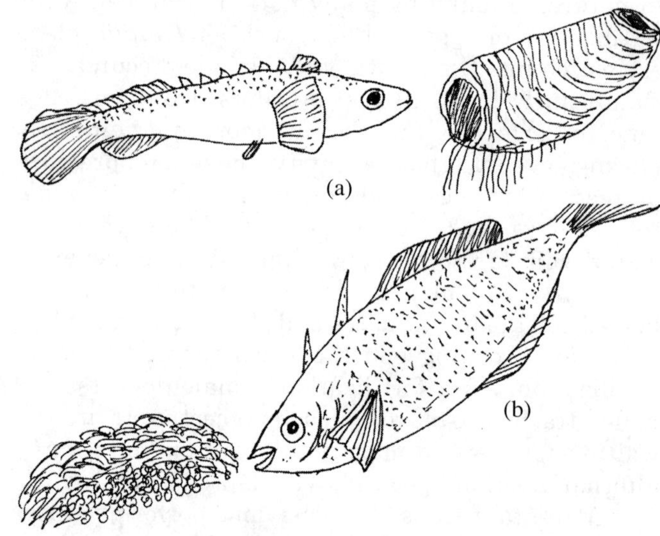

FIGURE 11.11 Fanning by male stickleback at entrance of its nest. (a) Nest of ten-spined stickleback, (b) Nest of three-spined stickleback.

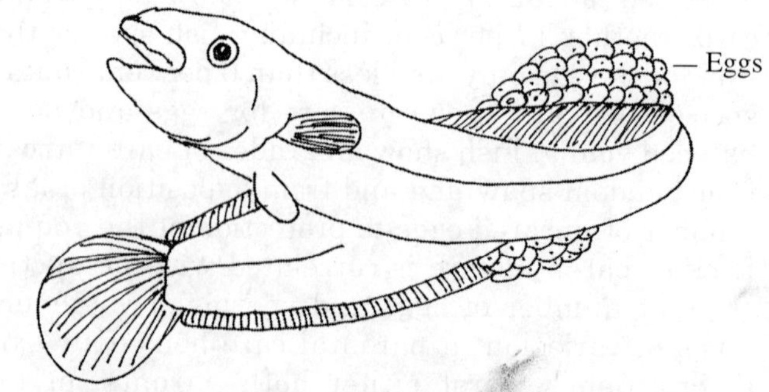

FIGURE 11.12 A butterfish coiling around the egg.

FIGURE 11.13 Male siamese fighting fish defending his floating nest.

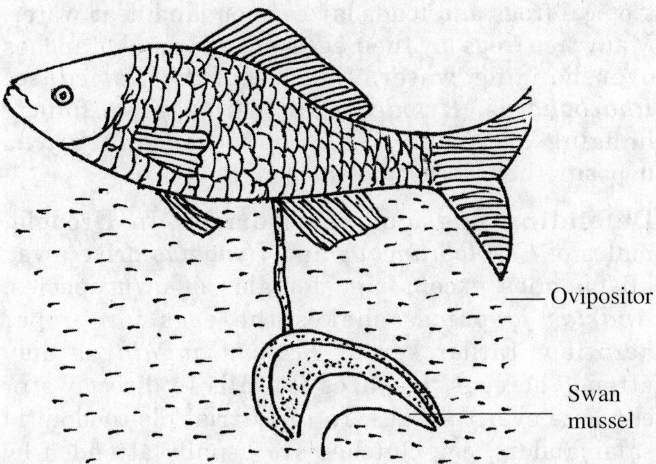

FIGURE 11.14 Oviposition by European bitterling in swan mussel.

FIGURE 11.15 Mouth breeding (oral incubation) in *Tilapia*.

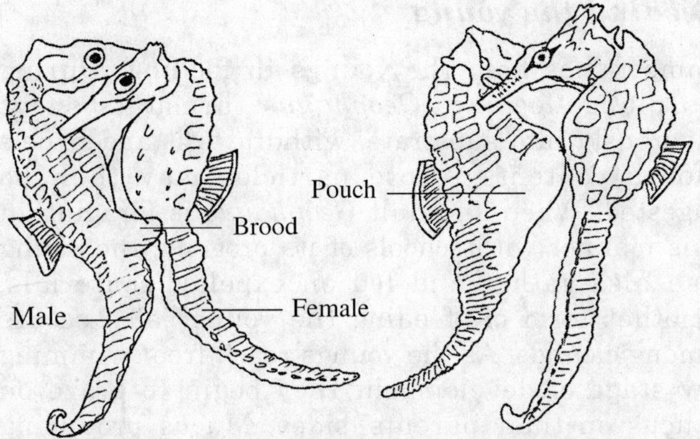

FIGURE 11.16 A male seahorse (left) receives eggs from his mate, which fills his brood pouch and swells his abdomen substantially.

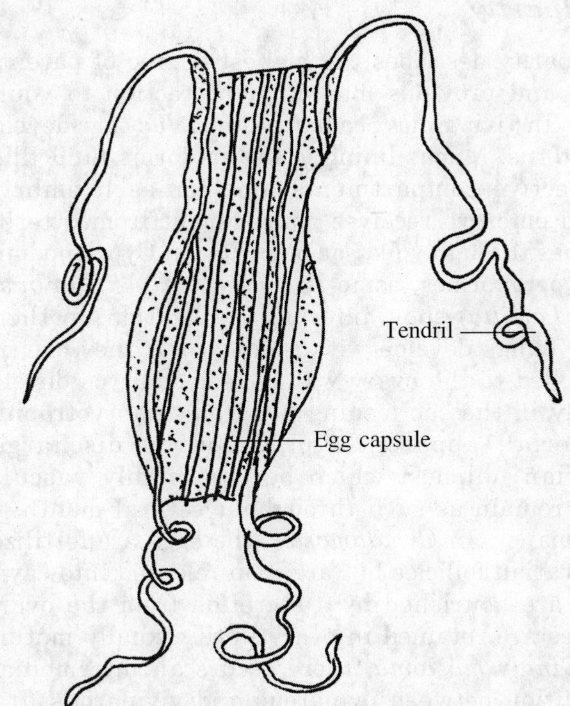

FIGURE 11.17 A horny egg capsule of cat-shark.

on belly of male. These eggs are carried by males until their hatching. Eggs become embedded in folds on brood pouch for exchange of respiratory gasses, a sort of placenta is formed. In male *Syngnathus acus*, a brood pouch is formed by two flaps of skin on underside of the body on which eggs are placed by the female. Brood pouch, develops as an inner spongy lining is richly supplied with blood vessels. Eggs get nourishment until hatching. Fry may return to pouch when in danger.

Attachment of egg to body

Male nursery fish (*Kurtus*) carries eggs held in a cephalic hook.

By putting eggs in integumentary cups

In siluroids (*Aspredo* and *Platystacus*), eggs are pressed into soft and spongy skin of female and carried about.

Egg capsules

In rays and cat sharks (*Scyllium* and *Raja*), fertilized eggs are laid inside protective horney egg capsules, called **mermaids purse**. This capsule remains attached to aquatic weeds by their tendrils. Development proceeds inside the capsule until yolk has been used up. Young hatch out, after rupturing off the egg case.

Feeding the young

Some fishes help the youngs in feeding. During its plunge-feeding, *Geophagus* thrust its snout vigorously into substrate, withdraws it and expels mouth contents. Loose particles may then be ingested. When an adult *Geophagus* is foraging in this manner with schools of its progeny, the young fish also gather and fed on expelled materials. Another form of "feeding the young" also occurs among cichlids. As the youngs reach free swimming fry stage of development they begin to graze on mucus on their parents' sides. Mucus producing cells in adult's epidermis are most numerous at this stage of reproductive cycle.

Viviparity

Viviparity describes the highest degree of parental care and provides maximum protection to young ones. In viviparous *Scoliodon, Mustelus* eggs develop in uterus. Mucus lining of uterus forms fluid filled protective compartments, one for each embryo. Each embroyo receives nourishment from uttering tissues through yolk, sac placenta. In Cyprionodonts and perciformes, some species (*Zoarces, Gambusia* and *Poicilla*) show internal fertilization, in them, the youngs develop within ovary, but they are not attached to the ovary wall. These embryos develop freely in the sac feeding upon an "embryotrophic" material, apparently produced by discharged ovarian follicles, which become highly vascular and remain as such throughout several months of pregnancy. In *Cymatogaster*, the eggs are fertilized in ovarian follicles but are soon released into cavity and are nourished by a secretion from the ovary. Males are retained in ovary until sexually mature. In *Gingly mostoma,* there occurs an intermediate condition between oviparous and viviparous. It is ovoviviparous. Its eggs are covered by a horny case and development of embryo occurs in uterus. Fully developed youngs are hatched out by breaking the shell inside uterus.

11.11.6 Parental Care in Amphibians

Perhaps no group shows more diversity in parental care than amphibians. Anurans show much greater diversity than urodeles and apodans. Parental care is associated only with those species that place their eggs in single clusters. Methods of parental care generally fall under two broad categories: (1) Protection by nests, nurseries or shelters, and (2) Direct caring by parents. In apodans and salamanders, parental care consists only of attendance of the eggs, mostly by the female. In anurans, eggs are usually guarded by the male, and parental care involves transportation of eggs and larvae also.

A. *Protection by nests, nurseries or shelters*

Amphibians have evolved various method for giving protection to their defenseless eggs and larvae.

Selection of site: Many amphibians lay eggs in protected sites, most of which are on land. Many tropical frogs and toads lay eggs on land near water. Many tree frogs lay their eggs on leaves and branches over hanging water. Species of *Phyllomedusa, Rhacophorus, Hylodes* glue their eggs to foliage hanging over water. *Rhacophorus malabaricus* deposits their spawn on trees.

Defending eggs, or territories: In Urodela, males of *Cryptobranchus* and *Hynobius* drive away all enemies except ripe females of own species. *Andrias japonicus* shakes the eggs for proper aeration. Either sex of *Proteus anguineus* may attend the eggs. It waves its tails to direct water currents over the eggs. In terrestrial ple-thodonitid salamanders, egg clutches are usually attended by the females. The female periodically rotates the eggs in clutch, in order to increase aeration, and to prevent adhesive malformation.

In anura, male green frog *Rana clamitans* and other species maintain territories and attack small intruders to defend eggs. In *Manlophryne robusta*, the male actually sits over and holds in hands the elastic gelatinous envelope containing eggs numbering upto 17. Some tree frogs laying eggs above water and may sit besides the eggs, or rest on top of them. Males of nest buildings gladiator frog, *Hyla rosenbergi* are highly territorial and aggressive. They dig a shallow pit near the water. Eggs are laid as a surface film on water in nest. If surface tension is changed, eggs sink to the bottom of the nest and die. Male moves around the nest and attack intruders whose entry to the nest may disrupt surface tension. Terrestrial eggs of *Dendrobates auratus* are periodically mositended by males by periodic urination on egg clutch. The female of *Hemissus marmoratum* provides moisture to developing eggs. She sits on clutch in an underground chamber and digs a tunnel leading to water. Tadpoles escape to water form the nest chamber.

Direct development: In some terrestrial tree frogs, like *Eleutherodactylus*, *Arthroelpits hylodes* and *Hyla nebulosa*, eggs hatch directly into little frogs, thus, avoiding larval mortality. In *Plethodon cineresuf*, hatchlings are miniatures of the adults.

Defending the larvae: Female *Leptodactylus ocellatus* provides protection to the tadpoles. The tadpoles move around the dense masses. Female remains with the tadpoles and attack birds attempting to fed on tadpoles. Tadpoles of *Dendrobates* develop in water filled with leaf axils of various kinds of plants or tree holes, etc. Females of *D. pumilio D. histrionicus* and *D. lehmanni* carry individual tadpoles to separate water filled with leaf axils and protect them. Female *Hyperolius obstertricans* remains with their eggs on leaves overhanging the streams. Upon hatching, the females kick tadpoles out of leaves into the water.

Foam nests: Many amphibians convert copious mucus secretions into nests for their youngs. In Japanese tree frog *Rhacophorus schlegeli,* the mating couples dig a hole or tunnel into which eggs are left in a frothy mass to avoid desiccation. During rains hatching tadpoles are washed down the slopping tunnel into pond, or river water for further development.

The female South American tree frog *Leptodactylus mystacinus,* stirs up a frothy mass of mucus, fills it in holes near water and lays eggs in them. The tadpoles developing in these foam nests can readily enter the water. Some anurans lay eggs in the nests of foam floating on water. Male *Adelotus brevis* actively defends the eggs laid in the foam nest. The males contain long tusks on the lower jaw which are used in fighting.

Mud nests (nurseries): In the Brazillian tree frog *Hyla fabre,* the male digs a little crater like hole or nursery in mud in shallow water, in which the female lays her eggs. Tadpoles hatch within these mud nest and develop up to stage, when they are large enough to defend themselves.

Tree nests: South American tree frog. *Phyllomedusa hypochondrales* lays eggs in a folded leaf nest with margins glued together by a cloacal secretion. The tadpoles when developed fall straight into water below. Another tree frog *Hyla resinfictrix,* lines a shallow tree cavity with bees wax obtained from the hives of certain stingless bees. Females lay eggs, when this cavity is filled with rain water. Here the youngs develop relatively free from predators.

Gelatinous bags: In *Phrynixalus biroi,* large eggs are enclosed in a sausage shaped, transparent, gelatinous, membranous bag, secreted by female and left in mountain stream. *Salamandrella keyserlingi* also are known to deposit 50 to 60 small eggs in a gelatinous bag which is fastened to aquatic plants.

Communal nests: *Nectophyrynoides malcolmi* prepare communal nests in which eggs are deposited by several females. This nest is reported to be guarded by a single male.

B. *Direct caring by parents*
(Figures 11.18 to 11.22)

Coiling around eggs: *Ichthyophis glutinosa* lays eggs in a shallow hole near the water and coils her around the gelatinous egg mass. Attendant female eats the infected eggs, and thus, saves other eggs from infection. Another apodan, *Idiocranium russeli* coils around the eggs in a dense mat of grasses. In salamander Congo eel *Amphiuma,* the female lays large eggs in burrows in damp soil and carefully guards them by coiling her body round them until they hatch. Female salamander *Plethodon* also coils round the eggs which are laid in small packages

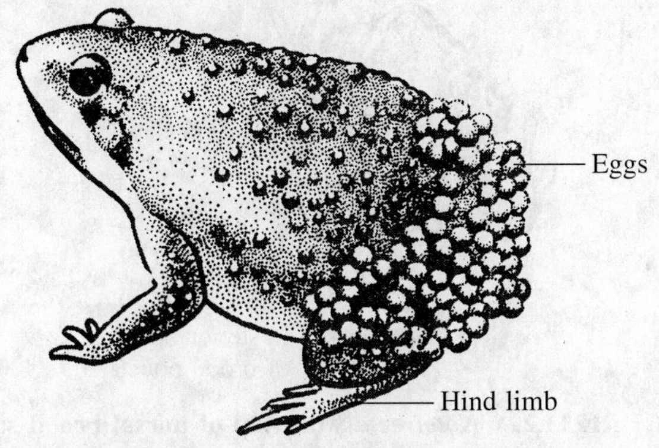

FIGURE 11.18 Male *Alytes* (midwife toad).

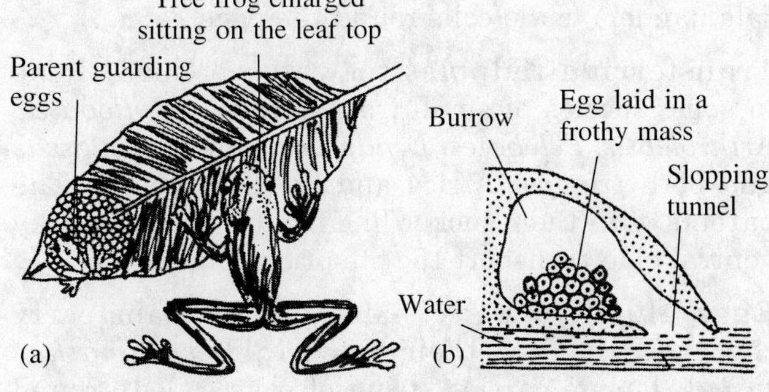

FIGURE 11.19 (a) A tree frog guarding eggs, (b) Foam nest of *Rhacophorus*.

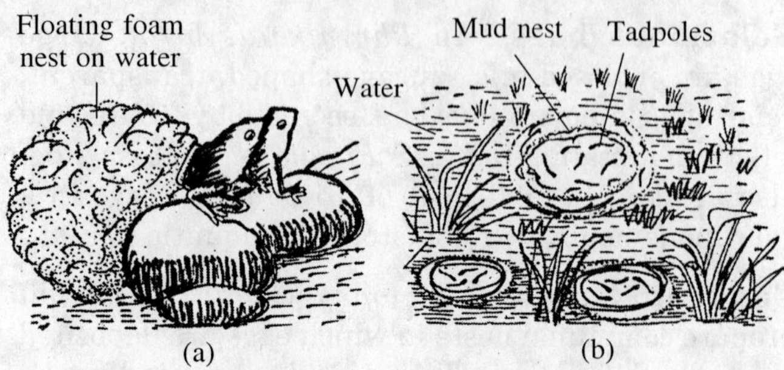

FIGURE 11.20 (a) Floating from nest (b) A mud nest of *Hyla*.

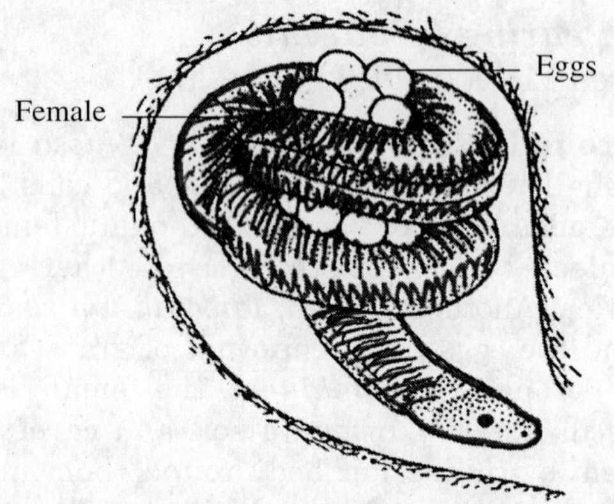

FIGURE 11.21 Female *Ichthyophis* coiling around eggs.

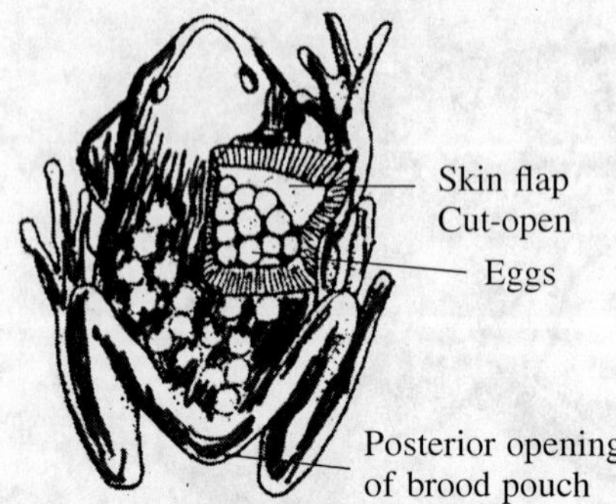

FIGURE 11.22 *Nototrema* with flap of dorsal brood sac.

in the hollow of a rotten log or beneath a rock. In *Megaloba-trachus maximus* (Japanee giant salamander), male coils round the eggs.

Transferring tadpoles to water: This is done in some species of small frogs such as *Phyllobates, Arthroleptis, Pelobates, Dendrobates,* and *Sooglossus* and both tropical Africa and South America. The parents with their sucker like mouth or their sticky ventral side transport the offspring to water.

Eggs glued to body: Many amphibians carry the eggs glued to their body. In *Desmognathus fuscus*, female carries string of eggs coiled around her neck, until they are hatched. Female places her chin on egg clutch, thereby subjecting egg to slight vibrations caused by pulsations of throat. *Desmognathus* aggressively defends her eggs. *Rhacophorus* reticulates the eggs which are glued to the belly of female. When female of European midwife toad *Alytes obstetricans*, lays eggs, the male entangles them around his hind-legs. He carries them with him until they are ready to hatch. On hatching, he releases the tadpoles into the nearest water.

Eggs in back pouches: In marsupial frogs or toads, female carries eggs on her back, either in an open oval depression—a closed pouch or in individual pockets. Eggs develop into miniature frogs before they leave their mother's back. In *Hyla goeldii* (*Cryptobatrachus evansi*), posterior part of back of female forms a sort of incipient brood pouch in which eggs remain exposed. In *Nototrema* (*Gastrotheca*) marsupium, eggs are covered by skin forming a single large brood pouch on the back, which opens posteriorly in front of cloacal aperture. In *Pipa pipa,* during breeding season skin of female's back becomes thick, vascular, soft, and gelatinous. Male presses fertilized eggs against females back, where they sink into individual pits. A hinged cover forms over each egg enclosing it in a small capsule. Complete metamorphosis occurs within capsule and tiny, tailless toads leave mother but do not enter water.

Parental care and egg size in frogs: Shine (1978) proposed an influential, yet controversial hypothesis: that evolution of parental care creates a 'safe harbour' that causes selection to favour increases in egg size. This hypothesis predicts that evolution of parental care will typically precede evolution of large egg size. Shine's hypothesis was based on a model that assumed a constant instantaneous mortality rate for juveniles, unaffected by the egg size variation. Sergent et al., (1987) incorporated size-dependent survival and growth rates into Shine's model. Removing assumption of constant mortality and, some anomalous predictions were made under Shine's original model, such as prediction of a bimodal distribution of egg size. Nussbaum (1987) proposed that evolution of large egg size typically precedes evolution of parental care, rather than reverse. Using salamanders as a specific example, he proposed that selection for ability of hatchling juveniles, to consume large food items in lotic environments favour large hatchling size. Nussbaum argued that larger eggs produced by this process of selection take longer to develop, and will be produced in smaller numbers. These two factors should increase egg mortality, leading to selection in favour of behaviours that reduce this increased mortality, such as parental care.

A model suggested that whichever evolved first, parental care and egg size should tend to coevolve, thereafter, making it difficult to disent-angle selective forces driving any correlation between parental care and egg size and there should be no clear trends in order of parental care and egg size evolution. Shine (1989) suggested that a third factor, such as certain forms of predation might simultaneously select for both large egg size and parental care, leading to a correlation between these two characteristics that is not due to their interaction. Other than Nussbaum's (1987) comparative investigations of parental care and egg size in salamanders, there have been few if any empirical investigations of predictions of safe harbour hypothesis. Reproductive diversity in anurans is associated with a spectrum of parental behaviours, including production and defense of foam nests or burrows, and transport of eggs or tadpoles among many others. Some comparative studies have indicated that egg size increases in presence of parental care. Egg size is significantly associated with parental care in frogs. Reconstruction of ancestral state of parental care across phylogenetic tree indicate that no parental care is the most likely state. Assumption that small egg size is primitive is most favourable for Shine's (1978) original hypothesis, which proposed that evolution of parental care in species with small eggs led to evolution of larger eggs (Summers et al., 1999).

11.11.7 Parental Care in Birds

Birds have internal fertilization. Females lay egg. On hatching, young either forage for themselves, or are given food. There are fewer constraints on evolution of male parental care. When young forage for themselves upon hatching (precocial species), parental care takes form of protecting youngs form predators, leading them to good foraging areas. Neither parent can provide parental care through feeding the young. Benefits to a male's fitness of helping female protect their offsprings, are generally small, and relative to advantages of finding additional mates. Parental care are forced on females. Species, in which youngs are born helpless, and need to be fed for an extended period (altricial species), a male's fitness must generally be higher, if it helps feed the hatchling, rather than invest its time and energy in finding additional mates. This is because addition of male's parental care should approximately double the surviving of the offsprings. Female can rear about ½ offsprings. To achieve same fitness, a polygynous male providing no parental care would have to obtain 2 mates. This require a greater investment by polygynous male, with a less certain return, than parental care, as he would require a territory twice as large. Altricial bird species tend to have monogamous mating system with biparental care. Traits of altricial offspring, monogamy, and biparental care probably evolved together. About 92% of all the birds form pair bonds, and that 2.5% of all the birds breed cooperatively. A recent review increased the number of falconiformes known to breed cooperatively to 14%. Identifying a common pattern between groups, in which males are predominant carers is difficult. In many waders, parental care is dynamic. Either male or female can abandon care, to other at various stages of nesting cycle. This group provides insights into the evolution of uniparental care. Male only care is primarily found in resident Gondwana taxa that lack female only care. Transitions are from male only care to biparental care, and have occurred twice in ratites, once in jacanas, and once in painted snipe. Three transitions involve a decline in incidence of male care, with increasing distance from the Equator. Some kiwis have male only incubation. Biparental incubation is more prevalent in populations from colder, southern sites, and extreme southerly populations are cooperative breeders to point that dominant male may not incubate at all. In both jacanas and rostratulids, biparental care occurs (Figure 11.23).

The common selection pressures have driven convergent evolution of female only care that occurs in a number of clades, with precocial nidifugous young. Numerous origins of female only care among

FIGURE 11.23 Parental care in Kangaroo.

taxa with nidicolous, altricial young is suggested, along with that female only care has evolved in birds that feed largely on tropical fruit and nectar. Many transitions in the families are associated with frugivory and nectarivory. Because tropical fruit and flowers can be massively abundant, yet availability can be patchy on short-term spatial and temporal scales, males may gain advantage from defense of fruiting trees or geographical locations that females frequently traverse, in order to find fruiting, or flowering trees. From female perspective, limitation on reproduction is likely to be associated with the ability of the young to extract nutrition, from abundant but low quality food. Hence, male care is of limited value, allowing females to choose freely among males for good genes, rather than for direct benefits from the male, such as a high quality territory, or paternal provisioning. Given the strength of this association between frugivory and female only care, it is profitable to examine exceptional cases, in which female only care has evolved in primarily insectivorous taxa, in which male care should be at a premium. Slow growth of chicks in family Menuridae could reduce cost of female provisioning, and hence, increase value of good genes, relative to paternal care. Many insectivorous taxa with female only care occurs in dense nesting aggregations in rich marshlands, where high abundance of food occur, because of seasonal irruptions of aquatic insects. This reduces need for females to obtain care and together with high female densities, facilitates evolution of polygyny. A variety of taxa cannot be explained via this approach, particularly some insectivorous denizens of rainforests. It is suggested variously that predation might be important in these species because as originally suggested for frugivores, males might enhance detection of nest by predators, as any attempt by males to guard a single female against extra-pair mating would impose impossible costs, from sit-and-wait predators that predominate in rainforest interiors, and because intrinsic mortality schedules of long-lived tropical species may make parents reluctant to take risk during reproduction. Further investigations of these cases will be extremely valuable.

Cooperative breeding is sufficiently pervasive. A useful first stage of comparative analysis may be to identify those taxa, in which cooperative breeding is extremely rare. Factors that determine presence or absence of cooperative breeding may not be the same, as those that lead to variation in number of cooperative species, within cooperative clades. Cooperative breeding is rare in clades with precocial young, in contrast to taxa with altricial young. Only clades, in which there is a high frequency of cooperative breeding, combined with precocial young are Rallidae, small families Psophiidae and Mesitornithidae and monospecific Rhynchocetidae and Anseranatidae. The complex cooperative systems have developed among rails, but many species contributing to prevalence show a simple level of cooperation, characterized by chicks of first brood provisioning later broods in same season. Low levels of cooperation among the precocial taxa are unsurprising, for same factors that often allow one sex to provide exclusive care, probably reduce benefits that both dominants and super-numeraries obtain from the care of offsprings. It is clear that cooperative breeding does not always reflect dependency on care. Some cases, in which care by more than one parent is most important, are not associated with cooperative breeding. Almost all species foraging at sea, yet breeding on land, form strong pair bonds, and exhibit exclusive biparental care. Lack (1968) argued that such species face distinctive selective forces and should be analyzed separately from other birds. Nine families involved share habit of feeding on marine prey, but often breed in dense colonies on offshore islands or cliffs. Species feeding in marine environments must return to the land to breed, yet still obtain food from the sea, in which resources are often distributed patchily, necessitating prolonged departures to forage. Greatest concentrations of food occur in areas of oceanic upwelling. These forces conspire to enforce constant nest attendance. Although, additional birds might be able to provide assistance, failure to coordinate such care may pose unacceptable risks to parents, and promote exclusive pair bonds.

Cooperation is well known and more common among residents than migrants. Migratory and island faunas are more likely to be comprised of pair breeders, as colonization is facilitated by dispersal of both sexes. A very large proportion of all oscines are cooperative breeders. There is repeated evolution of cooperative breeding from pair breeding and vice versa. It is unlikely that there is a simple ecological or life history explanation for this difference. Both clades have diversified into an enormous range of niches, and show overlapping variation in life history. Low prevalence in suboscines is unlikely to be a result of environment they occupy. Many of oscine taxa have a high incidence of co-operation (New World jays, mimids, emberizids, icterids and wrens). Among New World oscine clades only vireos, polioptilds and parulids are poor in cooperative breeding. Some have argued that foraging on ground is particularly conducive to evolution of cooperative breeding, which runs counter to empirical pattern in this case. Fore-going examples are a small subset of unexplained evolutionary contrasts in avian parental care. Hopefully, this new data compilation will help focus on these and other problems, and allow us to proceed to development of predictive models.

11.11.8 Coiling of Eggs in Female Pythons

Parental care is a complex behavioural trait with high costs. Reproductive individuals face substantial risks, and spend considerable energy in raising their offspring. Higher parental investment may increase offspring's probability of survival, or its subsequent reproductive success. Parent provides direct protection against pre-dators, or nutritional input, and benefits to offspring fitness. The benefits of energetically expensive forms of parental care that do not involve protection or nutrient transfer to neonates are less obvious. Parental care of eggs is relatively rare among squamate reptiles. A distinctive form of this behaviour is seen in pythons. Female pythons remain tightly coiled around their eggs, throughout incubation. In cool climates, brooding females maintain high and constant temperatures within clutch by shivering thermogenesis that entails high energetic costs. Brooding females do not feed during incubation. Most likely benefits involve effects of maternally controlled incubation regimes on embryogenesis. Developmental trajectories in reptile embryos are highly sensitive to physical conditions encountered during incubation. Eggs that experience conditions that are too dry or too wet, or too hot or too cold, either may die before hatching, or may hatch, but produce inferior hatchlings with a lowered probability of subsequent survival. This sensitivity has favoured evolution of careful nest-site selection in egg-laying species, and careful thermoregulation by pregnant females of viviparous taxa.

Plausibly, same kinds of selective forces have acted on maternal nest-attending behaviour. That is, benefit of maternal attendance might involve control over physical conditions, experienced by incubating eggs, in ways that enhance egg survival, and/or hatchling pheno-types. Only obvious alternative hypothesis is that maternal attendance functions to reduce predation on eggs, but this could be achieved by mother by simply remaining near eggs, rather than coiling around them and twitching. This simpler form of parental care is seen in many other squamate species. Reproducing females of water python (*Liasis fuscus*) display facultative nest attendance, depending upon thermal regimes inside burrows, where they lay their eggs. Maternal thermogenesis might substantially enhance offspring fitness in some, but not all natural nest sites. This thermal effect manifests both via increased hatching success, and via modifications to phenotypic traits of hatchlings.

It has shown that brooding females spent very little energy over 2 month incubation period. Such expenditure was independent of fecundity, challenging notion that intensive parental care necessarily entails major energy costs. In ball pythons, parental care over a prolonged period strongly affect not only hatching success of eggs, but also phenotypic traits of hatchlings that emerged from viable eggs. Female ball pythons coil so tightly around the clutch, that eggs are completely hidden. This ball surrounding clutch could create a saturated microclimate around the eggs, substantially reducing evaporation. If optimal incubation conditions generate optimal hatchling phenotypes, there will be strong selection for maternal behaviours that expose embryos to such conditions. Mechanism that generated phenotypic variation among hatchling pythons is one that has not attracted previous interest. Yolk coagulation due to egg desiccation reduces amount of resources available to the embryos.

11.11.9 In Mammals

They have internal fertilization, true viviparity, with an extensive period of embryonic development. Neonates are nourished by mother via lactation, for an extensive period, before they become independent. They are predisposed to parental care, provided care by the female parent. There is little that male can do in terms of parental care that enhance their fitness, as much as, would attempt to mate with additional female. They tend to have polygynous mating systems, and parental care is provided exclusively by the female. In some species, males may provide some parental care, but these are the exception (Figures 11.24 and 11.25).

In mammals, parental care is very complex and consists of multiple activities. Female is responsible for most of the parental care. The universal parental care in mammals is lactation. All the female mammals lactate to feed their very young offsprings. Maternal care is the predominant form of parental care. The mother-infant bond is normally achieved within the first 24 hours of birth,

FIGURE 11.24 Parental care in Leopard.

FIGURE 11.25 Parental care in Chimpanzee.

which is usuaully triggered by signals emitted by the newborn. In artiodactyls, precocial infants travel with mother within a large herd. Mother licks the infant within the first few hours after birth. If the licking does not occur early enough, the bond and recognition do not form. In pinnipeds, mother recognizes the infant after returning from feeding at the sea. In humans, the range of parental care is all time high. The monotremes lay eggs, which are incubated in an abdominal pouch in echidnas, and in a nesting burrow by platypus. The monotremes also lack nipples—milk oozes from skin in vicinity of mammary glands soaking the fur, which is then licked by the offsprings. The marsupials brood their young in a marsupium in the form of either a pouch or a fold of skin. These contain the mammary glands. The youngs of these species are born at an extremely early stage of development (superaltricial), make their way into marsupium, and grab hold of a teat. Teat then swells in their mouth, making it extremely difficult to dislodge the offspring. The placental mammals exhibit a broad range of offspring dependency after birth.

In many precocial species, such as artiodactyls, hares, and cavies, youngs are active soon after the birth and can feed for themselves. Other than lactation, the only parental care provided to them is protection from predators. The altricial species, like carnivores, most rodents, and rabbits require greater care and must be kept warm and sheltered. Lactation periods vary depending on condition at birth, and among terrestrial mammals increases with increasing body size and continues even after the young begin to consume solid foods. It tends to be rather short in the pinnipeds and whales.

In many mammals, youngs continue to associate with the mother, even after they are weaned. This occurs in most primates, carnivores, and ungulates. The extended period confers additional protection from predators, and allows the youngs a period of time to hone their foraging skills. In the primates, this juvenile period allows the offspring to develop important social skills. Paternal care occurs in mammalian orders primates, carnivores, and rodents, typically involve provisioning the youngs (carnivores), brooding the youngs (rodents), and guarding against predators. Males can also be important for teaching various skills (primates). Communal care of young occurs rarely, as in a few carnivores, like cape hunting dog, wolves, and lions. In such cases, individuals other than the direct parents help provision and care for the youngs. Typically, these are kin groups. Communal care of young is associated with cooperative breeding, where several individuals assist a focal pair in their breeding efforts. Cooperative breeding is relatively uncommon and is known in over 150 species of birds, over 25 species of mammals, and in a few species of fish. Packs of wolves typically consist of a central breeding pair (male and female), and several other adults, who are typically siblings or previous offspring of breeding pair. These other adults help the breeding pair to raise the current litter.

Men attempt for producing the greatest quantity of offsprings. Women are biologically the quality controllers. Fathering possible 500 babies a year for 60 years would enable a man to father 30,000 babies in his lifetime. Evolution of monogamous relationship and moral system perhaps contribute a lot in population control, in addition to costly parental care. Human body can be viewed as survival machine for sperm and ova, which are basically housings for DNA. The sexual act is the step by which DNAs are transmitted to the next generation. Most activities we do in our lives can be viewed biologically as steps towards this end. Every activity of daily life attempts to either protect and preserve our DNA or as a step toward its transmission. Eating, breathing, earning, sleeping, and saving up energy for the next day's round of survival activities are common. The typical forms of sexuality is the result of "genetic error". Reduction of high sexual tensions and craving for orgasmic gratification is DNA's main method of bringing about human reproductive activity that explains abnormal sex act and by extension, its passage to next generation.

Male care in mammals probably evolved, as male care sometimes improves offsprings survivorship to such an extent that the benefits of parental investment outweigh the costs of lost mating opportunities. Male California mouse, *Peromyscus californicus* exhibit extensive care of the young both in field and in the laboratory. Males display all the components of parental behaviour shown by mothers and to the same extent except lactation. Male parental care may not be indispensable for offspring survival, it is sufficiently important to outweigh the benefits of alternative mating strategy (Gubernick and Teferi 2000).

Lactation in male mammals

Regarding male parental care, humans being a case in point. Some degree of male care is not uncommon in primates, carnivores and perissodactyls. Until quite recently, there was no evidence for male lactation in wild mammals. Now, it has been reported in two species of old world fruit bats—*Dyacopterus spadecius* and *Pteropus capistratus*. Lactation in men was observed in World War II prisoners of war camps, when malnourished detainees were later liberated and provided with adequate nutrition (Greenblatt 1972). During the period of limited food supply, the prisoners suffered liver, testicular and pituitary atrophy. After postrelease increases in nourishment, the testes and pituitary gland rapidly began producing estrogens and androgens. However, the liver was slower to recover from stress of starvation and could not metabolise these products. The result was an imbalance of hormones that led to male lactation. Lactation in virgin and non-pregnant female sheep *Ovis aries*, as well as in some males has been observed, without signs of tumours or poor nutrition (Adams 1995). It is suggested that species with strict monogamy, little pre-copulatory male-male competition, and/or obligate biparental care, and perhaps extreme food unreiability, high male relatedness within groups and low predation, offer the potential of male lactation (Kunz and Hosken 2008).

Role of prolactin in parental care

Prolactin (PRL) appears to be an omnipotent hormone (*omnipotin*). It plays important role in production of milk proteins and is essential for physiology of lactation in female and maternal care. As PRL prepares the physiology of female for lactation and maternal behaviour, it is often referred to as *the hormone of maternity*. PRL also plays a significant role in paternal behaviour. Exclusive paternal care occurs in some fishes, amphibians, and birds, which means that the father is the only caretaker. Direct paternal care includes all actions of fathers towards their young that exert an immediate physical influence on them and are believed to increase their survival rate. Indirect paternal care includes acts of the father that are advantageous for the youngs, but are shown by the male independently of the presence of the youngs.

In two fish species with exclusive paternal care, a relationship between PRL and paternal care is known. In *Lepomis macróchirus*, males construct nests in colonies. Females visit nests to spawn and then disappear, leaving the males alone with the offsprings. Paternal care shown by *Lepomias* includes fanning and protecting the eggs and larvae from predators, until the fry leave the nest. Inhibition of PRL secretion significantly reduces paternal care and reproductive success in males. Most avian species are socially monogamous, with both mothers and fathers showing paternal care. Paternal care in birds can be divided into two phases: prehatching phase (incubation) and posthatching phase (brooding and feeding the young). In 25 species of 9 orders ranging from penguins to songbirds, relationship between PRL and paternal care is recorded. In male birds, PRL generally increases during incubation, either being elevated during brooding, or gradually decreasing after hatching.

Only 3–5% of mammalian species are socially monogamous. Not all monogamous mammals exhibit paternal care. Some carnivores like wolves show a communal breeding system, where mother (alpha female) nurses the young and other females, which experience pseudo-pregnancy with elevated

PRL levels. Even males, including father (alpha male), participate in caregiving: regurgitation of food, licking pups, playing with pups, and defending them. Parturition followed by parental care occurs in spring, coinciding with a seasonal PRL peak in both sexes. In wolves, both fathers and helpers have enhanced PRL levels during the period of infant care. Among primates, callitrichidae is well known for extensive paternal care. Callitrichids (marmosets and tamarins) live in family groups with one breeding pair and offsprings of successive births, all of which help rear the youngs. Father carries offsprings on his back and shares food with them. In common marmosets (*Callithrix jacchus*), fathers have significantly higher plasma PRL levels than non-fathers. In a study, PRL was measured in urine samples collected with a stress-free method from cotton-top tamarins (*Saguinus oedipus).* Not only did fathers have significantly higher PRL levels than non-fathers, but PRL levels already began to increase 2 weeks before birth and reached a maximum in 2 weeks after birth.

Hormones can directly modulate behaviour, when they act in the Central Nervous System (CNS). Two modes of action of PRL in CNS are conceivable: (1) organizational effects changing the neural substrate, and (2) activating effects changing neural action. Organizational effects could explain the continuation of paternal care even after PRL level begins to decrease (some birds, callitrichids), although this hypothesis is not yet tested. Because PRL is a protein with ~200 amino acids, the question arises regarding its transport from the blood into brain despite the blood-brain barrier. Several studies showed that PRL is able to pass from blood into brain. Choroid plexus, which shows specific binding sites for PRL play a major role in this process. In female rodents, Medial Preoptic Area (MPOA) is known to be important in controlling maternal care. In the biparental California mouse, virgin males have a larger MPOA than virgin females, although virgin mice of both sexes show no parental care. This sex difference in volume of MPOA is no longer found in parentally behaving California mice, because of increase in soma size of MPOA of females. PRL is known to facilitate maternal care in female rats, when implanted into MPOA, but nothing is known about PRL sensitivity of MPOA in the California mouse or its role in parental care. In combination with its role in maternal care, it might in fact more properly be called the *hormone of parenthood*.

Although relationship between PRL and paternal care is established, it is clear that PRL cannot be the only factor. Caregiving behaviour is a very important factor, as it contributes directly to representation of father's genes in next generation. It is unlikely that paternal care would depend on a single factor only, as this would render it highly susceptible to defects. Life-sustaining processes generally depend on numerous factors as a kind of insurance, so that they will continue to operate, albeit at a reduced level, after breakdown of one factor. Other possible factors influencing the occurrence of paternal care in addition to PRL are neonatal hormone levels, changes in steroid hormone concentrations, experience of male, and stimuli emanating from pups (Schradin and Anzenberger 1999).

Remarks

Parental care begins prior to fertilization, with investment made by females and can be divided into internal parental care and external parental care. Forms of internal parental care range from oviparity, through various levels of ovoviviparity, to viviparity.

Two important changes occur along this continuum:

1. Offsprings are retained longer by the females.
2. Primary source of nutrition for developing embryo shifts from egg yolk to nutrients supplied directly from female.

Most vertebrates are oviparous—a few sharks, all skates, most bony fishes, most amphibians and reptiles, all birds, and monotreme mammals. In oviparity, developing young derive all nutrients from the egg itself until hatching. Oviparous female must store up reserves ahead of time to produce eggs or exploit temporary, but rich resources during egg production. This is generally true for some ovoviviparous species, in which eggs are simply retained in females until they hatch. However, in many ovoviviparous species, females provide nutrients to developing offsprings. Some sharks secrete uterine "milk" from walls of oviduct, which is absorbed by embryos which allows the females to spread out the demand for resource need to produce the youngs. It may also allow the youngs to be born at a more advanced, and more competitive stage of development. Oviparity does have the advantage of relieving the female of the burden of carrying the developing young. In many viviparous species (mammals and snakes), females become slow and vulnerable during late pregnancy. Oviparous species can typically reproduce more frequently than viviparous species, because they do not have to wait for the young to develop in order to produce a new clutch of eggs.

Some viviparous or ovoviviparous species can partly overcome this problem by carrying several broods at different stages of development simultaneously. Ovoviviparous fishes of genus Poecilidae can carry several broods at different stages of development. This strategy slightly decreases the mean number of young produced per brood, but greatly decreases time interval between parturition events. Ovoviviparity is characteristic of most sharks and all rays, a number of teleosts, a few anurans, a salamander and several species of reptiles. True viviparity occurs in most mammals and also in a few species of sharks, several species of snakes and some skinks. An extreme case of viviparity occurs in surf perch, in which males are sexually mature at birth.

Viviparity and ovoviviparity is, as a rule, associated with internal fertilization, which involves some form of intromittent organ (penis, hemipenis, gonopodium). Oviparity is associated with external fertilization. But exceptions being some amphibians, many reptiles that have internal fertilization. One of these is tailed frog, which uses its tail as an intromittent organ. Reptiles have a hemipenis for internal fertilization. Urodelids also have a form of internal fertilization without use of an intromittent organ. In these species, male deposits a spermatophore on the bottom of the pond, which the female picks up and inserts into her cloaca. Female seahorses deposit eggs into male's brood pouch, where they are fertilized. All birds are oviparous, and all have internal fertilization.

Condition of mammals at birth and birds at hatching can be broadly divided into two categories. In altricial species, youngs are in a undeveloped state—typically blind, naked, and incapable of locomotion. In precocial species, youngs emerge as active and, at least partly self-sufficient. Condition of youngs at birth seems to be correlated with the foraging style. Altricial youngs are favoured in species with very active foraging styles, where prey must be pursued and subdued (carnivores). Mammals that consume food that is easily handled and readily available tend to have precocial young (artiodactyla). The amount of external parental care varies a lot among vertebrates.

11.11.10 Sexual Cannibalism

Sexual cannibalism in *Scorpion mantises*, spiders, midges, and horned nudibranchs have attracted researchers since Darwin. Females eat males after mating events in sexual cannibalism. Females always prefer to select the best fathers to pass on potential genes to their offspring. Male species attempt to reduce their risk of being eaten but ultimately surrender to their mates, perhaps to increase their reproductive success. Varied evolutionary pressures produce sexual cannibalism. The male bodies nourishes the mothers. Such nourishment would help their offspring and raise the chances of hatching and growing up their offspring. Thus, enhances the rate of carrying on the father's genes, the ultimate goal of life Female Chinese mantises eat a lot of males, and even up to 63 percent of the diet of females is made up of male mantids. Male redback spiders of Australia pluck the strands of web and court females for about eight hours. A male somersault onto her fangs at the start of mating. Male mate and the female feeds on male. The male crawls a short distance away to court the female again, and initiate second mating event. Male flips onto her fangs. Female becomes dead at the end of 2nd mating event.

During mating males find extra time to put a plug in a sperm receptacle of female. The plug also is the barrier to prevent other males to mate with the female. thus, ensuring that the first male's sperm will be placed in a sperm receptacle. Male redback spider has evolved for cannibalism to mate twice, with their two sexual organs. Two receptacles of a female spiders, in turn is get fertilized by separate male organs. Male redback spiders develop a pinched abdomen long before the males touch the females. The male's death may prevent other males from mating with the female. Because in death, male's sexual organ remains stuck in the female's receptacle. Even if she feeds on the rest of his body, the organ remains behind. This prevents her from receiving more sperm. In some species, males have few opportunities to mate. Chinese mantises are complicit in their own deaths. Females Chinese mantises eat a lot of male mantids. But that does not mean that male's benefit. Females sometimes grab males. So, the males may have waited out of caution. Male golden orb spiders reduce scope of being eaten by jumping onto females as they eat something else. When they survive mating, they enhance reproductive success by standing guard over their females. Female selects the males they will cannibalize. Female will let males' mate for a time period before eating them. Female mantises for their diet depend very much on male mantises. Natural selection favours females that attract more males. Females by the odor of pheromones draw the male mantises. Within spiders and scorpions, sexual cannibalism takes many variegated forms. The reproductive strategies of males and females often differ. This results in unsymmetrical financing of reproduction by the two sexes. This led to a mismatch between the sexes (Schneider and Lubin, 1998). Females obtain nutritional benefits to enhance fecundity by eating their male mates (Andrade 1998). Males sanction future mating scopes and fitness benefits by being eaten by females. Sexual cannibalism perhaps evolved as a form of paternal investment for males. If a male's sacrifice enhances the quality of offspring, sexual cannibalism would become an adaptive male strategy (Parker 1979). Buskirk et al., (1984) described the life history conditions under which sexual cannibalism would be selectively worthwhile for males. This model also suggests that natural selection favours sexual cannibalism. Cannibalism sometimes increases the number of eggs fertilized by the cannibalized male's sperm. It happens when the number of matings over a male's lifetime is low. This happens in situations in which cannibalism without sperm transfer result. Cannibalism occurs in about 65 %t of mating (Andrade 1998). It is unchanging while the male is in the somersault posture. Complicity in cannibalism results in two paternity advantages for this species. Females that cannibalize their mates are less likely to mate again. Cannibalized males mate longer and fertilize about double the eggs than those not killed in copulation. The male's mass is less than 2% of that of females (Andrade 1996).

Elgar and Nash (1988) proposed male rejection to explain incubation of premating sexual cannibalism. Females assess potential mates, cannibalize or mate based on phenotypic features that vary among male species. Females should consider to mate with larger males to produce bigger offspring. Females should opt to cannibalize smaller males to increase their rate of production of offspring (fecundity). Study of the spider *Araneus diadematus*, in which sexual cannibalism surpasses copulation help to develop a model. According to Newman and Elgar, (1991) cannibalism evolve through foraging. The ecological factors for the incubation of sexual cannibalism are important. Females cannibalize males when encounter rate by male is high, and the midpoint supply intake rate is low in comparison to other prey. When volitional prey items are abundant, cannibalism decrease. Reversed sexual cannibalism, in which the male *Gammarus pulex* consumes sexual conspecifics supported this theory (Dick 1995). In *Gammarus pulex*, males cannibalize freshly moulted females, which are ready for copulation and at their vulnerable stage, at low rates than they cannibalize other conspecifics. When other prey organisms decrease the reversed sexual cannibalism increases, as suggested by the Newman and Elgar in their "Economic" model of sexual cannibalism (1991).

Cannibalism is a possible disease transmission route in several species, including humans. cannibalism is widespread among the animals and is now considered as a major mortality reason of numerous species (Van Schaik & Janson 2000). It differs from other predator–prey interactions Because in cannibalism both the prey and the predator belong to the same species. This intra-specific predation result in population dynamics that are different from that found in without cannibalism systems (Claessen et al., 2004). In past human societies, individuals were either actively killed and consumed following an attack on neighbouring villages or tribes, or individuals were consumed as part of a funeral ceremony after death (Volhard 1968). I restrict use of the term cannibalism to the former case (where there is active killing and consumption of living prey) and refer to the consumption of dead conspecifics as 'intraspecific necrophagy'. Cannibalism in the past has been a common practice in many human societies worldwide (Lindenbaum 2004), in spite of some of the cases being exaggerated (Wendt 1989). The connection between cannibalism and disease has received considerable attention in the case of Kuru, a degenerative prion disease transmitted through cannibalism and necrophagy among the Fore people in Papua New Guinea (Collinge et al., 2006). Cannibalism is the predominant transmission mode in few species. However, specific examples of cannibalistic transmission are noted in many groups of animals and in many pathogens. Indeed, cannibalism was implicated as the major transmission mode for only two cases: prion transmission in humans (Lindenbaum 1979) and transmission of the protozoan *Sarcocystis* in lizards (Matuschka & Bannert 1989). In all other reported cases of cannibalistic disease transmission, alternative disease transmission modes such as necrophagy or heterospecific trophic transmission were of greater importance. While searching for the characteristics of diseases transmission by cannibalism, a classical susceptible–infected model was combined with a model of a cannibalistic predator–prey system. In many species including humans, another form of cannibalism is the consumption of dead conspecifics which is called intraspecific necrophagy. In the animal kingdom, this is a common transmission route for many diseases.

It has been shown that group cannibalism by multiple individuals on one victim is a necessary precondition for disease spread through cannibalism. in the animal kingdom, cannibalism is a one-on-one interaction in which a strong, large individual kills and then eats s a small and weak conspecific (Polis 1981). Under these conditions, cannibalism is likely to be an ineffective mode of disease transmission. Thus, it is assumed that cannibalism-independent population regulation is absent, and that infection itself does not influence the probability of being cannibalistic or being cannibalized and that there is no age structure. Intuitively, it is clear that in one-on-one cannibalism, the R0 of a cannibalistically transmitted disease has to be less than unity until the infected individuals get some compensatory advantage. An infected individual can transmit the disease to other individual. The victim is killed in the transmission process. This is analogous to the case of vertical transmission. In vertical transmission one female parent produces on average only one offspring and the spread of vertically transmitted parasites occurs under special conditions. That is when the infected individual has a higher fitness than the uninfected one. This is consistent with the fact that diseases spread through cannibalism are rare in natural populations. Cannibalistically transmitted parasites usually have one or more alternative transmission strategies. This rarity of cannibalistic transmission is especially notable. This is because trophic transmission itself (disease transmission from prey to predator) is a common place of parasites with life cycles involving alternate hosts, especially found in the digenean trematodes and cestodes (Odening 1976), and in many nematodes (Anderson 2000). It appears that within-species trophic transmission via cannibalism would evolve easily than cross-species trophic transmission, because in the former case the host species and hence the internal environment for the pathogen would remain unchanged.

An interesting 'exception that proves the rule' is the case of cannibalistic transmission of several species of the protozoan *Sarcocystis* in lizards of the genus *Gallotia*. *Sarcosystis* species normally have alternate hosts, but in several species of *Gallotia* that live on the Canary Islands, parasitic transmission is directly through cannibalism (Matuschka & Bannert 1989). However, in the lizards, cannibalism is frequently partial, taking place by consumption of autotomized tails. Thus, one individual host can infect several cannibals over the course of its lifetime following tail regeneration. The process therefore has the dynamics of group cannibalism, even though there is no simultaneous consumption by a group.

There are otherwise only a few records of group cannibalism in which several smaller individuals kill and together consume one larger conspecific. This has been seen in social Hymenoptera and Isoptera (Polis 1981) and in some species where the mother serves as a food source for its offspring (Evans et al., 1995). Cases where several adults kill and consume an infant together have also been recorded in social mammals. This occurs in lions, for example, when a group of new males acquires another male's harem (Bertram 1975). It also occurs in chimpanzees where male groups commonly attack conspecifics (Wilson et al., 2004). Humans have also had the required social structure and social practices that promote disease spread by cannibalism. Historically, members of families or a village often shared captured individuals in ritualized meals (Volhard 1968; Sahlins 1983; Wendt 1989), and the group size of individuals sharing one victim was often very large. Additionally, some human societies practiced cannibalism across groups and necrophagy within the group (Conklin 2001). This seems to have been the case for the Fore people of Papua New Guinea, in which both intraspecific necrophagy and cross-group cannibalism were common (Rumsey 1999). Thus, in the case of Kuru, necrophagy could maintain and spread the disease within a village, while cross-group cannibalism could be the mechanism that promoted the disease spread on a larger meta-population scale. Cannibalism in humans has been a common and widespread practice that dates back at least to the Neanderthals (Marlar et al., 2000). Several authors have argued that cannibalism has been a part of the natural ecology of human societies owing to the substantial nutritional gain (Darnstreich and Moren 1974).

Results suggest that the occurrence of a disease such as Kuru in humans is most probably the result of the frequent occurrence of group cannibalism and/or group necrophagy. Additionally, the persistence of Kuru into modern times may in part be explained by the pathogenic agent's resilience to cooking, although, in many of the cannibalistic rituals, raw human flesh was often consumed (Lindenbaum 1979). Because cannibalism is no longer a regular feature of human populations, it is not possible to assess the degree to which other diseases may have had this transmission mode. However, it has been speculated that the transmission and establishment of tapeworms in humans may have been aided by cannibalism (Hoberg et al., 2001). And it is quite conceivable that a number of blood-borne infections may have been regularly transmitted in the same way.

Although many pathogenic agents have the potential for transmission from prey to predator, diseases transmitted predominantly by cannibalism are rare because the epidemiological conditions necessary for its spread, especially group cannibalism, are rarely met in natural populations. However, we suspect that such transmission occurs in animals such as chimpanzees where cannibalism occurs by socially organized groups of males (Arcadi and Wrangham 1999; Mitani et al., 2002; Wilson et al., 2004). Cannibalistic transmission may therefore play an important role in the maintenance of blood-borne infections like simian immunodeficiency virus (Santiago et al., 2002) or different types of hepatitis (Birkenmeyer et al., 1998) in these populations. Given the importance of social primates as disease reservoirs for humans (Wolfe et al., 1998), this possibility deserves further investigation.

Review Questions

Short Answer Questions

1. What is asymmetric sexual reproduction?
2. What is symmetric sexual reproduction?
3. What is known as Muller's ratchet?
4. State the role of chemical communication in reproduction.
5. What is mate choice mechanism?
6. State the characteristics of k-selected taxa.
7. State the characteristics of r-selected taxa.
8. Define monogamy.
9. Define polygamy.
10. What is female defense polygyny?
11. What is lek polygyny?
12. What is prostitution polyandry?
13. What do you understand by sexual promiscuity?
14. What is lek?
15. Define fecundity.
16. Define viviparity.
17. Who first suggested the Lottery Principle?
18. State the Lottery Principle.
19. State the tangled bank hypothesis.
20. State the red queen hypothesis.
21. State the DNA repair hypothesis.
22. Comment on the good genes model of sexual selection.
23. What do you understand by asymmetric reproduction?
24. Comment on "Muller's ratchet".
25. What is the function of MOE?
26. What do you understand by sexual responsiveness?
27. What is the function of "fertilization efficiency gene"?
28. What do you understand by the term "sperm quality"?
29. Comment on the resource defense polyandry?
30. What is operational sex ratio?

Long Answer Questions

1. Describe sexual selection theory in your own words.
2. Write notes on the Red Queen hypothesis and the lottery principle.
3. Discuss the phenomenon of sperm competition in light of modern study.
4. Write a note on mating system.
5. Write an account of homosexual behaviour in animals.
6. State the hormonal mechanism in pair bonding with examples.
7. Give an account of parental care in fishes with suitable diagrams.
8. Give an account of parental care in amphibians with suitable diagrams.

Chapter 12

Behavioural Ecology and Conservation Biology

The Nobel Prize winning ethologist, Nikko Tinbergen first outlined two main questions, for any behavioural phenomenon. Proximate questions focus on behaviour's development or immediate causation. Ultimate questions focus on behaviour's evolutionary history or current adaptive utility. Thus, by definition, conservation behaviour is a field of study, including multiple levels and incorporates disciplines, like genetics, physiology, behavioural ecology, and evolution. This is a burgeoning field at interface between conservation biology and animal behaviour and necessarily not provides solutions to all conservation problems. However, by using conservation behaviour, conservation benefit can be gained to a better extent. Addressing both proximate and ultimate questions of conservation behaviour are essential to improve our management decisions.

A movement towards integrating behavioural ecology and conservation biology is indicated by four recent publications (*Animals Behaviour and Wildlife Conservation,* 2003; *Behaviour and Conservation,* 2000; *Behavioural Ecology and Conservation Biology,* 1997; and *Behavioural Approaches to Conservation in the Wild,* 1997). Some scientists suggest that behavioural ecology can be used in areas like baseline behavioural ecological data and conservation problems, baseline behavioural ecological data and conservation intervention, mating systems and conservation intervention, dispersal and in-breeding avoidance, and human behavioural ecology. Animal behaviour studies provide valuable information to conservation. For example, knowledge of mating system is helpful in determining harvest seasons, and limits. Dispersal is recognized as a key process in ecology, evolution, and conservation biology. Mechanisms of imprinting related to mate and kin recognition along with food and habitat selection, aids in captive breeding, and re-introduction programmes. Thus, major lines of current interest in behavioural ecology and conservation biology are:

- Dispersal behaviour
- Changing organisms in changing landscapes
- Mating system and conservation
- Resource based habitat concept and application
- Predator-prey interactions
- Cooperation and conflict

12.1 PREDATOR-PREY INTERACTION

Predation (+/– interaction) is a biological interaction, where a predator feeds on a prey. The predator lowers fitness of its prey and reduces prey's chances of survival, reproduction, or both. A true predator kills and eats prey. Predators, like jaguar kill large prey and chew it, prior to eating. Some predators eat their prey as a whole, as in case of bottlenose dolphin. Prey may die in the digestive system of the predator as in baleen whales. In seed predation, predator does not eat prey entirely, as in grazing. Many predators hunt only one prey species. Some like dog, kills and eats almost anything. Predators are usually well suited to capture their preferred prey. Preys are often equally suited to escape predator. This is called evolutionary arms race, which tends to keep populations of both predator and prey in equilibrium.

Predators are often another organisms' prey, and likewise preys are often predators. Birds prey on insect. Snakes in turn prey on birds. A predator at the top of any food chain is called **apex predator** (example tiger). Many predators eat from multiple levels of food chain. A carnivore may eat both secondary, and tertiary consumers, and its prey may itself be difficult to classify for similar reason. The organisms showing both carnivory and herbivory are called **omnivores**. The predators may increase biodiversity by preventing a single species from becoming dominant. Such predator is called **keystone predator**.

Act of predation may be broken down into four stages viz. detection of prey, attack, capture, and conservation. One adaptation helping both predator and prey to avoid detection is camouflage, a form of crypsis, where species have an appearance which helps them blend into the background. Background upon which organism is seen may be its environment (praying mantis resembles dead leaves). Zebra's stripes blend in with each other in a herd, making it difficult for a lion to target a particular one. Mimicry is a related form of deception, where an organism has a similar appearance to another species, as exemplified in drone fly, which resembles a bee, although it cannot sting. Predators may use mimicry to lure their prey. When hunger is not an issue, most predators generally not attack prey, since the costs outweigh the benefits. Treatment of consumption in terms of cost-benefit analysis is known as *optimal foraging theory*.

12.1.1 Predation Theory

Mathematical models of predation are amongst the oldest in ecology. The Italian mathematician Volterra developed his ideas of predation from watching rise and fall of Adriatic fishing fleets. Predator-prey models examine the interaction between two species. The earliest models considered the populations of lynx and snowshoe hair, due to availability of extensive data collected by the Hudsons Bay Company in 19th century. Predator prey models can be applied to herbivores (predator) eating grains (prey). So the model is not limited to carnivores.

In the Lotka–Volterra model there are two species.

$x(t)$ = Population of prey
$y(t)$ = Population of predator

The model makes the assumptions that in absence of predation, prey population grows exponentially, and prey captures are proportionate to predator-prey encounters, which in turn are proportionate to relative density of each species. These two assumptions imply that rate of change in prey population is given by:

$$x(t) = x(t) - x(t)y(t)$$

This model does not consider limiting environmental factors other than predation. Assumption that prey captures are proportionate to predator/prey density, is borne out by empirical studies. The constant β is rate of predation. For predators, assumptions are that in absence of predation, prey population dies off exponentially (no food supply), and predator growth rate is proportionate to prey captures, which are proportionate to predator prey encounters, which in turn are proportionate to relative density of each species.

These assumptions lead to an equation for predator population.

$$y(t) = x(t)t(t) - y(t)$$

The predators are assumed to have only one source of food (the prey) and no other environmental limitation.

12.1.2 Predation with Logistic Growth

Classic predator-prey model evolved from Lotka–Volterra equations has two major short comings (Berryman 1992). The model predicts that an enrichment of system causes an increase in equilibrium density of predator, but not in that of prey and destabilizes community equilibrium. It also predicts that there can not be a situation of both a very low and a stable prey equilibrium density. Neither of these predictions are supported by field observations.

The predators need consideration in conservation programme. Introduced predators cause harm to populations by not coevolving with them and may even lead to possible extinction. This depends largely on how prey species can adapt to new predator species, and whether or not predator can turn to alternative food sources when prey population fall to minimum level. If a predator can use an alternative prey, it may shift its diet towards that species in a behaviour known as *functional response*. Prey species may be able to survive, if predator has no alternative prey. In this case, their populations necessarily crash following decline in prey, allowing some small proportion of prey to survive (Figures 12.1 and 12.2).

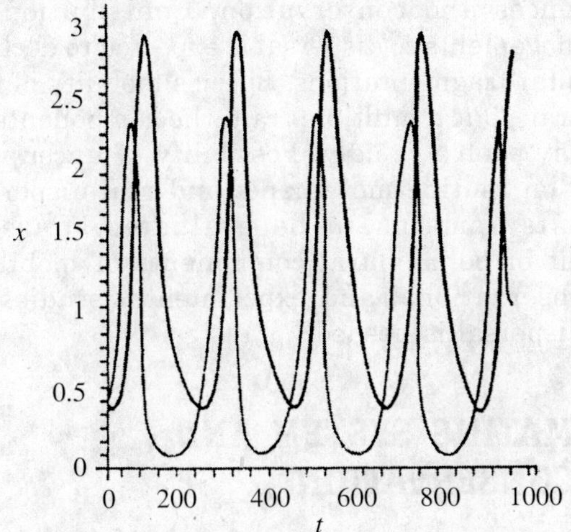

FIGURE 12.1 Predator and prey evolution over time.

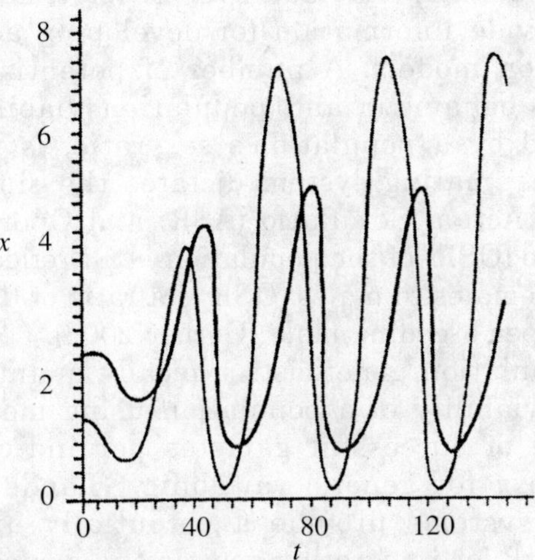

FIGURE 12.2 May system, predators and prey vs. time.

The predator-prey systems are potentially unstable as observed in laboratory condition, where predators often extinguish their prey and then starve. In nature, at least 3 factors promote stability and coexistence. Due to spatial hetero-geneity in environment, some preys are likely to persist in local pockets, where they escape detection. Once predators decline, prey can fuel a new round of population increase. Prey evolves behaviours, armour and other defenses to reduce their vulnerability to predators. Alternative prey may provide a kind of refuge because once a prey population becomes rare, predator may search for a different prey species. Predators may be used in conservation effects to control introduced species. Although aim in this situation is to remove introduced species entirely, keeping its abundance down is often the only possibility. Predators from its natural range may be introduced to control populations, though in some cases, this has little effect and may even cause unforeseen problems.

12.1.3 Predator Recognition and Avoidance

Recently Ian McLean has attempted to demonstrate how naïve wild and captive animals can be taught to recognize and avoid predators simply, quickly and economically enough to be part of every introduction program. The conditioning of animals to avoid predators highlights the importance of an ethological perspective in appreciating necessary stages and likely pitfalls of task. Rufous hare-wallaby (*Lagorchestes hir-sutis*) went extinct in mainland Australia in 1991, but persist today on two islands and in captivity. The species decline is attributed to human disturbance, introduced predators (foxes and cats) and ill-timed droughts. Despite intensive predator control at release sites, initial release of the species failed quickly due to predation. It has been shown that hare-wallabies responded warily, when novel stimuli were placed in their enclosures, particularly fox and cat models, suggesting an innate avoidance of unknown animals that could be and was accentuated by training. Successful training involved (1) sudden bursting of model predator from a box in association with a loud noise, and recordings of hare-wallaby alarm calls or (2) a model that 'lept' at hare-wallabies as they approached and a simultaneous squirt from a water gun.

12.2 DISPERSAL BEHAVIOUR

Dispersal is the spreading of individuals, away from each other. Dispersal consists of three components viz. emigration (crossing habitat boundaries), transience or traversing through a landscape (like movements in non-habitat or resource poor matrix), and settlement (immigration, or colonization). Behavioural differences may exist depending on motivation to disperse, like forced or voluntary emigration. Studies have focused mostly on dispersal propensity. Importance of behavioural variation in two other stages is mostly unknown. Dispersal patterns vary among sexes, populations, landscapes, or species. Mechanistic approaches have recently been advocated to gain better understanding of dispersal. Conservation biology benefits from such understanding to improve the ability to manipulate patterns for conservation. Recognition of different behavioural movement contributing to dispersal study has become particularly significant in the field of habitat fragmentation.

Various life processes along with their associated movement occur at different spatial scales, from food searching in a local patch to migratory movements on regional or continental scales (Ims 1995). Dispersal can be realized in two different ways. First, dispersal can be a byproduct of a search for daily resources by routine movements, which are called trivial or daily movements. Such routine movements are associated with foraging, mate location, seeking shelter. Second dispersal can also be realized by special movements that are designated for net displacement and settlement at some distance from natal site. Geometry of habitat in a landscape or basically of resources determine the probability at routine dispersal. In continuous landscapes with regularly spread resources, routine movements are predicted to contribute more to dispersal than in highly fragmented landscapes with resources in discrete patches, at considerable distance relative to scale of space used by average individual of target species. Results of experimental and empirical studies are in line with this prediction (Mennechez et al., 2003). Animals showing dispersal polymorphism and displacement movement are behaviourally distinguishable form routine movement. Distinct morphological variation provides an indicator of dispersal propensity of individuals, and frequencies of dispersive individuals can be compared among populations and landscapes (Dyck and Baguette, 2005).

All individuals engage in routine movements. Propensity of special dispersal movements may vary among populations and individuals. Individuals making long-distance movements represent a non-random sample of a population, with respect to morphology, physiology, and behaviour as exemplified by butterflies, fishes, lizards and naked mole rat. Evidence for a heritable basis for long distance dispersal in insects and birds exists. Significant variation among individuals, with respect to dispersal is not necessarily reflected in their routine, explorative movements. Colbert et al., (2004) predicted that species with least environment-sensitive dispersal strategies are specialized species in terms of habitat requirements. Habitat fragmentation alters resource distribution from widespread and abundant to localize in discrete patches with clumped resources. Contribution of routine movements to dispersal is expected to decline with degree of habitat fragmentation.

Dispersal is important to understand with more biological realism and with evolutionary consequences and conservation. Contribution of routine movements to dispersal is expected to decline with habitat fragmentations. Species mobility is not a static trait, but a multiple trait, the components of which may evolve rapidly. Possibility of uncoupled selection on routine movements and real dispersal movements remains to be evaluated. Careful treatment of behavioural components of mobility within observational and experimental studies of animal dispersal is needed.

12.3 MATING SYSTEM AND CONSERVATION

Individual behavioural strategies influence animals' response to perturbation. So, understanding factor that influence behaviour over short time scales may provide information for developing accurate population models. A number of potential links between behaviour and population viability, as mediated by a population's sex ratio, is known. A species' mating system dictates the similarity between Actual Sex Ratio (ASR) and Operational Sex Ratio (OSR) of the population. ASR reflects true ratio of females to males. OSR is a ratio of breeding females per breeding male (Gerber 2006).

Population geneticists typically quantify the genetic viability of a population using mean and variance of successful gametes per individuals, as a proxy for genetic variability (Lande 1994). Mating systems provide a potentially unifying framework for integrating genetic and demographic perspectives on population viability. While the effect of small population size on mating success has been considered in the context of Allee effects for small populations; few viability models include behavioural variation (Anthony and Blumstein 2000). Extent to which biases in OSR determine population growth is even less clear. In monogamous animals, female biased sex ratios are clearly a demographic liability, whereas in polygynous animals, this may actually improve the viability of a small population, relative to a population of the same size with 1:1 sex ratio. However, even for animals that have adopted mating systems, with strongly biased sex ratios social dysfunction and reproductive collapse may occur, when bias exceeds critical thresholds (Milner–Gulland et al., 2003). This suggests a more complex relationship between demographic consequences of bias in sex ratios and viability of a population. While sex ratio is an important parameter in demography of animal population, influence of individual variation on population dynamics remain poorly understood. Territorial behaviour exhibited by male animals during breeding season greatly influences the number of males copulating with females, and thus, siring young. At extreme biases in male sex ratio and small population size, number of males in a population is related to female fecundity. Gerber (2006) suggested that effects of sex ratio on viability of small populations depend strongly on mating system of local species. Mating system dynamics may influence a population's vulnerability to harvest and may determine recovery, as well as, extinction of populations. Exceeding the upper and lower thresholds in a population's sex ratio results in a reproductive collapse (Milner–Gulland et al., 2003).

Behavioural options for reproduction are diverse (Table 12.1). Generally mating systems are distinguished based on (1) number of individuals with which a male or female copulates; (2) whether a male and a female form a pair bond and cooperate in parental care, and (3) how long the pair bond is maintained (Alcock 2001). The number of combinations of these variables is substantial.

Among diversity of mating options, territoriality is one strategy that may have a profound impact on a population's sex ratio (i.e. if only a small fraction of males are breeding). Territoriality may also influence other parameters that are important for population viability. For example, among yellow bellied marmots, there is considerable variation in number of females occupying a male territory, such that the average annual reproductive success of a females actually declines as group size increases (Downhower and Armitage 1971). In polygynous mammals, such as sea lions, male reproductive success is largely determined by number of females present in their territory; males compete for males, and females invest in parental care (Trivers 1972). Territorial behaviour is likely to be of particular relevance for population dynamics in polygynous populations resulting in biased OSRs.

12.4 CONFLICTS IN ANIMAL SOCIETIES

Conflicts are ubiquitous in living world. Origin and resolution of many subtle and complex conflicts found especially in social animals are difficult to understand. Conflicts are of many kinds and may occur between prey and predators, between parasites and their hosts, and even between members of same species competing for limited resources.

TABLE 12.1 Overview of Mating System Theory

Mating System	*Behavioural Strategies*	*Which Sex has Higher Variation in Fertility?*	*Examples*
Male assistance monogamy	Male assists mate	Equal	Most birds, wolves
Male guarding monogamy	Male guards mate	Low	Black headed gull
Sperm replenishment polyandry	Females mate with multiple males to secure additional sperm	Females	*Drosophila*
"Prostitution" polyandry	Females mate with multiple males to gain access to sperm	Females	Humming birds
Resource defense polyandry	Males assume most parental duties and compete for mates	Females	Spotted sand piper
Lek polyandry	Females compete for dominant position because males select high ranking females	Females	Northern phalarope
Female defense polygyny	Male defends harem	Male	Gorillas
Scramble competition polygyny	Non-territorial search for mates	Male	Wood frog
Resource defense polygyny	Male defends resources	Male	Desert wood rats
Lek polygyny	Males defend symbiotic territory	Male	White bearded manakin

Since evolution by natural selection is based on competition and survival of the fittest, existence of these kinds of conflicts is not surprising. Conflict, even prolonged conflict is a natural state in wild animals for much of their lives. Sometimes decisions are taken at once, for example, escaping from an approaching predator. In many occasions, animals wait and collect more information, before a decision is made, which is particularly useful in social situations, where animals take considerable time to 'assess' each other. Red deer stags take half an hour or more to assess each other's fighting ability, before deciding whether to attack or flee. Roaring matches and parallel walking are used by stags as indicator of likely outcome of a real fight and mean that animals remain in a state of continuous motivational conflict, while they gather relevant information. Even escape from predation may be preceded by a period of lengthy assessment. Prey animal do not flee every time as predator appears. Prey animals weigh up risk that a predator will attack them against loss of feeding time they would incur by moving away. Zebras watch hyaenas carefully and are much more likely to flee on some days than others, apparently sensitive to small differences in hyaenas, behaviour and size of hunting group that indicates what prey hyaenas are after on any one occasion. Such assessments are highly adaptive, but take time to make. Motivational conflict is resolved only, when new information received tips the balance of causal factors in direction of one behaviour, or another.

Second reason for prolonged motivational conflict, or remaining in a state of high motivation for a considerable length of time, is that it is for some reasons prevented form carrying out the behaviour that it has 'decided' to do. An animal may be stimulated to drink, and all set to give top priority to dinking behaviour over anything else, only to find that its water is covered with an immovable glass lid. Such an animal is called **thwarted**. Alternatively animals may be highly motivated to drink, but unable to find any water at all. Such an animal is called **frustrated**. Both thwarting and frustration are commonly, but not exclusively seen where animals are kept in unnatural conditions. Understanding what happens when animals remain highly motivated for long periods of time is critical in study of animal welfare and to understanding the suffering and stress in animals. Animals exhibit symptoms of stress, when kept in unnatural environments arise as distortions of decision making mechanism (Manning and Dawkins 2002).

12.5 DISPLACEMENT ACTIVITIES

Tinbergen (1951) described a behaviour typically observed in conflict situation, which they called displacement activities. Common characteristics of such activities are their apparent irrelevance to situation. In middle of a fight, two cockerels would turn aside briefly and peck at the ground, sometimes picking up stones or grains and allow to fall again. Fighting gulls exhibit grass pulling in middle of a fight. A male stickleback courting an unreceptive female would suddenly swim to his nest and perform characteristic parental fanning movements which ventilate the nest with fresh water even though there are not eggs. An incubating tern in its nest makes preening movements just before it takes off at approach of an intruder. A thirsty dove prevented from getting to its water bowl by a sheet of glass may preen itself.

In these examples, animals are either thwarted, or in a conflict between two opposing motivations. Feeding in middle of a fight in cockerels is surprising, as it seems irrelevant to conflict. When a tern is faced with decision of whether to escape from a predator, or stay and incubate its eggs, preening seems irrelevant. Irrelevant behaviour is likely to occur, when animals remain in prolonged conflict and could not take decision. Experiment was performed on chaffinches by flashing a light, when they approached their food dish. Birds were hungry, but avoided light and were in a conflict between whether to feed or to escape. There were several different perches in birds' aviaries, so that they could either get as far way as possible form light, by using most distant perch or approach food dish, by using nearest perch. Chaffinches did most displacement preening and bill-wiping on intermediate perches. Displacement activities mostly occur at a balance point between two conflicting tendencies, but it is not conclusive. Since birds paused for longer on intermediate perches same result would be obtained, if preening simply occurred at a fairly constant rate, wherever a bird made a pause and was doing nothing else longer the pause, the more preening. The Exact causal basis for displacement activities is still controversial. The Displacement activities may not be such an unusual category as was once thought. If they occur in conflict situation and enable animals to gather a bit more information about their opponent or a potential predator that may be described as irrelevant, but may not be irrelevant at all to the animals involved.

12.6 CONFLICT AND DISPLAYS

Many other displays are seen as arising form unresolved motivational conflicts. These displays were initially called **threats**. Present day workers emphasize information about sender that a signal conveys (status, fighting ability)rather than its threatening or intimidating qualities. Tinbergen (1952) observed many signals resulting form dual motivation, conflict between simultaneously aroused tendencies to attack and escape, when neither can find separate expression. Threat often occurred at boundaries between territories, where both tendencies to fight and to flee are aroused simultaneously. This might be called evidence from the situation or context. Linked with this conclusion, mechanism of threat, which comes from independent manipulation of levels of attack and escape tendencies is also evident. Blurton–Jones had a group of completely tame Canada geese, which ignored him if he was dressed in familiar old clothes; if he wore a white coat, the geese attacked him uninhibitedly, whilst if he appeared carrying a broom they would flee. Familiar threat postures of goose appeared only, when Blurton–Jones combined wearing a white coat with carrying a broom. Tinbergen gave a detailed analysis of threat postures of lesser black backed gull. Bird moves towards its rival, with neck stretched upward and slightly forward, and head and bill turned down. Wrist joints of wings are lifted clear of body and plumage is slightly raised. Now, gulls normally launch an attack by beating with wings and attempting to peck down at their opponent. Raising of wings and down pointing bill look like actual attack. Elements of escape can be seen as two rivals come very close. Now, head moves increasingly back, bill becomes lifted and plumage sleeker. Bird may turn sideways to its opponent and more parallel not towards it. This turning aside looks like elements of escape. Gulls draw back their heads and sleek plumage preparatory to taking off. Not all threat postures can be interpreted as mixtures of attack and escape and animals may use quite different displays, as only honest way of signalling their fighting ability, rank and so on.

Developing a model of fighting behaviour of desert spider *Aegelopsis* is interesting. This spider is very aggressive, and fights over access to the web sites. In a model there are two variables fear and aggression, which fluctuate during course of an encounter between two spiders, according to such factors as who owns the disputed web site, relative body weight of two, territory quality, and so on. Some genetic factors involved for it are known. Some spider populations are more aggressive than others, even when external situation is kept constant. On this model, if fear greatly exceeds aggression, animals simply withdraw. If fear is equal to or less than aggression, then contest continues, but what behaviour is performed depends on absolute levels of two variables.

The model accounts well for many observed facts about spider's fighting behaviour, such as fat that owners of web sites tend to win; larger spiders tend to win and so on. A model with two variables is simplest one that will account for all observed data. It does appear then that traditional idea of agonistic encounters as extended conflict has stood test of time, although there has perhaps been a shift of emphasis. Animals may display, when they are in a state of conflict not necessarily to signal that they are in a conflict, but by displaying and seeing displays of others, to get themselves out of the conflict. They make their decision on whether to attack or to flee from a rival, or even a predator on the basis of new information received during the course of encounter. Their decision to perform one or other behaviour is then made in an adaptive way that would not have been possible, without the period of extended conflict and exchange of information it made possible.

12.7 COOPERATION

Cooperation is also known as helping behaviour. Termites, ants, bees and wasps show extreme helping behaviour. There is usually one reproductive female, the queen, and large number of sterile workers. Both males and females in termites, but solely females in ants, bees and wasps perform all tasks, such as foraging, rearing the young, nest construction and defense, and do not reproduce at all themselves. In *Hymenoptera*, such self-sacrifice of sterile workers does not appear to have arisen independently. Ants, bees and wasps exhibit "haplodiploidy". Males are haploid and develop from unfertilized eggs. Females are diploid and develop in normal fashion from fertilized eggs. All sperm from one male are therefore identical. When queen bee fertilizes eggs with this sperm, all resulting daughters received same paternal chromosomes and have half their genes in common. They share on average half the genes inherited from their common mother and so their degree of relatedness $r = 0.75$ (0.5 from father plus $0.5 \times 0.5 = 0.25$ from mother). Thus, although workers are sterile, many genes that they share with young queens (which are also their sisters) will be passed on to next generation. The sterile workers benefit, not because they help other workers, which are closely related to them and are reproductive dead-end, but because they care for small number of their sisters, which will develop into young queens. In terms of gene, it is more advantageous for female Hymenopterans to stay and help to rear their closely related reproductive sisters, than to leave and attempt to rear less closely related daughters of their own.

Haplodiploidy appears to predispose Hymenoptera to high degree of sociality they show. However, it is clearly not the only factor because a similar degree of sociality is also shown by termites. They have an ordinary diploid mating system. Sterile workers of both sexes have degree of relatedness to the young reproductives of only 0.5. King and queen termites are long lived and monogamous. A queen lay up to 36,000 eggs a day, and in some species live for 60–70 years. King and queen together may have literally millions of offsprings in their lifetime, vast majority of which will never reproduce at all. Understanding why so many termite workers should be sterile, we have to follow that there are three variability in Hamilton's equation. A high r predisposes towards helping, because it increases effective benefit, but even a relatively low value of r would lead to helping if benefits were high enough and costs low (Figures 12.3 and 12.4).

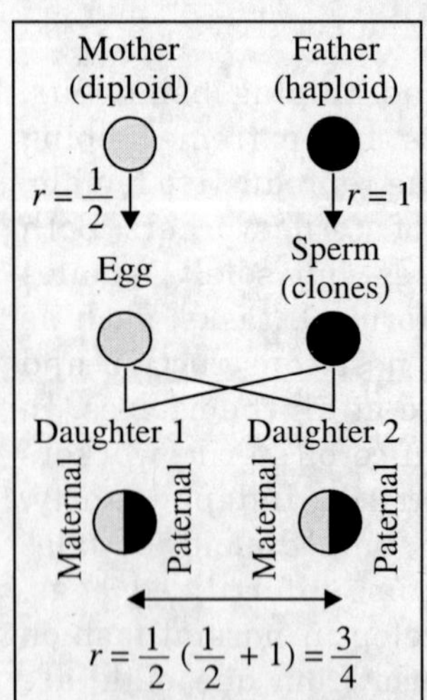

FIGURE 12.3 Genetic relatedness under haplodiploidy, as seen in the Hymenoptera. Notice that the relatedness (r) between two full sisters is 0.75.

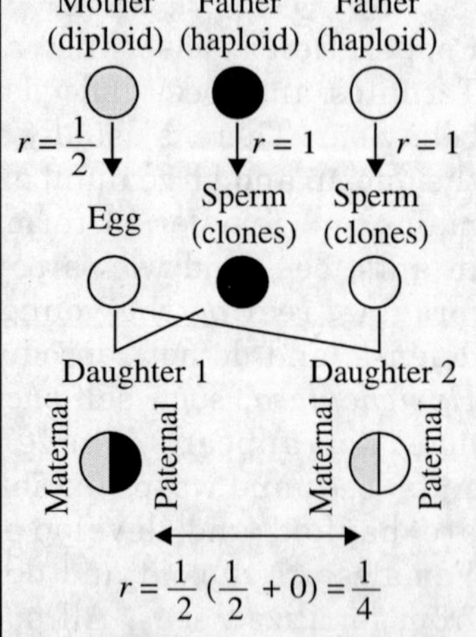

FIGURE 12.4 Genetic relatedness under haplodiploidy, with multiple mating. Notice that the genetic relatedness between two full sisters is 0.25.

Cost to each worker of being sterile is loss of these offsprings, it would have had, if it had not been a helping colony. Given that a pair of termites on their own would probably not survive, let alone reproduce even modestly, costs of helping must also be small (no hope of reproducing means no cost to losing it). Monogamy in termite breeding system means that workers are guaranteed a long series of full brothers and sisters to take care of. Since their own parents offer no care to them, without workers, young termites would die. Workers, therefore, make a substantial difference to survival chances of close relatives even though benefit b, accruing to each worker is only a fraction (because it is shared with other workers, which have helped) of output of each reproductive. Nevertheless, it appears that benefits of helping are significant, costs low and r at least as high as between diploid parents and offsprings. Termites often live in deserts, or in very dry regions and can only do so, because mounds built by collective labours of millions of workers enable them to create their own microenvironment. A single pair of termites removed from this specialized microenvironment would stand no chance and this fact effectively reduces costs and increases benefits of helping the royal pair in home colony (Manning and Dawkin 2002).

12.7.1 Cooperative Hunting in Sand Tiger Sharks

Despite popular conception of sharks as "long killers", many species feed in groups. Coles reported astonishingly coordinated manner in which a group of hundreds or more sand tigers of Cape lookout North Carolina, systematically surrounded school of blue fishes and herded them into shallow water. Similar instances have found at Seal rocks, New South Wales from where a diver reported a pack of sand tigers, known locally as Grey Nurse Shark, herding a small shoal of juvenile yellow tail king fish (*Seriola lalandi*) by whipping their tails to generate sharp under water pressure waves, which sound very like reports of a shotgun. Cooperative herding behaviour by using of bottom topography to strand schooling fishes has often observed in bottlenose dolphins (*Trusiops truncates*) feeding on mullet at several coastal wetland location of Southeastern United States.

12.7.2 Cooperative Hunting in Wild Dogs

A lone wild dog can successfully kill a medium sized ungulate, such as an impala, but group hunting increase overall hunting success and allow the group to capture larger prey. Hunting success increased from 42% for a group of three adults to 67% for a pack with 20 adults. The mass of prey killed also increased as pack size increased (21 kg for pack of three dogs to 68 kg for a pack of 20 dogs), while chase distance decreased for larger packs. An important reason for cooperative hunting is defense from kleptoparsitism. A lone wild dog would never be able to defend its kill from a pack of scavenging hyenas but a pack of wild dogs can often chase off a group of hyenas. A small pack may have lower hunting success and may often lose their kill to kleptoparasites. But usually each dog has more to eat at a successful kill. Similarly, a large pack may have greater hunting success and may rarely lose kills to kleptoparasites, but each dog will get less food at a given kill. Benefits and costs of cooperative hunting change with differences in pack size, age and experience of pack members and habitat, but in end, cooperative hunting is a crucial element of wild dog's behaviour.

12.7.3 Cooperative Breeding in Birds

The term cooperative breeding described situations in which adult individuals (helpers), in addition to genetic parents regularly aid in rearing of the young. This has been reported in at least 3.2% of extant bird species (308 cooperatively breeding species, out of a total 9672 species) Helpers are reproductively capable younger individuals that after attaining maturation delay dispersal, remain with and aid their parents in rearing of a later brood. These types of helpers are actually termed as primary helpers. Again there are second type of helpers, which after attaining maturity disperse from their natal family, join and aid distantly related/unrelated breeders. These are called **secondary helpers**.

12.7.4 Increased Success of Breeding Attempt

Simplest way to assess the importance of helpers on success of individual reproductive attempts is to compare production of youngs from the nests tended by breeding parents alone, with that from nests tended by three, or more adults. Many simple corelational analyses of this sort show a seemingly positive influence of helpers. In white fronted bee eaters, the linear trend of increasing fledging success, with increasing helper number was maintained, when environmental conditions, as well as, age and experience of the breeding pair were treated as covariates. Action of helpers has a major impact in increasing fledging success in this species. It appears that helpers have a moderate effect in enhancing breeding success in large number of species and a strong effect in few. There are number of species in which no influence of helpers on success of individual's reproduction attempts has been found.

12.7.5 Cooperation Economics

No one really seems to know why individuals should cooperate. Some economists and scientists have argued that cooperation is not a rational or logical behaviour for species individuals, since energy or other resources must be expended in effort with no direct benefit to individuals. Scientists have rejected such thinking. They argue that logic applies not just to ends, but also to means during economic decision making, and commented that "there is nothing irrational in being altruistic". General conceptual framework discussed is largely based on avian and mammalian studies. Works on cooperation among animals belonging to other species taxa are also required (Aich et al., 2006).

12.8 COOPERATION vs. EXPLOITATION

The principal prey of seabirds in shetland is lesser sandeel (*Ammodytes marinues*). In 1974, an industrial fishery for sandeels was started in this area. Annual fishery catches peaked in 1982, and then declined rapidly with catch in 1989, equalling only 3% of that in 1982. During this same period, breeding success of several seabird species like Arctic tern (*Sterna paradisea*), black legged kittiwake (*Rissa tridactyla*), shag (*Phalacrocorax aristotelis*), and common murre (*Uria aalge*), dropped sharply. Distribution and abundance of sandeels was conducted. A major oil spill occurred in study area, after in the Braer oil tanker sank on nearby rocks. These circumstances lead to an opportunity to examine behavioural and demographic data on both seabirds and fish population before and after an environmental perturbation.

Monaghan's group observed several reproductive and behavioural parameters of seabirds including egg laying date, clutch size, average breeding success, body conditions, foraging trip duration, foraging trip distance, and number of dives, and dive duration (for shag and murre). Sandeel population parameters were measured, using a combination of acoustic surveys and net hauls. The results showed that foraging behavioural data were much better indicators of status of sandeel population than reproductive data. Furthermore, comparison of changes in foraging among different species (diving vs. surface feeders) provided the best indication of fluctuations in sandeel distribution, abundance, and age structure.

The play behaviour can be a sensitive indicator of habitat quality in langur monkeys (*Presbytis entellus*). Frequency and duration of play behaviour in two bands of Hanuman langurs in habitats that differed in availability of food and water was compared. The langurs particularly immature individuals in low quality habitat played less often and for shorter periods of time than those in the higher quality habitat, which demonstrate that behavioural studies not only contribute to understanding of focal species, they can also serve as an indirect assessment of integrity of environment in which these species live.

12.9 REDEFINING THE HABITAT IN TERMS OF RESOURCES

Habitat is necessarily the location, where an organism completes its life cycle. It may be possible to map the bounds of a habitat in terms of life history requirements. Species require a set of resources and conditions in order to function. An adult butterfly requires resources for egg laying, mate location, resting, roosting, feeding and predator escape. Each stage can be treated similarly and the resources can be mapped. Habitat is then logical extension of this reasoning, defined by intersection and union of these resources. A functional definition of habitat is linked to Hutchinson's concept of a hyper dimensional niche. Habitat is to real ground conditions (e.g. occupied space) as niche is to biological space (vectors of influential agents). Accurate recognition of habitat is a prerequisite for determination of niche, which otherwise can only be notional (Dennis et al., 2003).

Southwood identified five conditions imposing selection for tactics in organisms within habitats: physical conditions, predation, food availability, mating and lethal conditions. Despite these distinct characteristics, habitat is modelled very much as a particulate, indivisible entity. Habitats, though varying in "favourableness" or quality, are envisaged to have discernible boundaries, where favourableness equals to zero. Impact that viewing habitats as particulate entities of life history strategies can be gauged by focusing on movements from one location to another (Baker 1978). Southwood described variation in mobility among arthropods, as being a function of differences in habitat longevity and disturbance regimes. He correlated habitat disturbance (temporariness) with extent of migratory movement, arguing that increased mobility enables a species to keep pace with space and time shifts in habitats. Baker extended this model by evolving geographical gradients to explain linear flight and inter-habitat distances and spatial isolation of larval and adult resources to account for differences in migratory distances. Baker (1969) recognized that resources need not physically overlap or be contiguous, but could be isolated from one another. The elements of a habitat are found in different locations, required successively, as well as, synchronously, and therefore, an organism's habitat cannot always be in one place. A clear example of spatially separated breeding and foraging habitats in a population of the butterfly *Leptidea sinapis* is recorded.

12.9.1 Habitat and Resources Geography in Population Biology

Example of behavioural options for increasing population viability with hypothesized effect on population growth for sea lion has been summarized in Table 12.2. Emphasis in population studies has shifted to a larger extent from single unitary, local populations to complexes of linked populations or metapopulations. Studies at these contrasting scales have accrued substantial knowledge on population dynamics. Reliance on particulate habitat model has affected these studies in important ways, including studies of both unitary populations and metapopulations.

Species subject to metapopulation analysis in UK are *Plebejus argus* (L.), *Hesperia comma* (L.), *Thymelicus acteon* (Rottemburg), *Melitaca athalia* (Rottemburg), *Coenonympha tulllia* (Muller), *Erynnis tages* (L.), *Eurodryas aurinia* (Rotttemburg), *Aricia artaxerxes* (Fabricius), *Aricia agestis* (Denis and Schiffermuller), *Lysandra bellargus* (Rottemburg), *Lysandra coridon* (Poda), *Polyommatus icarus* (Rottem-burg) and *Melitaea cinxia* (L.). Other species have been studied in a metapopulation context, but found not to accord to strict metapopulation modelling, e.g. *Aphantopus hyperantus* (L.), *Maniola jurtina* (L.) and *Pyronia tithomus* (L.). Butterflies described as having typical meta-population structures have host plants

TABLE 12.2 Example of Behavioural Options for Increasing Population Viability

Attribute of Population Viability for Males	*Behavioural Options for Increasing Population Viability*	*Hypothesized Effect on Population Growth for Sea Lions*
Female encounter rate	Increase proximity to female resources.	Low
	Increase proximity to sites ideal for displaying to attract females.	High
	Search for females (randomly or using cues).	Low
	Remain with females once encountered.	Moderate
Fertilization rate	Exclude other males by territoriality.	High
	Exclude other males by dominance.	Moderate
	Tolerate other males but control order or timing of insemination.	Low
	Guard female after insemination.	Low
Female fecundity	Select preferred mates using size or age as cue.	Moderate
Juvenile survival	Select females of appropriate age, size, and status.	Low
	Invest in territoriality, parental care, mate survival.	High
Adult survival	Defer breeding to later ages.	High
	Reduce investment in activities that increase risk.	Low
	Defend territory for own access to resources.	High
	Form social liaisons to share risks and costs.	Low

TABLE 12.3 Characteristics of Resource-based and Habitat-based models for Population Studies in Butterfly Ecology

Characteristic	*Metapopulations*	*Habitat-based Landscape Ecology*	*Resource-based Landscape Ecology*
Habitat zone/boundary	Yes	Yes	Yes—where feasible
Resources within patches	No	No	Yes
Resources in matrix	No	Yes-some	Yes
Topography, relief and drainage	No	Yes	Yes
Vegetation and land use	No	Yes	Yes
Microtopography	No	Probably not	Yes
Local climate	No	Yes	Yes
Paths of movements	No	Yes	Yes
Biotope dynamics	Yes	Yes	Yes
Patch dynamics	Developing	Yes	Yes
Target organism resource dynamics	No	No	Yes

with higher *S* strategy (Adversity tolerant) Scores (Correlation $r_s = 0.48$, $P < 0.001$) they are also less mobile $r_s = -0.47$, $P < 0.001$, $N = 60$ species in both correlations (Dennis et al., 2003).

Table 12.3 highlights the resource variables inducing movement and migration in butterflies. Initial conditions include agent as vegetation succession, human management, weather and climate. These influence five basic attributes of resource distributions, in turn, over different space-time frames influence the tendency to migrate. Resource disturbance refers to vegetation changes and host plant dynamics (generation time). Complementary resources refer to non-substitutable resource outlets. Some degree of dependence occurs among resource variables. No attempt is made to expand on individual (e.g. lifespan) and population influences on movement and migration distances in this simple process-response model, other than to indicate a link to resources, nor on the direct influence of conditions (e.g. weather). In Baker's (1978) initiation factor model, probability of an individual initiating migration depends on its migration threshold being exceeded. According to this model, both population density and environmental conditions interact to effect mobility, including changes to resources generated by individual resource use.

Sufficient resources would not make a habitat (a 'patch'), when a population of target organism is absent. Without supporting data from detailed autecological studies of target species, an area that contains host plants can not be categorized as either a vacant or non-suitable 'patch'. Organisms have different resource requirements in different places. Such requirements also change with time. The latter case introduces a problem arising from weak definition of habitat or patch. The emphasis on organism dynamics at the expense of patch dynamics is serious, as each resource has its own dynamics and a number of these resources, if not all regularly extend into landscape matrix. The metapopulation view is that matrix is considered empty of all resources. Movement is regarded as being linear and homogenous, and patch connectivity regarded as a decay function of distance and patch occupancy. Assumption that matrix has a neutral impact on migrating between patches is now known to be wrong (Rickttes 2001).

Metapopulation models do not fit a number of species, those with patchy and migratory dynamics despite increasing sophistication of metapopulation models. Where landscapes are not seriously degraded, it is not always clear that model applies to species believed to fit model elsewhere. In some cases, metapopulation patch-works have been drawn up for extremely complex landscape. However, no exercise has been undertaken to determine, whether independent surveys in such landscape can repeatedly identify same patch boundaries. Empirical problems of patch definition are reported to have serious repercussions for metapopulation modelling. Research on butterfly is being done in the field of landscape ecology. Landscape ecology is complex in structure, function and change. As metapopulation analysis increases sophistication, it would eventually and inevitably merge with landscape ecology and its enlarged focus on species diversity.

12.10 BEHAVIOUR AND CONSERVATION

Land use change has occurred rapidly during recent decades, particularly in tropical areas. Research, since 1980, has studied the impact of land use change including habitat fragmentation on distribution and abundance of animal taxa, including fish, arthropods, birds, and mammals. Although behaviour is intimately linked with species ecology and affects survivorship and population densities, behavioural responses to land use change are rarely investigated. Animal behaviour and conservation biology have not been integrated to the detriment of both fields and the species to be conserved.

12.10.1 Behavioural Information in Conservation Efforts

External light sources (photo pollution) affect nesting behaviour of seaturtles that nest on beaches and behaviour of young turtles, after they hatch. Laying and emergence behaviour occur at night. Adult females ready to lay eggs, often not emerge from the sea onto the beach, if too much external light exist. Even if emerge and lay their eggs, excess light may make it very unlikely that young turtles make their way to the sea. The hatchlings use visual cues to find the ocean. They balance visual inputs, heading towards the brightest direction, which would usually be the direction to the ocean. Artificial light disrupts their ability to do this and so they often wander in land, get hit by cars, preyed upon by beach crabs or birds or become exhausted, dehydrated and die in beach vegetation.

Strategies that have been effective in lowering negative effects of artificial light on turtle behaviour include:

(a) Focusing light towards ground so less of it "escapes".
(b) Using light shields and light designed to have little "stray light".
(c) Using light towards to yellow-red end of spectrum, although they still affect turtle behaviour to some extent.
(d) Using flashing rather than constant lights.
(e) Preserving and encouraging natural dune vegetation to diminish to glow from distant urban areas.

Behavioural information aids in prioritization of habitats that is necessary to maintenance of particular communities. Organisms like turtles, whales, birds and insects engage in short and long distance, one way and round trip movements between habitats. Fragmentation may or may not impede their movements, and thus, value of habitats. Thus, fulfillment of need from habitat determines to a large extent of their ability to survive.

Patterns of pollination, seed dispersal, herbivory and parasitism may be affected by fragmentation, because of changing patch size, edge effects, and plant species diversity changes and their associated effects on animal behaviour. Example to date of inter-specific interactions influenced by land use change is that of brood parasitism by brown headed cowbirds. Studies report higher rates of cowbird parasitism in fragmented vs. less fragmented landscapes (Robinson et al., 1995) and investigation of cowbird movement patterns revealed that interspersion of agricultural and forest land contributes to their ability to parasitize forest dwelling species.

Humans inevitably fragment habitats during development works. Conversion of forest and other native vegetation types for agricultural purposes are especially mentionable. This fragmentation and its effects on numbers and types of species have received a great deal of attention. The edge effects also need consideration. Temperature increasement, humidity decreasement, and wind velocity increasement occur near edges. These effects may extend tens of metres in fragments. In northern hemisphere, south facing edge will be warmer, drier, and wider than north-facing edges. Tree-mortality rates are higher along windward edges. Forests edges may contain more pioneer species, those adapted for dry conditions compared to interior forests, and species richness may be higher at edges. Species composition of forest may change, when these pioneer species disperse their seeds into forest interiors. Forest edges in suburban areas are affected by human activities, such as dumping of yard clippings and gathering of firewood.

Smaller fragments usually contain fewer species, than larger fragments probably because local extinction is more common in small fragments in accord with island biogeography theory. Mammals, amphibians, insects, birds, and herbaceous plants diminish in numbers in smaller fragments of habitat. Particular life history traits predispose certain species to extinction, or survival in smaller fragments. Species with poor dispersal ability and specialized resource needs are particularly susceptible to extinction in small fragments. Species with good dispersal ability and generalized resource needs and that can reproduce asexually, in contrast may be less prone to extinction in small fragments.

Nest predation and brood parasitism of birds are higher in small forest fragments, compared to large ones. Some migratory songbirds in the U.S. have declined by 50% since the 1960s. Early workers attributed the declines to habitat loss on their breeding grounds, or loss of wintering grounds in tropical America. Robinson et al., (1995) reported that fragmented landscapes in Midwest had very high rates of nest predation and cowbirds parasitism. The raccoons, opossums and cowbirds all use edge habitat. In addition grain fields have caused large increase in cow-birds population. Population decreases in small fragments leading to decreased seed production in plants.

Habitat fragmentation may take various forms:

(a) Native habitat may shrink over time, but remain as a single piece of habitat.
(b) Native habitat may be bisected by a road for example.
(c) Native habitat may be divided into many pieces.
(d) Native habitat may be perforated.

12.11 HUMAN ECOLOGY AND BIODIVERSITY CONSERVATION

Anthropogenic extinction has been continuing for at least 50,000 years. Small groups of humans armed with stone age weapons and fire were believed to be effective killers. They caused extensive changes in large ecosystems in short time. For example, megafauna's extinction on the North American continent approximately 10,000 years ago has been correlated with arrival of small groups of human armed with spears. However, some scientists have noted a strong correlation between indicators of high species diversity and diversity of small scale human communities especially in Africa (Mittermeier et al., 2004) in recent years.

During last past 500 years, rate of human-caused extinctions has increased exponentially. Sailors seeking spices, wood, whale blubber and other resources used in international trade; released goats, pigs, sheeps, and rats on remote islands and on Australia, which had never been home to these species before the age of global trade. These species took over habitats of many endemic species, causing extinction of some native species. Near the end of

20th century and beginning of 21st century many human activities are creating a cumulative effect that scientists call "crisis of extinction".

Marine animals are equally threatened and endangered by human activities as terrestrial animals. For example, populations of Pacific leatherback seaturtles has plunged 95% in last 22 years, primarily due to impact of inappropriate fishing activities, with fewer than 5000 nesting females remain in the Pacific. It is also estimated that pacific leatherback turtles will become extinct within 30 years. Some people deny these anthropogenic threats to biodiversity. Others hope for rescuing native biodiversity from a combination of interacting variables including the economic globalization and overharvesting of forest plants and animals in an effort to increase short-term profit through modern technology. Unsustainable population growth in some regions create expanded demands for resources to serve both vital needs and increasing per capita consumption. In many regions demands exist on limited supplies of fresh water, pasture for domesticated animals and arable farm lands. Some regions are experiencing massive growth of urban areas with their demands for fossil fuel drawn increasingly from remote region of the Earth (Ehrlich and Ehrlich 2004).

Supporters of the deep, long range ecology movement emphasizes that species have a right to exist apart form its use for benefit of humans. Humans have no right to cause extinction of native species across their habitat. Such movement of conservation of biodiversity is ongoing since 1960s (Sessions 1995). Some economists argue that "natural capital" of wild nature including "capital" of native biodiversity contributes to humans well being by providing clean water, clean air, soil and genetic diversity utilized to feed humans (Alexander et al., 2001).

Strategies for biodiversity conservation involve relationship between local residents, who live in critical habitat of threatened and endangered species, government agencies, international organization, and conservation biologists. In some cases local people have become involved in managing their homelands as reserves, parks or other protected areas. Outcome of such collaborations between NGOs and local people has been mixed. Some local tribes and communities want oil and gas development, new roads, schools and access to a variety of consumer goods. Others want to maintain as much as possible the old ways in their communities. Traditional tribal people when provided with rifles can effectively kill whatever animal species is sold for cash on market either as bush meat, trophies, skins, animal parts for medicines and other uses. They can also capture and sell animals as pets. Tribal people, peasants, miners, army or paramilitary all seems to have a hand in killing wild life. An example of interactive effects of these factors has been documented in creating a Tiger reserve in Myanmar (Robinowitz 2004).

Poverty eradication and "sustainable development" programs have undermined conservation efforts in some regions. Wild nature is debated in terms of economic value as part of "economic growth" and usually "sustainability" is defined by major donors such as the UN and World Bank as "sustainable economic growth" (Sanderson 2004). Optimistic hope and goal is that in which local populations will sustain their vital needs, using primarily local resources to serve their needs for food, shelter, water and clothing, without over exploiting local resource bases. In situation of dire emergencies for human populations including severe drought, flooding, warfare, tsunamis and earthquakes, the international community will develop institutional response to provide for vital needs of local humans population (Devall 2006).

Native biodiversity protection is dependent not only on relatively large, self-willed land and water areas, but also on changing cultural practices in industrialized and developing nations. Mammals like wolves, coyotes, bears and mountain lions are changing their habits as they survive and thrive in the urban/rural interface of vast suburbs in megacities of North America. The state of New Jersey allowed a bear hunt in 2003, in the suburbs of New York because population of bears was proliferating and bears were rapidly adapting to suburban habitat.

As we continue developing new technologies, we fell less dependent on nature for our survival. One result is that we find ourselves in the paradoxical position of making moral distinctions between animals that are permissible to eat or kill and those whom we bring into our lives on a more intimate and social basis (Cronin 1991). Even so, by seeking a relationship with nature, often through interactions with other animals, we may be able to connect with what is often a spiritual sense of wonder at being part of vast interconnected network. It has been suggested as human's endeavour to affirm their unity with nature. Natural world gives man a point of reference and support and to some people, an assuagement of existential anxiety. It appears that possession of pets symbolizes this unity with nature, and thus, satisfies some deep human needs.

We Must Think about Uncontrolled Population Growth and Behaviour

FATIK B. MANDAL

The world's human population has increased exponentially for the last 200 years. Human population of the world was estimated to be 5 million about 10,000 years ago, which had reached 6.7 billion in 2008, and is predicted to stand at 9.3 billion by 2050 as per the data of the World Bank and the United Nations Organization.

Soil, air, water, flora, and fauna constitute the life-support system on the earth. To fulfill our demand of livelihood, or/and to enhance our economic base, we are continually developing tools for commercial exploitation of natural resources.

We have changed the land use pattern dramatically causing deforestation, desertification, water logging, and groundwater depletion. Increasing pollution of air, water, and soil is threatening the survival of many species along with ours.

Our knowledge in science, society, and culture during the last 10,000 years has reached a significant height. We have devised birth control strategies as well as immunization program. Globalization has resulted in the homogenization of culture. Isolated human societies have also influenced seriously through globalization.

Now, we can reduce our population simply by opting one child per couple, thanks to the discovery of family planning program. Decrease in population has not been happening due to several reasons, including political will. Family planning norms are still not preferable, or accessible to a significant portion of world's population. Poverty rates in developing countries, based on an income threshold of $1 per day, declined from 29% of the population to 23%. Inequality in the distribution of wealth is a major barrier towards achieving sustainable development.

Standard of living of a huge portion of the world's population is not at all satisfactory now. Modern agricultural practices, combined with the chemical fertilizers and indiscriminate usage of pesticide, are matters of serious concern. Depletion and pollution of freshwater resources affect aquatic organisms. Degradation of forest and water is now contributing to pathogen pollution.

Use of fossil fuel, along with resultant global warming, is predicted to increase the sea level, inundate many places of the world, and seriously affect biodiversity with newly emerging diseases. Anthropogenic extinction of species and ecosystem degradation are threatening our survival.

Thus, it appears that human population growth is the root cause of many present-day problems of human civilization. Perhaps, the second most important factor in creating crises is the human behaviour.

The earth is perhaps incapable to fulfill our greed. Overconsumption and poverty, both, are unsustainable. Professor Paul Ehrlich first sounded the alarm about the impacts of over-population and unscientific resource exploitation in his 1968 book *The Population Bomb*. *The Millennium Assessment of Human Behavior* is his latest initiative to aware people about the danger of uncontrolled human activities.

The author acknowledges the information taken from the article: Ehrlich, P.R. 2010. The MAHB, the Culture Gap, and some really inconvenient truths. *PLoS Biol*, 8(4): e10000330.

At an early point in our history, humans were closer both physically and psychologically to natural world than they are now. Many argue that it was enlightenment and industrial revolution that provided the final coffin nails for the concept of humans and the nature's unity. We have become alienated form natural world and animals in their natural state by factors like industrialization and urbanization. Attitudes towards animals is reported to be changed radically at the time of enlightenment. It argued that immediately prior to this time, attitudes towards animals were characterized by theological view point that humans had right to use animals as they wished. Thomas believes that development of scientific disciplines, such as biology and astronomy changed this perception by causing humans to see themselves as part of a wider universe rather than centre of it. Increasing study of natural history inspired greater interest in nature out of sheer intellectual curiosity. Despite the fact that an intrinsic value of animals was becoming more recognized Cronin (1991) argued that twin forces of industrialization and urbanization contributed to a psychological split between humans and animals by physically removing humans from daily and routine contact with animals. Human relationships with animals that were based on proximity to them gradually changed to relations based on separateness. This feeling of separateness combined with a sense of intrinsic value of animals ultimately led to a more sentimental and emotional attitude towards them that is represented by increase in pet keeping and animal welfare movements that persist to this day (Vining 2003).

Review Questions

Short Answer Questions

1. What is predation?
2. What do you understand by the term dispersal?
3. What is ASR?
4. What is OSR?
5. What is called displacement activity?
6. Define threat.
7. Define habitat.
8. What is conservation biology?
9. What is conservation behaviour?

Long Answer Questions

1. Describe the evolutionary basis for behavioural ecology.
2. Give an account of predator-prey interaction along with Lotka–Voltera model.
3. Give an account of dispersal behaviour.
4. Conservation requires an understanding of mating system, discuss.
5. Write an account of conflict in animal societies.
6. Give an account of displacement activities.
7. Mention the significance of cooperation with examples.
8. Mention the relationship between human ecology and biodiversity conservation.

Chapter 13
Human Behaviour

Humans are intensely social species. Our social nature defines, what makes us human, with consciousness, along with large brains (Adolphs 2003). Attempts have taken to explain "why we behave the way, we do". Human behaviour often attracts media attention. Anthropology, biology, psychology, and other disciplines deal with human behaviour. Evolutionary theory, behavioural ecology, and biocultural approaches are also mentionable. Many aspects of human behaviour are controversial and perhaps misunderstood. Human behaviour study is required for a broad understanding about origins and characteristics of issues relating to humanity. Behavioural biology deals with interactions among the brain, mind, body, and environment. The human brain is shaped by evolution, constrained or freed by genes, and early experiences, modulated by hormones, and produce a range of behaviours. Little can be understood, considering any one influence, because combination of influences always work. The brain regulates hormones, and hormones influence the brain and behaviour. Hormones, evolution, genes, and behaviours are sensitive to environment. The part of brain relating to behaviour, is frontal cortex, which plays a central role in decision-making, and other functions. Frontal cortex makes one to do harder things. In primitive societies, people are more bound to land. Limited human activities were directed to self-defense and survival. Social structures were typically rigid, leadership was hierarchical, and traditions allowed reproduction of inherited instructions without alteration. Subsequent changes in behaviour have been occurred in due course of time. Innate social aptitudes of man have also been reviewed (Hamilton 1975).

Human behaviour study is important for a number of reasons. Recently, Professor Paul R. Ehrlich, Bing Professor of Population Studies, Stanford University, Stanford, during addressing various problems to humanity has written that "the central problem is clearly not a need for more natural science (although in many areas it would be helpful), but rather a need for better understanding of human behaviours and how they can be altered to direct humanity onto a course towards a sustainable society before it is too late."

13.1 HUMAN NEEDS

Human needs vary with respect to urgency. Motives for learning are built as a 'hierarchy of needs', which is described in terms of varying degree of urgency. In order of urgency, first are physiological and physical needs for safety and survival. Second are basic psychological needs for security. Third are needs for care and affection or 'unconditional love'. Fourth are basic psychological needs for self-respect and self-esteem. Fifth are needs for spiritual growth and development of social brain or 'socialization', or 'metaneeds'. Motivation provides an individual with a sense of direction and energy to carry out a particular task. Various needs are interrelated. Each individual is instinctively motivated to relate to other individuals, and to acquire things for survival, to 'socialize', to 'assimilate' work, and for defense, which are intrinsic. Degree of an individual's mental health is determined by balance of 'healthy' and 'unhealthy' psychological components. Mental health determines individual's level of awareness and perception of realities of changing social environment.

13.2 PURPOSE OF LIFE

This is cognitive awareness for achieving a goal, which guides decision making. Decision making is choosing appropriate actions, within a number of strategies. Purpose serves to achieve better set of conditions from the previous state. This change is due to motivation. Purpose is central to human life. Helen Keller stated that happiness comes from "fidelity to a worthy purpose", and Ayn Rand wrote that purpose must be one of the 3 ruling values of human life (others are reason and self-esteem). Personal and highest individual purpose of life is pursuit of soul level joy, which follow individual from birth. In most cases, it begins with desire for acceptance. From scientific view, purpose of evolution is progression of genes. Richard Dawkins stated that purpose is that which "grows up in the universe". Man has complex genetic make-up that allows him to choose a purpose. Dalai Lama, states in *The Art of Happiness* that purpose of life is pursuit of happiness. Happy people tend to be altruistic and less egotistic. So, choice of altruistic purpose leads to happiness. On most fundamental level, goal of an autonomous or autopoietic system is survival, that is, maintenance of its essential organization. This goal is built in all living systems by natural selection. Primary goal system has various subsidiary goals, such as keeping warm or finding food, that indirectly contribute to survival.

13.3 HAPPINESS

According to Maslow, relationship, acceptance and sexual intimacy is fundamental to meeting human needs and conducive in building happiness. Dalai Lama views sexual intimacy, as not necessarily conducive to happiness and fulfillment, but serves to provide temporary gratification and desire for a committed bond. Happiness is a direct sign for fitness. Meaning of life is to "learn and develop", "actualize our potentialities", "improve the balance of pleasure and pain", "enjoy ourselves" or "simply be happy", "love and be loved", or "promote cooperation and togetherness". Fitness for individuals require fitness for group and cooperation and "love", rather than selfishness and hostility." "Increasing fitness" is "implicit goal" governing evolution. Society strives for "the greatest happiness for the greatest number". Creating such happiness remains a question. Natural selection shapes our needs and emotions. Value of fitness is actually built in genes.

Organisms, whose genes do not maximize fitness, may be eliminated. We feel good in food, safety, curiosity, love, and social support. Happiness is an overall sign of satisfaction of basic evolutionary needs and can be viewed as a strong indicator of fitness. Utilitarian value of "the greatest happiness for the greatest number" is largely synonymous with evolutionary value of maximizing fitness of humanity. Main difference lies in time-scale. If it is possible to make every-body perfectly happy at this moment unsustainably, in longer term, evolutionary values would dictate for rejecting this state of affairs. Achieving self-actualization is accompanied by personality characteristics, such as openness to experience, empathy and tolerance towards others, creativity, high self-confidence and self-esteem. On individual level, differences in happiness between people living in the same society depend on their situation and personal characteristics. Happy people are much less likely to fall ill than unhappy people. This directly reflects strong correlation between happiness and biological fitness.

13.3.1 Psychological Characteristics of Happiness

Happy people are able to control their situation. Unhappy people believe that they are toy of fate. This reflects subjective competence. Happy people are more psychologically resilient, assertive, empathetic, and open to experience. These features accompany the perceived competence to satisfy needs. Social position happiness is common among those that have intimate ties and participate in various organizations. This reflects the degree to which people manage to satisfy their social needs, and get better control of their own situation by relying on support of others. Occupationally, happiness tends to be more common among professionals and managers, who are in control of work; they do, rather than subservient to their bosses. Life-events happiness is correlated with presence of favourable events and absence of troubles. These events on their own signal success or failure to reach one's goals, and therefore, control one's happiness. These observations cannot prove that perceived competence to satisfy needs is necessary and sufficient for happiness. These confirm basic tenets of evolutionary-cybernetic theory of happiness, and clarify how happiness can be promoted in practice by promotion of wealth, education, freedom, equality, health, personal control, self-actualization, and intimate relations.

"Pleasure consists in serenity of mind and absence of fear and are obtained only by those who has prudence and foresight and are ready to reject immediate gratification for the sake of permanent and tranquil satisfaction." Spencer pointed out parallel between biological function of pleasure and social evolution of man, and proposes that "remolding of human nature into fitness for requirements of social life must eventually make all the needful activities pleasurable, while it makes displeasurable activities at variance with these requirements." Happiness is man's greatest achievement. It is response of his total personality to a productive orientation towards himself and the world outside."

13.3.2 Physiology of Happiness

Feelings of pleasure and well-being are highly predictive of future good health, and originate in limbic system, which is involved in many homeostatic feedback loops. This is a kind of feedback from limbic system. Each person develops a unique kind of partnership based on a self-concept which is learned in the childhood between their conditioned behaviour and healthy or unhealthy self-control. Some develop a habitual pattern of always ignoring feelings, and curbing natural behaviour. Their life is driven entirely by obligations and duties that take priority over their natural feelings. The people with low pleasure and well-being index scores habitually ignore important emotional feedback from the limbic system. With this crucial feedback blocked, behaviour tends to drift into increasingly unhealthy and destructive patterns. Ability to feel pleasure and well-being is strongly influenced by mental habits and attitudes which are learned in childhood and continues to develop throughout the life. Enjoying life is a habit which is developed or lost through use/disuse. People attend cultural events for pleasure. Placebo effect proves that mental states directly affect health, but understanding of exact mechanisms is still at an early state.

Beta-endorphins are opioids, produced by Hypothalamic-Pituitary-Adrenocortical (HPA) axis, which reward instinctive behaviour, like sexual satisfaction by producing feelings of euphoria. It is the basis of "runner's high" and also of good feelings after meditation. Beta-endorphins increase Natural Killer (NK) cell activity by 30%, and increase maximal effector cell recycling capacity by 170%. Naloxone inhibits this increase. Since NK cells are important in defense against cancer, viruses and bacteria; this could be an important link between good health and feelings of pleasure and well-being. The decreased NK cell activity related to psychological stress is also known. Effectiveness of immune defense varies, as primitive emotional responses range between despair and ecstasy. Strong correlation between positive emotions and health, in no way conflicts with present understanding of disease. Bacteria, viruses, carcinogens, and external stresses are still actual causes of diseases. Mental factors affect health only by altering effectiveness of body's natural defenses against these insults.

13.3.3 Evolutionary-Cybernetic Theory of Happiness

Basic value is fitness, the capacity to survive and reproduce. Cybernetics add fitness is in first place achieved through control, that is, capacity to counteract deviations from goal state in which system optimally survive. Such deviations are lack of nutrients, high or low temperature, and damage to organism. When organisms deviate much from the goal, it cannot survive. Different variables, defining optimal state, are viewed as intrinsic needs. The better the control an organism has over its situation, more perturbations it can survive, and thus, higher its fitness. Control not only takes into account the present situation, but its likely evolution occurs by anticipating further deviations. Anticipation requires knowledge of cause and effect relations, and therefore, control is basis for cognition. Momentary happiness is defined as pleasant feeling or subjective experience of well-being. Long-term happiness corresponds to preponderance of pleasant feelings over a prolonged period. Biologically, feelings function to orient an organism away from dangerous situations, and towards positive situations. Positive feelings indicate that organism is approaching the optimal state. Happiness is viewed as an indication that a person is biologically fit, and cognitively in control and can satisfy all basic needs, in spite of perturbations from the environment. Such control,

Anatomy of Happiness

FATIK BARAN MANDAL

Genes and environment contribute to happiness. Neurotransmitters pass information from one nerve cell to another via synapse. Free-floating neurotransmitters create our happiness or misery. The theory of "central affect program" can explain happiness. Relaxation and meditation are helpful in increasing positive emotions, and controlling negative ones. Individual propensity for happiness depends on genetic heritage. Both innate temperament and negative early experiences in life are extremely influential in governing happiness. A healthy person returns to previous levels of happiness by a homeostatic mechanism that maintains the normal level of happiness. Evidence suggests that the brain is hardwired for happiness via goal-seeking behaviour.

Genes and environment contribute to happiness. Genes carry the messages for production of neurotransmitters, their receptors; also regulate reabsorption portals along with the storage and release of neurotransmitters. In short, genes determine the prevalence, scarcity, and activity of neurotransmitters like serotonin and dopamine. Dopamine is associated with positive emotions, activation of the reward system, and sets in motion the neural circuits for motivation. The gene for controlling the D4 dopamine receptor, which in turn regulates the amount of dopamine binding, is reported. In fact, a specific connection between gene(s) and people's levels of happiness has been established. Howeever, we are perhaps built not to be happy ones, rather to pursue happiness. Attainment of Darwinian goals (sex, status, and so on) makes one happy at least for a while. The absence of happiness forces us for pursuing it, and to be productive. Surveys on 1.1 million people shows that most people are pretty happy. Although happiness fluctuates within a set point, it is concluded that broad heritability of happiness ranges between 40% and 50%.

Of more than 300 recorded neurotransmitters, dopamine governs pleasure, while reduction of serotonin leads to misery. Neurotransmitters pass information from one nerve cell to another via synapse. The bulbous end of nerve cell releases neurotransmitter, when an electrical impulse moving along the nerve reaches it. Then, they cross the synapse to dock at the other nerve cell's receptor, and prompt or inhibit the impulses along the second cell. The first nerve cell reabsorbs excess neurotransmitters although not necessarily the total amount of them. The remaining free-floating neurotransmitters create our happiness or misery.

The theory of "central affect program" can explain happiness. In the 1950s, pleasure centers in the brain were discovered. Nucleus accumbens (NA) was found to be the most active in processing motivational input. Stimulation of NA elicits smiling, laughter, happiness, and even euphoria. The existence of a "reward system" in the brain of many mammals including human is known. The cortex of brain processes a sensory stimulus, and when it is a reward, it sends the signal to the ventral tegmental area (VTA) of the midbrain. VTA then releases dopamine into NA, septum, amygdala, and prefrontal cortex, which is connected through medial forebrain bundle (MFB). In the 1970s, endorphin, a kind of internal morphine, is discovered. Receptors for endorphins called opiate receptors are found in several parts of the brain. When released by the pituitary and neurons in the hypothalamus, endorphins suppress pain. Pleasurable feelings involving laughing, touching, meditating, singing, and listening to good music are partially attributed to endorphins. Endorphins increase the dopamine level.

Desire and pleasure are two distinct processes with interrelated neurochemical systems in the brain. Opiate receptors and endorphins are the key to the effect of emotions. Happiness has a functional pattern distinctly different from sadness. Happiness activates the right posterior cingular gyrus, left insula, and right secondary sensor motor cortex. Sadness decreases activation in these regions. Structures like pons become activated in sadness but not in happiness. Voluntary control of negative emotions resists sadness and depression. Sadness causes changes in right ventrolateral prefrontal cortex, anterior temporopolar cortex, affective division of the cingular cortex, and the insula. By contrary, the voluntary suppression of sadness activates the right dorsolateral prefrontal cortex and right orbit frontal cortex.

Ventrolateral prefrontal cortex is the part of the circuits, which process information from the body, when activated by emotion. Dorsolateral prefrontal cortex suppresses sadness, and is involved in the willed suppression of positive emotions. General set point of happiness can be modified downward by depression, or upward by meditation. A healthy person returns to previous levels of happiness by a homeostatic mechanism that maintains the normal level of happiness.

Positive and negative feelings in human have been shaped by evolution. We share basic emotions and neural substances with other mammals. Happiness exists in the form of hardwired circuits; otherwise, it probably could not be selected during evolution. In a study conducted by Lykken on identical twins separated at birth, part of the happiness was attributed to genetic factors. Lykken argues that "the laws governing happiness were designed not for our psychological well-being but for our genes' long-term survival prospects." Happiness involves contentment and well-being with positive feelings such as good humor, joy, laughter, hope, enthusiasm, and without negative feelings like sadness, worry, anxiety, anger, irritability, and despair. Happy people appear to be healthier, and live longer. Relaxation and meditation are helpful in increasing positive emotions, and controlling negative ones. Meditation accompanied by a clear increase of activity in the left prefrontal cortex is known to be related to positive emotions. Equally effective hobby for many is achieving pleasurable sensations through music, color, and touch.

Individual propensity for happiness depends on our genetic heritage. Both innate tempera-ment and negative early experiences in life are extremely influential in governing happiness.

Happiness acts as a good motivator, an internal reinforcer which is important for our survival. Evidence suggests that the brain is hardwired for happiness via goal-seeking behaviour. For human beings, happiness is a mixture of nature and nurture which is largely influenced by cultural experiences and learning.

Source: http://expertscolumn.com/content/anatomy-happiness

over one's situation, has three components—material competence, cognitive competence, and subjective competence. Problem of promoting happiness, then simply reduces to promoting material-, cognitive-, and subjective-competence.

13.3.4 Empirical Confirmation of Theory

Veenhoven has created an extensive world database of happiness by collecting results of many studies, in which people were asked about their happiness with their life. He studied main factors that correlate with resulting happiness scores and concluded that happiness is not dependent on a purely subjective outlook, rather accurately predicted on the basis of objective "liveability" of society, and on the basis of one personal profile. Characteristics of societies, where people are happy is wealth. However, correlation between purchasing power and happiness becomes less important for more wealthy societies, implying that once basic material needs of nutrition and shelter are satisfied, further prosperity adds little to happiness.

13.4 NERVOUS SYSTEM AND MIND

Human behaviour is the collection of activities, performed by human beings, under the influence of culture, attitudes, emotions, values, ethics, authority, rapport, persuasion, and/or coercion. Many factors affect human behaviour. A person, usually is a part of a society, which consists of community. A number of groups form a community. Community, in which a person develops, plays a significant role, in his/her behaviour. Until 50,000 years ago, modern *Homo sapiens* are reported to be existed only in Africa, and continued to live as a hunter-gatherer until about 10,000 years ago. Agriculture suddenly began and rapidly changed prosperity of humans. Certain genes, related to brain size, changed between 5800 and 37,000. Human brain size has increased 3 times, in last 2 million years with more connections and less nerve cells. About 3 million years ago, *Australopithecus afarensis,* had a brain size of about 400 cc, while modern human brain size is about 1400 cc. Large brain is costly, and 18% of energy expenditure is consumed by the brain. This suggests the significant advantages to possess a larger brain, which outweighed the expense.

Accumulated knowledge played a crucial role in allowing our ancestors, to develop a rich and varied diet, which required language and memory. The prolonged retention of juvenile features allowed development of a larger brain as a secondary effect of a longer growth period. Machiavellian hypothesis proposes that main evolutionary pressure for increased intelligence was competition with other people. Human evolution is based on highly developed communicative ability which associated with increasing brain size. Need to guess the likely actions of other individuals require capability to imagine others' minds. Theory of mind was a key factor in development of self-consciousness. Some argue sexual selection as driving force behind explosion in our brain size.

The human brain perceives external world through senses and individual is influenced greatly by experiences leading to subjective views of existence, and passage of time. Humans possess consciousness, self-awareness, and a mind. These correspond to mental processes of thought. Extent to which mind experiences outer world is a matter of debate. Dennett argues that there is nothing as "mind". But there is a collection of sensory inputs and outputs. Skinner has argued that mind is an explanatory fiction that diverts attention from environmental causes of behaviour. What are commonly seen as mental processes may be better conceived as forms of covert verbal behaviour. Human brain contains neurons. On average, each neuron is connected to other neurons through about 10,000 synapses. Actual figures vary greatly. Network of neurons forms a massively parallel information processing system. Being built with very slow hardware, brain has remarkable capabilities. Partial recovery from brain damage is possible, if healthy units can learn to take over functions previously carried out by damaged areas. It performs massively, parallel computations extremely efficiently, and supports our intelligence and self-awareness.

Neural networks attempt to bring computers a little closer to brain's capabilities by imitating certain aspects of information processing in brain. Brain is not homogeneous. Cortex, mid-brain, brainstem, and cerebellum are distinguished. Each being subdivided into many regions and areas, according to anatomical structure of neural networks or functions performed by them. Overall pattern of projections of neural connections between areas is extremely complex and only partially known. A neuron has dendrites, cell body and axon. A neuron receives input from other neurons. Once input exceeds a critical level, neuron discharges a spike—an electrical pulse that travels from the body, down the axon, to next neuron(s). This spiking event is called depolarization which is followed by a refractory period during which neuron is unable to fire. Axon endings almost touch dendrites, or cell body of next neuron. Transmission of electrical signal from one neuron to next is effected by neuro-transmitters released from first neuron and bind to receptors in second. This link is called a **synapse**. Extent to which signal from one neuron is passed on to the next, depends on many factors, e.g. amount of neurotransmitter available, number and arrangement of receptors, amount of neurotransmitter reabsorbed, etc.

Motivation is based on emotions specifically, on search for satisfaction, and avoidance of conflict. Positive and negative motivation is defined by individual brain state, which is influenced by social norms. Emotional experiences, such as love, admiration, joy are perceived as pleasant in contrast with those perceived as unpleasant, like hate, envy, or sorrow. A person may be driven to self-injury, when his brain is conditioned to create a positive response to such action. Motivation is involved in performance of learned responses. Conflict avoidance and libido are primary motivators. Refined emotions are socially learned, and survival oriented emotions are innate. Emotions commonly develop in reaction to superior survival mechanisms and intelligent interaction with each other and the environment.

13.4.1 Synaptic Learning

Brain learns by altering strengths of connections between neurons, and by adding or deleting connections between neurons. It learns "online" based on experience. Efficacy of a synapse change as a result of experience and provides both memory and learning through long-term potentiation. This happens through release of more neurotransmittes. An enduring (>1 hour) increase in synaptic efficacy results from high-frequency stimulation of an afferent (input) pathway. "When an axon of cell A excites cell B and repeatedly or persistently takes part in firing it, some growth process or metabolic change takes place, in one or both cells, so that A's efficiency as one of the cells firing B is increased", as postulated by Hebbs.

13.4.2 Nature and Origin of Mind

The nature and origin of consciousness and mind are debated. Multiple drafts model holds that consciousness can be explained by neuroscience, through the workings of brain and neurons. Electromagnetic field generated by brain may be actual carrier of conscious experience. There is disagreement about implementations of such theory relating to other works of mind. Dennett emphasizes 3 key factors in evolution of consciousness-tool use, ability to predict future, and ability to take intentional stance. Tool uses require intelligence to recognize, use and maintain a tool, and tools confer intelligence to organism, which possess them. Most important tool of human use is the mind-language tool. Dennett says "the process of evolution by natural selection has no foresight at all, but has gradually built beings with foresight. A mind is fundamentally an anticipator, an expectation generator. It mines the present for clues, turning them into anticipations of future. And then, it acts rationally on those hardwon applications." Ability to take intentional stance is a key attribute of human. Ability to take intentional stance towards others and towards self provides an example of clever behaviour that can occur in presence of cleaver thoughts.

Dennett argues that journey towards conscious-ness involves step taken between first and second order intentional systems. A first order intentional system is one that has beliefs and desires about something, but not about beliefs and desires themselves. A second order system is one that has both. He proposes that our ability to take intentional stance towards others may have come before our ability to apply it to ourselves and they could be prerequisite for self-consciousness. A novel approach was employed to identify specific genes that may influence emotional stability, using mouse as a model system, and defined emotionality by covariance of a set of four measures. Using these measures, three candidate regions that influence emotionality were identified. Evidence suggests that genetic basis of emotionality in mice is similar to that in other species and it may underlie the psychological trait of emotional instability in humans. Discovery of QTL in mouse would provide molecular characterization and may lead to identification of genes influencing human emotional stability. The improvement of analytical methods, refinement of genetic and physical maps and results from QTL analyses in model systems have led to possibility of dissecting genetic component of complex traits.

13.4.3 The Rule of 150

Gladwell stresses importance to groups either smaller or bigger than 150 persons. Adding just a few members around this threshold can make a big difference. This principle of group size has important social policy implications (Ridley 1996). Belief in the potential magic number 150 stems from research by Dunbar in his Harvard University Press book, *Grooming, Gossip, and the Evolution of Language*. Brain can be divided into 3 parts. Neocortex is mostly associated with primates' "smartness" (Dunbar 1996). Dunbar found a strong relationship between the species' neocortex size and the size of groups in which the species lived and suggested that 150 persons was somehow an optimal size for human groups and that groups larger than this were somehow too unwieldy to sustain as effectively. This group size corresponds approximately the maximum level of social complexity that human minds have evolved to be able to handle. Such groupings "appear to be the largest grouping in which everyone knows everyone else, in which they know not simply who is who, but also how each one is related to the others". Dunbar discovered that 150 is the typical clan size. Dunbar's figure is challenged by Jarvenpa (1993) who criticizes him for relying on old sources.

13.5 EMOTION

Most neuroscientists recognize six basic emotions—anger, disgust, fear, joy, sadness, and surprise. Facial expressions associated with these emotions are universal and almost certainly plumbed in genetically. LeDoux studied fear, an emotion that mankind shares with other species. Wager's map shows that many emotional pathways converge on structures called **amygdalas**; the part of limbic system. LeDoux demonstrated their importance through experiments carried out initially on rats. He confirmed that amygdalas are most active part of brain, when subject is afraid. He also produced fear by stimulating neurons of amygdalas with electricity. Subsequent work suggests that amygdalas have same role in people. Lose parts of them, as happens sometimes as a result of disease or surgery, causes inability to experience or recognize fear. The amygdalas orchestrate fear. They do so in role of conductors, as much as players. Certainly this emotional orchestra cannot play without conductor, but other instrument's functions are shown in Wager's map. The amygdalas also conduct other emotions, like anger, sadness, and disgust. Joy involves hypothalamus.

The common factor in criminal members of a family turned out to be absence (due to a faulty gene) of an enzyme called **monoamine oxidase A**. This enzyme regulates a group of neuro-transmitters including serotonin, and dopamine. Serotonin- and dopamine-based neurons are important for emotional responses. Finding about monoamine oxidase A was widely reported as discovery of "a gene for violence". Violence is an expression of anger.

13.5.1 Understanding Emotional Intelligence

Emotional intelligence is a multidimensional ability of a person. In simple terms, this is "a positive and proactive attitude towards all aspects of life" or "the ability to get along with people and situation" which reflects one's ability to deal with daily environmental challenges and helps predict one's success in life, including professional and personal pursuits. It refers to a person's maturity in dealing with the unavoidable ups and downs be either external or internal of life. External changes could be due to increased competition that one encounters now-a-days at every stage. Internally, these may be due to failures or disappointments. Emotions play a very significant part in our decision making process. Three essential or basic components of EQ are:

(a) Motivating oneself
(b) Motivating others
(c) Empathizing with others

Motivating oneself involves our own feelings and thoughts that help us to remain in control and also inspire ourselves. Motivating others and empathizing with others pertain to interpersonal skills. One needs to match other people's wave-length to ensure best results while dealing with complex situation or handling personal relations. EQ pertains to our ability or skill to observe, communicate, empathies, counsel, understand and cope with changing circumstances. Human mind has two important parts. We operate from the rational mind as well as the primitive mind, which is mainly the emotional mind. While the rational mind enables us to care, love and take moral decisions, the emotional mind is the source of sorrow, anger, shock, fear and disappointments. A healthy mix of emotional and rational mind mainly constitutes emotional intelligence which now emerges as the key to success in life.

Emotional development of an individual is associated with his/her development in the early years of life. Improper emotional development may adversely affect subsequent behaviour. Emotional intelligence makes us capable of having a good insight into the behaviour of others as well. This helps us immensely in adapting to new situations and also interacting with all kind of peoples in our society.

13.5.2 Significance of Emotional Intelligence

The failure to achieve even our self-defined goals, such as getting rid of a bad habit results in considerable disappointment leading to disgust and manifestation of undesirable behaviour. EQ encourages a person to aim high and ultimately become more result oriented in life with discernible emphasis on human relations. EQ makes an individual to manage him as well as others in a meaningful and proactive manner to achieve higher goals.

Emotional intelligence also plays a prominent role in group effectiveness of an individual. In the present day context when almost all jobs invariably entail monumental team effort, with an ever-increasing time and resources crunch, a person must be competent enough to motivate and influences his team members adequately to optimize results. A person must be equipped with sufficient emotional intelligence to become an effective motivator for himself as well as for others around him.

EQ has tremendous potential to contribute in:

- Controlling your negative thoughts
- Becoming optimistic
- Developing trust in others
- Inspiring yourself and others around you
- Cultivating an empathetic attitude towards others
- Controlling anger
- Decision making under complex situations
- Managing stress effectively
- Building and maintaining a cohesive team, etc.

13.6 SENSE OF SMELL

Importance of human sense of smell is largely underestimated. Humans and other primates are believed to be microsmatic, equipped with highly developed powers of vision that supposedly make humans "visual creatures." This concept needs reconsideration, as olfaction plays a very important role in human reproductive biology. Nasal mucosa is functionally divided into two areas—respiratory and olfactory regions. In nose, olfactory region is found on both sides of nasal septum and in upper nasal conchae. Ability to discern between different odours suggests that specific receptors exist in sensory cells. Excitation of axons from sensory cells occurs, when an odour molecule "docks" with a receptor protein in membrane of olfactory cilia. That superior cognitive power allows us to better use olfactory input compared with other mammals is suspected. Axons of sensory cells enter olfactory bulb. Sensory input is then projected via olfactory tract into olfactory lobe. From there, olfactory input is projected via thalamus to neocortex, and to limbic system. This pathway allows olfactory stimuli and to be consciously detected interpreted and olfactory stimuli to directly influence the neuroendocrinology of emotions.

'Affective primacy hypothesis' asserts that positive and negative affective reactions can be evoked with minimal stimulus input and virtually no cognitive processing. Olfactory signals seem to induce emotional reactions, whether or not a chemical stimulus is consciously perceived. Chemical cues allow humans to select for and to mate for. The traits of reproductive fitness cannot be assessed simply from visual cues. Ontogenetic link between olfaction and hormones become evident in patients suffering from X-linked Kallmann's syndrome. They show under-developed gonads, lacking secondary sexual characteristics, and both male and female patients are anosmic, i.e. unable to detect odours. This syndrome results from underdevelopment of olfactory bulb in the embryo. Gonadotropin Releasing Hormone (GnRH) neurosecretory cells of hypothalamus originate in olfactory placode and migrate into hypothalamus. In Kallmann's syndrome, this migration does not occur, accompanied by underdevelopment of olfactory bulb and minimal, if any, secretion of hypothalamic GnRH.

Preliminary evidence suggests that people with Kallmann's syndrome do not respond to putative human pheromones. Effect of human pheromones on hormones, like GnRH and on the behaviour is conditioned in presence of other sensory input. That mammalian neuroanatomical pathways link vision and olfaction is suggested. Kohl proposed that LH is measurable link between sex and human sense of smell. Kohl detailed reciprocity in olfactory-genetic-neuronal-hormonal-behavioural relationships that appear to link nature and nurture of human sexuality. A more complete overview of nongonadal, non-hormonal, influences on sexual differentiation, and influence of sensory stimuli, especially chemosensory stimuli on human sexuality is on record. The affect of chemosensory stimuli on behaviour was integrated with tactile cues. This is suggested that male pheromones and tactile cues lead to increase in GnRH-immunoreactive (GnRN-ir) cell numbers, which are correlated with LH modulated estradiol levels and with sexual behaviour. This is also recorded that sexual arousal of males as classically conditioned. Sexual interest in males might result from Pavlovian conditioning, and likely to be odour induced. GnRH directed conditioning of LH release may be used to evoke functional changes in mammalian neuroendocrine pathways that mediate release of T and E, with, or without visual awareness of any associated stimuli.

13.6.1 Vomeronasal Organ (VNO)

VNO also termed Jacobson's organ, is a special part of olfactory system(s). Recent data shows that VNO exists in adult humans. This is also recorded that adult human VNO responds to picogram amounts of human skin pheromones with depolarization, suggesting that human VNO may function as a pheromone detector, as in other mammals. However, there is no evidence that human VNO is connected to a functional accessory olfactory system.

13.7 LEARNING

The human ability to think has demonstrated that productivity of basic material resources, such as land, water, and fuel can be multiplied. Various forms of technology are organized, and replicable. Thinking enable us to convert sand into bricks, glass, fiber optic cables, micro-processors, and to convert petroleum into lamp oil, plastics, clothing, and life-saving pharmaceuticals. Humans learn both from experience and from theorizing, by which people develop attitudes, opinions, values, and other mental resources determine how creatively and effectively a society responds to challenges and opportunities. Human mind and spirit appear to be ultimate resources that determine usefulness and productivity of other resources. Systematic knowledge transfer through family and formal education followed by life experience includes what happens, what is dreamed and imagined. This elevates productive, adaptive, and even prophetic responses to needs.

13.7.1 Language

Unlike call systems of other primates that are closed, human language is far more open and gains variety in different situations. Language has quality

Human Language in Eco-region Concept

—FATIK BARAN MANDAL

Modern human is believed to be originated about 160,000 years ago. The simplification of human linguistic diversity is represented by the term, "monoculture of the mind". Out of 7,000 languages, 80–85% is spoken by indigenous peoples.

Modern human is believed to be originated about 160,000 years ago in East Africa. Mega-droughts in East Africa, between 135 and 75 thousand years ago, have caused human migration out of Africa. India civilization, being an ancient one, now includes 532 tribes, 72 primitive tribes and 36 hunters and gatherers. Human language might have originated from gestures rather than vocalizations. The anatomical and neural changes necessary for the production of articulate sounds probably have occurred late in human evolution, with the ability to sustain autonomous speech development only after the emergence of *Homo sapiens*, about 200,000 years ago.

Armstrong and Wilcox support this claim by referring to evidence that the *FOXP2* gene, involved in vocal articulation, has undergone a mutation within the past 100,000 to 200,000 years, a final, crucial step in the evolution of human language. This evolution would have been aided by the increase in brain size in response to the enhanced dependence on cooperation and social communication due to drastic ecological changes in the Pleistocene era.

Local societies, whether indigenous peoples or rural communities, and other small-scale societies, co-evolved with the ecosystems, in which they lived. By nurturing the environments, they nurtured them.

A total of 895 eco-regions have so far been identified, of which the WWF estimates 238 (referred to as the Global 200) have outstanding international importance as about 4675 linguistic groups reside in these eco-regions. Approximately half of all spoken languages are used by communities of 10,000 speakers.

The trend towards the simplification of human linguistic diversity and cultural uniformity has been represented by the term, "monoculture of the mind".

Some 420 languages are known as extinct. UNESCO in 2002 estimated that at least 3,000 languages are endangered in various parts of the world, and up to 90% of the word's languages in the course of 100 years would be extinct. Out of 7,000 languages spoken today, 80–85% is spoken by indigenous peoples. Most languages are spoken by very small communities. Their languages are threatened to replacement by majority languages.

As rapid cultural and linguistic change takes place, the "inextricable link" between people and the environment begins to break down. A number of initiatives to sustain and restore biocultural diversity, for example, Terralingua have gathered information about several of them in the volume *Biocultural Diversity Conservation: A Global Sourcebook* (Earthscan, 2010) (Related posts: Restoring human cultures to the web of life and Talking to the clouds and listening to the trees.)

Further information is available in www.sott.net. www.unescco.org. www.nationalgeographic.com

Source: http://www.digitaljournal.com/print/article/301048

of displacement, using words to represent things, and happenings that are not presently, or locally occurring, but occurring elsewhere, or at a different time. Data networks are important to continue development of language. Speech is a defining feature of humanity, possibly predating phylogenetic separation of modern population. Language is central to communication between humans and central to sense of identity that unites nations, cultures, and ethnic groups. Invention of writing systems, at least 5000 years ago, allowed preservation of language and was a major step in cultural evolution. Science of linguistics describes structure of language and relationship between languages. Approximately 6000 different languages currently in use include sign languages. Many thousands of the languages are considered extinct.

13.8 HELPING

Helping is mostly provided to children (Figure 13.1), parents, spouses, or other close relations, and less likely, when it involves cousins, or unrelated neighbours. Helping decreases with kinship distance, and occurs only, when sacrifice is outweighed by advantage. Haldane told "I'd gladly give my life, for three of my brothers, five of my nephews, nine of my cousins". This is called kin selection, altruism based on genetic selfishness. "Sociopathy," is found in a sizable portion of population. Helping decrease with kinship distance and in fact, it occurs when the sacrifice one makes is out-weighed by advantage and the genes behind the sacrifice are shared with those relations. If we feel empathy towards a person, who needs help, we are likely to help them without any selfish thoughts. Otherwise, we will help them only, if rewards of helping them outweigh costs. Rewards of helping can be many and various including relief from distress of seeing another in trouble. Beggars live totally off empathy and can be expert at putting themselves in situations to increase this, such as using children and animals, to find empathetic people, or create an empathetic situation.

FIGURE 12.1 A human baby.

13.9 AGGRESSION

Some think aggression as a great virtue, while others as a symptomatic mental illness. Aggression serves to enhance self. Its positive version called **assertiveness** act to enhance self without implication that we are hurting someone else. The negative version is called **violence**. The resources are important for "fitness" and relevant to individual reproductive success. Male's aggressiveness is mediated by testosterone. People get aggressive with that happened long ago and that they think will happen in future, or that they have been told is happening. People get angry with that happen to their houses, communities, nations and religious establishments. People get frustrated with interruption of ongoing behaviour, or by a delay of goal achievement by disruption of ordinary behaviour patterns. Behaviours are intelligent, if they are conducive to social cooperation. Injurious behaviour causes pain and beneficial behaviour causes pleasure. Evolution of social intelligence involves pleasurable behaviours because they are conducive to social cooperation. 'Human solidarity' as 'universal brotherhood' or 'love' is the instinctive motivation for peace, and the community form the basis for human civilization.

13.9.1 Aggression in Male

Richard Wrangham presents an interesting analysis of male violence. He argues that chimpanzees and humans are the only species in which groups of males hunt and kill members of their own species. Goodall discusses implications of lethal intraspecific aggression in chimps. Murder is not a unique 'culturally determined' behaviour. Some forms of violence involve an accurate assessment of risk of injury. Warfare is a unique behaviour. In a battle, both sides suffer casualties regardless of who finally wins. Consequently, battles involve a failure to assess true costs of combat by both sides. Wrangham suggests that this failure is due to 'positive illusions' by each set of combatants that they will emerge victorious.

13.9.2 Aggression in Female

Until recently, relatively little attention was focused on female aggression. Campbell (1999) argues that "lower rates of aggression by women reflect not just the absence of masculine risk-taking, but are part of a positive female adaptation, driven by critical importance of mother's survival for her own reproductive success." Campbell reviews evidence that women show greater fear of physical harm, compared to men. Women show more fear of open spaces, dogs, snakes, insects, and rodents, than men .Women are less likely to engage in hazardous sports, dangerous driving, military combat, and drug abuse, than men. Women are more afraid of being victims of crime, involving aggression and are more likely to visit a doctor to seek advice on preventative care, than men. Women commit fewer violent crimes than men. Women show less concern for status.

13.9.3 Neurotransmitters and Aggression

Increased serotonergic activity tends to reduce aggressive behaviour in rodents. Serotonin levels are affected by dominance rank. 5-HT level is higher in dominant, than subordinate male vervet monkeys. Removing dominant male changes dominance hierarchy within remaining animals. New dominant male shows increase in his 5-HT level and aggression.

13.9.4 Aggression and Sexual Selection

Females make a higher parental investment than males. Males compete with each other for access to females. Males use their dominance and resources to deter rivals and attract females. Higher rate of aggression in men shows crucial importance of status to male reproductive success. Women can show aggressive behaviours. Females compete with each other. Men complete for resources that enable them to successfully raise their children. Over 80% of homicides are committed by men. Most victims are also men. Common cause of homicide is due to escalation of a relatively trivial disagreement over status that starts with words and escalates into lethal violence. Men resort to violence to protect or gain status and honour. This sex difference is found across all the cultures. Criminal violence is most likely to occur between ages of 14 and 24. Psychologists argue that boys are trained to be aggressive and girls learn to be passive. Dyson–Hudson (1995) found that: "'low-conflict societies' with affectionate socialization and aversion to interpersonal confrontation have high rates of violent deaths. In contrast, Turkana pastoralists (East Africa) are taught to fight as children and most men having participated in interpersonal fights intended to cause injury, having engaged in recreational within-group fighting, mimicking warfare, and having taken part in raids on neighbouring Pokot. Yet demographic data indicate that within-group homicide rates among the 'violent' Turkana are lower, than those reported for 'low-conflict' societies. It may be that Turkana rules, which require bystander intervention and adjudication by elders are effective in preventing within-group aggression and violence, from escalating into lethal fights."

13.10 FORAGING THEORY

Dietary selectivity is a fundamental area of interest and has strong implication for structuring social relations among hunter-gatherers, group size, mobility patterns and understanding economic dimensions of human evolution. Earliest application of behavioural ecology came with application of Optimal Foraging Theory (OFT) in study of hunting, fishing and gathering (foraging behaviour). OFT allows to develop a large set of fundamental hypotheses that predict, which food resources foragers will pursue when encountered during search (diet breadth) or where foragers will travel to search for resources (patch choice models) and how long they will stay in these places before moving to other areas (marginal value theorem).

Assuming that forager's goal is to maximize their net rate of return while foraging, as by doing so they are ultimately able to maximize their fitness. Rate of return is indexed by a currency, such as calories, or another attribute of food value (Hill 1988). This assumption is reasonable, as food resources are fundamental to survival and reproduction. When environment is poor, humans can be said resource limited. Maximizing one's net rate of return allows foragers to acquire greatest amount of food. When environment is relatively rich, humans may be time limited so that efficient acquisition of food resources leaves foragers with more time to engage in alternative fitness enhancing activities, such as child care (Winterhalder and Smith 2000).

A number of qualitative and quantitative tests of the optimal foraging models applied to problem of diet breadth exists (Hill and Kaplan 1992). Qualitative test predict directional tendencies in prey choice. This will expand, or contract, depending on encounter rates of highly profitable species (Winterhalder and Smith 2000). Quantitative rates predict the precise number of species, a forager will pursue under a variety of environmental conditions, which species will enter, or drop out of optimal diet breadth and exact rank-ordering of species. Diet breath generally shows a division of labour between men and women and children and adults (Hill and Kaplan 1992). Although OFT was designed to account for foraging behaviour in animals and later extended to human foraging, recent applications have been made in areas of origins of agricultural and pastoral food prediction with considerable success (Winterhalder and Smith 2000).

13.10.1 Group and Resource Transfer

Movement of goods and to some extent services between families and individuals have received considerable attention. Traditionally, this is subsumed under terms of exchange or reciprocity. Behavioural ecologists term such activities as resource transfer as it neutrally characterizes resource movement between groups and/or individuals, compared to traditional terms like exchange or reciprocity. In analysis of transfers, one assumes giving a resource that one has acquired will be done, such that acquirer of resource gains something in return, or avoids a cost in defending that resource. As a result, models of transfer assume that benefits of giving are greater than costs of monopolizing resource. Model of transfer developed, thus, include scrounging, risk minimization, trade or exchange, and show-off hypothesis.

At one extreme, coercive models of scrounging assume that resource acquired cannot be easily or effectively defended against those who lack resources and that benefit of giving, or yielding a portion of resource presents a higher payoff than costs of defending it against another, who are in greater need, and thus, more willing to contest the resource. Models of sharing which assume that individuals give resources to others, contingent upon an expectation that something of value will be returned to the giver in future. When resources cannot be predictably acquired, risk minimization or pooling strategies are employed through mechanism of reciprocal altruism. Individuals, who acquire resources, share with unlucky individuals with expectation of return in future.

13.11 ADOLESCENCE

Adolescence is a transitional stage of development between childhood and adulthood during which an individual experiences a variety of biological and emotional changes. Ages of adolescence vary by culture and ranges from preteens to 19 years. According to WHO, adolescence covers period between 10 and 20 years of age. Adolescence is often divided into 3 phases—early, mid, and late adolescence. This is a specifically turbulent and dynamic period, in which people develop abstract thinking abilities, become more aware of sexuality, develop a clearer sense of psychological identity, and increase their independence from parents. Hall denoted this period as one of "Storm and Stress".

Conflict is normal, and Margaret attributed behaviour of adolescents to their culture and upbringing. Freud saw it as "genital phase" of psychosexual development, where child recaptures sexual awareness of infancy. Piaget focused on cognitive development, seeing adolescence as "formal operative stage", where young person draw conclusions from information available. Erikson's theory of psychosocial development identified identity crisis as central to adolescence. Adolescents report that they are far happier, spending time with similarly aged peers, as compared to adults. Conflict between adolescents and their parents increase at this time from sense of independence. Peer pressure is very prevalent. Males tend to exhibit less interest in infants and are prone to recklessness and risk-taking behaviours, which can lead to substance abuse, car accidents, unsafe sex, and crime. Risk-taking is biologically driven, caused by social and emotional part of brain, the amygdala, and develop faster than frontal cortex. Most adolescents are psychologically healthy.

13.12 GESTURE

Gestures, the movement of arms and hands, are different from other body language. They tend to have a far greater association with speech and language. While rest of the body indicates general emotional state, gestures have specific linguistic content. Gestures have three phases— preparation, stroke, and retraction. Real message is in stroke, preparation, and retraction elements consist of moving arms to and from rest position, to and from start, and end of stroke. Emblems are specific gestures with specific meaning that are consciously used, and understood. They are used as substitutes for words and are close to sign language than everyday body language. Holding up hand with all fingers closed in, except index and second finger which are spread apart, can mean 'V for victory', or 'peace'.

The iconic gestures are closely related to speech, illustrating what is being said. Iconic gestures are used to show physical, concrete items, and are useful as they add detail to mental image that the person is trying to convey. Timing of iconic gestures in synchronization with speech show, whether they are unconscious or are being deliberately added for conscious effect. In an unconscious usage, preparation for gesture start before words are said, while in conscious usage, there is a small lag between words and gesture. When using metaphoric gestures, a concept is being explained. Gestures can also be used to display emotion from tightening of a fist to many forms of self-touching and holding self. Holding hands or whole body indicate anxiety as person literally holds themselves. Self-preening shows a desire to be liked and indicate desire of another. Beat gestures are rhythmic beating of a finger, hand or arm. They can be as short as a single beat, or as long as needed to make a particular point.

13.13 BEAUTY AND MATE CHOICE

Skin's overall homogeneity and colour saturation received little attention by scientists. Skin tone and luminosity gives visual cues about a person's health and reproductive capability and may be a major signal for mate selection, attractiveness, and perceived ages. When one thinks of women, irrespective of their nationality, culture, religion, class, and social situation, beauty comes in mind. A desire to be sexually attractive causes us to aesthetically alter how we appear. Sex and beauty go hand in hand. Women with hourglass figure are more fertile due to higher level of female hormones, a condition that may subconsciously condition male's mate choice. Secret of beauty and attractiveness has been a quest of human. In fact, the human brain has a special part called the fusiform, located in the back of head near spine. It is needed to recognize faces of family, friends, and people. When it is damaged, the patients can not recognize anyone, even people they have just met. Facial symmetry is one of the best indicators of good genes and healthy development. Fluctuating Asymmetry (FA) indicates the presence of genetic disturbances. FA increases with exposure to environmental perturbations during development, indicates a genetic weakness, and less than optimum health. Bilateral symmetry is equated with heterozygosity, and resistance to infection. Women, for marriage, usually react to a man with a wide smile, small eyes, a big nose, and a large jaw which indicate a strong testosterone level, a potentially good provider and protector of family life. Attractive female face is characterized by two measurements, i.e. the distance from eyes to chin, which is stronger and size of lips which is fatter.

Men's physical traits commonly associated with masculinity are greater height and broader shoulders. Women's feminine traits include larger hips, longer and more slender necks. Female features most positively correlated with attractiveness are the neonate features of large eyes, small nose, and small chin. Maturity features are prominent cheekbone and narrow cheeks, while the expressive features are high eyebrows, large pupils, and a large smile. Maturity features may have social structural implications, as these tend to be non-sexed typed and have greater status and cross-cultural significance (Cunningham 1986). Women are attracted to an optimal combination of neonatous, mature and expressive facial features (Cunningham et al., 1990). Such studies can be linked to power, status, maturity and sociability. Rhodes (2006) demonstrated the relationship of symmetry, averageness, and sexual dimorphism to facial attractiveness. Each of these characteristics may be considered ideally attractive as they are related to judgments of physical health, also known as the "good genes" approach. The "good-genes" approach is based on the idea that the attractive features have evolved to represent freedom from parasites and infectious diseases. Kalick et al., (1998) showed that people are blinded by beauty.

13.13.1 Facial Signals

When we communicate with others, we look mostly at their face. This is not a coincidence, as many signals are sent with the 90-odd muscles in the face. The way the head tilts also changes the message. Eyes are particularly important, and when communicating, we first seek to make eye contact, then break and reestablish contact many times during discussion. Eyebrows and forehead also add significant signals from surprise to fear mouth, when not talking can be pursed down turned or turned up in a smile. Facial symmetry may be a marker for health associated with 'good-genes'. During development, the body is subject to stressors that cause deviations from symmetry. Some individuals are able to withstand this stress and show less asymmetry. Humans have evolved a preference to mate with people with symmetrical facial features. People with asymmetric features are judged as less healthy and less attractive as partners (Buss 1999). They may become sick or die, and therefore, not provide resources for their children. They may carry diseases, which pass on to their children.

Although prototypical facial expressions (Figure 13.2) reliably signal the so-called basic emotions, such as fear, or happiness, human viewers are surprisingly adept at making reliable judgments from impoverished stimuli, such as faint changes in facial expression or a few seconds of full-body interpersonal interactions. Not only we are exceedingly sensitive to social signal themselves, but we are also sensitive to details of context in which they occur. The social judgment that we make from faces is attractiveness, which can be manipulated by specific properties of faces. For instance, faces are perceived to look more attractive, the more average, or symmetrical they are, or with greater exaggeration of robusticity and neoteny features, all of which have proposed to signal differential fitness. Moreover, such preferences by women can vary across different phases of the menstrual cycle (Adolphs 2003).

FIGURE 12.2 A unique facial expression.

13.13.2 Thinness and Fatness

Female body fats have great evolutionary significance as it has historically served several fitness enhancing functions, including insulation, storage of calories, and fertility regulation (Anderson et al., 1992). However, preferential attitudes regarding female fatness is not static, rather vary with food supply. Specifically, men with scarce resources tend to prefer heavier women, whereas men with abundant resources prefer thinner women. Ugandan participants assign higher attractiveness ratings to more obese male and female than do British participants (Furnham and Baguma 1994). A direct relationship between socioeconomic status and female obesity rate in developing societies and an inverse relationship in developed nations is recorded (Sobal and Stunkard 1989). Resource availability may thus be a driving force in determining mate preference (Tovee et al., 2007). Nelson and Morrison (2005) used financial satisfaction and hunger as variation of resource scarcity and found evidence that men who feel either poor, or hungry prefer a potential female partner who is heavier compared to man who feel rich or satiated. Importantly, female hunger state and economic situation did not influence potential male partner weight preferences.

13.13.3 Waist-to-Hip Ratio

Singh has advocated an evolved male preference for a waist-to-hip ratio (WHR) of .70. The WHR of healthy, premenopausal women in developed societies typically ranges from .67 to .80 (Singh 1993a). Men prefer the attractiveness of female figures that varied in both WHR and total amount of fat. Regardless of the total amount of fat, men find women with low WHRs the most attractive. Women with a .70 WHR are seen as more attractive than women with .80, who in turn are more attractive than women with .90. Low WHR preference is known to be one of the best documented instances of male adaptation in mate preference (Singh 1993b). However, "every culture tested so far has been exposed to the potentially confounding effects of western media (Yu and Shepard 1998). Yomybato rated women with a .90 WHR more attractive than women with a .70 WHR. Shipetiari rated women with a .70 WHR as attractive, but their preference for low WHR women was weaker than a sample of men from the United States. Wetsman and Marlowe (1999) examined male WHR preferences in a foraging Hadza society who live in a partly savanna environment in Tanzania. They found that Hadza strongly preferred heavier women over thinner women, and they did not systematically prefer either a .70 or .90 WHR suggesting that mate preference could be sensitive to food scarcity.

According to Westman and Marlowe "WHR may only become relevant when food resources are plentiful, enough that the risk of starvation during pregnancy and lactation for women is minimal". "A foraging lifestyle means that surplus food is rarely available" and that foraging societies are "the closest link to societies in which WHR preferences evolved". A Low WHR woman has advantages for health and fertility, which makes her a lucrative mate from Darwinian standpoint than a high WHR woman. Relationship between WHR and reproductive potential can be inferred from findings that body-weight-matched girls with lower WHR exhibit earlier pubertal endocrine activity, as measured by high levels of lutenizing hormone, follicle-stimulating hormone and steroid activity.

Direct relationship between WHR and fertility has been reported recently. Married women with higher WHR and lower Body Mass Index (BMI) reported difficulty in becoming pregnant and have their first live birth at a later age than married women with lower WHR. Difference in WHR between women who did and did not have difficulty in getting pregnant was 0.838 versus 0.840 (Kaye et al., 1990). Difference in WHR between women who had their first live birth before 30 versus those having their first birth after 30 was 0.838 versus 0.843. Obesity is typically presumed to have been rare in our evolutionary past, especially among reproductive-aged women. If correct, then the consequences of different kinds of obesity could not have been a target of selection.

13.13.4 Beauty and Intelligence

Human perceive attractive others to more intelligent, good, competent in general and also competent at specific tasks as piloting a plane (Webster and Driskell 1993). People expect physically attractive others to be more intelligent than less attractive others. Children hold the perception that better looking teachers are more intelligent (Zebrowitz et al., 2002). Men like young and attractive women, and women like rich and powerful men (Buss 1994). Infants (12-month old) exhibit more pleasure, more play involvements, less distress, and less withdrawal when interact with strangers wearing attractive masks than with stranger with unattractive mask. They play longer with facially attractive dolls than with unattractive dolls. Thus, judging the standard of beauty might be innate not learned (Langlois and Roggman 1990). Standard of beauty is also culturally universal and invariant (Cunningham et al., 1997). Beauty is an indicator of genetic and developmental health. Physically attractive people are healthier than less attractive people (Langlois et al., 2000). Bilateral symmetry, averageness, and secondary sexual characters characterize the attractive face (Little et al., 2002). FA of faces increases with exposure to toxins and genetic disruptions (Parsons 1990, 1992). Individuals with attractive faces are less likely to be homozygous on deleterious alleles (Thornhill and Gangestad 1993).

Higher status persons tended to be taller than lower status people throughout the evolutionary history. In the ancestral environment, many competitions for status were physical, although alliances and coalitions were also important (de Waal 1982). Since 1776, only two U.S. Presidents (James Madison and Benjamin Harrison) were below average in height, and the taller candidate almost always wins the Presidential election (McGinnis 1988). Evidence shows that man who occupy higher status position are more intelligent. The 1970 data from the U.S. Department of labour shows the following mean IQs for selected occupations: engineer (130), accountant (118), teacher (114), bookkeeper (110), photographer (108), stenographer (106), machinist (104), carpenter (99), labourers (92) and stock clerk (84) (Jencks 1972).

Men prefer to mate with physically attractive women and women prefer to mate with socially dominant men. Among both black and whites, upwardly mobile women are more physically attractive than others (Urdy 1977). Women's attractiveness has a significantly positive effect on their household income, although it has no effect on their income (Urdy and Eckland 1984). Wang and Oakland (1995) confirms that general intelligence is largely heritable. Indirect evidences suggest that beauty is heritable. There is a significant genetic component to developmental stability and fluctuating asymmetry, and hence, to beauty. Fluctuating asymmetry, hormone markers in the face and facial structure in general are heritable. So, available evidence indicates that physical attractiveness is also heritable (Kanazawa and Kovar 2004). Elder (1969) notes that middle class girls have higher IQs, and are more attractive than working class. Facial attractiveness significantly correlates with IQ among both men and women throughout the life course except late adulthood. The significant effect of physical attractiveness on income is documented (Biddle and Hamermesh 1998).

13.13.5 Mate Choice

Evolutionary characteristics that led to an increase in women's reproductive success include a mate who is able to invest resources in her, and her children, physically protect her, and her children, and show good parenting skills, along with sufficiently compatible goal and values to enable strategic alignment without inflicting too many costs on her, and her children (Buss 1994). Worldwide, women and men wanted mates who are intelligent, kind, understanding, dependable, and healthy. Similarly, mutual attraction/love emerged as a most valued qualities in a spouse, worldwide (Buss 2002).

Men ideally like 18 mates in their lifetime, whereas women average 4.5 (Buss and Schmitt, 1993). Greiling and Buss (2000) formulated a number of hypotheses about potential benefits of women from short-term mating. These include resource, genetic, mate switching, and mate skill acquisition hypotheses. Resource acquisition and mate switching hypotheses have received most empirical support. For example, women expect jewelry, money, free dinners or clothing in short-term mating. According to mate switching hypotheses, women find interest in a sex partner who is interested in making a commitment to them, willing to spend a lot of time with them, and able to replace her current partner.

In addition to ambition, industriousness, and social dominance, women rate the emotional stability and family orientation of prospective marriage partners more highly than men do (Oda 2001). They seem to prefer men with whom they feel physically safe and who are protective for them (Geary and Flinn 2001). Many women prefer men with whom they can develop an intimate and emotionally satisfying relationship (Buss 1994), although this appears to more of a luxury than a necessity (Li et al., 2002). Women prefer somewhat taller men with an athletic and symmetric body shape and wider shoulders (Oda 2001) and find attraction in quality physical, and genetic health of men (Gangestad and Simpson 2000). Women rate prominent cheek bone, as this is related androgen/estrogen ratios during puberty (Fink and Penton–Voak 2002). Men with asymmetric faces and body features have higher basal metabolic rates, lower IQs and fewer sexual partners than their more symmetric peers (Manhing et al., 1997). Physically smaller and less robust men are less likely to be chosen as marriage partner than taller and more robust men (Nettle 2002). Immune-system genes are signalled through pheromones. Women are sensitive to and respond to these scents especially when they are more fertile (Gangestad and Thornhill 1997). The best outcome for offspring occurs when there is high variability in immune-system genes (Hamilton et al., 1990). Couples with dissimilar immune system genes conceived more quickly (2 versus 5 months) and fewer spontaneous abortions than the couples with similar genes (Ober et al., 1997). Most men want a long-term marriage partner, and many women only want a long-term partner (Miller et al., 2002). Men are predicted to be nearly as choosy as women and show both similarities and differences in the criteria used to choose long-term mate (Geary et al., 2004). In long-term relationship, men prefer intelligent marriage partner with whom a compatible and cooperative relationship can be established (Li et al., 2002).

13.14 CULTURE

Culture is a set of distinctive material, intellectual, emotional, and spiritual features of a social group, including art, literature, sports, lifestyles, value systems, traditions, rituals, and beliefs. Link between human behaviour and culture is often very close, making it difficult to clearly divide the topic into one area or other. As such, placement of some subjects may be based primarily on convention. A culture's values define what it holds to be important, or ethical. Closely linked are norms, expectations of how people ought to behave bound by tradition. Artifacts, or material culture are objects derived from culture's values, norms, and understanding of world.

13.15 HAPTIC COMMUNICATION

Haptic communication is communicating by touch, used in a number of contexts, often intimate and can be used as an act of domination or friendship, depending on context and who is touching who, how and when. Young children and old people use more touching, than people in middle ages. Touch provides a direct contact with other person. This varies greatly with purpose and setting. Some jobs require touching in some way, very typically in medical profession, or other caring jobs. Touch can be negative and positive and is a common part of many greeting rituals, from shaking hands to cheek-kissing to full-body hugs. Such communication contains subtle symbolism. Degree of touch ranges from a gentle touch on arm, to an arm around the shoulder to a full-body hug. Friends touch more during greeting and in many spontaneous touch one communicates with another.

More about Human Voice

Fatik B. Mandal

Behavioural biology is really interesting with a variety of survival strategies exhibited by animals and humans. Our understanding about behavioural biology is astounding, although not yet complete. We can now explain various social behaviours in terms of gene. The topics like sociobiology, sociogenomics, chronobiology, and biomimicry have achieved significant success within a very short time.

Human behavioural study is also steadily advancing. Studies on human behaviour can explain why we are so careful for our sons, or daughters. The beauty of human behavioural studies lies in our communication ability through language, unique facial structure, facial symmetry, and even in the attractiveness of voice. Development of language along with an increase in brain size is an important contributor in the evolutionary history of human.

Behaviour is known to be influenced by gene, nervous system, endocrine system, and environment. New disciplines are coming out using the knowledge of behavioural biology. Now let us consider the attractiveness of voice in human.

A recent (2004) study published in the journal, *Evolution and Human Behavior* (Vol. 25, pp. 295–304; doi:10.1016 /j.evol.hum.behav.2004.06.001), states that voice of a person may convey important information about his/her socioeconomic status, personality traits, age, height, and weight. Individuals with symmetrical morphological traits have more attractive voices. Deviation from bilateral symmetry decreases rating of voice attractiveness.

Estrogen and progesterone shape the female voice. Testosterone shapes the male voice. These hormones are also associated with male/female body configuration features. Individuals with attractive voices are perceived more favourably and generally have more desirable personality.

13.15.1 Attribution Theory

Human attributions are significantly driven from emotional and motivational drives. Blaming others and avoiding personal recrimination are real self-serving attributions. Attributions are made to defend attacks. We point to injustice, tend to blame victims for their fate, seek to distance ourselves, from thoughts of suffering, tend to ascribe less variability to other people, than ourselves, seeing ourselves as more multifaceted and less predictable, than others. This may because, we see more of what is inside ourselves. We often tend to go through a two-step process, starting with an automatic internal attribution, followed by a slower consideration of whether an external attribution is more appropriate.

13.15.2 Politeness Theory

We maintain two kinds of face—positive face, when others respect and approve us; negative face, when others cannot constrain us in any way. Both of these may be threatened. Positive face is satisfied, but negative face may lead them to think they can take advantage of us. Reverse is also true, as defensive talk threaten positive face. Conformance to social rules of politeness is taking a central and safe path, which neither threatens nor signals that one may be threatened. Politeness means acting, to help save face, for others.

13.15.3 Prosocial Behaviour

It occurs, when someone acts to help another person, particularly, when they have no goal, other than to help a fellow human. Kin selection is a genetic response to support broader gene pool. Social conditioning also has been a cause, and prosocial parents lead to prosocial children. Reciprocity norm may have an effect, where people help others, knowing that one day, they may want someone else to help them, in same unselfish way. Prosocial behaviour varies with context, as much as between people. Men tend to be chivalrous for short periods. Women work quietly for longer periods. People, who are in a good mood, are more likely to do well, in comparison to people, who are feeling guilty. People in small towns are more likely to help, than those squashed together in cities. Evidence abounds of people helping others, without asking for anything in return.

13.16 SELF-EFFICACY

Self-efficacy is an individual's impression of their own ability to perform a task. This impression is based upon factors like the individual's prior success in task, individual's physiological state, and outside sources of persuasion (Bandura 1977). Self-efficacy is thought to be predictive of amount of effort an individual will expend in initiating and maintaining a behavioural change. So, although self-efficacy is not a behavioural change theory per se, it is an important element of many theories including learning.

13.17 BEHAVIOUR ANALYTIC THEORIES OF CHANGE

Learning theories state that complex behaviour is learned gradually through modification of simpler behaviours. Imitation and reinforcement play important roles in these theories, which state that individuals learn by duplicating behaviours that they observe in others and that rewards are essential to ensuring the repetition of desirable behaviour. Simple behaviour is established through imitation and subsequent reinforcement. When verbal behaviour is established, organism learns through rule governed behaviour, and thus, not all action needs to be contingent shaped. Critical role of imitation (what is termed echoic behaviour) in the learning of language was recognized. Behaviour analytic theories of change are quite effective in improving human conditions.

13.18 SOCIAL LEARNING

According to the social learning theory (social cognitive theory), behavioural change is determined by environmental, personal, and behavioural elements. Each factor affects the other. For example, in congruence with principles of self-efficacy, an individual's thoughts affect his/her behaviour, and an individual's characteristics elicit certain responses from the social environment. Likewise, an individual's environment affects development of personal characteristics, as well as the person's behaviour, and an individual's behaviour may change their environment, as well as the way the individual thinks or feels. Social learning theory focuses on reciprocal interactions between these factors, which are hypothesized to determine behavioural change.

13.18.1 Theory of Reasoned Action

Theory of reasoned action assumes that individuals consider the behaviour's consequences before performing particular behaviour. Intention is an important factor in determining behaviour and behavioural change. The intentions develop from an individual's perception of a behaviour as positive or negative, together with individual's impression of way, their society perceives same behaviour. Thus, personal attitude and social pressure shape intention, which is essential to performance of a behaviour, and consequently behavioural change.

13.18.2 Theory of Planned Behaviour

In 1985, Ajzen expanded upon theory of reasoned action and formulated theory of planned behaviour, which emphasizes role of intention in behaviour performance and is intended to cover cases, in which a person is not in control of all factors which affect actual performance of a behaviour. As a result, new theory states that incidence of actual behaviour performance is proportional to the amount of control that an individual possesses over behaviour and strength of the individual's intention in performing the behaviour.

13.18.3 Transtheoretical Model

According to transtheoretical model (Stages of change model), behavioural change is a five-step process. Five stages between which individuals may oscillate before achieving complete change, are precontemplation, contemplation, preparation, action, and maintenance. At precontemplation stage, an individual may or may not be aware of a problem and has no thought of changing their behaviour. From precontemplation to contemplation, individual develops a desire to change a behaviour. During preparation, individual intends to change behaviour within next month, and during action stage, individual begins to exhibit new behaviour consistently. An individual finally enters maintenance stage, once he/she exhibits new behaviour consistently for over six months.

13.19 BIORHYTHM IN HUMAN

A biorhythm in humans is a theoretical process, by which human body and mind are regulated according to set patterns. Biorhythms are usually separated into 3 distinct groups—emotional, mental, and physical cycles. Classically, lengths of these cycles are 28 days, 33 days, and 23 days, respectively. The biorhythms are thought to follow a regular wave, oscillating back and forth over length of cycle. On first day of one's cycle, one is thought to be at an optimal functioning level, while at low point, a person is viewed as being at his or her worst possible level of functional capability. The practice of tracking body's biorhythm dates back to end of 19th century, when a number of doctors began observing, what they perceived as repeating cycles, in a number of ailments and immune system weaknesses. The lengths of the biorhythm cycles date from this period and are based on early observations of these physicians. There is scientific evidence for a number of biological cycles, such as a woman's menstrual cycle. It is highly unlikely that a biorhythm would be same for all people. Everyone needs different amounts of sleep, but general consensus is that adults require between six and eight hours of sleep per night. Aspects of mood disorders are linked to biological rhythms with 3 different periodicities—daily, menstrual and annual. The rhythms are related to mood disorders in several ways. Some conditions are expressed cyclically; they have a period linked to an identified internal or external periodicity, such as menstrual and annual rhythms of depression. Other conditions, like cases of bipolar illness express cycles which are not tied to any identified periodic stimulus.

13.20 SELFISH GENE

Genes are unit of inheritance on which natural selection operates. Dawkins characterizes genes as "selfish," not saying that genes have wishes, or greedy impulses. The gene that causes functional changes in organism proliferate in subsequent generations and increases in relative frequency in population. Genes exhibit interest in the well-being of relatives of bearers. Closer blood relation between individuals is exhibited by higher number of genes they have in common. A gene increases its representation in the next generation by enhancing reproductive success of its bearer and bearer's relatives. Pinker (1994) writes: "Genes are not imprisoned in bodies; the same gene lives in the bodies of many family members at once. The dispersed copies of a gene call to one another by endowing bodies with emotions. When a parent wishes she could take the place of a child about to undergo surgery, it is not the species or the group or her body that wants her to have that most unselfish emotion; it is her selfish genes". "Child-specific parental love" may inhibit parents to respond angrily to their children's bad behaviour. This is designed to work for one's own biological children, and might explain the abuse of step-children (Daly and Wilson 1996).

Genes are no longer treated as static entities, frozen in time. Genes evolve as they negotiate the interspecies journey and they evolve in humans to maximize reproduction, survival opportunities, and copes with exogenous stimuli. Culture itself placed certain genes under selective pressure, thus, spawning new alleles (Lumsden and Wilson 1981). The D4DR dopamine receptor has come under selective pressure, enhancing *Homo sapiens*' wanderlust (Olson 2002). All of us share 99.9% of our DNA sequences. Polymorphisms are stretches of DNA we do not share, thus, ensuring that we are not all clones. Genes that exhibit different alleles in different people hold the secret of phenotypic variation. That is they hold the secret of political attitude and behavioural heritability. The human genome is now estimated to contain 9 million SNPs, four hundred thousand of them reside in exons. SNPs responsible for amino acid composition shifts could number two hundred thousand promoter region. SNPs can have a marked influence in gene expression levels (Carmen 2007).

13.20.1 Genetics and Politics

The years 2004 and 2005 have witnessed the dawn of a new subfield, a new focus for political sciences, "Genetics and politics". During the two years period, Alford and Hibbing published two salient reports developing the genetics-politics nexus, while Carmen published a monograph staking out the broader lineaments of the new subfield. Some argue for a strong relationship between political ideology and psychological profiles (Jost et al., 2003); others disagree (Greenberg and Jonas 2003). At one glance, one can spot the importance of the neurotransmitters serotonin and dopamine in influencing sociality. The 5-HTT promoter on chromosome 17 controls serotonin transporter junction. Normal version of the promoter, containing 16 sequences repeats, approximately 20 base pairs per repeat, effectively clears serotonin deposits; while the shorter version, containing 14 of these repeats, permits serotonin accrual. A difference than stands at 40 nucleotide sequence. High correlation between the short version regime and neuroticism exists and 3 manifestations—anxiety, angry hostility, and impulsiveness are significantly related to short version. Neuroticism has an heritability quotient of .50. Political scientists would consider neuroticism and its subtraits 5-HTT promoter configurations of subject leaders-followers or winners-losers which would help in defining political behaviour (Carmen 2004).

The misnamed "novelty gene", actually the D4DR gene is located on chromosome 11 and makes a dopamine receptor protein. Dopamine, a chemical, needs to be at equilibrium in the typical case and its overload correlates with risky behaviour like too much gambling, too much sex and too much drinking. The D4DR gene contains a series of 48 letter repeats. Average number of repeat runs from 4 to 7; those with 2, or 3 are extra-ordinarily effective in clearing dopamine, whereas those with 8, or more (the ceiling is 11) are not very effective at all. If a subject has 2 longs or a long and a short, the correlation with novelty seeking is greater than for a subject exhibiting 2 shorts (Hammer 2004). A proliferating literature in political science stresses the need to address decision making as a neuro-scientific phenomenon (McDermott 2004). Neurophysiologic rules and processes capture the brain's several structures as action systems during the play of political contexts. Data were recruited by employing functional MRI technology as subjects participated in a wide variety of these contests.

MAOA gene on X chromosome break down neurotransmitter deposits to help smooth communication among neurons. The MAOA promoter is a polymorphism of 30 base pair repeatable from 2 to 5 times. The shorter versions alone may possibly precipitate impulsivity, even aggression (Caspi et al., 2002). For certain, the shorter versions slow enzymatic expression, thus causing neurotransmitter inefficiency, and "cleanup" deficits like serotonin. Researchers from University of California and University of Los Angeles essayed a tentative first step when, during the 2004 presidential election, they used the MRI technology to measure partisan reactions to facial images of President George Bush and Senator John Kerry. They found an intriguing dance between the emotional centres and cognitive centres, as both republicans and Democrats fought to convince themselves of their candidate's manifest superiority. The investigators proceeded to spoil the party by jumping to conclusion that the red state/blue state divide is fictional artifact of our own self-deception (Freedmen 2005).

The short arms of chromosome 6 and 8, and the long arms of chromosome 13 and 22 seem to be prime locations for mutations involving various disorders. With the latter, comparing the presence of candidate genes in those displaying sundry antisocial pattern against those who are immune from these stresses in the context of genomic investigation has produced difference as found by Bouchard and McGuee (2003). Micro-arrays become a truly robust vehicle for socio-genomic investigation, when the experimental data take the form of human SNPs.

13.21 HUMAN SEXUALITY

Status of sexual selection between males lie for possession of females. Result is not death to the unsuccessful competitor, but few or no off-springs. Offspring of males with more eyespots are bigger at birth, and better at surviving in wild than the offspring of birds with fewer eyespots. Sexual selection operates through two mechanisms—(1) Intersexual selection: competition between members of opposite sex, e.g. females and males choosing to mate with 'attractive' mates, which is equal to 'selective mate choice' and 'male sexual proprietariness' over females to protect against undetected cuckoldry. (2) Intrasexual selection: competition between members of same sex, e.g. males competing with each other for access to females. Satin Bower bird inspects an avenue of twigs a foot or so apart constructed by male which has bright shiny blue feathers. Male arranges blue objects in front of this avenue, in order to attract a mate. A brightly coloured male bird courts a relatively dowdy female.

The intrasexual selection includes stags fighting during rutting season, creation and maintenance of 'dominance hierarchies' and possibly tendency of young men to engage in risk taking and intermale aggression. Frequently inter- and intrasexual selection behaviours are exquisitely interwined. Darwin's theory of sexual selection was developed further by Trivers (1972). Trivers argued that because of parental investment, sex that invests greater resources in offspring will evolve to be choosier sex in selecting a mate. In contrast, sex that invests fewer resources in offsprings evolves to be more competitive with its own sex for access to high-investing sex. Normally females are limiting sex and invest more in offsprings than males. Because males tend to be in excess, males tend to develop ornaments for attracting females or engaging other males in contests. Buss (1999) provides a clear description of application of Trivers' theory of parental investment to human mating. "The differences between men and women, in terms of fitness costs of making a poor mate choice are profound. An ancestral man, who made a poor choice, when selecting a mate could have walked away, without incurring much loss. An ancestral woman, who made a poor choice, when choosing a mate might risk, becoming pregnant and perhaps having to raise the child alone, without help."

Buss (1999) showed that males prefer young, physically attractive and chaste mates. Females place greater emphasis on earning capacity and ambitiousness-industriousness of potential partners, than men do. These psychological differences between sexes evolved because adaptive problems posed by reproduction are different for men and women. If we examine adaptive problems posed by long-term mating, there is remarkable symmetry between problems faced by both sexes. Both men and women need to select mating partners with 'good biology', i.e. to maximize fitness. Both sexes need to mate with fertile, healthy partners who are likely to produce fertile, healthy children. Both men and women need to mate with partners, who will remain 'faithful'. Women select men, who are able and willing to provide resources to promote survival and reproductive success of their child(ren).

The cuckold indicates a man, whose wife has sex or becomes impregnated by a man other than himself. Daly and Wilson (1998a) have suggested that because of paternal uncertainty and cost of paternal investment, male fitness will be reduced if a male is cuckolded. Male sexual proprietariness is an adaptation that has evolved through natural selection to overcome this risk. Cuckoldry is a real potential risk to male fitness. Both men and women can experience strong emotions if they believe their partner has formed a romantic relationship within another. Although male sexual proprietariness is emphasized, a comparable module in females is also predicted. Predictions of 'risk factors' that may increase male sexual violence, like degree of local competition between men for access to women, poor economic and health prospects, size of age cohort are on record. Some aspects of men's behaviour may be directed towards controlling female reproductive behaviour. 'Male sexual proprietariness' is seen as part of an adaptation to reduce risk to male reproductive fitness posed by cuckoldry, which reduces returns from paternal investment.

A small, but significant proportion of women in long-term relationships engage in short-term mating. Some selective advantage for women to engage in short-term mating is suggested. Buss (1999) distinguishes between different types of explanations for female short-term mating. 'Sexy son hypothesis' allows female to bear a son with her genes, who will attract females in the next generation. 'Mate-switching' may be due to dealing with a long-term resource-poor partner. It would advantage inclusive fitness of both sexes to engage in short-term mating, where partner offers 'good biology', and offspring would be sufficiently resourced to ensure their survival to the reproductive age. A man can increase his inclusive fitness by engaging in short-term mating, provided his mating partner has 'good biology', and access to sufficient resources to promote survival and reproductive success of his child. Short-term mating can increase an unpartnered woman's inclusive fitness, provided mating partner has 'good biology', and she has access to sufficient resources to promote survival and reproductive success of her children. This analysis suggests hypotheses about the attitudes of the genetic relative to short-term mating.

Besides ensuring biological reproduction, sexuality creates physical intimacy, bonds, and hierarchies among individuals and may be directed to spiritual transcendence. Sexual desire is experienced as a bodily urge, often accompanied by strong emotions like love, ecstasy and jealousy. Sexuality is seen in hidden ovulation, external scrotum, and penile design suggesting sperm competition, absence of an os penis, permanent and secondary sexual characteristics, forming of pair bonds based on sexual attraction as a common social structure and sexual ability in females outside of ovulation. Complex sexual behaviour has a long evolutionary history. The sexual choices are usually made in reference to cultural norms, which vary widely. The restrictions are sometimes determined by religious beliefs or social customs. Freud believed that humans are born polymorphously perverse, which means that any number of objects could be a source of pleasure. Human pass through five stages of psychosexual development. For Alfred Kinsey, people can fall anywhere, along a continuous scale of sexual orientation. People bear with one sexual orientation, or another. Common variations in sequence of DNA impact sexual desire, arousal and orgasm intensity, and lead to differences and diversity of sexual phenotype. Humans are particularly adept at utilizing communication for self-expression, exchanging of ideas, and organization. Complex social structures composed of many cooperating and competing groups from families to nations. Social interactions result in a variety of traditions, rituals, ethics, values, social norms, and laws, which together form the basis of human society. A marked appreciation for beauty and aesthetics, combined with desire for self-expression, has led to cultural innovations like art, writing, literature, and music. The notable features are desire to understand

and influence the world, seeking to explain and manipulate natural phenomena through philosophy, art, science, mythology, and religion. Humans are as they cook their food, clothe themselves, manipulate and develop technologies, and pass down their skills and knowledge to the next generations. Descent of larynx and hyoid bone make speech possible in a human. Evolution of hidden estrus or concealed ovulation may have coincided with evolution of behavioural changes, such as pair bonding.

Helen Fisher defines that "human beings have a common nature, a set of shared unconscious tendencies or potentialities that are encoded in our DNA and that evolved because they were of use to our fore-bearers millions of years ago. We are not aware of these predispositions, but they still motivate our actions" (*Anatomy of love*). Understanding sex require knowledge about males and females similarity, and dissimilarity. Both differ physically in external/internal sexual structures, chromosomes (XX/XY), hormones, and secondary sex characteristics.

Sex differentiation creates males and females. SRY gene on Y-chromosome starts a cascade of events in developing foetus that leads to development of the male. If that gene is absent, a female body results. SRY switches on gene SOX9, which does work switching on and off all sorts of genes in testis and brain, causing switch on and off production of hormone. Hormones alter body and affect other genes. SRY is "an archieve, recipe, switch, interchangeable par, or health-giver of maleness". Genes are sensitive to external experience, reacting to diet, social setting, learning and culture. Variations in expression of human sexuality are the result of learned behaviour. Individual variations in many aspects of human sexuality are genetically determined. D4 receptor is responsible for producing dopamine receptor protein (DRD4). Dopamine, being a neurotransmitter is linked to pleasure. Some forms of variants in this gene have a depressing effect on sexual desire, arousal, and orgasm intensity, while other common variants have opposite effect—an increase in sexual desire.

Behavioural genetics explains behaviour in terms of gene function. Genetic differences are created through evolution. Genes and environment act to determine structure and function of the brain. Separation into two sexes justifies in ensuring variation that would prevent destructive gene competition within one organism and allows evolution of new genes for improved change of survival. Through sexual selection, males compete with other males for females, and the females choose their partner. The greater the difference in reproductive pay off, the greater the difference in aggressiveness among males. Female aggression is more subtle, less direct, defensive and reactive. Personality differences may be explained by differences in our so called "Junk DNA" (95% of our DNA that is not used to make protein). Because a significant portion of genome consists of junk DNA, and due to way microsatellite DNA expand and contracts over the time, microsatellite may represent a previously unknown factor in social diversity.

In humans, genetic information is stored in chromosomes and each chromosome is made up of two parts called **alleles**. The researchers found that the 334 allele of a common AVPRIA variation seemed to have a negative association with men's relationship with their spouses. Allele 334 has also been shown to be associated with increased activity in amygdala, a brain region involved in regulating emotions. Gene variant related to the hormone vasopressin appears to be associated with how a human makes bond with their partners or wives (Figure 13.3).

FIGURE 13.3 The bond between a man and his wife is gene regulated.

13.21.1 Extreme Sexual Behaviour

Sophisticated theories of sexual assault assert that there is no "gene" that causes men to rape. Existence of a predisposition to rape may be a consequence of evolution. Men, who are predisposed to rape, may have more reproductive success. Over long periods of time, this reproductive advantage results in a widespread predisposition to rape among males. Predisposition to rape may be a side-effect of reproductive adaptations, such as, pursuit of a number of partners. For women, sexual activity with a limited number of partners is desirable, and

Evolution of Fatherhood

FATIK B. MANDAL

Male mammals attempt to maximize their reproductive success by mating with as many females as possible. In the absence of extra mating opportunities, males might support their kids, when they are sure of paternity. Males invest more in mating than in parenting to spread their genes far and wide. Female mammals invest more energy in parental care. However, that males avoid fatherly responsibilities may be wrong. Being a mammal, in human fatherhood is extremely variable. In some societies, men provide little parental care; while in others, fathers provide economic and emotional support to children. It has been reported that men who care children more are more attractive to women.

Behaviour evolves rapidly. Two closely related species may exhibit remarkable differences in behaviour, while quite distantly related species may show behavioural similarities. The caring father feeds his kids, brings them to school, and extends other fatherly support. Some situations make it easier for men to support children. In some society, more than one man claims paternity of the same child, as in some groups of peoples of the Amazon basin. In such society, fathers contribute crucial calories to their children, and protect them from intergroup violence. This exemplifies that fathers are so important that kids benefit from having more than one.

In hunter-gatherer society, during the first year of an infant's life, when mothers are least able to forage for themselves, fathers bring home extra calories including meat, honey, and plant foods. Hunter-gatherer men are known to spend about 5% of their time holding infants having intimate relationships with them than do men from horticultural, agricultural, and herding societies. In these latter societies, father's relationship with the offspring is distant with main roles of father in child's discipline, status, and training. The fathers do not always provide sufficient economic support. In 68% of the societies, the death of a father is recorded having no impact on his children's survival. Grandmothers were rather more important. Hunter-gatherer men show higher involvement with offspring because females contribute a large portion of calories for survival. Most marriages were monogamous. Men spend less time engaged in war, and similar work. Men did not accumulate much wealth. When they did so, they wished for multiple wives, and engaged in warfare to defend their wealth. Men contributed less paternal care, when they were not married to the children's mother. Men with doubted paternity are less likely to support the kids. In cultures, where men are less likely to be the genetic fathers of their wives' children, men channel their wealth to their nephews or nieces.

Oxytocin and prolactin hormones produced in the hypothalamus are associated with childbirth. Oxytocin acts on uterus muscles to promote labor, and to move the foetus for delivery. During pregnancy, prolactin level is increased 10–20 times to develop the mammary glands. Following delivery, prolactin stimulates milk production, while oxytocin promotes the release of breast milk. Fatherhood is also associated with hormonal changes. Men involved with the care of children show a drop in testosterone level, and may experience elevated level of prolactin, the hormone that stimulates parental behaviour in a variety of mammals. Fathers with higher prolactin levels are more responsive to babies. Evidence suggests that living with an infant inhibits a man's sex drive.

Higher prolactin levels in human father compared to non-father similarly suggest a role in father-infant bonding. Increase in levels of both hormones in father is very similar to that in mother. Fathers with high oxytocin levels are seemed to be more attached to their children, and men with higher prolactin levels are seemed to be more alert and responsive to their infant's discomfort. Father's oxytocin level is boosted by physical contact with children, suggesting the hormonal rise in males is stimulated by close contact with the infant. Interestingly, in a recently published paper of Kunz and Hosken (2008), lactation in men has been reported in World War II prisoner. Male lactation has also been recorded in Dayak fruit bats and the masked flying fox although the reason of such lactation is not clear.

Source: http://www.amazines.com/article_detail.cfm/2161292?articleid=2161292

thus, women have evolved to resist rape. Experience of "trauma" associated with sexual assault is a reproductively successful response because women, who experienced such trauma subsequently avoid rape. Some argue that because men cannot control their irresistible impulses to rape, it is women's responsibility to avoid dressing provocatively. Women, who are being raped, learn to avoid rape. Finally, biological theories do not explain why men rape women, who cannot bear children, or why they rape their intimate partners and spouses.

Thornhill and Palmer (2000) did not decide regarding any adaptations that male mind contains and evolve specifically to facilitate rape. They proposed that specific adaptations could modulate the psychological pain of rape victims. Individual rapists "learn" rape from experiences. Rapists learn to dominate women with the belief that rape is an effective means of achieving this domination. Rape point to a proximate asymmetry in sexual desires of men. Males typically covet sexual access to a number of females. Women are choosier than men regarding targets of their sexual desire. The minimal "start-up cost" to bring a child into being is larger for women than men, i.e. ejaculation for men versus an egg, 9 months of gestation and a risky childbirth for women. Breast-feeding impose a significant tax on a mother's caloric resources, delays the ability to give birth, or care for other child. Men have little reason to be discriminating in their mating behaviour. For women, it makes sense to be highly discriminating, as becoming pregnant by a man implies a heavy subsequent investment. All else being equal, selection is expected to favour men who follow less discriminating mating strategies, and women who follow more discriminating strategies. This difference alone produces a condition in which a male desire for very often thwarted sex. Since females provide the limiting resource for offspring, keen competition can be predicted among men for females, with the result that some men end up fathering children by multiple mates and other men end up never getting to mate at all (Symons 1979). Rape is an extreme and risky tactic for obtaining sex from unconsenting females, with possibility of subsequent retaliation and punishment (Thornhill and Palmer 2000a). This tactic is employed by those who cannot attain mates through other means. This hypothesis is called the mate deprivation hypothesis, or the rapist-as-loser hypothesis (Malamuth and Heilman 1988).

The rapist-as-loser hypothesis does not imply anything about whether the male mind contains rape-specific adaptations. A possible psychological mechanism could operate by means of a threshold, below which the male desires to rape. Such threshold could be raised or lowered according to availability of alternative mating opportunities. Rapists on the whole may actually be successful than non-rapists in obtaining sex through non-coercive means. Rape by wealthy males is explained by "a combination of impunity and hypothetical adaptation pertaining to evaluation of a victim's vulnerability". Women sexually desire attractive men in exchange of material benefits from them. Females display to attractive males function as a signal to male that the female is discriminating about mates, and may result in getting more material benefits from male, than she would get without display. If a woman's display is truly effective, a man who achieves copulation with her will perceive that he achieved it by force. Women have a short-term desire for sex with attractive men, but also know that seeming "too easy" can harm their chances for winning long-term interest. A woman might feign resistance to sexually attractive men into believing that they have raped her. Female choosiness explains far more about male sexual behaviour than rape.

Weinberg says that the men interviewed by him fall into one of two broad types. The first type fit the zoophile-as-loser hypothesis, i.e. men having difficulty in finding affection or sexual gratification from women in their youth, try to find such gratification from pets or livestock. For those in the first group, inability to get sex from women might explain the initial breach of interspecies sexual boundary, but the continuation of behaviour is harder to explain. The second type is pansexual swingers who do sexual experimentation of many kinds along with bestiality. Members of this second group may have more sexual contact with women in course of their pansexual pursuits. Bestiality may persist even when the man has obtained a human sexual partner. Weinberg suggests that the same could be said for men who molest children. No doubt the inability to obtain sexual access to females of suitable age explains some of initial instances of this behaviour, but once the boundary has been crossed, the behaviour can persist on a trajectory that cannot be explained by unavailability of age-appropriate mates (Kelley and Byrne 1992).

Sexual functions of six main brain regions including 3 sub-cortical and 3 cortical regions are identified. Sub-cortical regions comprise septal region, mediating pleasurable response and orgasm; hypothalamus, mediating neuroendocrine and autonomic aspects of sexual drive with possible role in sexual orientation; and ansa lenticularis and pallidus, mediating sexual drive. Cortical regions comprise frontal lobes, mediating the motor components of sexual behaviour and control of sexual response; parietal lobes, specifically the paracentral lobule, involved in genital sensation, and temporal lobes particularly the amygdalae, implicated in sexual orientation, sexual drive and sexual disorders. Sexual behaviour can be conceptualized in terms of various facets of sexual response cycle defined by Kaplan (1979), who proposed a triphasic model: sexual desire, excitement, and orgasm. These three phases are interconnected, but governed by separate neurophysiological systems. Associations between the key brain regions are known and phases of sexual response cycle can be made. The initial phase of sexual desire is mediated by subcortical structures specifically the hypothalamus, ansa lenticularis, and pallidum. Temporal lobes specifically the amygdalae play a significant role in this initial phase. Sexual excitement (phase 2) mediated by cortical structures, namely the parietal and frontal lobes, control genital sensation and motor aspects of sexual response respectively, sustaining sexual activity until progression to orgasm (phase 3), for which the septal region is implicated. It is likely that there is an interconnected neural network where various brain regions and associated phases of sexual response are simultaneously activated and produce the physical and psychological manifestations of human sexuality.

"True hypersexuality" is more commonly associated with limbic and temporal lobe injury (Zasler and Horn 1990). Hyposexuality may occur following injury to prefrontal convexity as part of a general tendency towards apathy and indifference (Benson et al., 1972) or in relative isolation as in temporal lobe epilepsy. Septal self-stimulation elicit orgasm. Chemical stimulation of septal region elicited similar sexual responses (Heath 1972). Septal injury markedly increases sexual activity and "the septal nuclei form one locus on an amygdaloid–hypothalamic–septal circuit mediates sexual behaviour". Septal region appears to be intrinsically involved in pleasurable response and orgasm. Isolated hypothalamic lesions are reported to reduce sexual drive (Miller et al., 1986). Alterations in sexual behaviour and/ or orientation can occur in association with various combinations of limbic structure lesions, including the hypothalamus. Hypothalamus appears to mediate neuroendocrine and autonomic aspects of sexual drive, with focal lesions resulting in a reduction or abolition of sexual drive. In contrast, lesions including but not restricted to hypothalamus have resulted in increased sexual drive. This may reflect the possibility that different nuclei within hypothalamus serve opposing functions. Hypothalamus may serve to trigger an overt sexual response. This is a key brain region in mediation of human sexual drive (Ferretti and colleagues 2005). Interruption of a bundle of myelinated fibres known as ansa lenticularis is found to effect sexual function. Ansa lenticularis and pallidum play a role in mediation of sexual drive.

Some authors consider disinhibited sexual behaviour as a byproduct of general behavioural disinhibition occurring after frontal lobe damage (Zasler and Horn 1990). Effect of frontal leucotomy on sexual behaviour has been assessed (Hutton 1947) and some researchers reported cases of hypersexuality and others described no changes after leucotomy. Frontal lobes mediate the motor components of sexual behaviour, in addition to control of sexual response that may be disinhibited after frontal lobe damage in context of general behavioural disinhibition. The individuals are typically conscious and alert during the episodes and can recognize their sexual nature without necessarily finding them pleasurable or erotic (Hopkins and others 1995).

13.21.2 Fertility, Parental Investment, and Parent-offspring Conflict

Fertility is most direct contributor to an organism's fitness. All else constant, any increase in fertility increases an organism's fitness. An organism can increase its energy capture rate in future by growing, and thus, increasing its future fertility. Organisms, typically, have a juvenile phase, in which fertility is zero, until they reach a size, at which some allocation to reproduction increase fitness more, than growth. Among organisms that engage in repeated bouts of reproduction, some energy during reproductive phase is diverted away form reproduction and allocated to maintenance, so that it can live to reproduce again. Natural selection on age of first reproduction and on adult reproductive rate tends to maximize total allocation of energy to reproduction over life course. Natural selection on offspring number and investment per offspring tend to maximize long-term production of descendents. This may be estimated by number of offsprings that survive to reproduce themselves during an organisms life time, or if fertility affect production and survival of grandchildren by more distant effects.

Two major decisions, whether to reproduce now or later, and amount of care to invest in each offspring depend on costs and benefits of mating and parenting effort, which may be influenced by environmental risk. This risk is 'unpredictable variability in outcome of an adaptively significant behaviour'. In life history theory, extrinsic mortality is risk of dying at a particular age that is shared equally by all members of a population and is 'not sensitive to changes in reproductive decisions'. Predation is greatest source of extrinsic mortality for most organisms, but there may be others. Mortality need not be only source of risk relevant to life history. Any unavoidable factor that can reduce an organism's reproductive value is called **'extrinsic risk'** and it may have effects similar to extrinsic mortality. Extrinsic risk, like extrinsic mortality is independent of parental care, which should usually benefit offspring and creates substantial diminishing returns to parental effort.

Trivers used the phrase "parent-offspring conflict" to describe the divergence in the genetic interests of parents and children. This conflict would occur, when parents try to shape children's personalities in ways that serve the parents' interests, but not the child's. In humans, parent-offspring conflict may begin in the womb. In a prenatal tug-of-war, foetus extracts more maternal resources than mother's wishes. Foetal cells embed themselves in artery walls, restrict the mother's ability to control blood flow to placenta, and placenta also secretes a hormone that tries to siphon off a larger ration of the mother's blood sugar. Obtaining extra resources is foetus's advantage even if it slightly increases the mother's risk of dying in childbirth.

Trivers writes that "socialization is a process by which parents attempt to mold each offspring in order to increase their own inclusive fitness, while each offspring is selected to resist some molding." In his 1985 book *Social Evolution*, Trivers argues that personality and conscience are formed early in socialization, probably during the first 5 years of a child's life. Trivers present adolescence as a time when children "reorganize their personalities in a way to reflect their own self-interest more exactly" (Trivers 1985). Harris's conclusions about the minimal influence of parents are far from accepted wisdom within psychology (Begley 1998). Failures of repression occur during adolescence and for this reason repressed childhood experiences are often revealed in adulthood. Repression acts as a psychological arbitrator maintaining a tenuous equilibrium between the fitness interests of parent and offspring during an extended childhood. The childhood wishes and motivations that are acceptable to the parent and maintain family cohesiveness remain in a child's consciousness, whereas those are not congruent with parental views are repressed and exist primarily at unconscious level. At adolescence, the individual's self-interest begins to emerge again as it gains independence. A sexually mature organism is able to pursue its own fitness interests through production of offsprings. Children repress manifestations of their genetic self-interest. They need to do so to remain in good graces of the parent on whom they are dependent. But when they grows into adolescence, they have a diminishing need to repress things and are increasingly able to assert their own genetic interests more strongly. Financial expenditures may greatly enhance children's future, but children get benefit from greater personal involvement by parents in their education (Carter and Wojtkiewicz 1997). In rational choice theory, effort spent in monitoring and supervising the activities of other persons is a cost, no different than spending money (Hechter 1987). Parental supervision is positively associated with improvements in children's self-reliance, happiness, sense of social responsibility, and educational success (Baumrind 1989).

There is a point (S_{max}) beyond which fitness does not respond to additional parental care. Child outcomes are determined by chance, once the parents reach 'saturation' point at S_{max}. If extrinsic risk is high, then parents reach S_{max} at lower levels of effort. Channelling resources away from parenting into mating effort or additional offspring should enhance fitness, when extrinsic risk is high. If extrinsic risk is low, then saturation point of parental care is 'high', and responsive parenting can have important influence on children's survival and ultimate success. Thus, tradeoffs between mating and parenting effort, or offspring quality and quantity depend on, whether environmental hazards can be avoided by increasing parental effort. S_{max} (saturation point of parental effort) is a function of extrinsic risk. S_{max} is a point, at which additional parental effort does not improve offspring fitness. Maternal care increase as pathogens increase to moderate levels, then decrease at higher pathogen levels. This suggests that effects of pathogens on child outcomes may depend on parenting at low to moderate levels, but at high levels, pathogens may present extrinsic risk. Warfare and famine appear to be hazards that are not remedied by parental care. Parental care may improve child outcomes at low to moderate levels of pathogen stress. Weaning age and maternal care increase at low to moderate pathogen levels, but higher pathogen levels are associated with earlier weaning and less maternal care. Multiple linear regression shows associations between environmental risk and parental effort.

In suburban North America, parents do not sleep with children. They maintain distant proximity throughout most of the day, and make substantial indirect parental investment through schools and clothing. Negative association between paternal involvement and patrilineal inheritance of movable property suggests tradeoffs between paternal effort in childhood and later investment. Pathogen stress show variable relations with parental effort. At low pathogen levels, parental effort is positively associated with pathogen stress, but parental effort decreased at higher levels. Responsive parental care may buffer children from chronic psychosocial stressors and improves immune function. The children's stress reactivity and hormone levels (cortisol) are positively associated with unresponsive parenting.

Pathogen stress may increase benefits of genetic diversity in offspring. If so, then pathogen stress should be positively correlated with indicators of multiple mates cross-culturally, which appears to be the case. This is not clear that whether fathers reduce parental care, as they are attempting to produce more offspring to 'beat the odds', or to increase genetic diversity in offspring or both. Certain behaviours, over others, depend on their perception of environment'. Such phenotypic plasticity is hallmark of human adaptation. Other hazards, such as substance abuse, HIV infection, crime, imprisonment, etc. may be insensitive to parental care, and hence, present 'fitness cliffs' that may affect allocation of effort among mating, parenting, and self-preservation. Thus, these findings suggest an important role for environmental risk in allocation of human parental effort. Maternal and paternal care appears to have somewhat different relationships with environmental hazards, presumably, owing to sex-specific benefits of parenting and mating effort.

Females provide all investment to offspring in greater than 95% of mammalian species, but males provide similar amounts, or more total investments among most altricial birds, male brooding fish and some insects, such as katydids (Clutton-Brock and Parker 1992). One that invest more sex results in operational sex-ratios. The more investing sex is selected to be choosy about, when and with whom to mate. Less investing sex is selected to possess characteristics that increase its mating opportunities. Male resources are wasted on costly display, or handicaps, or on fighting, rather than in offspring production.

13.21.3 Divorce

Helen Fisher found a cross-societal tendency of divorce around the 4th year of marriage. This four-year peak in divorce rates is an evolved disposition for mates. She puts it a "planned obsolescence" of the human pair bond (Fisher 1992). It is often advantageous for an individual to seek a new mate, once a child from an existing union has passed through infancy. Particular neurochemical mechanisms may facilitate the transition from one mate to another. The ideal lifelong marriage is a holdover of a special case adjustment for agriculture, and the tension between "till death do us part". Fisher's proposed "four-year itch" reflects the evolved dispositions those were selected for in our pre-agricultural past. "Four-year itch" is part of a broader, evolution-minded theory of monogamy that was developed in social, scientific (Fisher 1998) and more popular forums (Fisher 1996). The "planned obsolescence" proposes that a variety of factors in our ancestral past led humans to evolve a tendency to enter into monogamous pair-bonds. Tendency is not for lifelong monogamy, rather for a period to conceive a single child and rear her or him through infancy. As births during our ancestral past were spaced about 4-years apart, evolution has left humans to pair off for roughly the same period. Divorces in most societies tend to peak around the fourth year of marriage.

13.21.4 Son or Daughter

Daughters are more valuable to parents in poor "reproductive condition", while sons to parents in above average condition, due to evolved differences in parental investment. Parents in poor condition treat their daughters better, than they treat their sons, while parents in good condition do the opposite (Trivers and Willard 1973). Wealthier parents favour their sons, while those who are less wealthy favour their daughters. A man's sexual behaviour is sensitive to society's sex ratio variations. Men pursue multiple sexual relationships, when there is a surplus of women relative to men, and they commit to a single relationship, when there is a shortage (Buss 1994). Depending on sex ratio, societies may appear to be relatively "promiscuous" or "monogamous." Prevalence of parasites, scarcity, or reliability of food, frequency of warfare, and other external threats could potentially interact with evolved mechanisms to produce cross-societal differences in sexual behaviour.

13.21.5 Birth Order Effect

In human, same sex individuals compete for the best mates. Competition occurs between mothers and fathers over parental investment. Competition takes place between siblings over the parental resources (Sulloway 1995). Such competition leads children to develop divergent attitudes, personalities, and tendencies to maximize the parental resources. Firstborns having a consistent edge compete with their younger siblings at least in the early years. They tend to be larger, stronger, intellectually developed and occupy the dominant ("alpha") position among siblings. They develop conservative, tough-minded attitudes, and more jealous, conscientious, and dominant personalities. Laterborns differentiate themselves from the firstborn to increase their share of parental resources and in their efforts to secure a family niche develop more liberal, compassionate attitudes and more rebellious personalities. Laterborns are innovators; while firstborns are traditionalists. Sulloway assembled data on over 3000 scientists which indicated that laterborns were twice as likely as firstborns to adopt a scientific innovation early. Firstborns, usually the ambitious achievers have higher IQ's, higher academic achievement, and a greater probability of achieving eminence than laterborns. "First-borns are overrepresented among political leaders, including American Presidents and British Prime Ministers". Laterborns on average do worse but occasionally rise up. Sulloway cites the work of Zajonc's well-known "confluence model" which hypothesizes that children are affected by the intellectual milieu of their home environment, which in turn is influenced by the ages of other family members. The model predicts that for most societies the firstborns will outperform laterborns. Arwine and Patterson considered the achievement of various leadership positions in society and found no systematic correlation between birth order and eminence. In some cases, laterborns performance was better than firstborns (Zajonc and Bargh 1980). Rowe (1997) presents sibling data in which the correlations of IQ among siblings adjacent and removed in birth order did not vary and in which IQ also did not significantly differ by birth order. Retherford and Sewell (1991) also found no effect of birth order on IQ, or academic achievement.

13.22 BODY ODOUR

Pheromone production is primarily linked to apocrine glands of the skin, and also to other glandular secretions and to skin flora present in moist areas of the body, like axillae, mouth, feet, and genitals. Apocrine glands are found in genital area, around navel, chest, breasts, areola, and are concentrated in axillae. Apocrine glands are associated with hairs. High concentration of apocrine glands found in armpits led to term, "axillary organ", an independent "organ" of human odour production. Apocrine glands have a tubular, coiled structure, and are about 2 mm in diameter. Apocrine glands develop in embryo, and become functional only, with onset of puberty. Concentrations of C2–C5 aliphatic acids that are secreted from vaginal barrel, referred to as "copulins," vary with menstrual cycle phase.

Odour of copulins and its behavioural effects also appear to vary with menstrual cycle. Copulins are also referred to as pheromones. In sufficient quantity, pheromones are consciously detected as natural human body odour. Link, between apocrine gland function and puberty, reflects that function is closely linked to the levels of sex steroid hormones that increase with onset of adrenarche and puberty. Freshly produced apocrine secretion has no odour, and is transformed into odorous products by micro-organisms. Odours of skin, saliva, urine, and genital secretions contribute to the amount and hedonic quality that is characteristic of natural human body odour. Pheromonal communication, typically, occurs without consciousness. Pheromones, when produced in high concentration, may have both conscious and aversive effects on others.

Pheromones of women regulate ovulation in other women, presumably by affecting levels of LH and FSH. It is suggested that a progesteronic pheromone alters LH pulsatility in men. Human pheromones elicit change in hormones. Juette showed that an aqueous mixture of five ovulatory fatty acids evoke increased saliva T levels in men. Pheromone androstadione is shown to elicit behavioural changes. Shinohara and others showed that axillary pheromones from women, either in follicular, or in ovulatory phase of menstrual cycle, differentially modulate pulsatile LH pulse frequency in other women. Preti and others recently showed that male axillary extracts effect LH, and mood in female recipients, and suggested that LH response may be used to determine precisely the compound involved in this pheromonal effect, which is a typical mammalian female response to pheromones from a male conspecific.

Most studies focused on the 16-androstenes, metabolites of characteristically male sexual hormones, and the androgens, which are secreted by apocrine glands. Dorfman assumes that

16-androstenes develop with metabolism of testosterone. Two of these androstenes, androstenol, and androstenone have odorous characteristics that bear a similarity to smell of male axillae. Androstenol has a musk-like scent, while andro-stenone smells urinous. Odours arise only via activity of microorganisms—aerobic bacteria *Corynebacterium sp.*, which transforms odourless precursor's androstadienol and androstadienone into odorous androstenone. If axillae are treated with antibacterial detergents, production of and-rostenone decreases significantly. Male axillary sweat contains approximately five times more androstenone, than female sweat. This sex difference can be explained by sexually dimorphic levels of blood androgens, and by sex differences in colonization of microorganisms. Jackman and Noble showed that in most men axillae were dominated by *Corynebacteria sp.*, whereas in women they found bacteria *Micrococcaceae*.

It was presumed that human pheromones had potential to facilitate inter-sex communication. Influence of pheromones on social behaviour may pale by comparison to influence that pheromones may have on reproduction. Olfactory cues are essential in animal, especially in mammalian sexual behaviour. These olfactory cues are difficult to isolate and related discussions are controversial. Humans are capable of discriminating between males and females by olfactory cues alone. Pheromones also influence human menstrual cycle. Male pheromones have a regulatory effect on menstrual cycle. Many authors have speculations that both androstenone and androstenol are male pheromones, raising questions of whether and how females perceive them. Maiworm found that females perceive males positively, under exposure to androstenol, and negatively under exposure to androstenone. Finding that females are emotionally more affected by androstenone and androstenol, than by control substances, like rose water, led to hypothesis that both androstenone and androstenol might be male pheromones.

Role of androstenol in any hypothetical signalling system is clear, since it promotes female sexual attraction towards males. However, problems arise in attempts to determine function of androstenone, which induces negative female emotions towards males. Odour of androstenone prevails, whereas fresh sweat odour of androstenol disappears quickly. Production of attractiveness-enhancing androstenol inevitably produces repellent androstenone, and makes it difficult to propose a definite advantage for sender of such chemical signals, compared to a non-sender. If a male repels females with androstenone, this would contradict hypotheses, which assert male promiscuity on evolutionary basis. A less odorous male could out reproduce a more odorous male, simply because he could approach more females, in less time and with less energy. This only holds, if cost of more odorous androstenone production are greater than benefits reached through producing more sexually attractive androstenol. As androstenol oxidizes to androstenone, initial attractive signal becomes repellent. Because this effect takes place within 20 minutes, a less odorous male would be better off, since repellent smell of androstenone is the long-term prevailing signal. If androstenone is a signal for females, then what advantages do more odorous males have? Schneider proposed that females have a higher olfactory acuity at ovulation and Doty et al., showed a direct correlation, between estrogen levels, LH levels, and heightened olfactory sensitivity. These changes in olfaction, during menstrual cycle, extend well to odour of androstenol, and in general to more "musky" odours typical of males. Overall, results suggest existence of two different olfactory signals: androstenol, which induces female attraction to males, and androstenone, which induces negative emotions in females. Functional assessment of such a positive-negative mood-inducing signal requires consideration of a set of evolutionary hypotheses.

13.22.1 Pheromones and Sex

Parental investment theory predicts that females, who look for long-term relationships seek out and choose males, who are ready to invest resources in their offspring. This minimizes female investment, but maximizes overall investment through added male assistance. In contrast, males are expected, either to attempt copulation frequently with as many fertile females as possible, or to develop a pair bond. This helps to ensure that either a large number of offspring survive without significant paternal investment, or that paternal investment occurs primarily when another male does not father offspring. It is adaptive for females and males to develop and use cognition in mate selection, which takes into account biological constraints. Thus, mate selection is a task of information processing, and evolution would favour individuals, who quickly and reliably process information that allowed them to make appropriate mating decisions.

Adaptive cognition could be expected to lead to optimal decision making under a wide spectrum of socioeconomic constraints. Existence of ubiquitarian sex specific differences in mate selection criteria attests that male and female cognition is adapted to biological constraints of mate selection. For example, neither males nor females consciously perceive human ovulation. Since, ovulation is associated with a number of overt physiological and behavioural changes; it is surprising, that it is not consciously detected. Olfactory perception, an "unconscious" mechanism is associated both with physiological and behavioural changes of menstrual cycle. Alexander et al., 2001 have argued that concealed ovulation evolved as females need to trick males into forming a bond. Males, who were not aware of optimal (ovulatory) female fertility would remain bonded to ensure impregnation and paternity. A female, who provided cues to ovulation might risk losing paternal investment, due to paternal uncertainty and limited temporal reproductive interaction. This hypothesis implicates male fear of cuckoldry as an evolutionary pressure. One evolutionary outcome would be that female's ability to secure paternal care is affected by mechanisms that increase temporal aspects of pair bond and enhance male confidence of paternity.

Concealed ovulation is a mechanism that fits this hypothesis. Symons (1979) has proposed an alternative evolutionary scenario, where concealed ovulation evolve to increase chances of successful cuckoldry by females, so they "can escape the negative consequences of being pawns in marriage games". Once monogamy is established, a female's best strategy would be to copulate outside the pair bond, because she could then obtain superior genes, with a certain expectation of paternal investment, and increase survival of genetically superior offspring. These two hypotheses imply different impacts of heritable traits. Change in female attitude towards male body odour can be expected, to impact mate selection and perhaps self-initiated copulations by females. With regard to the androstenol-androstenone signalling system, situation for androstenol seems clear—it makes males more attractive to females. But females are less likely to act on this olfactory-based attraction; unless fitter males produce more androstenol. A non-producing male could do quite well in a population of producers because females would not be repelled by his body odour. Thus, attractiveness-enhancing component of smell does not seem to be main, or at least only function of signalling system.

Similarly, pheromones from female elicit changes in hormonal milieu of male that affect his behaviour, by chemically signalling him that female is ovulating. Females faced with an evolved male strategy to detect concealed ovulation would be likely to develop a counter strategy. One possible strategy could be to manipulate male cognition, and thus, adaptive male information processing in mate selection. Studies on menstrual cycle fluctuations in fatty-acid composition of women's vaginal fluids indicated that a similar type of estrogen-based chemical signalling system might also exist in humans. Human vaginal secretions have a composition that is similar to vaginal secretions of female rhesus monkeys. Application to ovariectomized female rhesus monkeys, either of human, or rhesus vaginal secretions, induced similar activation of rhesus male sexual interest. Copulins may act as putative human pheromones, and provide beautifully balanced "strategic weapons", in the "battle of the sexes" and the "war of signals" resulting from, sex differences in parental investment theory. It is not necessary to view these "battles" or "wars", only from the perspective of parental investment theory.

Persky and others commented on observed ovulatory increase in T levels of human males, and suggested that, somehow, female was signalling male that she had ovulated, though neither of these studies specifically mentioned human pheromones, affective reactions were present both in male and in female, and pheromones are most likely cause of affective reactions. For example, increased T in male can readily be linked to increased intercourse, whether or not increased T was an observable event. Finally, Singh and Bronstad (2001) showed that human males find natural body odour of ovulatory females to be most pleasant, when compared to natural body odour during other phases of menstrual cycle. Male's hedonic rating of pleasant ovulatory odour increased T and increased intercourse collectively offer significant support for concept that chemical communication is more important to properly timed reproductive sexual behaviour than visual or other sensory input. Because the sexual activity is not limited to the ovulatory phase of the menstrual cycle, human sexual behaviour is considered to be more complex than that of other mammals that depend upon properly timed reproductive sexual behaviour for species survival. There are other cues besides chemical cues that are involved.

However, it is remarkable that many people consider visual cues to be more important than olfactory cues, when consideration is given for mammalian mechanisms that ensure properly timed human reproductive sexual behaviour. It is presumable that human scent, apart from above-mentioned functions, could like other cues in mate selection and also signal aspects of reproductive fitness. Several studies have found that bodily and facial symmetry plays a role in attraction, and thus, in choice criteria for human mating. Symmetry is believed to signal developmental stability, which refers to an individual's ability to cope with genetic and environmental perturbations during early development. Recent research has focused on the significance of developmental stability as mate choice-criterion. Sex steroid hormone dependent human body odour could transmit information about an individual's developmental stability as an additional redundant olfactory signal. Since olfactory and visual cues have different physiological roots, signalling errors are likely to be uncorrelated.

Thus, taking information of both signals into account reduces error, and allows much more reliable mate choice decisions. Thus, pheromones could indeed be regarded as honest signals for human mate choice based on testosterone-immunocompentence-developmental stability link to pheromone production. In humans, female olfactory preferences also seem to induce disassortative mating for components of the Major Histocompatibility Complex (MHC) as is observed in other mammals. In other words, olfactory cues may be able to reflect parts of an individual's genome, and body odour seems to influence female mate choice, in order to find a partner, who possesses fitting MHC-dependent immune system components. Simply put, ovulatory women seem to prefer scent of genetic diversity. Indeed, both women, who are not taking oral contraceptives and men rate similar genetically determined odours as less attractive, than dissimilar genetically determined odours. Thus, not only are men and women able to distinguish among genetically distinct, self versus non-self odours, they prefer the scent of non-self (genetic diversity). Men and women with shared markers of genetic diversity also select perfumes that may amplify body odour that is linked to their genetic diversity. Male facial attractiveness is mediated by hormones and generally support a hormonal theory of facial attractiveness dependent interaction between visually displayed hormone markers and hormonal state of viewer. There is no biological pathway that directly links visual input, either to neuroendocrine function, or to hormonal state of viewer, and male and female visual systems are not sexually dimorphic. Accordingly, means and biological mechanisms by which sexually dimorphic, hormone-dependent facial features become attractive have yet to be detailed. However, olfactory pathways link hormonal state of "viewer" to chemical signals of reproductive fitness that correlate well with degree of hormone-dependent, sexually dimorphic facial features. For example, higher T levels correlate with the visual appeal of a "stronger" jaw. Interaction of these visually displayed hormone markers of reproductive fitness and effects of hormones on pheromone production and distribution suggest that effects of pheromones on reproductive neuroendocrine function might provide a critical, well-detailed, mammalian link between hormone-mediated facial signals and perception of facial attraction.

Sex steroid hormones control regional fat distribution, which interacts with reproductive control mechanisms. For example, fat tissue converts androgens to estrogens. Circulating E levels appear to lower WHR, while circulating T levels appear to increase WHR, which is believed to signal reproductive fitness in women and perhaps in men. In addition, high levels of LH and FSH, as well as, estradiol levels are linked to lower WHR and to earlier pubertal endocrine activity of females. However, conscious or unconscious mechanisms linked to perception of WHR and its link to physical attractiveness have not been detailed. Presumably, these mechanisms exist cross-culturally, but they have defied explanation. Conditioning of visually perceived physically attractive WHR by association with steroid hormone-dependent chemical cues (human pheromones) seems to be a very likely explanation for increased desirability of men and women, whose weight and height are proportionate. Each example above, of symmetry, genetic diversity, hormone-mediated facial attraction, and of WHR, has some as yet undetermined link to what we visually and consciously perceive to be attractive. Human life and interactions are influenced by pheromones, whether or not affect or effect are part of our consciousness. The affective hormonal reactions caused by olfaction and pheromones dominate social interaction and these affective reactions may be primary influence on social interactions. Human pheromones have more potential, than any other social environmental sensory stimuli to influence physiology and behaviour.

Human's babies and children are nearly helpless and require high levels of parental care for many years. One important parental care is the use of mammary glands to nurse the baby. Natalism is promoting reproduction, while antinatalism promotes birth control as a solution to overpopulation and its effects are continuing. Parenting is the process of raising a child from birth, or before, until adulthood. It is usually done by the biological parents, although governments and society take a role. In many cases, orphaned or abandoned children receive parental care from non-parent blood relations. Others may be adopted, raised by foster care, or be placed in an orphanage. Parental cares provided for a child's physical needs, protect him/her from harm, and impart in him/her skills and cultural values, until he/she reaches adulthood. Degree of attention parents invest in their offsprings is largely inversely proportional to the number of offsprings produced.

13.23 LIFE HISTORY EVOLUTION

Species level specializations result in a number of life history characteristics, which are known to be arranged on a fast-slow continuum. Among mammals, species on fast-end exhibit short gestation times, early reproduction, small baby size, large litters, and high mortality rates with species on slow end having opposite characteristics. Organisms are capable of slowing down, or speeding up their life histories, depending on environmental conditions, such as food availability, density of conspecifies and mortality hazards. Allocation to reproduction, as measured by fecundity and fertility, vary over short-term in relationship to food supply and energetic output, among plants, birds, and humans.

Birds under variable conditions adjust clutch sizes, in ways that tend to maximize number of surviving young produced the life course. Calories restriction of rats at young ages tends to slow down growth rates and leads to short adult stature, even when food becomes abundant later in juvenile period (Shanley and Kirkwood 2001). Rates of senescence vary across different populations of grasshoppers, with those at higher altitudes and earlier witness, senescing faster than those at lower altitudes, as a result of differential selection on genotypes. Male and female parental investments vary in relation to local ecology. Among katydid, males provide females with a "nuptial gift", to support offspring production. Specialization and flexibility are fundamental in human life histories and mating systems. Large human brain supports the ability to respond flexibly to environmental variation, and to learn culturally. Four major factors affecting the timing of reproduction in life course and hence reproductive rates and parental investment for each sex. Important resources are consumed and utilized in reproduction. Production process by which these resources are obtained.

- Risks of mortality and technology of mortality reduction.
- Extent of complementarity between sexes in production of offspring.
- Degree of variation in resource production and capital holdings among individuals and within individual over-time.

Neural tissue monitors organism's internal and external environment and induces physiological and behavioural responses to stimuli. Brain has capacity to transform present experience into future performance. Cerebral cortex specialized in storage, retrieval and procession of experiences. Major investments in brain results in increased reproductive rate during adulthood. The longer the time spent growing and learning prior to reproducing, the more natural selection favours investments in staying alive to reap the benefits of these investments. Similarly, any investment of that produce increased energy capture rates later in life, such as learning select for additional investment to reach those older ages.

13.24 BEHAVIOURAL ECOLOGY

Human behavioural ecology examines adaptive design of traits, behaviours, and life history in an ecological context. This field attempts to explain variation in behaviour as adaptive features to competing life history demands. Human behavioural ecology has its base on general evolutionary theory, like natural selection, sexual selection, kin selection, inclusive fitness, and middle levels evolutionary theories, like theory of parental investment, parent offspring conflict, and theory of receiprocal altruism, Trivers–willand hypothesis, r/k selection theory, evolutionary game theory and evolutionary stable strategy. Other principles of HBE are ecological selectionism, piecemeal approach, conditional strategies, phenotypic gambit and modelling.

13.24.1 Ecological Selectionism

This assumes that humans are highly flexible in their behaviour and various ecological forces select for various behaviour which optimize human's inclusive fitness in their given ecological context.

13.24.2 Piecemeal Approach

This approach holds view that by taking complex social phenomenon (marriage patterns, foraging behaviours) and then breaking them down into sets of components involving decisions and constraints, models may be created and human behaviour may be predicted. This may be exemplified through examining marriage systems by examining the ecological context, mate preferences, distribution of particular characteristics within population and so on. Behavioural strategies, according to this approach, are the most adaptive strategy in one environment, and may not be the most adaptive strategy in another environment.

13.24.3 Phenotypic Submit

This refers to the assumption that humans posses a high amount of phenotypic plasticity and attempt to discover correlations between variations of ecological contexts and variations of human behaviour.

13.25 SCIENCE AND HUMAN VALUES

In "developed" societies, social structures are more flexible and adaptive in having greater social mobility, competition, and opportunity for individual initiative, outside of established patterns. Some people display an increasing capacity for change, response to external opportunities and challenges, and to successes and failures of their own and other people's experience. It is typical of more "developed" societies to grant a higher social value, to brainwork, and spread opportunity for mental development much more widely. Excess energy pours into development of complex technological organization of material processes, social organization of life processes, mental organization of information, knowledge, even intuition and wisdom. Mental society displays a wider range of adaptive responses and creative initiative. "More developed" society has people, who try to resist mobility and change, discourage creative imagination, suppress deviations from norm, lead by command and control, cling to tradition, and turn aside from new choices and chances. Social development tend to generate unanticipated and unwanted excesses, side effects, and untoward reactions, like environmental damage, over-population, destructive applications of technology, economic crises, and social conflicts. So, it needs to consider, whether these problems in biology and society have a common source, and whether at least part of difference is that biological systems have been better than human systems at self-correcting. Values are something qualified as good and are prepared to set as goals in life.

13.26 BIOLOGICAL THEORY

Biological theory can be broken down into evolutionary theory, which focuses on why aspect of theory, and biological processes, which explains how people are affected by biology, thus causing differences in personalities. Focal point of evolutionary theory is idea that personality is inherited from parents through particular genes. Biological processes are concerned with genes influence on behaviour, through chemicals within brain. "Each individual personality is embedded in make-up of their body" and many aspects of personality are inherited from parent genes, and by passing inherited characteristics through time, many people share aspects of their personality with numerous others. It explains differences in people, and mode of sharing of personality characteristics with others throughout the world.

Influence of genetics on personality occurs, through biological processes. Biological processes deals with how workings of body influence personality and behaviours of each person, processes by which inherited genes influence behaviour. Gray performed research on monkeys and discovered that some monkeys learn from negative reinforcement, while others learn from positive reinforcement. From this experiment reinforcement sensitivity theory evolved. There are two different systems describing behaviour, within theory-Behavioural Approach System (BAS) and Behavioural Inhibition System (BIS). BAS is associated with left frontal lobe of brain and responds to positive reinforcement. Those with a sensitive BAS are receptive to positive conditioning and rewards. People with a responsive BAS are more likely to be outgoing and extraverted. BIS acts as 'break petal' of brain; with increased threat of danger, BIS begin to respond, lighting up right frontal lobe of the brain. Those sensitive BIS do not seek rewards, rather, they respond to negative emotions and threat of danger or punishment. These people are most closely linked to neuroticism.

Although parents are a major influence in personality of child, environment also plays a major role in the behaviour individual's display: "Genes and environment account for variation in human behaviour". Inherited genes from parents working with different situations of person create personality of an individual. Inherited genes, may strongly suggest a behaviour to be exemplified by offspring, situations experienced may lead to a totally new behaviour. According to evolutionary theory, male should portray characteristics used by men, for thousands of years to spread their own seeds.

Psychologists are "taking one aspect of personality out to describe a person as a whole, while all aspects of personality work together to form the person" when attempting to describe a person using only one theory. Each theory is separate from the other, yet each theory cannot exist without others; all theories are strengthened by each other. Evolutionary theory and biological processes produce, what is known as biological theory, that works with inherited genes and affect on behaviour. Biological perspective attempts to wholly break down characteristics of human personality. Because of complexity of composition of a person they will never be able to be perfectly and fully expressed, with only one typing. It is the role of biological theory to "examine how biological functions influence human action, being very much a 'process' approach to personality".

13.27 HUMAN EVOLUTIONARY PSYCHOLOGY AND ANIMAL BEHAVIOUR

Human Evolutionary Psychology (HEP) is an integral part of ordinary animal behaviour science (Daly and Wilson 1999). Many prominent contributors to the development of HEP came to the subject from backgrounds in non-human animal behaviour, ecology and evolutionary biology. A number of contributors to HEP have published on the behaviour of other species or usually continuing their non human research in parallel with their work on human behaviour and often focusing on more or less the same issues in both. So, HEP is just the ordinary pursuit of animal behaviour research.

Happiness in Life Expectancy

FATIK BARAN MANDAL

A definite correlation between happiness and health has been established by scientists. The cross-country average of the life expectancy gap between men and women is about five years. Higher life expectancy of women is attributed to the compensatory influences of the second-X chromosome, longer telomeres, strong immune system, better capability of protection against oxidative stress, and the protective effects provided by estrogen, the details of which have been documented in a research article published in the Journal, *Gender Medicine* (3: 79–82) in the year 2006. Cross-society differences (for example, 13.3 in Russia and 0.46 years in Zimbabwe) in the life expectancy between male and female have been mentioned. Lifestyle, smoking, drug, alcohol consumtion, etc. are indicated as contributing factors behind such gap in an excellent review done by Junji Kageyama in the year 2009, which has been published by Max Plannck Institute for Demographic Research. Kageyama has concluded that happiness influences life expectancy even at national aggregate level. Various factors like stress responsiveness, composition of marital status, and economic resources are correlated with happiness and life expectancy. Happiness may explain the differences in the life expectancy gap between countries as well as between sexes. It may be concluded in summary that nature favours women in regard to survival span.

Source: http://expertscolumn.com/content/happiness-life-expectancy

Review Questions

Short Answer Questions

1. What is motivation?
2. What do you understand by the term intention?
3. Define culture.
4. What is haptic communication?
5. What are the purposes of life?
6. What is happiness?
7. Which are basic human emotions?
8. What is emotional intelligence?
9. What is self-efficacy?
10. What does the term cuckold indicate?
11. Write a brief note on selfish sene.

12. What is the full form of GnRH?
13. Name the hormone associated with aggression in male.
14. What is the full form of OFT?
15. What do you understand by adolescence?
16. What do you understand by gesture?
17. What do you understand by culture?
18. What is biorhythm?
19. What is pheromone?
20. Define fertility.
21. What do you understand by parental investment?
22. What do you understand by parent–offspring conflict?

Long Answer Questions

1. Give an account of happiness in humans.
2. Give an account of human emotion.
3. Write a note on human aggression.
4. Describe the role of body odour in humans.
5. Describe "four-year itch" according to Helen Fisher.
6. Write an essay on parental investment and parent–offspring conflict in light of modern study.
7. Discuss how genes influence political activities.
8. Discuss the importance of human sense of smell.
9. Write a note on facial signals.
10. Discuss the relationship between beauty and intelligence in light of recent study.
11. Discuss the concept of selfish gene.
12. Write an essay on fertility, parental investment and parent–offspring conflict.
13. Discuss the significance of studying human behaviour.

Glossary

Acquisition: Strengthening of association between stimulus and behaviour after pairing of CS and US.

Adaptation: Decline in amplitude of a sensory response in presence of a constant stimulus.

Aestivation: The condition of dormancy or torpidity.

Aggression: Antagonistic behaviour that normally occurs among members of the same species.

Allee effects: When the growth rate of a population is affected by population size. In small populations, birth rate may decline because mates may be hard to find.

Allele: An alternate form of a gene at the same locus.

Alpha animal: Animal that takes a lead role and occupies the dominant position in the group.

Altricial: Young are born in a relatively underdeveloped state in which they are unable to feed or care for themselves independently for a period of time after birth/hatching.

Altruism: Reduction in reproductive potential of "helper" individual while increasing that of "helped" individual.

Anadromous: Migration in which adults spend lives at sea and return to freshwater to spawn.

Analogy: Resemblance in characteristics (body form, behaviour) as a consequence of independent adaptation to same or similar environmental conditions and not due to common ancestry.

Animal behaviour: Wild ways in which animals interact with each other, with members of other species and with environment.

Anthropomorphism: Attribution of human characteristics to inanimate objects, animals, forces of nature, and others.

Aposematic: Refers to colour or other signal that any organism might use to advertise the danger of preying upon it.

Arousal: Activation of the nervous system generally.

Associative learning: A form of behavioural plasticity in which an animal forms an association between a stimulus that initially evoked little or no response and one that normally evokes a reflex response. Association is formed as a consequence of paired presentation/occurrence of two stimuli.

Attitude: Predisposition to behave in a certain way.

Autonomic nervous system: Division of the vertebrate nervous system serving internal organs such as the heart, blood vessels, lungs, intestines and also certain glands.

Aversive stimulus: A stimulus against an organism acts to escape or avoid which can result in some problematic secondary effects, such as aggression, countercontrol, social disruption and emotional escape/avoidance behaviour. Averser is sometimes equated with punisher.

Avoidance: Refers to behaving so that the aversive stimulus is not experienced at all.

Axon: Elongated process for long-distance signalling. Part of neuron that conveys information away from the cell body.

Balanced polymorphism: Maintenance of two or more alleles for a trait in a population at a more or less constant frequency for the selective advantage of heterozygotes.

Bateman's principle: Variance in net reproductive success will be greater in the sex with the smaller per-offspring investment. This sex is likely be subjected to more intense sexual selection.

Behaviour: All coordinated actions and responses of an individual organism/group/society.

Behaviourism: Movement in psychology that advocates use of strict experimental procedures to study observable behaviour in relation to environment. Behaviourism deals with learning under strictly controlled conditions.

Biological clock: A timekeeping mechanism of the organism for measuring the approximate length of a day, even without external environmental cues.

Biomimicry: Refer to study of nature's most successful development and then imitating these design and processes to solve human problems.

Bipolar disorder: A mood disorder characterized by fluctuations between manic and depressive states.

Bipolar neuron: Neuron with two dendrites, one extending from each end of cell body which is typical of many invertebrate sensory neurons.

Brain: Supervisory centre of the nervous system located in the head.

Bruce effect: Possible form of post-copulatory competition. In some animals, a strange male can cause a female to abort (previous male's foetus) and make the female receptive.

Camouflage: Colouration and/or shape that renders an organism inconspicuous in its environment.

Capturing: Reinforcing a behaviour when it occurs naturally, as opposed to prompting it.

Carrying capacity: An environment's maximum persistently supportable load.

Central nervous system: Main organ of nervous system, made up of brain and spinal cord in vertebrates.

Channel: Central pattern generator group of neurons in central nervous system that due to properties of neurons and connections between them can produce a specific pattern of output in absence of sensory feedback.

Chemical synapse: Type of synapse in which communication is achieved by release of chemicals from pre-synayptic membrane, diffusion across synaptic cleft and binding to receptors on post-synaptic membrane.

Chemoreceptor: Sensory receptor that responds to binding of a specific class of molecule.

Circadian clock: Inherent timekeeping mechanism with a capacity to drive or coordinate a circadian rhythm.

Circadian rhythm: Biological rhythms with a periodicity of approximately 24 hours.

Classical conditioning: Behaviour changes when an organism comes to associate one stimulus with another, a reflexive or automatic response transfers from one stimulus to another. It requires an unconditional reflex where an unconditional stimulus brings about an automatic, unlearned response.

Coding: Means by which information about sensory quality or strength is conveyed to central nervous system.

Coefficient of relatedness: The proportion of genes shared by two individuals. Also known as degree of relatedness, and symbolized by 'r'.

Coercion: Use of punishment and the threat of punishment to get others to act, and to reward people just by letting them escape from punishments and threats.

Coevolution: The evolution of adaptations in two species due to their interactions with one another.

Cognition: Identifying and acting upon information derived from environmental or internal information to selectively acquire, organize, and apply information. Sensory information is transformed into perceptions which are then categorized and organized, stored, recovered, abstracted and used. Cognition includes the use of memory to guide behaviour and thinking.

Cognitive neuroscience: A branch of neuroscience involving the study of the neural mechanisms of cognition, but sometimes it is seen as part of a wider interdisciplinary study of cognitive science.

Cognitive science: Usually defined as the scientific study either of mind or of intelligence, and often said to consist of, take part in, and collaborate with psychology, artificial intelligence, philosophy and neuroscience.

Collective behaviour: Social behaviour that is spontaneous and a departure from the norm, but is not deviant.

Communication: Intentional information transfer between sender and receiver, usually advantageous to both.

Compass: Mechanisms that allow finding a particular spatial direction.

Complex System: A group of relatively simple but interconnected elements which causes the emergence of a higher order phenomenon.

Concept: Any class, the members of which share one or more defining features.

Conditioned response: Response elicited by a conditioned stimulus, often, but not always, similar to unconditioned response.

Conditioned stimulus: A previously neutral stimulus that has been paired with an unconditioned stimulus and now elicits reflexive behaviour.

Conditioning: Reflexive or automatic response transfer to an unrelated stimulus/behaviour.

Connective: Relatively large nerve trunk in invertebrates consisting of axons that connect one ganglion of central nervous system to another.

Consciousness: Consciousness is a graded integration of multiple cognitive functions yielding a unified representation of the world, our bodies, and ourselves.

Contingency: Functional relationship between behaviour and the environment. It includes the behaviour itself as well as the antecedent that evokes it and the consequence that influences the behaviour's strength.

Continuous reinforcement: A schedule of reinforcement in which every response results in reinforcement.

Cooperative breeding: When more-or-less closely related relatives assist in raising young.

Coprophagia: Eating of faces.

Core area: Part of an animal's home range of heaviest use.

Culture: Sum of characteristics, ways of living of a particular group, learned from one another and passed down to generations.

Current: Moving charge in an electrical circuit.

Dark adaptation: Adaptation to decreasing levels of ambient illumination by an increase in receptor sensitivity.

Decision making: Cognitive process of selecting a course of action from among multiple alternatives.

Declarative memory: Capacity to retain information about the past.

Dedicated circuit: Neural configuration of particular sensory, intersensory, and motor neurons responsible for generating a specific behaviour.

Dendrite: One of the branching or small neural processes carrying signals to cell body of a neuron.

Depolarization: Change in cell membrane potential where interior becomes less negative.

Direct fitness: Probability of reproductive success through one's own offspring.

Directional selectivity: Most vigorous neuronal response to stimulation from a particular direction.

Discrimination: Process of behaving, or coming to behave, differently in the respective presence of different stimuli.

Discriminative stimulus: An immediate antecedent stimulus that evokes a behaviour as part of a contingency.

Dispersal: Travel of individuals which as an ecological process affects distribution of individuals; as a genetic process affecting geographic differentiation and variation.

Displacement activity: Behaviour patterns seemingly unrelated to behavioural context in which it occur.

Dominance (Social dominance): Social relationship which addresses the management of social conflict including the allocation of limited resources, through the exertion of control and influence in a way that minimizes the risk of overt aggression by way of the use of conventionalized ritual display behaviours and involves a cost–benefit evaluation of the benefits of seeking to win a particular social conflict versus the likely associated cost.

Drive: An internal factor that determines how likely an animal is to perform a behaviour.

Echolocation: The use of ultrasonic sounds to produce an auditory image of nearby surroundings.

Ecological factors: Aspects of the environment that contribute indirectly to the contingencies.

Electrical synapse: Type of synapse in which communication is made by electrical synaptic transmission. Potentials are transmitted by direct flow of ions through gap junctions.

Emergent property: A property that could not have been predicted on the basis of knowledge of its constituent parts.

High-order feature that arises from synaptic interactions of neurons or group of neurons that is not intuitively obvious from an analysis of its individual elements.

Emotion: Emotions are coordinated states, shaped by natural selection, that adjust physiological and behavioural responses to take advantage of opportunities and to cope with threats that have recurred over the course of evolution.

Emotional behavior: Behaviours which are respondently conditioned and changed only via respondent conditioning. They may act as motivating operations or antecedent conditions for operants.

Empathy: Ability to recognize and understand the emotion of another.

Entrainment: Organism's response to environmental cues such as light, that enables resetting of the clock.

Environment: Collective term for the conditions in which an organism lives.

Environment of evolutionary adaptedness: Often applied to the environment in which a particular species evolved, more usefully applied to a particular trait.

Epigenesis: Interaction of genes and their environment in determining the expression of traits during development, including behavioural traits, and involves specific effects of environment on activation of specific genes.

Escape: Refers to behaving to cease an existing aversive stimulation. After learning escape contingencies, animals will often learn to avoid the aversive stimulation. It is argued that avoiding a stimulus requires exposure to a stimulus that evokes the avoidance behaviour and that must be aversive to evoke the behaviour and therefore, the behaviour functions to escape that aversive experience.

Escape behaviour: Behaviours that function to allow an organism to stop or diminish an aversive experience that has already commenced.

Ethogram: Comprehensive compilation of the different behaviours exhibited by a species.

Ethology: Biological study of animal behaviour comprising study of proximate causes of behaviour, including external stimulation and internal physiological mechanisms and states; development, including ontogenesis and genetics of behaviour; biological function, including immediate effects and ultimate consequences.

Eusocial: A specific social system that shows cooperation in caring for young, reproductive castes cared for by non-reproductive castes, and overlapping generations such that offspring assist parents in raising siblings.

Evolution: Any change in the proportions of different genotypes in a population from one generation to the next.

Evolutionary stable strategy: This is a strategy which, if adopted by most of a population, cannot be displaced by any alternate strategy, and will therefore tend to become established by natural selection.

Evolutionary psychology: Behavioural ecology plus socio-biology.

Excitatory postsynaptic potential: Potential that is generated in a postsynaptic neuron by activation of an excitatory synapse, usually depolarizing.

Excitatory transmission: Synaptic transmission that has a depolarizing effect on postsynaptic neuron and which increases probability that postsynaptic neuron will fire an action potential or amount of neurotransmitter release.

Experience: This is the basis for trial and error learning.

Extinction: Extinction is "the process of decreasing difference between the antecedent and postcedent environmental conditions in a three-term contingency featuring a behaviour that was previously effective in the setting and that was maintained at its previous rate by reinforcement alone or by a combination of reinforcement and punishment."

Facilitation: Increase in amplitude of post-synaptic responses as a result of repetitive activation of synapse.

Fear: An emotional response involving the release of certain hormones into the bloodstream and certain neural behaviours that tends to motivate escape operants.

Fight-or-flight response: Rapid activation of the sympathetic nervous system, characterized by either flight from an aversive stimulus or attack of it.

Fitness: The reproductive success of an organism.

Fixed action pattern: Term used to describe stereotyped patterns of movement common to all members of a species, often performed in response to specific sign stimuli. Once initiated, FAPs continue until completion.

Flight distance: The distance at which an animal will commonly flee from a particular aversive stimulus.

Fundamental niche: Set of resources and physical factors required for survival and reproduction of organism.

Game theory: A branch of applied mathematics that uses models to study interactions with formalized incentive structures. This theory encompasses decisions that are made in an environment where various players interact strategically.

Gap junction: Channel formed by proteins that allow ions and other molecules to pass between adjacent cells.

Gene: Segment of DNA that specifies structure of proteins.

Gene flow: The exchange of genes between different groups.

Genetics: Branch of life science that deals with various aspects of gene.

Genotype: Aggregate of specific genes an organism possesses or an organism's genetic composition.

Geomagnetotaxis: Response of an organism to earth's magnetic field.

Habituation: Gradual decline in response to recurrent stimuli without specific significance. Tendency to become familiar with a stimulus of no consequence after repeated exposure to it.

Hamilton's rule: The idea that the genetic relatedness of the performer of an altruistic act to the recipient is a key consideration in a costs/benefits analysis of its likeliness to be performed and in its evolution. The "rule" can be expressed as $b/c > 1/r$ (where b = benefits, c = costs, and r = coefficient of relatedness).

Handicap hypothesis: The idea that apparently harmful traits in males become attractive to females as they indicate the male's capacity to cope with them.

Hardy–Weinberg law: Gene frequencies in a population remain constant as long as there is no chance of fluctuations or outside influences.

Heritability: Refers to the total proportion of genetic variance in a trait

Hierarchical selection: Refers that selection can act at any level in the biological hierarchy, from single genes to entire ecosystems.

Hierarchy: When there are multiple levels of organization, control, or information flow; higher levels emerge out of or are supported by lower ones and lower levels are then at least partly controlled by the higher ones.

Home range: Area in which an individual pair or group occupies or regularly returns to.

Homing: Ability to return to a home site after being displaced.

Homology: Resemblance in characteristics due to common ancestry.

Hormones: Chemical messenger which are produced by the ductless glands and are carried by the circulating blood to another parts of the body where they evoke systemic adjustment by acting on specific cell, tissues and organs.

Imprinting: Capacity to learn specific types of information at certain critical periods in development.

Inbreeding: Preferential mating between relatives, mating between sibs, half-sibs, and parent-offspring.

Indirect fitness: Probability of reproductive success through non-descendant relatives.

Inhibitory postsynaptic potentials: Potential generated in a postsynaptic neuron by activation of an inhibitory synapse, usually hyperpolarizing.

Innate: Refers to built-in, inborn, hereditary.

Insight learning: Learning that requires the manipulation of mental concepts to arrive at adaptive behaviour.

Instinct: Capacity of an animal to complete a complex behaviour in response to a first-time encounter with a given situation; an inborn behavioural mechanism that manifest itself in an orderly movement sequence. An inherent disposition of a living organism towards a particular behaviour.

Intelligence: Refers mental capabilities such as the ability to reason, plan, solve problems, think abstractly, comprehend ideas and language, and learn.

Intention: Behavioural patterns that precede or prepare for other behaviours.

Interspecific interactions: Interactions among organisms of different species.

Intraspecific interactions: Interactions among organisms of the same species.

IQ: A score derived from a set of standardized tests that are developed for the purpose of measuring a person's cognitive abilities in relation to their age group.

Iteroparity: Repeated production of offspring at intervals throughout the life cycle.

Key stimulus: Any event that initiates an instinct.

Kin selection: Selection for traits that lower an individual's personal fitness, but raise a relative's fitness.

Kinesis: Non-directional orienting reactions in presence of a particular sensory stimulus. Animals which suddenly find themselves in an unfavourable environment may change direction by trial and error.

Klinokinesis: Kinesis where rate of change of direction increases in proportion of increase in light intensity.

Klinotaxis: Taxis in which successive comparison of stimulus intensity is taken as reference.

K-selected traits: K is an attribute of an environment. K represents the "carrying capacity" of that environment. A trait that is "K-selected" is one that helps an organism to maximize its capacity to survive and thrive very close to that carrying capacity.

Language: Humans have an inborn capacity to extract word meanings, sentence structure, and grammatical rules from the complex stream of sounds they hear.

Larynx: The portion of the air passage between the pharynx and the trachea.

Law of effect: Responses that produce a satisfying effect are likely to occur again in that situation. Responses that produce a discomforting effect are less likely to occur again in that situation.

Learning: An adaptive change in behaviour resulting from experience.

Lek: Communal mating area in which individuals hold small territories solely for courtship and copulation.

Limbic system: A diverse group of brain structure mostly in lower forebrain.

Logic: Study of formal structures of valid arguments.

Major histocompatibility complex: Proteins located on cell surface that identify the cell as self.

Maternal effects: Influences of the mother on an offspring's behaviour that is not due to direct genetic inheritance.

Mechanoreceptor: Sensory receptor for mechanical energy.

Meme: A unit of information that can be communicated to human minds, possibly altering their behaviour, and thereby propagating themselves.

Memory: Capacity to retain an impression of past experiences.

Menotaxis: Mechanism for maintaining a constant direction by steering a course at constant angle to incidence of stimulus.

Migration: Seasonal movements of animals from one defined place of residence to another.

Model: A selective, simplified representations of the things important about a phenomenon.

Monogamous: Mating system in which each male or female individual has only one mate.

Mood: A person's mood can consist of a combination of emotions. In normal functioning, moods are largely adaptive and influenced by external events.

Motivation: An internal factor that determines how likely an animal can perform a behaviour.

Processes that account for direction, intensity, and persistence of effort expended to meet needs.

Natural selection: Process by which environmental effects lead to varying degrees of reproductive success among individuals of a population of organisms with different hereditary characters or traits. Characters that inhibit reproductive success decrease in frequency from generation to generation.

Navigation: An animal moves about using external cues to determine its position relative to a goal.

Negative reinforcement: A behaviour change process in which a decrease of a stimulus to the environment during or immediately following a response results in an increase in the strength of the behaviour across subsequent occasions.

Negative reinforcer: Any stimulus that, when removed following a behaviour results in an increase in the strength of that behaviour.

Nervous system: System that coordinates the activity of the muscles, monitors the organs, constructs and processes input from the senses, and initiates actions.

Neurohormone: Compound that is released at a synapse and diffuses across the synaptic cleft to act on a receptor located on the membrane of a postsynaptic cell, which may be another neuron, a muscle cell or a specialized gland cell. It is released from nerve endings by nerve impulse activity at morphologically distinguishable synaptic junctions producing suitable changes in the excitability of the postsynaptic membrane.

Neuromodulator: Compound that is released within a localized region of CNS, the receptor for which is not necessarily sited on an anatomically apposed postsynaptic cell.

Neurotransmitter: This is an endogenous chemicals which relay, amplify, and modulate signals between a neuron and another cell. Neurotransmitters are packaged into synaptic vesicles that cluster beneath the membrane on the presynaptic side of a synapse, and are released into the synaptic cleft, where they bind to receptors in the membrane on the postsynaptic side of the synapse. Release of neurotransmitters usually follows arrival of an action potential at the synapse, but may follow graded electrical potentials.

Neutral stimulus: A stimulus that does not elicit a response.

Niche: All aspects of an organism's environment that enable it to survive and reproduce. Functional position of an organism in its ecosystem.

Non-associative learning: A change in behaviour as a result of exposure to stimuli that is not associated with positive or negative consequences.

Ontogeny: The progressive expression of programmed developmental change, from conception to death.

Operant conditioning: A form of associative learning in which behaviour changes as a result of an action that leads to a specific consequence.

Operant: Behaviour that operates on the environment to produce consequences.

Orientation: Positioning of individual with regard to external cues.

Orientation response: A reflex in which an organism orients their attention to a change in their environment.

Orthokinesis: Kinesis where speed of locomotion is related to stimulation intensity.

Outbreeding: Preferential mating between non-relatives.

Pack: Pack as a stable social unit hunts, rears young and protects a communal territory as a group.

Paradigm: A set of beliefs that complement each other to create a model about works of world or even a comprehensive world-view.

Parental investment: Behaviour towards offspring that increases the chances of the offspring's survival at the cost of the parent's ability to invest in other offspring.

Peck orders: Animals social position within a group.

Perception: Interpretation of sensory signals within the CNS where it produces an internal representation of electrical activity from sensory organs. This is largely dependent on which part of the brain receives the signals.

Perceptual learning: A change in the way in which events are perceived and may involve associative and non-associative learning.

Peripheral nervous system: Complete set of nerves and ganglia carrying information towards or away from central nervous system.

Personality: Personality describes the unique emotion, thought, and behaviour patterns of a person.

Persuasion: Usually used in distinction to coercion.

Phenotype: External expression of an individual.

Phenotypic plasticity: Ability of an organism to adjust to an environment during the life time.

Pheromones: Chemicals that are released by individuals to influence the behaviour of other members of their own species.

Philosophy of mind: philosophical study of the nature of the mind, mental events, mental functions, mental properties, and consciousness.

Photoreceptor: Receptor cell specialized to respond to light energy.

Phylogenetic: Phylogenetic change involves changes that are genetically based and have evolved over time.

Phylogeny: Origin and ancestral succession consti-tuting the evolutionary descent of a species or class of species.

Pleiotropy: Situation under which a gene may influences multiple phenotypes (there is no 1:1 correspondence between gene and behaviour).

Polyandrous: Mating system in which females mate with more than one male.

Polygamous: Mating system in which males mate with more than one female.

Polygenic: Phenotypic trait that is under control of multiple genes.

Polymorphism: One of alternative "morphs" or variations of a trait that is present within a population sharing a common genotype.

Post-copulatory competition: One consequence of mating can be deposition of a copulation plug, left in female's vagina to thwart subsequent copulators.

Postsynaptic neuron: At a synapse, neuron that receives a signal from another neuron.

Precocial: Youngs are born relatively mature and mobile from the moment of birth or hatching.

Predator: An organism that feed directly other organism in order to survive.

Presynaptic inhibition: Form of inhibition in which inhibitory effect is brought about by a reduction in amount of neurotransmitter released by target neuron, hence reducing probability that excitation of target neuron will produce an action potential in a postsynaptic neuron.

Procedural memory: Capacity to retain information about past experiences related to procedures and skill.

Proprioceptor: Sensory receptor that provides an animal with information about relative position or movement of its body parts.

Proximate explanation: Explanation those appeal to motivational variables, experience, and genotype as the cause of behaviour.

Psychology: An academic and applied field involving the study of behaviour, mind and thought, and the subconscious neurological bases of behaviour.

Punctuated equilibrium: Macroevolutionary changes occur as "sudden" bursts, separated by long periods of stasis.

Quantitative Trait Loci: Genes in multiple gene system.

Receptor potential: Electrical potential that appears in sensory neurons when a stimulus is applied.

Reciprocal altruism: Altruistic acts between unrelated individuals.

Reflex: The elicitation of an unconditioned response with an unconditioned stimulus.

Reinforcement: Process in operant conditioning where a response is strengthened by reward (positive reinforcement) or avoidance of punishment (negative reinforcement). Negative reinforcement is not same as punishment.

Reinforcer: A stimulus that, when presented or removed contingent on a behaviour, increases the future strength of that behaviour across subsequent occasions.

Relaxation: Relative minimal arousal.

Releaser: A sign stimulus emanating from a conspecific play.

Releasing mechanism: Functionally organized neural circuit that filters just the required features from sensory input and detects a specific sign stimulus and then initiates appropriate behavioural response. Previously, RM was referred to as innate.

Repression: Refers to the memory inhibition which is the ability to filter out the irrelevant information from the memory for recalling.

Resistance to extinction: The persistence of an operant behaviour after it is put on an extinction schedule.

Ritualization: Process by which a functional behaviour pattern or structure is transformed into a communication signal.

***r*-selected traits:** *r* is an attribute of an organism. *r* represents "the intrinsic rate of population growth." A trait that is "*r*-selected is one that helps an organism maximize expression of that intrinsic rate."

Runaway sexual selection: The process in sexual selection by which positive feedback between an ornamental trait in a male and the female preference for the trait can lead to very elaborate even burdensome ornaments.

Satiation: Decline in the effectiveness of a reinforcing stimulus due to excess exposure or repeated presentation of it.

Self: Refers to the conscious, reflective personality of an individual.

Semelparity: The occurrence of a single act of reproduction during an organism's lifetime.

Sensitization: Enhancement of response to most stimuli following exposure to a significant signal.

Sensory adaptation: Decreased signalling of a peripheral sense organ with continued exposure to a stimulus.

Sensory filter: Neural mechanisms that discriminate more-or-less salient stimuli that are detected by sensor organs; a low level mechanism of selective attention; may help prevent "sensory overload".

Sensory receptor: Cell endowed with ability to respond to a specific stimulus.

Sensory transduction: Chain of physiological reactions which convert environmental energy presented to the sense organs into electrochemical energy which can be transmitted by the nervous system and affect physiological variables such as the stability of polarized cells.

Sexual selection: Process by which changes in gene frequencies result from individuals that are better than others at either competing for or at attracting mate, i.e. evolution of traits based on differences in mating success.

Sign stimulus: Particular subset of features or qualities of a natural stimulus or complex environmental cues sufficient to elicit a behavioural response via a releasing mechanism.

Signal: Physical coding of information capable of transmission through environment.

Skinner box: Animal placed inside the box is rewarded with a small bit of food each time it makes the disired response, such as pressing a lever or pressing a key. A device outside the box records the animal's responses.

Sleep: A state of inactivity during which animals are not responsive to external stimuli.

Socialization: Sum total of social experiences that alter the development of an individual.

Society: Group of individuals of the same species that is organized in a cooperative manner extending beyond sexual and parental care.

Socio-biology: Interdisciplinary field of knowledge that explores the biological basis of social behaviour including morality.

Sociopathy: Disorders of empathy are associated with sociopathy.

Sperm competition: A form of post-copulatory competition which occurs when ejaculates from more than one male might be in female's reproductive tract.

Stereotypic behaviour: Behaviours performed in a repetitive manner for extended periods of time, usually to the point of distress.

Steroid hormone: Hormone with chemical structure similar to cholesterol.

Stress: A nonspecific response of the body when any demand is placed upon it.

Successful dispersal: Dispersal is successful if propagule obtains opportunity to breed and raise the young.

Survival strategies: Strategies that ensure survival and reproduction of organism.

Synaptic plasticity: Changes in excitability and transmitter release at synaptic junctions between neurons.

Taxis: Act of orientation towards some external stimulus or combination of stimuli.

Telotaxis: Taxis in which certain stimuli are kept at a fixed place on a receptor.

Territory: Any defended area; area of more or less fixed boundaries from which rival conspecifics are excluded.

Tonic receptor: Type of receptor cell in which adaptation is slow and the response of cell to the stimulus does not declines quickly.

Ultimate explanations: Explanations that refer to selection pressure and other factors as the cause of behaviour.

Unconditioned stimulus: Stimulus that elicits an unconditioned response.

Vomeronasal organ: A special part of olfactory system which may function as a pheromone detector in mammals.

Waggle dance: A form of communication about food source as seen in honey bees.

Warning colouration: Bright colouration that warns predators that the potential prey is distasteful.

Zeitgeber: An environmental stimulus that entrains a biological clock.

For List of References visit: www.phindia.com

Index